Studies in Classification, Data Analysis, and Knowledge Organization

Titles in the Series

H.-H. Bock and P. Ihm (Eds.)
Classification, Data Analysis and Knowledge Organization

M. Schader (Ed.)
Analyzing and Modeling Data and Knowledge

O. Opitz, B. Lausen and R. Klar (Eds.)
Information and Classification

Hans-Hermann Bock, Wolfgang Lenski,
Michael M. Richter (Eds.)

Information Systems and Data Analysis

Prospects - Foundations - Applications

Proceedings of the 17th Annual Conference
of the Gesellschaft für Klassifikation e.V.
University of Kaiserslautern, March 3-5, 1993

With 96 Figures

Springer-Verlag

Berlin Heidelberg New York
London Paris Tokyo
Hong Kong Barcelona
Budapest

Prof. Dr. Hans-Hermann Bock
Institut für Statistik
Rheinisch-Westfälische Technische Hochschule Aachen (RWTH)
Wüllnerstr. 3
D-52056 Aachen, FRG

Dr. Wolfgang Lenski
Forschungsstelle "Mathematische Logik"
der Heidelberger Akademie der Wissenschaften
Fachbereich Informatik
Universität Kaiserslautern
Erwin-Schrödinger-Str. 57
D-67653 Kaiserslautern, FRG

Prof. Dr. Michael M. Richter
Fachbereich Informatik
Universität Kaiserslautern
and
Deutsches Forschungszentrum für
Künstliche Intelligenz GmbH (DFKI)
Erwin-Schrödinger-Str. 57
D-67653 Kaiserslautern, FRG

ISBN-13: 978-3-540-58057-7 e-ISBN-13: 978-3-642-46808-7
DOI: 10.1007/978-3-642-46808-7

42/2202-5 4 3 2 1 0 - Printed on acid-free paper

Preface

This volume presents a selection of 42 refereed and revised contributions (including some invited lectures) which were presented at the 17th Annual Conference of the "Gesellschaft für Klassifikation e.V." (GfKl), the German Classification Society. This conference was held at the University of Kaiserslautern from March 3 – 5, 1993 and jointly hosted by the Research Center for Mathematical Logic of the Heidelberg Academy for the Humanities and Sciences, the University of Kaiserslautern, and the German Research Center for Artificial Intelligence (DFKI GmbH). The local organization was performed by Michael M. Richter, Wolfgang Lenski, Sabine Friedrich, Annette Klein, and Elisabeth Wette-Roch of the Computer Science Department of the University of Kaiserslautern and the Research Center for Mathematical Logic of the Heidelberg Academy for the Humanities and Sciences, respectively. The members of the Scientific Program Committee were Hans-Hermann Bock, Paul O. Degens, Matthias Herfurth, Hans-Joachim Hermes, Rüdiger Klar, Rudolf Mathar, Michael M. Richter, and Rudolf Wille.

Under the title " **Information Systems and Data Analysis: Prospects - Foundations - Applications**" the conference covered a wide range of topics related to information handling, classification and data analysis, either in a structural, theoretical, computational, mathematical or statistical framework, or by providing illustrative and typical applications from practice. Therefore this volume brings together several complementary perspectives from quite distinct fields such as archeology, computer science, conceptual structure theory, library science, linguistics, mathematics, medicine, molecular biology, philosophy, sociology, statistics, systematics – all focusing on aspects of classification and information. In fact, this diversity reflects the interdisciplinary character of the "Gesellschaft für Klassifikation" which provides since many years a broad forum for cooperation between different domains of research and application.

The scientific content of this volume has been classified into four parts with suitable subsections. They can be summarized as follows:

Part I: Information Processing, Classification-Based Approaches, Linguistic and Conceptual Analysis

The first part discusses issues related to **information systems** and **knowledge organization**. It comprises contributions on **learning** and **case-based reasoning** in the framework of classification, approaches based on **linguistic analysis**, work on **information processing** and the **design of information systems** as well as papers on **uncertainty** and recent progress in the field of **neural networks**. The last subsection deals with investigations on **concepts** and includes formal concept analysis, a philosophical treatise on concept formation, and the problem of organizing concepts in a suitable framework.

The first subsection is on **learning and case-based reasoning**. The paper of K.-D. ALTHOFF ET AL. clarifies the relationship between induction and case-based reasoning, points out the main differences, and pleads for an integrated approach. – CH. GLOBIG & S. WESS analyze principles of case-based classification and demonstrate interrelations between the distance measures and the sets of (symbolically) learnable concepts. – K. P. JANTKE argues for an extended view at similarity concepts in case-based reasoning which may also be used for inductive learning purposes. – F. PUPPE gives an overview over approaches to case-based learning techniques discussing the viewpoint of expert-systems, statistics, and neural networks and evaluates the performance of respective applications.

The second subsection is devoted to **approaches based on linguistic analysis**. R.-J. FISCHER discusses the use of phonetic transformations for automated indexing of medical texts which should eliminate spelling variants. – The paper of J. INGENERF introduces a terminological formalism to represent semantical relationships between medical terms which aims at overcoming syntactical ambiguities. – G. RAHMSTORF presents an approach to the conceptual representation of noun phrases to be used for an retrieval component as well as for a thesaurus as part of a knowledge based information system.

The third section is on **information processing and design of information systems**. The paper of H. CZAP deals with consistency of heterogenous distributed databases and provides formal conditions as well as practical suggestions to guarantee data integrity. – M. HERFURTH surveys methodological trends in sociology and sketches the possible assistance of information processing tools for epistemological progress. – The paper of G. MEYER, CH. BEIERLE, R. SCHEUBREIN discusses several types of integrations of logic programming systems and deductive databases and reports on implementations. – P. MUTSCHKE describes a technique to detect importance of, and structural relationships between, scientists within a specific research field out of bibliographic databases. – M. RUNDSHAGEN and S. MARX report on the progress made in the implementation of the object-oriented knowledge base system WIMDAS. – The paper of M. STAUDT and H.W. NISSEN shows how to use classes for queries in object-oriented databases and gives an approach to integrate query classes and deductive rules into one framework.

The next section deals with **neural networks** and **uncertainty**. The survey article of R. KREE reviews principles of a theory of neural networks. – S. PÖHLMANN gives a critical examination of approaches treating uncertainty in particular the Dempster–Shafer method in diagnostic systems. She furthermore introduces interval probabilities and mentions opportunities as well as problems.

The final section of part I is devoted to **conceptual approaches**. CH. THIEL outlines the history of concept formation in science and philosophy and gives an account on modern positions. – The contribution of F. VOGT and R. WILLE formalizes an extension of formal concept analysis by algebraic methods and shows that formal contexts can be embedded into algebraic or bialgebraic contexts. – The paper of E. WETTE-ROCH proposes a reorganization of concepts in mathematical logic for bibliographic documentation using an integration of terminological concept descriptions into a frame-based system.

Part II: Mathematical and Statistical Methods for Classification and Data Analysis

Mathematical and statistical approaches are the main theme of this second Part which combines methods for **classification and clustering** with articles on **scaling, optimization, three-way data matrices** and **spatial data analysis.**

The section on **classification** starts with the paper of TH. ECKES and P. ORLIK on the simultaneous clustering of the rows, columns and layers of a three-dimensional data matrix by generalizing WARD's well-known hierarchical clustering method. – E. GODEHARDT and A. HORSCH address the problem of testing homogeneity versus a grouping structure in the case of a (possibly binary) similarity matrix. Their approach is based on classical and new results in random graph theory, with special emphasis on random directed graphs and multigraph clustering (including dependence between different layers). – The paper of P. WILLETT deals with chemical information systems and methods of similarity searching for two- and three-dimensional chemical structures by an efficient combination of data-analytic, graph-theoretical and clustering methods. The paper focuses on atom-mapping methods where the similarity of molecules is described by the distance matrix of their atoms, but surveys other methods and computer programs as well. – H.-J. MUCHA explains the use of the interactive computing environment *XploRe* for clustering purposes, e.g. the k-means method, adaptive distance clustering, hierarchical clustering, multivariate plots and dynamic graphics.

Discriminant analysis is the main theme of K.-D. WERNECKE who provides a survey on numerous discrimination rules and computer programs and discusses their use for medical diagnostics (including practical examples). His emphasis is on the incorporation of mixed data, the estimation of error rates, and the evaluation of variable selection strategies by modified cross-validation. – D. KRAUSE and W.-D. RICHTER consider the problem of assigning a one-dimensional sample into one of two Gausian distributions from which training sets are available and obtain an explicit integration formula for the probabilities of a correct decision. – The paper of J. KRAUTH shows how jackknife techniques can improve the estimates of various indices of similarity between sites in ecological biology when using quadrat sampling. – Finally, CHR. WISOTZKY and F. WYSOTZKI propose a procedure for classifying time series which is based on the coefficients of a spline approximation.

The next subsection is devoted to problems of **multivariate data analysis** in a broad sense. The paper by S. KROLAK-SCHWERDT, P. ORLIK and B. GANTER starts with a three-way binary data matrix and looks for maximal (three-dimensional) blocks of 'ones' in the same way as two-dimensional binary arrays are analyzed in formal concept analysis or block clustering. – Three papers deal with **optimization methods** designed, in particular, for multidimensional scaling (MDS) where objects must be represented by suitable points in R^p such that a given dissimilarity matrix is optimally reproduced (minimization of a STRESS criterion). J. DE LEEUW presents a systematic classification of several quite commonly used optimization strategies and shows how block-relaxation methods split up into augmentation, majorization and EM algorithms, and in alternating least squares and ACE methods. – R. MATHAR considers MDS problems with Minkowski metrics instead of the Euclidean distance. – The best rank-r-approximation of a given (data) matrix is discussed by R. MEYER.

She obtains an extension of the classical optimality properties of the eigenvector solution to the case of orthogonally invariant norms and thereby unifies many investigations relating, e.g., to correspondence analysis and canonical correlation. – The next two papers are devoted to **spatial data analysis**: D. PFEIFER, U. SCHLEIER-LANGER and H.-P. BÄUMER present stochastic models from point process theory, stochastic geometry, and geostatistics in order to analyze the spatial dependence and structure of marine ecosystems. – F.-U. SIEWERT and P.O. DEGENS consider real-valued data sampled on a regular (rectangular, hexagonal) lattice in the plane and discuss methods for finding an additive decomposition, e.g., in a smooth, a periodic and an irregular component.

Part III: Genome and Molecular Sequence Analysis

This Part is devoted to the actual and rapidly developing domain of **modern molecular biology** and **genome or sequence analysis** including related methods for **phylogenetic tree reconstruction**. It contains various mathematical or statistical methods for the **analysis, seriation and classification of molecular data sets** (proteins, DNA). This field is surveyed in the paper of B. LAUSEN which focusses on (a) the mapping of genes by maximum likelihood methods (b) the alignment of DNA sequences or DNA fragments and (c) tree reconstruction with genetic distance data by using ultrametrics and additive trees. – W. SCHMIDT proposes a new distance measure for comparing sequences (amino acids) and shows its use for the construction of evolutionary trees and optimum multiple alignments (e.g. for 11 globin sequences). – M. SCHÖNIGER, A. JANKE and A. VON HAESELER investigate the phylogeny of nine vertebrates on the basis of 13 protein coding mitochondrial genes. They present, apart from parameter estimation methods and the models of Kimura and Hasegawa for nucleotide substitutions, a new method including properly weighted 'third codon positions'. The consideration of this special position is shown to be indispensible under certain evolutionary constellations. – The prediction of the hidden three-dimensional structure of amino-acids from the one-dimensional order and type of the amino acid sequence is the topic of J. SELBIG's paper. He uses the framework of machine learning and develops an appropriate (graph-based) pattern description language, suitable formal rules and generalization operators. – Finally, G. VOGT, D. FRISHMAN and P. ARGOS discuss the application of massively parallel processors for recognizing distantly related protein sequences in large databases by training a neural network to recognize suitably chosen motifs (net/motif search).

Part IV: Applied Data Analysis in Special Fields

The last part of this volume illustrates the practical use of classificatory, knowledge-based, or statistical methods in **economy, archeology, and medicine.** Two papers relate to **marketing studies**: R. DECKER presents the model- and knowledge-based data analysis system KVASS (within the WIMDAS system) for analyzing the consumer bying behavior on the basis of market diagnostics and panel data. – The paper of W. GAUL, U. LUTZ and E. AUST describe the analysis of a survey data set concerning the attitudes towards domestic or foreign products when people decide to purchase an automobile. After a brief survey on literature and previous work a four clusters segmentation of consumers is interpreted in detail. – **Archeological**

finds in Alamannic cemeteries are underlying the paper of P. IHM, B. SASSE and C. THEUNE. They propose a detailed classification system for (Merovingian) glass beads which is based on suitably coded information on material and technology, surface pattern, color, and shape. This leads to the construction of a hierarchical thesaurus for the indexation of beads. – The last two papers deal with screening problems in **medicine**: O. GEFELLER and R. MUCHE consider the evaluation of cholesterol screening strategies when classifying persons into low and high risk groups for cardiovascular disease. In addition to classical criteria such as sensitivity, specificity and predictive values, they use measures for the gain in certainty and inverse odds. – The paper of O. GEFELLER, J. WINDELER and H. J. TRAMPISCH discusses the evaluation of cancer screening strategies. They make the point that case-control-designs bear some unresolved problems as yet and advocate the use of randomized controlled trials.

The conference would not have been possible without the financial support of several institutions and private companies. First of all, we are very much indepted for the support provided by the Heidelberg Academy for the Humanities and Sciences and by our sponsors: Freundeskreis der Universität Kaiserslautern, Stadtsparkasse Kaiserslautern, TEHALIT (Kaiserslautern), Technische Werke Kaiserslautern, Knoll AG (Ludwigshafen), Hewlett-Packard GmbH (Böblingen), TecInno (Kaiserslautern). The Pfälzische Getränke GmbH (Kaiserslautern) provided refreshing drinks for all participants. The organizers also gratefully acknowledge the support by the commercial software presentations provided by LARS (Frankfurt) and GraS (Berlin).

The editors would like to thank the numerous (anonymous) referees whose help was invaluable in the difficult task of selecting among the submitted contributions. Finally, the organizers of the conference are most indebted to all the collaborators for their excellent work in running the conference. Special thanks is given to Edith Hüttel-Schwichtenberg, Dieter Hartmann, Christopher Wetzler who generously helped in various ways in the organization of the conference.

The editors hope that this volume will stimulate further research and communication among researchers and practitionners from the fields of knowledge processing, classification, and data analysis.

Aachen and Kaiserslautern, January 1994

Hans-Hermann Bock
Wolfgang Lenski
Michael M. Richter

Welcome address*:

The Conference Organizers' View of the Gesellschaft für Klassifikation

I would like to introduce this year's meeting of the Gesellschaft für Klassifikation, the German Classification Society, by sharing several thoughts that look at the themes to be raised at this conference from a unifying point of view.

We, the organizers of this conference, conceive the GfKl as an umbrella organization that brings together such different scientific disciplines as archeology, computer science, biology, psychology, mathematics, medicine and economics, to list just an arbitrary selection, because of a shared interest. I believe this also correctly describes the GfKl's (and its organs') understanding of itself.

Our approach of not losing sight of the central issues and the points of thematic convergence that unite the individual scientific disciplines, despite their great heterogeneity, put us in good company. Concepts such as 'unity', 'holism' and 'integration' are the subject of much discussion in the scientific and academic worlds. At a programmatic political level, they even take on a truly visionary dimension, being understood as standing in contradiction to a world that is disintegrating because of specialization and particularistic interests. If diffuse longings for unity are not to rob these concepts of all of their individual content, then it is necessary for us to reassure ourselves of the points of reference of the various forms of holism, three of which seem essential to me.

First, man himself is at the center of interest; his *personal 'wholeness'* is traditionally considered to be a unity of body, spirit and soul, and this is now once again the primary objective of the newest ideas about education. Second, there is the *wholeness of man's experiencing of the world*, which attempts to understand how objects and appearances are intertwined, or in modern terminology, linked as in networks. Third, and this seems to me paramount in our context, is the intersubjective effort concerned with the re-creation of an overall framework of understanding in which discoveries made in individual disciplines can be gathered and understood in a new, integrative sense. In the sciences, this general approach is called *interdisciplinary.*

Although we may reject the idea of science serving as a source of unity, we should not reject the idea of our attempting to work together on shared issues, each from the perspective and with the methods of his own scientific discipline. To us, such an issue would especially include the questions : What are the structures among the apparently unordered material (data) that provide unity? And how do we recognize them, use them, and represent them? Especially in the daily work of science, there is the danger of knowing more and more about less and less. Complaints have been raised about a tendency toward an isolation of research results (in which they cannot be understood outside the original discipline's paradigms) and a tendency against their use in other disciplines. From the point of view of striving for holism, we must work against this trend to particularization.

In saying this, I am by no means attempting to undermine the competence of specialists in various disciplines. On the contrary, one insight of interdisciplinary research is rather that the interdisciplinary approach invariably depends on competence in

*This is a slightly modified version of the welcoming address given in German language at the opening ceremony of the 17th Annual Conference of the Gesellschaft für Klassifikation (GfKl) e.V.. I am very indebted to Dr. Mike Wilson for his invaluable help in translating it into English.

various disciplines, since an openness for questions and answers from neighboring disciplines depends first and foremost on strength and competence within one's own discipline. One precondition of interdisciplinary achievements is to develop methods in one's own field that respond to issues raised in other disciplines, and to redirect problems that lie outside the methodical sphere of one's own sphere to other disciplines.

As an example of research where interdisciplinary collaboration is absolutely necessary, I would like to mention my own avenue of work. As a bibliographic enterprise of the Heidelberg Academy for the Humanities and Sciences, we set ourselves the task of cataloging one small field of mathematics in its entirety, namely mathematic logic, of presenting it as a complete and separate entity, of classifying it according to a basic structure, and of making its contents accessible. This programm of mapping knowledge in mathematical logic is a cultural task requiring the support of other scientific disciplines.

It is thus only natural that the first item on the agenda of a conference concerned with issues of relevance to many disciplines is that those participating meet in separate sections. That this can be more than simply working side by side, that is, achieve a high degree of true collaboration, is an insight I have gained in the GfKl at various occasions. This is one reason I was immediately attracted by the idea of, together with Michael M. Richter, organizing this 17th Conference here in Kaiserslautern.

We hope that this conference provides a forum promoting an exchange of information and helping scientists make personal contacts, a factor of such great importance for scientific progress. I wish that everyone attending this conference may experience personal enrichment and gain an impression of thematically related holism.

Wolfgang Lenski

Table of Contents

Part IV: Applied Data Analysis in Special Fields

Economy and Marketing

Archeology

Screening in Medicine:

Part I

Information Processing,
Classification-Based Approaches,
Linguistic and Conceptual Analysis

- Learning and Case-based Reasoning
- Approaches Based on Linguistic Analysis
- Information Processing and Design of Information Systems
- Uncertainty and Neural Networks
- Conceptual Approaches

Induction and Case-Based Reasoning for Classification Tasks

K.-D. Althoff[1], S. Wess[1], M. Manago[2], R. Bergmann[1], F. Maurer[1], E. Auriol[2], N. Conruyt[2], R. Traphöner[3], M. Bräuer[3], S. Dittrich[3]

[1]University of Kaiserslautern, D-67653 Kaiserslautern, Germany

[2]AcknoSoft, 58a rue du Dessous des Berges, F-75013 Paris, France

[3]tecInno GmbH, Sauerwiesen 2, D-67661 Kaiserslautern, Germany

Summary: We present two techniques for reasoning from cases to solve classification tasks: Induction and case-based reasoning. We contrast the two technologies (that are often confused) and show how they complement each other. Based on this, we describe how they are integrated in one single platform for reasoning from cases: The INRECA system.

1. Introduction

Induction and case-based reasoning are two technologies for the development of experience-based expert systems that have received considerable attention during the past decade. They provide methodologies for knowledge acquisition, validation of the knowledge base, and expert system maintenance. However, a confusion is often made between induction and case-based reasoning by tool vendors or even by academic researchers: Several systems presented with the label "case-based reasoning" are simply inductive tools and, on the other hand, some incremental versions of induction tools work in a case-based reasoning fashion. We distinguish between case-based reasoning and induction by considering that the first technique makes direct use of past experiences (cases) at the problem solving stage (diagnosis) while the second one only uses an abstraction of the cases. In other words: induction *compiles* past experiences into general heuristics which are then used to solve problems. Case-based reasoning directly *interprets* past experiences (cf. also Manago, Althoff et al., 1993; Wess, 1993a; Althoff, 1992).

Many systems are often at the frontier of the two approaches. For example, ID5 (cf. Utgoff, 1988) refers back to the cases in order to incrementally modify the decision tree. The question is, however, if such a system is purely an inductive system since it remembers past cases. We prefer to clearly distinguish the two kinds of systems in order to perform a cost and merit analysis which gives clues on how to integrate the two technologies such that they can indeed benefit from each other. Note that the fundamental distinction that we make between the two kinds of systems is not so much in the underlying technology. For example, information theory as in ID3 (cf. Quinlan, 1983) might be used to implement a case-based reasoning system.

The key distinction lies in how the technology is used. We believe that the integration of induction and case-based reasoning is one key issue for improving the development of diagnostic expert systems and will expand the set of applications that can be tackled. While both technologies in their own right are able to solve special instances of diagnostic problems, the combination of these approaches may result in more than "the sum of the respective single approaches". Up to now, no satisfying systems are available that base on a really deep integration of the underlying technologies.

We present the INRECA integrated learning system[1] which goes first steps into this direction. It includes inductive and case-based reasoning techniques. Currently, it is tested on two applications, name ly fault diagnosis of machine tools as well as the identification of marine sponges (cf. Manago, Althoff et al., 1993). While a more cooperative kind of integration of induction and case-based reasoning is described in Althoff, Bergmann et al. (1993), within this paper we focus on a deep integration of these technologies.

First, we motivate our approach on a more intuitive basis. Chapter 2 results in a more or less concrete guideline for integrating inductive and case-based reasoning based on mechanisms known from the field of information retrieval. We introduce multidimensional retrieval structures for associative search, especially k-d trees and describe the basic algorithms for tree construction and search. These basic data structures and algorithms are then extended to meet all the requirements of real complex diagnostic problems. Finally, we discuss our approach from several scientific points of view.

2. Inductive and Case-Based Reasoning

Case-based reasoning is a technology that allows to find analogies between a current working case and past experiences (reference cases). It makes direct use of past experiences to solve a new problem by recognising its similarity with a specific known problem and by, at least partially, applying the known solution to reach a solution for the actual new problem (cf., e.g., Kolodner, 1980; Schank, 1982; Althoff & Wess, 1991a+b).

Induction is a technology that automatically extracts knowledge from training examples (reference cases). It derives general knowledge from the cases: From an extensional description of concepts (i.e. the examples), it derives an intensional description of these concepts in the form of a decision tree, a set of most general rules (most general version of the concepts), or a characteristic description of the examples (most specific version of the concepts) (cf., e.g., Michalski, 1983; Quinlan, 1986; Manago & Kodratoff, 1987; 1990). This general knowledge is then used to solve new problems.

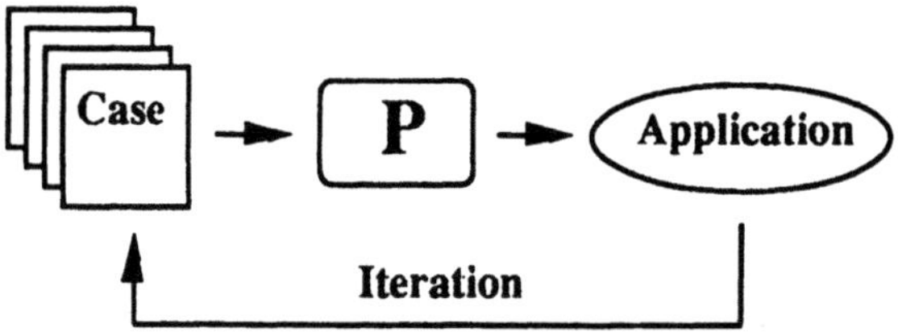

Figure 1: Inductive and Case-Based Processing of Cases

Induction and case-based reasoning both are processes that prepare cases for a certain application (cf. figure 1). If new cases occur, the inductive system has to update its derived concept descriptions. Therefore, the generation of concept descriptions (normally) has to be carried out again. For the case-based reasoning system, the

[1]This description does not necessarily reflect the official opinion of the whole INRECA consortium. Ongoing applications might change this.

consideration of new cases (normally) is no problem, because they only have to be included into the case base. But, the underlying similarity measure is, of course, not guaranteed to classify all new cases correctly. Thus, we may have to improve the measure based on the extended case base.

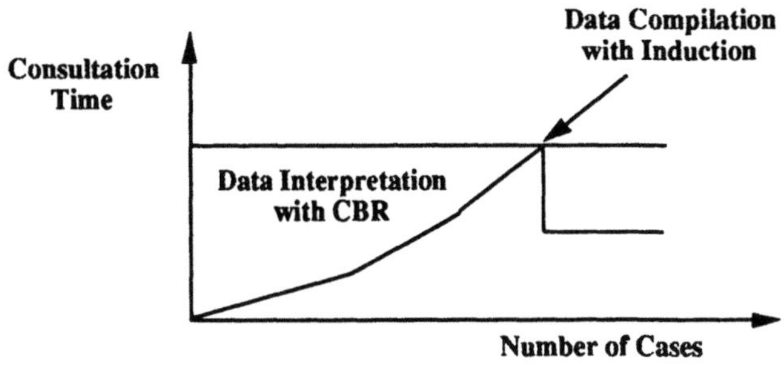

Figure 2: Combining Interpretation and Compilation

Having the above described commonalty in mind, we can use the above stated compilation-interpretation contrast to come up with an (abstract) idea for integration. This view is summarised in figure 2. Case-based reasoning is used as a flexible interactive mechanism to directly interpret the presented cases. If the number of cases strongly increases such that the time needed for consultation becomes too high, induction can be used as a compilation-like procedure which allows to reason with abstract knowledge being derived from the presented cases. If new cases are presented, case-based reasoning can be used again etc. Thus, we arrive at an interlocking of both reasoning schemes (cf. Althoff, Bergmann et al., 1993).

For the development of one single homogeneous architecture based on inductive and case-based reasoning technology that goes beyond this interlocking of the respective reasoning schemes, we generalise our view on induction. Through-out the following chapters, we consider induction as a technique which constructs abstractions from data for efficient processing. We denote the underlying construction mechanism by *fixed-order processing*, where the discovered structure within the given data corresponds to the "fixed order". With respect to the derived abstract structure, the procedure is static, i.e. inflexible in some sense. For instance, a TDIDT-like (top-down induction of decision trees; Quinlan, 1986) procedure derives a decision tree from the given cases. If certain assumptions are fulfilled[2] (no unknown or missing attribute values, no noise, no exceptions), such a decision tree enables an efficient consultation. Case-based reasoning does not apply such a kind of fixed-order processing. Therefore, its efficiency normally is worse, but it is more flexible in its reaction on data which do not meet the above mentioned requirements.

For instance, Althoff, Bergmann et al. (1993) describe the cooperation of a TDIDT-like inductive system and a case-based reasoner. The decision tree is used to preprocess the entered attribute values in a way that the number of interesting cases can be reduced. Thus, it works like a fixed indexing structure for the case retrieval where the induction and the case-based reasoning module are on the same level. We now suggest another kind of integration of induction and case-based reasoning by building

[2]At least to a high degree

a case-based reasoner that uses inductive techniques to improve its performance. The improvement will be in two different ways:

- reducing the average case complexity of the case retrieval step
- correcting misclassifications of the similarity measure

The main focus will be on the first kind of improvement (chapters 3-4), the second kind will be one major aspect discussed in chapter 5. We hope that the introduction of the fixed-order processing view helps to make transparent that using an efficient information retrieval technique, namely multidimensional retrieval structures for associative search, for case retrieval is a step towards the integration of induction and case-based reasoning. We will describe the basic retrieval algorithms in the next two chapters. To overcome certain restrictions of these algorithms, especially to keep the advantages of the case-based reasoning approach, we will introduce certain extensions for these algoritms. These extensions also allow the above mentioned second kind of inductive improvement, namely the heuristic adaptation of the (global) similarity measure to avoid misclassifications.

3. Multidimensional Retrieval Structures

We developed a retrieval mechanism that is based on a k-d tree, a multi-dimensional binary search tree (Wess, 1993b; Bentley, 1975; Friedman, Bentley & Finkel, 1977). This mechanism is built on top of an object-oriented data base (Öchsner & Wess, 1992). This leads us, e.g., to the following correspondences: case = entity/object, case base = data base, problem = query, similarity-based case retrieval = best-match search. Within the k-d tree an incremental best-match search is used to find the m most similar cases (nearest neighbours) within a set of n cases with k specified indexing attributes. The search is guided by application-dependent similarity measures based on user-defined value ranges. The used similarity measures are constructed according to Tverskys contrast model (Tversky, 1977), but the user is free to define other ones. He is only restricted to use ordered value ranges as well as monotonic and symmetric similarity functions, which is not a problem for many real applications. The k-d tree uses the inhomogeneity of the search space for density-based structuring. The balanced retrieval structure results in a small number of accesses to external media.

Every node within the k-d tree represents a subset of the cases of the case base, the root node represents the whole case base. Every inner node partitions the represented case set into two disjoint subsets. The next discriminating attribute within the tree is selected based on the inter quartile distance of the attributes' value ranges (cf. Koopmans, 1987). Splitting in the median of the discriminating attribute makes the k-d tree an optimal one (the tree is optimal if all leaf nodes are at adjoining levels).

Search in the k-d tree is done via recursive tree search and the use of two test procedures: BALL-WITHIN-BOUNDS (BWB) and BOUNDS-OVERLAP-BALL (BOB) (cf. figure 3). These procedures check whether it would be reasonable to explore certain areas of the search space in more detail, or not. Such tests can be carried out without retrieving the respective cases. The geometric bounds of the considered subspaces are used to compute a "similarity interval" whose upper bound then "answers" the question to explore, or not.

The average case effort (measured by the number of comparisons; cf. Jacquemain, 1988) for generating a k-d tree is $O[k * n * log_2 n]$, for the worst case $O[k * n^2]$. The

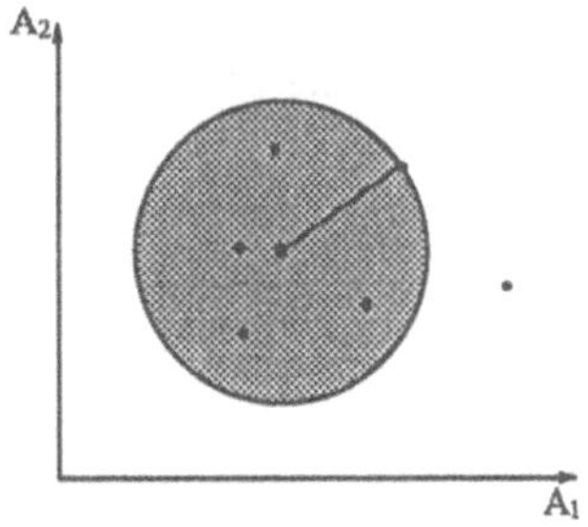

Figure 3: Bounds-Test for Nearest Neighbour Search

average costs for retrieving the most similar case are $O[log_2 n]$, if the tree is optimally organised. For the worst case, the retrieval costs are $O[n]$. The retrieval mechanism is correct in that sense that it always finds the m most similar cases. The costs for the reorganisation of the k-d tree (making the tree an optimal one again) are $O[l * log_2 l]$, where l is the number of leaf nodes belonging to the non-balanced subtree, i.e. the costs to rebuild the whole tree are $O[n * log_2 n]$.

4. k-d Trees

We describe the basic procedures for generating and searching a k-d tree. Here, we already include some modifications, e.g. changing distance to similarity measures. This is necessary because we need the notion of similarity for the case-based reasoning component. The similarity measure is split into local measures for each value range and a global measure that is composed from the local ones. We need the local measures during the construction of the k-d tree for selecting the next discriminating attribute. The global measure is used for searching the tree. Starting from this basic retrieval mechanism, we introduce several extensions that are necessary in the context of diagnostic reasoning.

4.1 Building a k-d Tree: Basic Algorithm

For generating an optimal k-d tree, we need as input the case base $CB := \{C_i \mid C_i := (c_{i1}, c_{i2}, \ldots, c_{ik}), i \in \{1, \ldots, n\}, c_{ij} \in R_j := R(A_j)\}$, the indexing attributes $A_1 \ldots A_k$, the value ranges $R_1 \ldots R_k$, the local similarity measures $\mu_1 \ldots \mu_k$, $\mu_i : R_i \times R_i \rightarrow [0, 1]$, and the bucket size b which defines how many cases are at most allowed to be included in one leaf node. Every case includes a distinguished attribute (called *diagnosis*) which is, of course, not used for indexing.

If $|CB| \leq b$ then only one leaf node is generated and the construction process terminates. Otherwise, an inner node is generated. For every attribute $A_i, i \in \{1, \ldots, k\}$, the quartiles $q_1^{(i)}$ and $q_3^{(i)}$ of its in CB occurring values are computed. The inter quartile similarity is defined as $iqr^{(i)} := \mu_i(q_1^{(i)}, q_3^{(i)})$. As a discriminating attribute d, which is attached to the generated inner node, we select that one of which the inter quartile similarity iqr is the lowest: $d := \{i \mid iqr^{(i)} \leq iqr^{(j)}\}$. This easily corresponds to the use of inter quartile distances where that attribute is selected of which the respective quartiles have the maximal distance.

Since every inner node should partition the case set into two equally-sized subsets, for every discriminating attribute d the respective median p for the value range R_d

is computed: $p := median\{a_j \mid (a_1, \ldots, a_k) \in CB, j = d\}$. Then optimal k-d trees for the partitions $CB_{\leq}$ and $CB_{>}$ are generated: $CB_{\leq} := \{(a_1, \ldots, a_k) \in CB \mid a_d \leq p\}, CB_{>} := \{(a_1, \ldots, a_k) \in CB \mid a_d > p\}$.

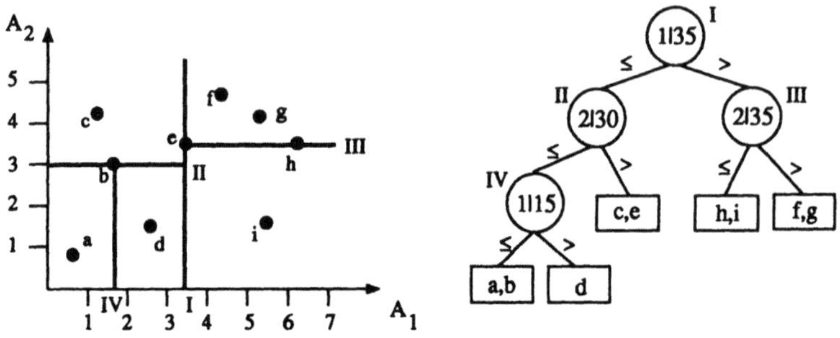

Figure 4: An Exemplary Two Dimensional Search Space and the Corresponding k-d Tree

4.2 Searching a k-d Tree

For finding the m most similar cases for a given working case (or query case)[3], we apply recursive tree search. Thus, as input we need the query case C_q, the number m of most similar cases, the k-d tree represented by its root node, and the global similarity measure $sim : [0,1]^k \rightarrow [0,1]$, and

$$sim(C_h, C_i) \rightarrow F(\mu_1(c_{h1}, c_{i1}), \mu_2(c_{h2}, c_{i2}), \ldots, \mu_k(c_{hk}, c_{ik})) \quad C_h, C_i \in CB$$

One simple example for the (monotonic) function F is:

$$sim(C_h, C_i) := F(\mu_1(c_{h1}, c_{i1}), \ldots, \mu_k(c_{hk}, c_{ik})) := \frac{1}{k} \sum_{j=1}^{k} \mu_j(c_{hj}, c_{ij})$$

During search a priority queue is continuously updated which includes the m most similar cases. If the recursive search procedure examines a leaf node, the similarity of all included cases is computed and, if necessary, the priority queue is updated. If the examined node is an inner node, then the search procedure is recursively called for that son node which should include the query case. If this call terminates, it is tested whether it is also necessary to examine the other son node by using the BOUNDS-OVERLAP-BALL test. It is TRUE if the cases of the actual tree node have to be explored.

The inner nodes are correct generalisations of the all the cases they represent in that sense that they include the geometric (upper and lower) bounds (for every indexing attribute) which correspond to the respective subspace. These geometric bounds are used to compute a similarity interval whose upper bound then answers the question to explore, or not. The closest point C_{min} within the actual nodes subspace is computed as the projection onto the actual nodes geometric bounds. C_{min} is on the actual nodes bounding box on the edge facing the query case C_q. If there is no overlapping in any of the k dimensions between the nodes bounding box and the k-dimensional ball

[3]For a query case the value of the distinguished attribute *diagnosis* is unknown

round C_q then C_{min} is a corner of the bounding box. If C_q is within the bounding box then $C_q = C_{min}$ (cf. also figure 3).

Before the recursive search procedure terminates the BALL-WITHIN-BOUNDS test is applied. It is TRUE if the k-dimensional ball round C_q is completely within the bounding box of the actual tree node. If this is the case, no overlapping with other bounding boxes is possible any more. Thus, the search is finished. Two cases $C_1^{(i)}$ and $C_2^{(i)}$ per dimension $i \in \{1, \ldots, k\}$ are generated (building an interval according to the geometric bounds of the actual tree node's bounding box) to test whether the m most similar cases are all within that bounding box.

5. Extensions

The associative search mechanism, as described above, is used for the basic memorisation and retrieval task in our case-based reasoner. But, there exist a lot of real diagnostic problems which cannot be handled satisfactorily up to now (cf. Althoff & Wess, 1991a; Manago, Althoff et al., 1993; Wess, 1993). Our approach is to introduce extensions for the global similarity measure, the k-d tree representation and search, as well as the overall similarity assessment process (e.g., use of domain knowledge). Within this paper, we want to focus on the integration of an adaptive learning mechanism to automatically improve the global similarity measure. It is the second kind of improvement of our case-based reasoner using induction. Another reason is that this learning strategy is already built on top of other important extensions which then can be introduced implicitly by this procedure. The used learning strategy is similar to competitive learning (cf. Rumelhart & Zipser, 1985) and has been described in Wess (1993) and Richter (1992). Here, we concentrate on the combination of this learning strategy with the above described basic memorisation and retrieval mechanism.

We now stepwise introduce all necessary extensions. First, we improve the global similarity measure using global and local weights. The latter are defined by use of a relevance matrix R which includes a special weight for every attribute/diagnosis pair. A local weight w_{ji} denotes the relative importance (relevance) of an attribute A_j for the diagnosis D_i. Such weights effect the ball tests BOB and BWB because (only) here the global similarity measure sim is used. The consequence for the k-ball round the query case C_q is that there is a tendency to exact matching on important dimensions, and that there is an increasing degree of flexibility for less important dimensions. The relevance matrix is defined as follows:

$$\mathrm{R} = \begin{pmatrix} & D_1 & D_2 & \ldots & D_m \\ A_1 & w_{11} & w_{12} & \ldots & w_{1m} \\ A_2 & w_{21} & w_{22} & \ldots & w_{2m} \\ \vdots & \vdots & \vdots & \ddots & \vdots \\ A_n & w_{n1} & w_{n2} & \ldots & .w_{nm} \end{pmatrix}$$

$$\sum_{j=1}^{n} w_{ji} := \sum_{j=1}^{n} w_j(D_i) := 1$$

For every value range R_j, we introduce the distinguished value of *unknown*. During the generation of the k-d tree it has the meaning of *don't care*, during retrieval that of a missing value.

We also introduce global, i.e. diagnosis-independent, weights for special groups of attributes. Such groups are defined using the distinguished values of unknown$^{(1)}\ldots$ unknown$^{(k)}$ as well as the additional information whether an attribute value is a pathologic[4] one, or not. Let $C_i \in CB, C_i := (c_{i1}, c_{i2}, \ldots, c_{ik})$, be some case of the case base and C_q a query case, $C_q := (c_{q1}, c_{q2}, \ldots, c_{qk})$, where C_i includes the diagnosis $D^{(i)}$ and C_q's diagnosis is not known (per definition). We distinguish the following sets of attributes:

- $E := \{j \mid \mu_j(c_{ij}, c_{qj}) > \Omega_j\}$ *Equivalent attribute values*
- $C := \{j \mid \mu_j(c_{ij}, c_{qj}) \leq \Omega_j\}$ *Conflicting attribute values*
- $U := \{j \mid c_{qj} = unknown^{(j)}\}$ *Unknown attribute values*
- $A := \{j \mid c_{ij} = unknown^{(j)} \land c_{qj}$ is pathologic $\}$ *Additional attribute values*

Note, that the decision whether two values are equivalent or conflicting, i.e. belong to E or C, is made by use of the respective local similarity measure μ_j as well as a range-dependent threshold $\Omega_j \in [0,1]$. Based on the above defined attribute sets, we introduce the following improved global similarity measure *sim*:

$$sim(C_i, C_q) = \frac{\alpha E}{(\alpha E + \beta C + \eta U + \gamma A)} \qquad \alpha, \beta, \eta, \gamma > 0$$

where E, C, U, and A denote the following expressions:

$$\begin{aligned} E &:= \sum_{j \in E} w_j(D_i) * \mu_j(c_{ij}, c_{qj})) \\ C &:= \sum_{j \in C} w_j(D_i) * (1 - \mu_j(c_{ij}, c_{qj}))) \\ U &:= \sum_{j \in U} w_j(D_i)) \\ A &:= |A| \end{aligned}$$

Practical experience led us to the use of $\alpha = 1, \beta = 2, \gamma = 1$, and $\eta = 1/2$.

Since we have introduced the distinguished values unknown$^{(1)}\ldots$unknown$^{(k)}$ as well as diagnosis-dependent similarity measures, we have to extend the k-d tree mechanism:

- Attributes are selected as discriminating attributes only if the percentage of occurring unknown values is not too high. Otherwise, they are not used for indexing at all.
- Every node within the k-d tree "remembers" which diagnoses are included within the cases belonging to the respective node's subtree.
- While searching the k-d tree the tests BOB and BWB are applied using the diagnosis-dependent similarity measures of all diagnoses which occur in the respective node's subtree.

---

[4]Pathologic (or abnormal) attribute values within a query case are very important and must be explained by a similar case in the case base. Thus, they are weighted maximally (=1)

If only a few attributes can be used for indexing (because of many unknown values), it might happen that the leaf nodes contain more than (bucket size) b cases. For such cases we have, of course, linear retrieval costs. Within the buckets, the cases are sorted by their included diagnoses. Thus, the BWB test can be performed more efficiently.

In real applications, we are not always interested in the most similar case(s), only if such case(s) are sufficiently similar. This leads us to the definition of diagnosis-dependent thresholds $\sigma(D_i)$, which must be exceeded by the global similarity measure *sim* in order to terminate the overall classification with a certain diagnosis D_i as output: $sim(C_i, C_q) > \sigma(D^{(i)})$ if C_i sufficiently similar to C_q.

5.1 Automatic Adaptation of the Similarity Measure

Experiments in our laboratory with given case bases of correctly classified cases CB_{cor} (iteratively selecting, and temporarily removing, one case for the use as query case) showed that the similarity measure often did not classify correctly, though only one case has been removed from the case base. But, this can be improved applying an adaptive learning process. The goal is to learn new weights, i.e. new entries of the relevance matrix R. This process has an initial and a learning phase, the training set is the case base CB_{cor}.

Initial phase: the initial weights w_{ji} are determined according to the observed frequencies in the base.

Learning phase: the query cases C_q are taken from the case base CB_{cor}, i.e. every case of CB_{cor} will be a query case once. Such a query case is then temporarily removed from case base. The system determines the most similar case C_{sim}. Since the query cases are selected from CB_{cor}, it is possible to compare the respective diagnoses of C_q and C_{sim}. If $D^{(sim)} = D^{(q)}$, then nothing will be changed. For $D^{(sim)} \neq D^{(q)}$ we distinguish two possibilities:

- C_{sim} contains less known attribute values than C_q, i.e. the known values of C_{sim} are a subset of the known values of C_q. Here, the diagnosis $D^{(sim)}$ was obviously only correct by accident and C_{sim} is eliminated from the case base.
- In all other situations C_{sim} remains in the case base but the weights are updated.

The numerical form of the learning rule is not of interest here (cf. Wess, 1991). The leading principles are the achievement of $sim(C_{sim}, C_q) = \sigma(D^{(sim)})$, i.e. C_{sim} and C_q are no more sufficiently similar, the increase of the weights belonging to the attributes in C and U, the decrease of the weights belonging to the attributes in E, as well as the normalisation of the weights. The weights belonging to the attributes in A remain invariant. Since there is a remaining degree of freedom in the underlying equation formula, we choose the following: high weights belonging to the attributes in E are highly decreased, low weights only to a low degree. In addition, low weights belonging to the attributes in C and U are highly increased, high weights only to a low degree. Here, the goal is to "support" attributes which had only a small responsibility for the misclassification, and vice versa.

5.2 Domain Knowledge

The overall similarity assessment process can be improved by the use domain knowledge. Default values can be used to increase the number of known attribute values.

Causal and heuristic determination rules can be used to derive new attribute values from known ones. Since such knowledge increases the available information, similarity is estimated on a broader basis. For the automatic generation (of a part) of that knowledge and its detailed use cf. Althoff (1992), Rehbold (1991), Althoff and Wess (1991a), and Wess (1991).

6. Discussion

The overall scenario we assumed is comparable to Gentner and Forbus' MAC/FAC model[5] (Gentner & Forbus, 1991; cf. figure 5). We used a fixed-order processing technique as the basic case retrieval mechanism which can be compared to the MAC phase. The described extensions (chapter 5) then correspond to the FAC phase.

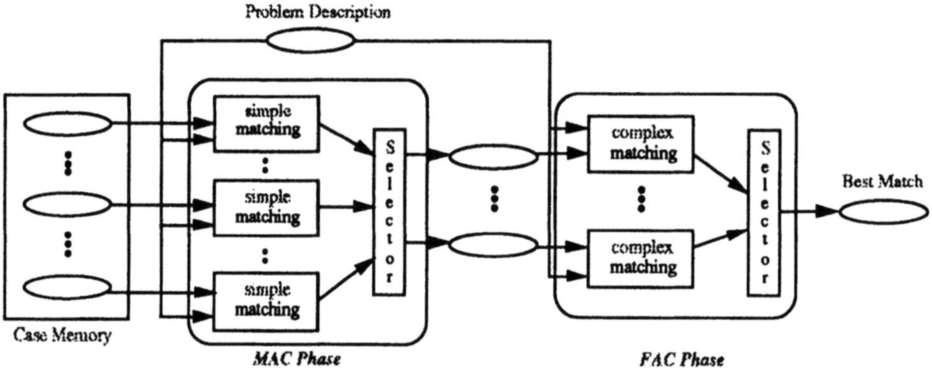

Figure 5: MAC/FAC Model

Characteristics of inductive learning have been summarised in, e.g., Jantke and Lange (1989). From this abstract point of view, case-based learning could be described as follows: From a given sequence of cases, learning hypotheses are incrementally generated. Such a hypothesis consists of a pair of a set of cases and an associated similarity measure. Questions to be answered are: Which cases will be entered into the case base, which will be removed from it, and how to realise the similarity measure. Since the criteria of Jantke and Lange can be analogously applied to case-based learning, it can be viewed as a special instance of inductive learning. A common theoretical framework is necessary to achieve results on this topic. First steps in this direction have been described by Jantke, Richter et al. (1991), Jantke (1992), and Globig and Wess (1993).

Both inductive learning and case-based learning have in common that they derive "global" knowledge from "local" observations (which, of course, are uncertain, respectively). However, they use different techniques to achieve this: Inductive learning bases mainly on logical concept descriptions ("logical reasoning"), whereas case-based reasoners often use analytic descriptions ("geometric reasoning") (cf., e.g., Richter, 1992). One consequence from this is that inductive learners mostly start with the "dropping of complete dimensions" in contrast to case-based reasoners which "decompose complete dimensions into intervals". It depends on the use of a learning

[5]Many Are Called but Few Are Chosen

result which particular technique is then the more successful one. Therefore, the INRECA approach integrates both learning strategies within a broader architecture for identification and diagnostic reasoning.

Up to now, much work has been done on the integration of different knowledge representation and processing schemes to improve knowledge acquisition. E.g., a comparative analysis as well as a proposed integration of models, cases and compiled knowledge have been given by van Someren, Zheng and Post (1990). The MOLTKE architecture also bases on these three schemes (cf. Althoff, Maurer & Rehbold, 1990; Althoff, Maurer et al., 1992; Althoff, 1992). The GRANUL system integrates several existing knowledge acquisition tools into one coherent system that supports several styles of knowledge acquisition (Aben, van Someren & Terpstra, 1992). The MOBAL system is an interesting example for the integration of manual and automatic knowledge acquisition methods (the balanced cooperative modelling issue, cf. Morik, 1991). Van de Velde and Aamodt (1992) have analysed the possible use of machine learning techniques within the KADS approach to expert system development. Rissland and Skalag (1989) introduced the notion of mixed paradigm reasoning for the integration of different reasoning schemes (reasoning from cases, rules, constraints, deep models etc.). Examples here are CABARET (Rissland, Basu et al., 1991), CREEK (Aamodt, 1991), PATDEX/MOLTKE (Althoff & Wess, 1991a; Richter & Wess, 1991), GREBE (Branting & Porter, 1991), and JULIA (Hinrichs & Kolodner, 1991), among others. A first suggestion for the integration of case-based reasoning and model-based knowledge acquisition is given in Janetzko and Strube (1992).

7. Conclusion

We have introduced basic parts of the architecture of the Inreca system that uses induction and case-based reasoning for solving classification tasks. INRECA is being applied to real world problems in the areas of technical maintenance as well as the pharmaceutical industry. Results from this applications might change the suggested architecture.

8. Acknowledgement

Funding for this research has been provided by the Commission of the European Communities (Esprit contract P6322, the INRECA project). The partners of INRECA are AcknoSoft (prime contractor, France), tecInno (Germany), Irish Medical Systems (Ireland), and the University of Kaiserslautern (Germany).

9. References

AAMODT, A. (1991). A Knowledge-Intensive, Integrated Approach to Problem Solving and Sustained Learning. *Doctoral Dissertation*, University of Trondheim

ABEN, M., VAN SOMEREN, M. W. & TERPSTRA, P. (1992). Functional and Representational Integration in Knowledge Acquisition. *Proc. International Machine Learning Conference, Workshop on "Computational Architectures for Supporting Machine Learning and Knowledge Acquisition" in Aberdeen*

ALTHOFF, K.-D. (1992). Eine fallbasierte Lernkomponente als integrierter Bestandteil der Moltke-Werkbank zur Diagnose technischer Systeme. *Doctoral Dissertation*, University of Kaiserslautern; also: Sankt Augustin (Germany): Diski 23, infix Verlag

ALTHOFF, K.-D., BERGMANN, R., MAURER, F., WESS, S., MANAGO, M., AURIOL, E., CONRUYT, N., TRAPHÖNER, R., BRÄUER, M. & DITTRICH, S. (1993). Integrating

Inductive and Case-Based Technologies for Classification and Diagnostic Reasoning. *Proc. ECML-93 Workshop on Integrated Learning Architectures (edited by E. Plaza)*

ALTHOFF, K.-D., MAURER, F. & REHBOLD, R. (1990). Multiple Knowledge Acquisition Strategies in MOLTKE. In: B. J. Wielinga, J. Boose, B. Gaines et al. (eds.), *Current Trends in Knowledge Acquisition* (Proc. EKAW-90). Amsterdam: IOS Press, 21-40

ALTHOFF, K.-D., MAURER, F., TRAPHÖNER & WESS, S. (1992). MOLTKE - An Integrated Workbench for Fault Diagnosis in Engineering Systems. *Proc.* EXPERSYS-92, *Paris*

ALTHOFF, K.-D. & WESS, S. (1991a). Case-Based Knowledge Acquisition, Learning, and Problem Solving in Diagnostic Real World Tasks. *Proc.* EKAW-91, *Glasgow & Crieff*

ALTHOFF, K.-D. & WESS, S. (1991b). Case-Based Reasoning and Expert System Development. In: F. Schmalhofer, G. Strube & T. Wetter (eds.), *Contemporary Knowledge Engineering and Cognition*, Springer Verlag

ALTHOFF, K.-D., WESS, S., BARTSCH-SPÖRL, B. & JANETZKO, D. (eds.) (1992). Ähnlichkeit von Fällen beim fallbasierten Schliessen. *Proc. of the first Meeting of the German Special Interest Group on Case-Based Reasoning*, Seki-Working-Paper SWP-92-11, University of Kaiserslautern

BENTLEY, J. L. (1975). Multidimensional Search Trees Used for Associative Searching. *Communications of the ACM 18*, 509-517

BRANTING, L. K. & PORTER, B. W. (1991). Rules and Precedents as Complementary Warrants. *Proc. AAAI-91*, 3-9

FRIEDMAN, J. H., BENTLEY, J. L. & FINKEL, R. A. (1977). An Algorithm for Finding Best Matches in Logarithmic Expected Time. *ACM Trans. math. Software 3*, 209-226

GENTNER, D. & FORBUS, K. D. (1991). Mac/Fac: A Model of Similarity-Based Retrieval. *Proc. of the 13th Annual Conference of the Cognitive Science Society*, 504-509

GLOBIG, CH. & WESS, S. (1993). Symbolic Learning and Nearest Neighbour Classification. *Technical Report*, University of Kaiserslautern

HINRICHS, T. R. & KOLODNER, J. L. (1991). The Roles of Adaptation in Case-Based Design. In: R. Bareiss (ed.), *Proc. 3rd DARPA Workshop on Case-Based Reasoning*, Morgan Kaufmann, 121-132

JACQUEMAIN, K. J. (1988). *Effiziente Datenstrukturen und Algorithmen für mehrdimensionale Suchprobleme.* Hochschultexte Informatik (Bd. 5), Heidelberg: Hüthig Verlag

JANETZKO, D. & STRUBE, G. (1992). Case-based Reasoning and Model-based Knowledge Acquisition. In: F. Schmalhofer, G. Strube & Th. Wetter (eds.), *Contemporary Knowledge Engineering and Cognition*, Springer Verlag

JANTKE, K. P. (1992). Case-Based Reasoning in Inductive Inference. *Proc. COLT-92*

JANTKE, K. P. & LANGE, S. (1989). Algorithmic Learning Theory (in German: Algorithmisches Lernen). In: J. Grabowski, K. P. Jantke & H. Thiele (eds.), *Grundlagen der Künstlichen Intelligenz*, Akademie-Verlag, 246-277

JANTKE, K. P., RICHTER, M. M., ALTHOFF, K.-D., LANGE, S. & WESS, S. (1991). IND-CBL - Vergleich ausgewählter Ansätze aus dem induktiven und dem fallbasierten Lernen. *DFG project proposal*

KOLODNER, J. L. (1980). Retrieval and Organisational Strategies in Conceptual Memory: A Computer Model. *Ph.D. Thesis*, Yale University

KOOPMANS, L. H. (1987). *Introduction to Contemporary Statistical Methods.* Second Edition, Duxbury Press, Boston

MANAGO, M. & KODRATOFF, Y. (1987). Model Driven Learning of Disjunctive Concepts. *Progress in Machine Learning (Proc. of the 2nd European Working Session on Learning)*, edited by Bratko & Lavrac, Sigma Press (distributed by John Wiley & Sons)

MANAGO, M. & KODRATOFF, Y. (1990). KATE: A Piece of Computer Aided Knowledge Engineering. *Proc. of the 5th AAAI Workshop on Knowledge Acquisition for Knowledge-Based Systems*, edited by B. R. Gaines & J. Boose, Banff, Canada, AAAI Press

MANAGO, M., ALTHOFF, K.-D., AURIOL, E., TRAPHÖNER, R., WESS, S., CONRUYT, N. & MAURER, F. (1993). Induction and Reasoning from Cases. In: RICHTER, WESS ET AL. (1993), 313-318

MICHALSKI, R. S. (1983). Theory and Methodology of Inductive Learning. In: R. S. Michalski, J. G. Carbonell & T. M. Mitchell (eds.), *Machine Learning: An Artificial Intelligence Approach*, Tioga Publishing Co.

MORIK, K. (1991). Balanced Cooperative Modelling Using Mobal - An Introduction. *Technical Report (Gmd-F3-Nachrichten AC Special Nr. 3)*, GMD, Sankt Augustin

ÖCHSNER, H. (1992). Mehrdimensionale Zugriffspfadstrukturen für das ähnlichkeitsbasierte Retrieval von Fällen. *Diploma Thesis*, University of Kaiserslautern

ÖCHSNER, H. & WESS, S. (1992). Ähnlichkeitsbasiertes Retrieval von Fällen durch assoziative Suche in einem mehrdimensionalen Datenraum. In: ALTHOFF, WESS ET AL. (1992), 101-106

PEWS, G., WEILER, F. & WESS, S. (1992). Bestimmung der Ähnlichkeit in der fallbasierten Diagnose mit simulationsfähigen Maschinenmodellen. In: Althoff, Wess et al. (1992), 47-50

QUINLAN, J. R. (1986). Induction of Decision Trees. *Machine Learning 1*, 81-106

REHBOLD, R. (1991). Integration modellbasierten Wissens in technische Diagnostik-Expertensysteme. *Doctoral Dissertation*, University of Kaiserslautern

RICHTER, M. M. (1992). Classification and Learning of Similarity Measures. *Proc. of the 16th Annual Conference of the Gesellschaft für Klassifikation e.V.*, Springer Verlag

RICHTER, M. M. & WESS, S. (1991). Similarity, Uncertainty and Case-Based Reasoning in PATDEX. *Automated Reasoning - Essays in Honour of Woody Bledsoe*, Kluwer Academic Publishers

RICHTER, M. M., WESS, S., ALTHOFF, K.-D. & MAURER, F. (eds.) (1993). *Proc. of the First European Workshop on Case-Based Reasoning*, Seki-Report SR-93-12, University of Kaiserslautern

RISSLAND, E. L., BASU, C., DANIELS, J. L., MCCARTHY, J., RUBINSTEIN, Z. B. & SKALAG, D. B. (1991). A Blackboard-Based Architecture for Case-Based Reasoning: An Initial Report. In: R. Bareiss (ed.), *Proc. of the 3rd DARPA Workshop on Case-Based Reasoning*, Morgan Kaufmann, 77-92

RISSLAND, E. L. & SKALAG, D. B. (1989). Combining Case-Based and Rule-Based Reasoning: A Heuristic Approach. *Proc. IJCAI-89*, 524-530

RUMELHART, D. E. & ZIPSER, D. (1985). Feature Discovery by Competitive Learning. *Cognitive Science 9*, 75-112

SCHANK, R. C. (1982). *Dynamic Memory: A Theory of Learning in Computers and People.* Cambridge, UK: Cambridge University Press

TVERSKY, A. (1977). Features of Similarity. *Psychological Review 84*, 327-352

UTGOFF, P. (1988). ID5: An Incremental ID3. *Proc. of the 5th International Conference on Machine Learning, Irvine, CA: Morgan Kaufmann*

VAN DE VELDE, W. & AAMODT, A. *(1992). Machine Learning Issues in CommonKADS.* Esprit-Project P5248, Technical Report KADS-II/TII.4.3/TR/VUB/002/3.0

VAN SOMEREN, M. W., ZHENG, L. L. & POST, W. *(1990). Cases, Models or Compiled Knowledge: a Comparative Analysis and Proposed Integration. In: B. J. Wielinga, J. Boose, B. Gaines et al. (eds.),* Current Trends in Knowledge Acquisition *(Proc.* EKAW-90*). Amsterdam: IOS Press*

WESS, S. *(1991).* PATDEX/2 - *ein System zum adaptiven, fallfokussierenden Lernen in technischen Diagnosesituationen.* Seki-Working-Paper SWP-91-01, *University of Kaiserslautern*

WESS, S. *(1993).* PATDEX - *ein Ansatz zur wissensbasierten und inkrementellen Verbesserung von Ähnlichkeitsbewertungen in der fallbasierten Diagnostik. In: F. Puppe & A. Günter (eds.),* Proc. of the 2nd German Conference on Expert Systems, *Hamburg, Springer Verlag*

Symbolic Learning and Nearest-Neighbor Classification[1]

Christoph Globig, Stefan Wess

Fachbereich Informatik
Universität Kaiserslautern
Postfach 3049
D 67653 Kaiserslautern, Germany
{globig,wess}@informatik.uni-kl.de

Summary: The Nearest-Neighbor Classification has a long tradition in the area of pattern recognition while knowledge-based systems apply mainly symbolic learning algorithms. There is a strong relationship between Nearest-Neighbor Classification and learning. The increasing number of cases and the adaptation of the similarity measure are used to improve the classification ability. Nowadays, Nearest-Neighbor Classification is applied in knowledge-based systems by a technique called case-based reasoning. In this paper we present first results from a comparison of case-based and symbolic learning systems.

1. Introduction

The Nearest-Neighbor Classification has a long tradition in the area of pattern recognition while knowledge-based systems apply mainly symbolic learning algorithms. Nowadays, Nearest-Neighbor Classification (Dasarathy, 1990) is applied in knowledge-based systems by a technique called case-based reasoning (Kolodner, 1991). A Nearest-Neighbor (or case-based) classifier consists of a finite set of already classified objects – the case base – and a measure of similarity.[2] Given a new object the system searches in the case base CB for the nearest neighbor (or the most similar case) with respect to a measure of similarity sim (or a distance measure d) and states its classification as the classification of the new object. If the case-based classifier gets more and more cases it builds a sequence of pairs (CB_j, sim_j), i.e. the classifier may change the case base and the measure of similarity. The aim is to get in the limit a pair (CB, sim), that needs no further change because it is a correct classifier for the target concept.

The two paradigms of symbolic (Michalski *et al.*, 1983) and case-based classification (Aha *et al.*, 1991; Aha, 1991) arise the question which one performs better than the other. For the area of Inductive Inference, Jantke (Jantke, 1992) proved the equality of the learning power of symbolic and case-based classifiers. The proof is based on the learner's ability to adjust the measure of similarity to the given problem. To use Nearest-Neighbor classification it is necessary to understand the possibilities to adjust the measure of similarity, or more generally to understand the use of information in the algorithms.

We want to illustrate the abilities of case-based classification by a comparison of a well-known symbolic algorithm (the Version Space (Mitchell, 1982)) and an equivalent case-based variant. In section 3 we examine the interdependence between the concept,

---

[1]The presented work was partly supported by the *Deutsche Forschungsgemeinschaft*, SFB 314: "Artificial Intelligence and Knowledge Based Systems" and the Project IND-CBL.

[2]It is possible to use a distance measure instead of a measure of similarity. We will use the term *measure* to capture both types. For the equivalence of distance and similarity measures see (Richter, 1992).

Version Space algorithm

1. Initialize G as the set containing only the most general concept $G = \{(*, ..., *)\}$ and $S = \{a_1\}$.
2. Assume the new example a is positive.
   Remove all concepts g from G which are not fulfilled by a. Search for the most specialized concept C in the Version Space which is fulfilled by all positive examples and set $S = \{C\}$.

   Assume the new example a is negative.
   Remove all concepts s from S which are not fulfilled by a. For every concept g from G which is fulfilled by a, search for the most general specializations, which are fulfilled by all known positive and no known negative example. Replace g by the found specializations.
3. If there is a concept g in G which is more specific than a concept in S, or $S = \emptyset$ then **HALT**(The examples do not fit any concept of the Version Space).
4. If $S = G$
   then **RETURN**(Found concept = S)
   else go to 2)

Figure 1: The Version Space Algorithm

the case base, and the measure of similarity in more detail. The fourth section discusses the notion of appropriateness of a measure for the efficient learning of a concept.

2. Example: Version Space

The Version Space algorithm is a simple and well-known symbolic learner (Mitchell, 1982). Because of its simplicity it is easy to show a lot of properties, which hold for many other learning systems, where it would be difficult to prove them. First we want to describe the Version Space.

Let $W_i := N_{k_i}$ $(i := 1, \ldots, n)$ be sets of values.[3] $U := W_1 \times \ldots \times W_n$ is the universe. A concept is a vector $C = (C_1, ..., C_n)$, where $C_i = *$ or $C_i = a_i$ $(a_i \in W_i)$. An object $a = (a_1, a_2, \ldots, a_n)$ fulfills the concept C (i.e. $C(a) = 1$), if for all $1 \le i \le n$ holds: $C_i = *$ or $C_i = a_i$. Otherwise $C(a)$ is set to 0. We demand for every concept $C_i \neq *$ for at least one i. These concepts are the pure conjunctions of atomic formulas $x_i = a_i$. All possible concepts can be arranged in a directed acyclic graph (the Version Space) where the concept at the end of an arrow specifies just one more attribute than the concept at the starting-point.

The algorithm presented in figure 1 gets a sequence (a^i) of positive and negative examples of the concept. With respect to the known examples the Version Space algorithm constructs two sets of concepts. The set S contains all concepts, which are fulfilled by all the known positive and by no known negative example and there is no more specialized[4] concept with the same properties. The set G contains the most general concepts which are fulfilled by all the known positive and by no known negative example. The algorithm describes the way to modify the sets S and G, when

[3] $N_k := \{0, 1, \ldots, k\}$

[4] A concept C_1 is called *more specialized* than C_2, if $\forall x \in U\ [C_1(x) \Rightarrow C_2(x)] \wedge \exists y \in U\ [C_2(y) \not\Rightarrow C_1(y)]$. The term *more generalized* is defined analogously.

a new example is presented. To define S and G properly we force the first example to be positive. The sets S and G characterize at every moment the classification ability of the Version Space. Under the assumption that the concept is a member of the Version Space every object which fulfills the concept in S must be positive. If the object does not fulfill any concept in G then the object must be negative. This observation leads to a decision function $VS : U \rightarrow \{0,1\}$ that can be used to classify new objects.

$$VS(x) = \begin{cases} 1 & \text{if } \forall C \in S \; [C(x) = 1] \\ 0 & \text{if } \forall C \in G \; [C(x) = 0] \\ ? & \text{otherwise} \end{cases}$$

As long as $S \neq G$ VS will not classify all objects of the universe. If an object is not covered by S but by G it is not clear, whether it belongs to the concept of not. So VS will not return an answer for those objects (this is the semantic of the "?" in the decision function).

2.1 A Case-Based Version Space

It is obvious that the main ability of the Version Space algorithm is to separate relevant and irrelevant attribute values. A value is called *relevant*, if it is part of the concept the learner has to learn. The following case-based variant (VS–CBR) will use this basic idea. For every attribute i a function f_i is defined which maps the set W_i to $\{0, 1\}$. If the concept $C = (C_1, \ldots, C_n)$ is learned then for every $x \in W_i$ holds:

$$f_i(x) = \begin{cases} c & : \; C_i = * \\ 1 & : \; C_i = x \\ 0 & : \; \text{otherwise} \end{cases}$$

where c is arbitrary but fixed. We will use $c = 0$ in the sequel. The constructed f_i will be combined to a function $f : U \rightarrow \{0,1\}^n$. The distance between two objects is then defined as $d_f(a, b) := |f_1(a_1) - f_1(b_1)| + \ldots + |f_n(a_n) - f_n(b_n)|$. During the learning phase the function f is learned by the algorithm presented in figure 2. It is obvious that every change of the function f will change the distance measure on the universe. Like the original Version Space the first presented case has to be positive to initialize the function f. The algorithm defines $f_i(x) = 0$ if $C_i = *$ holds.

Step 5) tests like the symbolic Version Space whether the known examples fit any concept which is learnable. If the learning is done the function f and the case base are used for classification. Given a new object c the set $F := \{x \in CB \mid d_f(x, c) \leq d_f(x', c)$ for all x' in the case base CB$\}$ is build up. If F contains more than one element the classification is determined by a fixed strategy. For example the strategy may state the lowest classification value (remember that we have defined a concept as a mapping $U \rightarrow \{0,1\}$. So 0 is the lowest value if there are positive and negative cases in F). This strategy is used in the following decision function.

$$\text{VS–CBR}(x) = \min\{class(y) \mid y \in CB \wedge \forall y' \in CB d_f(y, c) \leq d_f(y', c)\}$$

If the measure d is total the decision function is total, too. The cases in the case-base are used at classification-time as well as at learning-time.

2.2 Classification with VS–CBR

We now want to compare the classification abilities of VS and VS–CBR. In step 2) for all i exactly one $a_i \in W_i$ is mapped to 1. Step 4) occasionally deletes a 1. So, there is never more than one value of an attribute mapped to 1. Let us look at VS

Learning Algorithm for f

1. Define $f_i(x_i) = 0$ for all i, $x_i \in W_i$
2. If the first positive example is a $= (a_1, a_2, \ldots, a_n)$ define $f_i(a_i) = 1$. Define $CB = \{[a, 1]\}$
3. Let $b = (b_1, \ldots, b_n)$ be a new example. If b is negative, then store b in the case base: $CB := CB \cup \{[b, 0]\}$
4. If b is a positive example and for some i there is $f_i(b_i) = 0$ then set $f_i(x) = 0$ for all $x \in W_i$.
5. If there exist a positive case p and a negative case n in the case base with $d_f(p, n) = 0$ then **HALT**(Not a concept of the Version Space).
6. Delete redundant cases from the case base.[a]
7. If the concept is unequivocal go to step 8) otherwise go to step 3)
8. **RETURN**(The concept is learned. Classifier: (CB, d_f))

[a]A case r is redundant if there is a case s in the case base such that $d_f(r, s) = 0$ holds.

Figure 2: Algorithm to learn f for VS-CBR

and VS–CBR after the presentation of every object. It is obvious that $f_i(a_i) = 1$ holds if and only if the concept in S contains the value a_i for the attribute i. The function d_f forces that at every moment $d_f(a, b) = 0$ implies that a and b must be equally classified.

Based on these observations it is easy to verify that objects which can be classified by VS are equally classified by VS–CBR. But VS–CBR will give a classification to every object even if the classification is uncertain[5], because the distance measure is total and therefore the decision function. This arises from the totality of the decision-function of VS–CBR. It is possible to suppress this uncertainty by a test whether the classification must be correct under the assumption that the target concept is a pure conjunction. But this test is more ore less a variant of the original VS. If we call the hypothesis when the ith example is presented VS_i or VS-CBR$_i$, respectively, then VS-CBR$_i(x) \Leftrightarrow$ VS$_i(x)$ holds for all i and all $x \in U$. The information that is used by VS to change S and G is used by VS–CBR to change the distance measure. And we have seen that the resulting classifiers are identical. If it is clear, that the concept is representable in the Version Space both algorithms can be used to learn the concept.

We have seen that it is possible to rewrite the Version Space algorithm in a case-based manner so that the case-based variant behaves exactly as the symbolic algorithm. For other well-known symbolic learning algorithms like ID3, AQ, CN2 etc. (cf. (Globig, 1993)) it is possible to find case-based versions, too. Therefore, it is important to understand the implications of a measure to the set of representable concepts. This will be done in the next sections.

[5]Uncertain means that the current cases do not allow to decide whether the case at hand belongs to the concept or not.

3. Basic Issues of Case-Based Classification

In the last paragraph we have seen a simple case-based classifier. In this paragraph we want to discuss some basic issues of case-based classifiers and the related learning algorithms. First of all we have to clarify the conditions which must be fulfilled to learn a concept in a case-based manner. In a second part we present some examples to show the interdependence between the measure of similarity and the learning power. We must distinguish between the representability and the learnability of a concept. A concept is called representable by a measure sim, if there exists a finite case base such that (CB, sim) is a classifier for the concept. A concept is called learnable by a measure sim, if there exists a strategy to construct a finite case base such that in the limit (CB, sim) is a classifier for the concept.

A case-based classifier consists of a case base and a measure of similarity (or a distance measure). Neither the case base nor the measure is sufficient to classify alone. The knowledge about the concept is spread to both. Even in the VS–CBR you can extract the concept from the distance measure only because you know the way in which the measure was constructed. Therefore, the use of the cases is important in the application phase.

If we try to symbolize the relationship we can describe a case base system as a sum:

Concept = Case Base + Measure of Similarity

There are always multiple concepts which can be learned by a given measure. Because of the distribution of the knowledge between the case base and the distance measure it is clear that there are many tuples (CB, sim) which represent the same concept. If the hypothesis of the learner must be modified there are always three possibilities. Either to change the case base, to change the distance measure, or to change both. VS-CBR uses the positive cases to change the distance measure (by updating the functions f_i) while the negative cases are stored in the case base without changing the distance measure.

To illustrate the relationship between a case base and a distance measure we simplify the framework for a moment.

1. Let U be a finite universe
2. $\forall a, b \in U \; [d(a,a) = 0 \wedge d(a,b) = 0 \Rightarrow \forall x \in U \; [d(x,a) = d(x,b)]$
3. d is fixed.

The assumption 2 means that the relation $\sim$ defined as $x \sim y :\Leftrightarrow d(x,y) = 0$ is an equivalence relation. $\sim$ builds $|U/\sim|$ equivalence classes. It is obvious that a concept C can be represented and learned by a measure d if and only if for all $x, y \in U \; d(x,y) = 0 \Rightarrow C(x) \equiv C(y)$ holds, i.e. all elements of an equivalence class must have the same classification. On the other hand the equivalence classes can be classified without any respect to each other. Therefore, we can conclude that d is able to distinguish between $2^{|U/\sim|}$ different concepts. Each of these concepts can be represented by a case base with $|U/\sim|$ (appropriate) cases (i.e. one case in every equivalence class).

As a result we can state that for the learnability of a concept the only question is the definition of the distance 0. If we have two measures d and d' where $d(x,y) = 0 \Leftrightarrow d'(x,y) = 0$ and $d(x,y) \neq 0 \Leftrightarrow d'(x,y) = 1$ they can represent the same concepts.

Case based systems can be compared with respect to two important dimensions. The first dimension relates to the implicit knowledge in the measure.

Definition 1 *The similarity measure sim_1 of a case-based system (CB_1, sim_1) is called better informed than a measure sim_2 of a system (CB_2, sim_2), iff both systems are classifiers for the same concept C, $|CB_1| < |CB_2|$ holds and there is no $CB_i' \subset CB_i$ so that (CB_i', sim_i) is a classifier for the concept C.*

The second dimension relates to the set of learnable concepts.

Definition 2 *A similarity measure sim_1 is called more universal than a similarity measure sim_2 iff the set of concepts which are learnable by sim_2 is a proper subset of the set of concepts which are learnable by sim_1.*

To use a universal similarity measure stands in the way of a minimal case base. To minimize the size of the case base usually results in a less universal similarity measure. We illustrate the contrasting nature of these two aims in figure 3. It shows different distance measures in relation to the minimal size of the case base to select a certain concept and the total number of learnable concepts. For the table we use a universe with objects which consists of four attributes. Each attribute can take one value out of 16. So, the size of the universe is 65536. The concept which the measures try to learn fixes two attributes.

We can distinguish two extrem measures:

Identity of objects: The similarity is maximal if and only if the compared objects are identical. The measure is universal because it is able to recognize every binary concept in the given universe. But to do so it needs the whole universe as a case base.

Identity of classification: The similarity is maximal if and only if the classification of the compared objects is identical. Nearly the whole knowledge about the concept is coded in the measure by the definition of the concept. The case base is used only to choose between some trivial variations. $sim(x, y) = C(x) \Leftrightarrow C(y)$ can only distinguish four concepts (C, $not(C)$, $TRUE$ – i.e. all objects are positive, $FALSE$ – i.e. all objects are negative).

The other measures in figure 3 are between the extremes. VS_CBR$_1$ and VS_CBR$_2$ are neither maximally universal nor able to recognize a concept with a minimal case base. VS_CBR$_1$ is the distance measure, which is built for VS–CBR when the first case is presented. In every dimension exactly one value is mapped to one so that the universe is mapped into the vertices of an four dimensional cube. VS_CBR$_2$ is the measure, which is used, when VS–CBR has recognized the concept. It distinguishes only between the two relevant values of the concept and, therefore, builds up only four equivalence classes.

measure	minimal size of CB	number of learnable concepts
$sim(x, y) := (x = y)$	$65536 = 16^4$	2^{65536}
VS_CBR$_1$	16	$65536 = 2^{16}$
VS_CBR$_2$	3 (4)	$16 = 2^4$
$sim(x, y) := C(x) \equiv C(y)$	2	2^2

Figure 3: Measures in relation to the minimal case base and the number of learnable concepts

From the table we can draw the conclusion that the distance measure determines the space of possible target concepts and the case base selects one of them. In other words we can say that the choice of the distance measure is the bias of case-based classification. Its choice determines the set of target concepts which can be recognized and the efficiency of the learning process as we will see in the next section.

The results of this section can be applied to systems that change their measure two. These systems reduce the number of unfilled equivalence classes by changing the measure and by storing cases in the case base. In a typical case-based learner two processes – reducing the size of the hypothesis space and increasing the size of the case base – are performed. So it can be difficult to specify the influence of a single case.

The last measure in figure 3 indicates a simple way to rewrite any symbolic learning algorithm in a case-based measure, i.e. use the actual symbolic hypothesis to construct such a measure and store one positive and one negative case in the case base.

4. Appropriateness of Distance Measures

In the last section we have analyzed the learnability in a case based system in principle. Here we want to illustrate the influence of the knowledge in the distance measures to the efficiency of the learning process by an experiment. We therefore compare four distance measures (d_3, d_4, d_5, and d_6) while they are learning different concepts.

The universe is finite. It consists of objects with four attributes with sixteen values each ($W_i = \{0, 1, \ldots, 15\}, \quad i = 1, \ldots, 4$).

The measure d_3 is a statistical variant of the measure used in VS–CBR. Instead of mapping the values to $\{0,\ 1\}$ the interval $[0,\ 1]$ is the range of f_i. $f_i(a_i)$ is the relative frequency of a_i *is element of an positive case in the case base.* The functions f and d_f are build from the f_i as in VS–CBR. d_3 is the only measure which is adapted during the learning process. d_4 is a very specialized measure, tailored for a certain concept. $d_4(a, b) := |\sum_i a_i - \sum_i b_i|$. d_5 is a measure which is able to distinguish between all binary concepts, if it can use the whole universe as a case base. $d_5(a, b) := \sum_i |a_i - b_i|$. d_6 is similar to d_5. But d_6 computes the distance only on a subset of all attributes. If the concept is defined on the same attributes like d_6 the selective evaluation can be seen as a hint from a teacher that the other attributes are irrelevant. $d_6(a, b) := |a_2 - b_2| + |a_4 - b_4|$.

All measures fulfill condition 2 from page 5 so learnability can be discussed by analyzing the equivalence classes.

The concepts are all defined sharp, so it is possible to describe them as a disjunctive normalform. The probability, that a randomly drawn object is member of a concept is more or less 0.5.[6]

$$
\begin{aligned}
P_1 : C(x) = 1 &\Leftrightarrow \sum_i x_i \geq 30 \\
P_2 : C(x) = 1 &\Leftrightarrow (x_2 < 5) \vee (x_4 \, mod \, 2 \ = 0) \\
P_3 : C(x) = 1 &\Leftrightarrow (x_2 \geq 8 \wedge x_4 < 8) \vee (x_4 \geq 8 \wedge x_2 < 8) \\
\\
P_5 : C(x) = 1 &\Leftrightarrow ((x_2 \, mod \, 2 \ = 0) \wedge (x_4 \, mod \, 2 \ = 1)) \vee \\
& ((x_4 \, mod \, 2 \ = 0) \wedge (x_2 \, mod \, 2 \ = 1)) \\
P_6 : C(x) = 1 &\Leftrightarrow (x_1 \geq 8 \wedge x_3 < 8) \vee (x_3 \geq 8 \wedge x_1 < 8) \\
P_7 : C(x) = 1 &\Leftrightarrow ((x_1 \, mod \, 2 \ = 0) \wedge (x_3 \, mod \, 2 \ = 1)) \vee \\
& ((x_3 \, mod \, 2 \ = 0) \wedge (x_1 \, mod \, 2 \ = 1))
\end{aligned}
$$

[6]For all concepts you have to add C(x) = 0 else

4.1 Experiment

The experiment was done in the following manner. For every concept a case base of the given size was created and tested against 500 new test objects. The objects of the case base and the test set were drawn randomly. The rate of correct classification is the mean value of ten case bases. The test sets and the case bases for each measure are identical. If the set of nearest neighbors has more than one element, the lowest classification was selected, i.e. it is supposed the object is not member of the concept.

The results of the experiment are shown in figures 4 and 5. We now want to analyze the relationship between the concepts the measures, and the change of the rate of correct classification in more detail.

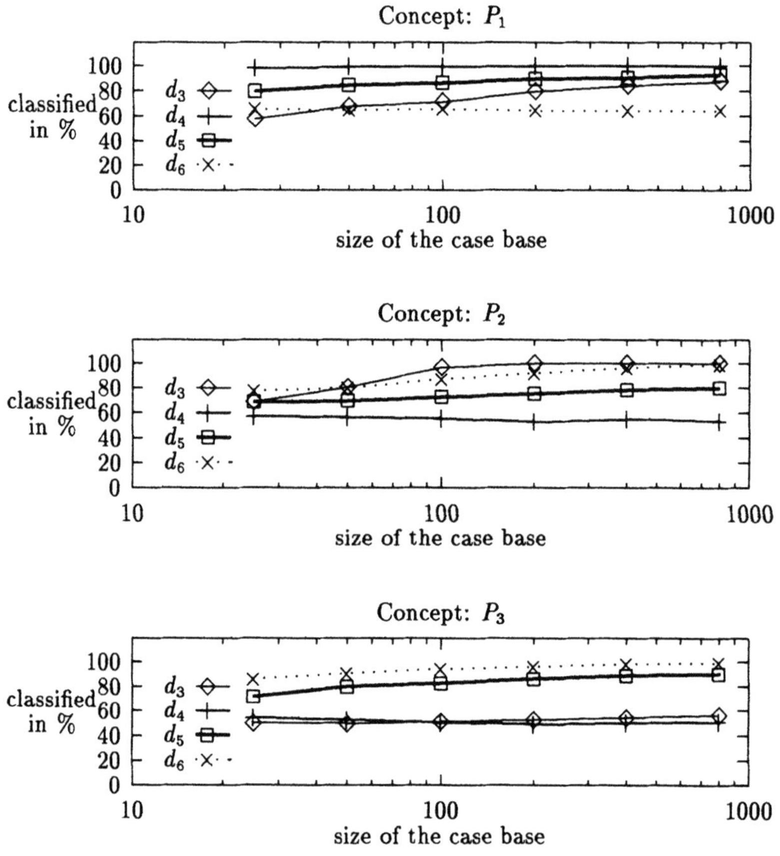

Figure 4: Results for P_1, P_2, and P_3

The concepts are selected in such a way that each distance measure has superior rates of correct classification for one of them. d_6 has compared to d_5 additional information about the concepts P_2 and P_3 because it knows which attributes are relevant for representing the concept. Of course, if the information about the relevant attributes would be wrong, d_6 would be unable to recognize the concepts even with the whole universe as a case base.

It should be mentioned that all concepts are learned well by those distance measures

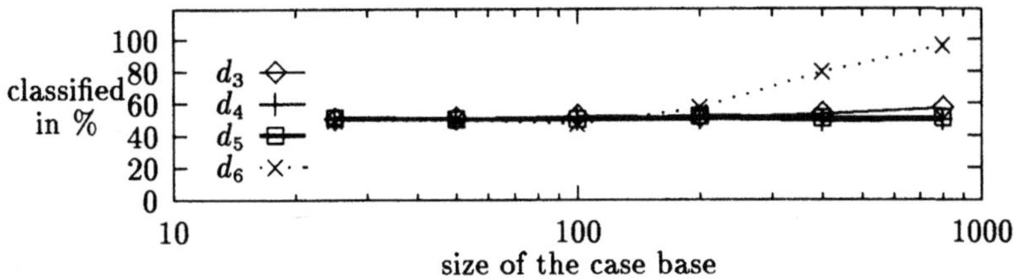

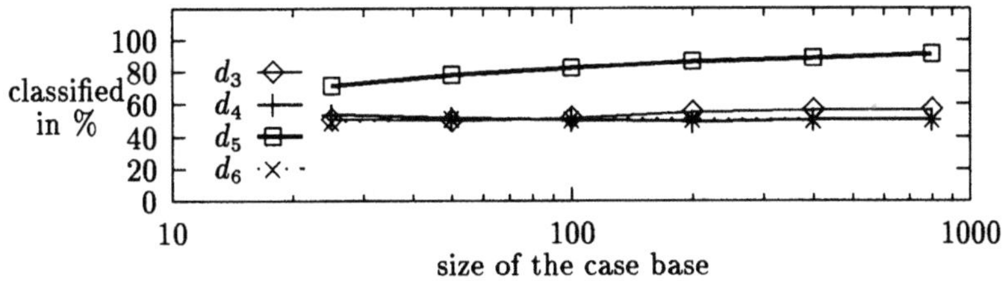

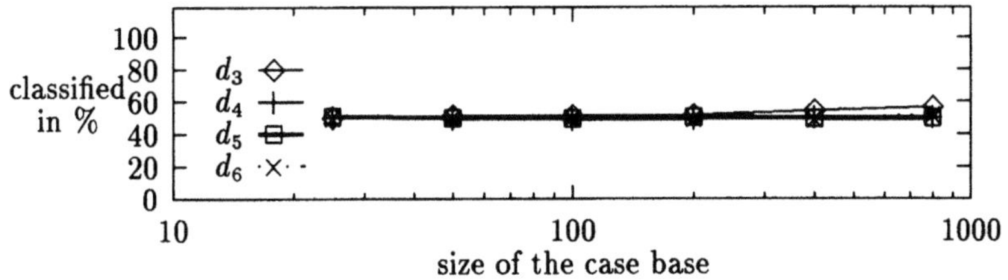

Figure 5: Results for P_5, P_6, and P_7

which have a meaningful distance value, i.e. a small value of the distance between a and b indicates a high probability that a and b have the same classification. The only exception is d_6 with the concept P_2. d_6 has only very few equivalence classes so that most of them are filled by a case base with 800 cases.

The concepts P_5, P_6, and P_7 illustrate what happens if the assumptions of the distance measures are not fulfilled by the concept. The difference between P_3 and P_5 is the selection of the relevant attribute values. The number of relevant values is the same. But the distance between objects is of no information about the classification of the compared objects. d_3 and d_4 are unable to recognize P_5. Of course d_5 and d_6 will learn the concept, because the concept is a union of their equivalence classes. But they need a very large case base to do so. Even short distances do not indicate that the classification of the objects is identical. P_6 is a variant of P_3, too. P_6 changes the relevant attributes instead of the attribute values. Consequently d_6 is unable to represent the concept, because of the misleading assumption that the relevant attributes are 2 and 4. d_5 learns P_6 as fast as P_3 because it did not assume that there are irrelevant attributes. P_7 combines the difficulties of P_5 and P_6. The consequence is that none of the given distance measures is able to classify a large amount of the objects in the universe with 800 cases in the case base.

The examples indicate that the guaranty that the target concept is in the hypothesis space of a given measure is not enough for efficient learning. The distance measure d_5 is able to learn every binary concept in a finite universe, because every object is an equivalence class of its own. But the case base must be huge even for relative simple concepts as P_2. To learn a concept with a small case base it is necessary to encode information about the target concept into the distance measure.

We have demonstrated three types of information in the distance measures. It is very important that a small distance between two objects indicates a high probability that the compared objects have the same classification (for example P_3 with measure d_5, d_6; P_1 with d_4; P_2 with d_3). In other words the distance measure should describe a neighborhood of equally classified objects. The second type of information is knowledge about the structure of the equivalence classes needed to learn the concept. A measure with only a few equivalence classes is superior if the target concept is learnable with these classes. A special type of information is the knowledge about the relevant attributes of the concept. If we compare d_5 and d_6 we can see the advantage of knowing the important attributes. The advantage becomes more important if the number of irrelevant attributes increases.

The examples show the interdependence between the concept and the distance measure. Because of this interdependence there cannot be a distance measure, which depends only on the universe and not on the set of the possible target concepts.

5. Consequences for Case Based Classification

We have analyzed the principles of case-based classification. The theoretical and experimental results reveal the strong interdependence between the distance measure and the set of learnable concepts. If we agree to the assertion that there is no measure which depends only on the universe and not on the set of target concepts we must conclude that we are confronted with a bias in case-based classification, too. The bias is the distance measure (resp. the method to change the measure). Like the bias in symbolic classification the distance measure determines which concepts are learnable and in addition the efficiency of the learning process.

It is therefore very important to select an appropriate distance measure with respect to the given problem. The measure must have equivalence classes which cover the target concepts and meaningful distance values, i.e. a short distance between two objects must indicate a high probability that the objects have the same classification. If we know that there is a distance measure with these properties case-based classification seems to be a good choice.

Given an appropriate distance measure, case-based classification has some other useful features. If there is some noise in the data and the effect of the noise is small according to the distance measure then case-based reasoning is a very natural way to implement a noise tolerant learner (Aha and Kibler, 1989). Instead of the classification of the nearest neighbor, it is possible to use the majority vote of the k nearest neighbors for some $k > 1$. But it is important to remember that the noise tolerance is a property of the measure of similarity and not a property of the use of a case-based algorithm.

To summarize we can say that there is no fundamental advantage of case-based classification *cf.* (Cost and Salzberg, 1993) compared to the traditional symbolic approach in the simple framework we have considered here. So the question which algorithm is better for a given task depends on the simplicity and adequacy of the representation of the given knowledge. Both approaches need a method to cut down the size of the hypothesis space is needed. While the symbolic approach uses this extraevidential knowledge to construct useful abstractions, the case-based algorithms need it to get appropriate measures of similarity.

Acknowledgement: The authors thank Michael M. Richter, Klaus Dieter Althoff, and the whole research group in Kaiserslautern for many helpful discussions and the anonymous referees for instructive comments.

References

AHA, D.W. (1991): Case-Based Learning Algorithms. *In:* Bareiss, Ray (ed), *Proceedings: Case-Based Reasoning Workshop.* San Mateo, California: Morgan Kaufmann Publishers, for DARPA. Washington, D.C., USA, May 8–10, 1991.

AHA, D.W., KIBLER, D., and ALBERT, M.K. (1991): Instance-Based Learning Algorithms. *Machine Learning,* **6**, 37–66. March 1991.

AHA, D.W., and KIBLER, D. (1989): Noise-Tolerant Instance-Based Learning Algorithms. *Pages 794–799 of: Proceedings of the 11th International Conference on Artificial Intelligence IJCAI-89.* IJCAI. Detroit, Michigan, USA.

COST, S., and SALZBERG, S. (1993): A Weighted Nearest Neighbor Algorithm for Learning with Symbolic Features. *Machine Learning,* **10**(1), 56–78.

DASARATHY, B. (1990): *Nearest Neighbor Norms: NN Pattern Classification Techniques.* IEEE Computer Society Press.

GLOBIG, C. (1993): *Fallbasiertes und Symbolisches Lernen.* Diplomarbeit, Universität Kaiserslautern.

JANTKE, K.P. (1992): Case-Based Learning in Inductive Inference. *In: Proc. COLT-92.*

KOLODNER, J.L. (1991): Improving Human Decision Making through Case-Based Decision Aiding. *AI Magazine,* **91**(2), 52–68.

MICHALSKI, R., CARBONELL, J.G., and MITCHELL, T. (eds). (1983): *Machine Learning: An Artificial Intelligence Approach.* Vol. 1. Palo Alto, California: Tioga.

MITCHELL, T.M. (1982): Gerneralization as Search. *Artificial Intelligence,* **18**(2), 203–226.

RICHTER, M.M. (1992): Classification and Learning of Similarity Measures. *In: Proc. of the 16th Annual Conference of the German Society for Classification (Gesellschaft für Klassifikation e.V.).* Springer Verlag.

Nonstandard Concepts of Similarity in Case-Based Reasoning

Klaus P. Jantke

HTWK Leipzig (FH), FB IMN
Postfach 66, O-7030 Leipzig, Germany
jantke@informatik.th-leipzig.de

Summary: The present paper is motivated by the author's work in learning, by his work on case-based learning and by his machine learning work in the case-based reasoning research project FABEL, in particular. The presentation is focussed on case-based classification, but the paper is more specific in one respect and more general in another. The core of the paper is more specific, as it is driven by investigations in learning, and it is more general, as the nonstandard concepts developed are of importance for case-based reasoning, in general. It seems that a much too restricted view at similarity is one of the basic drawbacks of recent research in case-based reasoning. A rigorous criticism is intended to set the stage for new approaches to similarity which may contribute to a considerable progress in the similarity-based use of cases. New generation approaches to similarity should allow both non-symmetric similarity concepts and similarity of more than two objects. Those concepts exist already in certain specific versions mainly in the cognitive sciences area. The present paper is introducing nonstandard concepts of similarity in a quite formal setting which allows an integration of well-known procedures like unification algorithms.

1. Introduction

The present paper is aimed at propagating new concepts of similarity more flexible and expressive than those underlying most case-based reasoning approaches today. So, it mainly deals with criticizing approaches in use, with motivating and introducing new notions and notations, and with first steps towards future applications.

The investigations at hand originate from the author's work in learning theory. In exploring the relationship between inductive learning and case-based learning within a quite formal setting (cf. [Jan92b]), it turned out that both areas almost coincide, if sufficiently flexible similarity concepts are taken into acount. This provides some formal arguments for the necessity of non-symmetric similarity measures. Encouraged by these first results, the author tried to investigate more structured learning problems from the view point of case-based reasoning. It turned out that an appropriate handling requires formalisms allowing similarity concepts of more than two objects. This tremendously extends the computational power of case-based devices, as it enables a system to detect the similarity of some case with respect to some case base, although there is no distinguished element of the case base individually carrying this information. Thus, similarity is lifted from a local property to become a global one. Starting with a brief description of the state of the art and referring to some typical references, there will be developed some of the new approaches announced.

1.1 Standard Concepts

The majority of case-based reasoning work today, if it is based on formalizations at all, invokes a small number of standard concepts. Even very recent workshops and publications do reflect this state of the art very well. The basic notions and notations are as follows. Cases are (represented as) attribute-value tuples with respect to a given finite number of attributes. Similarity is measured via some metrics. RICHTER is introducing several approaches (cf. [Ric92]) which are somehow closely related in spirit. ALTHOFF's Ph.D. [Alt93] contains a similar discussion of a number of standard approaches. POSTHOFF and SCHLOSSER, when investigating chess end games (cf. [PS92]), are explicitly invoking some HAMMING metrics. For measuring the distance of two tuples, one usually assumes some way to transform non-numerical values into numerical ones, and a vector of weights allows to calculate the distance $\delta(t_1, t_2)$ of two tuples t_1, t_2 as the weigthed sum of the distances of their corresponding attribute values. Assuming the attribute names $a_i (i = 1, \cdots, n)$, the weights $w_i (i = 1, \cdots, n)$, and using some notation from data base theory to denote a tuple's value at some attribute by t.a, the standard distance of two tuples t_1, t_2 is

$$\delta(t_1, t_2) = \sum_{i=1}^{n} w_i \cdot \delta(t_1.a_i, t_2.a_i) \tag{1}$$

Usually, distances are transformed to describe similarities. There is a standard way which seems to be used in most approaches:

$$\sigma(t_1, t_2) = 1 - \frac{\delta(t_1, t_2)}{1 + \delta(t_1, t_2)} \tag{2}$$

There are several versions of defining a similarity measure from some given distance. However, the particular way of encoding distances by figures called similarities does not matter.

In the author's opinion, there is absolutely no reason for a transformation as above. Instead of calling $\sigma(t_1, t_2)$ the similary of t_1 and t_2, one should speak about the distance of these objects. Such a poor view at similarity may sometimes do. But in many cases, it does not meet the necessities quite well. Furthermore, it is hiding the essence of the approach: measuring the distances of objects.

These arguments above are more methodological in spirit: If there is no deeper concept of similarity than measuring distances, one should admit to do so. But there is another problem. A similarity measure σ as defined by equation (2) above is inheriting essential properties from the distance measure δ behind. One is *symmetry*. Another one is the triangular inequality in its encoded version. The thoughtless adoption of approaches like the one sketched here is providing similarity considerations with properties which seem to be inevitable, but which are not, as the case of symmetry exhibits.

Note that there are very different viewpoints of researchers concerning similarity concepts, in general, and the problem of symmetry, in particular. Reflexivity and symmetry are used as *basic axioms* for one of the approaches considered in [Ric92]. KODRATOFF, for instance, when specifying the concept of a similarity measure (cf. [Kod88], Chapter 2, Section 2.1.2, page 189) is defining it by two conditions. One is logically incomplete (thus, wrong), and the other one is symmetry.

Some recent problems illustrate the necessity to abandon symmetry of similarity concepts. More flexible approaches will be motivated and introduced in the following chapter particularly emphasizing the problem of (non-)symmetry.

Before doing so, another problem has to be considered which is usually out of the scope of similarity approaches in use: The global knowledge represented by a collection of cases can rarely be reflected by comparing cases pairwise. The standard approach of case-based classification (cf. [Aha91], [AKA91] for a collection of typical algorithms) is to collect a number of cases in a case base, first. A new case is classified by searching a most similar case inside the case base and classifying the new one accordingly. This standard approach restricts the power of case-based classification considerably. Typical problems which require a different view at similarity are related to sets of texts representing text patterns (see [Ang80] for related learning problems), where the characteristics of a particular new case can only be recognized by comparing it to a couple of other cases simultaneously. This motivates lifting similarity from a local to a global concept.

A more general approach to overcome some of these problems will be developed below.

1.1 Nonstandard Concepts

Whereas the previous section was intended to criticize the restrictedness of most recent approaches to similarity, the present section is devoted to constructive arguments directly leading to a more general view at the problem.

Cognitive sciences are providing a huge amount of knowledge illustrating that human beings are reasoning based on cases and similarity in many ways. [Lak87] is a rich source in this regard putting emphasis on categorization. Thus, in the author's opinion, LAKOFF's work provides an important cognitive sciences background for classification. ROSCH, in particular, is providing experimental results and theoretical investigations bearing abundant evidence of the need of more sophisticated and flexible approaches to similarity. This is closely related to ROSCH'S concept of prototypes (cf. [Ros75]). One of the illustrative examples found in [Ros75] (see also [TG78]) is that humans when asked to classify countries considered Mexico to be more similar to the United States than the United States is to Mexico. (Here, one can not go into the details of circumstances for those results. Moreover, there is a considerable number of further experiments one should try to exploit.) In [Lak87], this phenomenon is abstracted to the following insight:

- *Less representative examples are often considered to be more similar to more representative examples than the converse.*

In the example mentioned, this may be expressed by

$$\sigma(USA, Mexico) > \sigma(Mexico, USA) \qquad (3)$$

where the first argument of the measure is considered to be the given one whereas the second one should be understood as the newly presented case to be decided about.

New generation similarity concepts should allow to model effects like the one reported here. This rules out encodings of metrices as in the chapter before.

Another motivation for nonstandard similarity concepts stems from inductive inference research. [Jan92b] is a first publication integrating case-based reasoning and inductive inference. Roughly spoken, the main result of [Jan92b] valid in the area of learning recursive functions from input/output examples is that every inductive inference device can be transformed in to a case-based normal form, if symmetry of similarity measures is not required any longer.

Finally, the participation of the author's learning theory research group in the BMFT joint research venture FABEL has faced the researchers with interestingly structured cases (cf. [Hov93b], [Hov93a] for more details) together with a number of ad hoc approaches to specify similarity. Deeper investigations have exhibited the need for structured case representations and similarity considerations taking structural properties into account. (The main application domain of FABEL is in industrial building design, where cases are highly structured in a natural way.) This led to structural approaches to similarity developed in the sequel. The first idea is to "measure" similarity of two objects by mapping into some partially ordered space. This allows to compute and compare similarity as in the standard approach. But under certain circumstances, it easily generalizes to mappings which allow to express the similarity of a couple (possibly containing more than two) of cases.

There is a key problem with new formal approaches, in general: They need to be justified by making a reasonable use of them. But this is always a little difficult in a first attempt. The present paper is attacking this problem as follows. First, there are results from recursion-theoretic inductive inference based on a kind of nonstandard approaches to similarity. These results are very abstract and far from real world applications, but they demonstrate with mathematical precision that the nonstandard approaches allow to derive new results, at least in principle. Second, there will be introduced an algorithm for case-based reasoning below which makes essential use of the pecularities of the nonstandard concepts introduced.

2. Similarity in Inductive Inference

Inductive inference is the theory of learning from possibly incomplete information. Frequently, the information to be processed is assumed in a form (examples, ...) which may be understood as a sequence of cases. This bridges the gap between inductive inference and case-based reasoning. [AS83] and [KW80] are excellent surveys on inductive inference. [Jan89] is an easy introduction. For the present paper, a brief introduction will do.

2.1 A Peep at Inductive Inference

There are several areas of inductive inference. Among them, learning recursive functions is most thoroughly studied. In recursion-theoretic inductive inference as invented by GOLD in his seminal paper [Gol67], total recursive functions are assumed as target objects to be learned from input/output examples only. In the general approach, for any recursive function f, any complete presentation of its graph is admissible as information fed into some learning device. Arbitrary partial recursive functions S may be used for learning any function f stepwise presented by finite initial sequences of orderings of its graph. If some ordering is denoted by $\mathcal{X}$, the corresponding initial segment of the first n input/output examples of f w.r.t. $\mathcal{X}$ is abbreviated by $f_{\mathcal{X}}[n]$. If $\mathcal{X}$ is some ordering of arguments

$$x_1, x_2, x_3, x_4, x_5, \cdots$$

then $f_{\mathcal{X}}[n]$ may be understood as the corresponding sequence

$$(x_1, f(x_1)), (x_2, f(x_2)), (x_3, f(x_3)), \cdots .$$

S is said to learn f in the limit, if and only if there is some n_* such that for all information covering the initial segment $f_{\mathcal{X}}[n_*]$, the hypotheses generated by S are identical to each other and represent f correctly.

Above, we have presented informally one of the basic definitions in recursion-theoretic inductive inference. The corresponding family of learning problems solvable according to this definition is called $\mathcal{LIM}$ in [JB81], for instance. In the early work of Latvian scientists, this so-called identification type is denoted by $\mathcal{GN}$, whereas it is called $\mathcal{EX}$ in the recent American literature. Certain authors are used to call those partial recursive functions S solving some learning problem an inductive inference machine (*IIM*, for short). We adopt this notation in the sequel. The reader is directed to [AS83] for further details.

There is a large variety of additional requirements which may be considered to be natural properties of learning strategies (cf. [JB81]). One cannot go into further details here.

The approach outlined illustrates the gist of inductive inference quite well. There is a class of target objects to be learned from possibly incomplete information within some limiting process. In many settings, the pieces of incomplete information presented may be understood as cases. From this point of view, one may search for a particular *IIM* which learns in a case-based manner by collecting cases, perhaps learning similarity concepts, and expressing its knowledge in these terms as specified in the sequel.

First, there will be introduced and discussed some fundamental notations. Assume that C is any space of case representations. There will be used a notation suggesting some relationship to attribute-value representations, although one may attach any other suitable semantics. For any given case $c \in C$, there are two objects denoted by $c.b$ and $c.s$. If b and s are vectors of attribute names, one may interpret $c.b$ and $c.s$ as the corresponding vectors of values. In other words, if a case may be understood as some tuple, it is a standard approach to divide it into two parts, where one called the body describes the problem presented by the case, whereas the other presents its solution. However, $c.b$ and $c.s$ may denote anything else. For notational convenience, one introduces $C.b = \{c.b/c \in C\}$ and $C.s = \{c.s/c \in C\}$.

Second, these notations may be used to define some formal semantics. A case base CB together with some similarity measure $\sigma : C.b \times C.b \rightarrow [0,1]$ may be interpreted as a classifier as follows. Given any problem description (a case body) $p \in C.b$, search for the most similar member in CB and return its value as solution $p.s$. Note that exactly this approach is underlying the 7 algorithms for case based classification presented in [Aha91] and [AKA91]. The approach may be expressed formally as follows, where $f_{[CB,\sigma]}$ denotes the classifier defined by CB and σ:

$$f_{[CB,\sigma]}(p) = c.s \quad : \quad \begin{cases} c \in CB & \wedge \\ \forall c' \in C(\sigma(c.b, p.b) > \sigma(c'.b, p.b)) \end{cases} \tag{4}$$

The reader may recognize some problems concerning the semantics above. Sometimes, there is no unique case c satisfying $\forall c' \in C(\sigma(c.b,p.b) > \sigma(c'.b,p.b))$ This leads to several formal approaches which turn out to be logically different. But the discussion of those details exceeds the present paper. [Jan92a] contains a more comprehensive discussion in this regard.

For the moment, it is sufficient to understand that collecting cases and learning similarity measures may be interpreted as learning (synthesizing, ...) classifiers. And this does not require to look at cases as tuples over some collection of attributes. One usually assumes that case bases are always finite. If the similarity measure σ used in equation (4) is computable, the defined classifier is computable, too.

The recursion-theoretic inductive inference provides techniques and further results about learning computable functions. This includes computable classifiers. Therefore, it is one of the basic problems whether or not inductive inference is more general than learning in a case-based manner as sketched above.

2.2 General Results

The key result (cf. [Jan92b]) is that every inductive inference strategy able to learn computable classifiers may be transformed into a case-based normal form. In other words (recall that $\mathcal{LIM}$ denotes the family of all classes of computable functions which are learnable as described above), the following theorem holds.

The following result becomes particularly interesting by the corollary attached.

Theorem 1

For every class U of computable classifiers belonging to $\mathcal{LIM}$, there exists a pair of computable devices M and S able to learn each classifier $f \in U$ in a case-based manner as follows:
Initially, there is some empty case base $CB_0 = \emptyset$. σ_0 is any given computable similarity measure. Being presented with any case c, M and S are updating the case base to $CB_{n+1} = M(CB_n, c)$ and the similarity measure to $\sigma_{n+1} = S(\sigma_n, c)$, respectively, such that

$$CB_n \subseteq CB_{n+1} \subseteq CB_n \cup \{c\} \tag{5}$$

and both sequences $\{CB_n\}_{n\in\mathbb{N}}$ and $\{\sigma_n\}_{n\in\mathbb{N}}$ converge to some CB resp. σ with

$$f_{[CB,\sigma]} = f. \tag{6}$$

It may be useful to rephrase the theorem: In the area of inductively learning recursive functions, every *IIM* can be transformed into a normal form which learns by collecting cases and tuning similarity measures. The reader may consult [Jan92a] for details of the proof. The construction invoked is quite instructive:

Corollary 1

The similarity measures used to prove the theorem above are nonstandard in the sense that they are usually only partially defined and non-symmetric.

This provides formally precise reasons to consider more general similarity concepts.

3. Structural Similarity

This chapter is devoted to the introduction of a particularly new approach to similarity. It will turn out that the approach adopted has some close relations to other work in artificial intelligence and theoretical computer science, although in no case there has been established an explicit relation, before. It is particularly close to first order unification theory and to term rewriting.

3.1 Notions and Notations

The following concepts are folklore in mathematical logics, theoretical computer science, and artificial intelligence. This allows to reduce the present introduction remarkably. Most concepts are only briefly mentioned. Only those notions and notations, which will play a particular role in the approach to similarity proposed, will be introduced in more detail.

There is assumed any finite, heterogeneous, and finitary signature Σ. This is taken as a basis for building terms and formulae, as usual. One needs a sorted family of variables called X. For simplicity, we assume all variables to be called x and y, perhaps with certain decorations. $T(\Sigma, X)$ denotes the term algebra over Σ and X. Sometimes, it is useful to consider terms as particular trees, where trees are mappings from some tree domain into some alphabet. The concept of a tree domain allows to associate with every place in a term the concept of a position in the corresponding tree. Positions are words over the alphabet of natural numbers. If $\mathbb{N}$ denotes the set of natural numbers, $\mathbb{N}^*$ contains all possible positions. The empty word ε is understood as the position of the root of any tree. Formally, a tree domain D is any subset of $\mathbb{N}^*$ such that it is closed in the following sense: First, for all $w \in \mathbb{N}^*$ and all $p, q \in \mathbb{N}$ it holds that $wp \in D$ and $1 \leq q \leq p$ implies $wq \in D$. Second, if $\sqsubseteq$ is the relation over $\mathbb{N}^* \times \mathbb{N}^*$ to denote by $v \sqsubseteq w$ that v is an initial part of w, for any two words

$u, v \in I\!N^*$, $u \in D$ and $v \sqsubseteq u$ implies $v \in D$. For illustration, the term $2 * sin(x + \pi)$ may be understood as a tree t over the tree domain $D = \{\varepsilon, 1, 2, 2.1, 2.1.1, 2.1.2\}$, where the letters of a word of natural numbers are separated by dots, for readability. t is mapping as follows: $t(\varepsilon) = *$, $t(1) = 2$, $t(2) = sin$, ..., $t(2.1.2) = \pi$. This formalism allows a quite flexible stepwise construction of terms, in the sequel. In computer science, one is mostly dealt with finite terms resp. trees, i.e. finite tree domains.

Based on the concept of positions, it is useful and quite easy to introduce the notion of a subterm of t at position p. The notation is $t_{|p}$. As trees and terms are mappings, this may be done as follows, for all positions u:

$$t_{|p}(u) = t(pu) \tag{7}$$

If t and t' are terms, p is a position in t, then $t[p \leftarrow t']$ denotes the term which results from inserting t' into t at position p. It is defined for any position u by

$$t[p \leftarrow t'](u) = \begin{cases} t(u) & : \; p \not\sqsubseteq u \\ t'(v) & : \; p \sqsubseteq u \wedge pv = u \end{cases} \tag{8}$$

For the reader's convenience, there is sketched an example. Assume t to be the example term above, and $t' = \psi - \varphi$. Then, the introduced concepts apply as follows: $t_{|2.1} = x + \pi$ and $t[2.1 \leftarrow t'] = 2 * sin(\psi - \varphi)$

Substitutions are mappings from X into $T(\Sigma, X)$. The set of all substitutions is denoted by Γ, and individual substitutions will be written as β or γ, usually. Substitutions are canonically extrapolated to map $T(\Sigma, X)$ into $T(\Sigma, X)$. The application of some substitution $\beta \in \Gamma$, where $\beta : T(\Sigma, X) \rightarrow T(\Sigma, X)$, to some term $t \in T(\Sigma, X)$ is denoted by $t\beta$. Substitutions may be understood as sets of assignments like $\{x_1 \leftarrow t_1, \cdots, x_n \leftarrow t_n\}$. As there are only finite terms taken into account, substitutions are always finite sets. The concatenation of two substitutions β and γ is written as $\beta\gamma$ and means that β has to be applied before γ.

For two given terms t_1 and t_2, some substitution β is called a unifier, if and only if $t_1\beta = t_2\beta$. Note that "=" denotes structural equality. (It is a far more complex problem to consider term equality in dependence on background knowledge. Some remarks concerning this extension will follow below.) β is called a most general unifier, if for every other unifier β' of t_1 and t_2, there is some substitution γ satisfying $\beta\gamma = \beta'$.

For structural unification, there is a naive standard algorithm. Furthermore, there is a considerable number of attempts to find more efficient versions like the algorithms of [PW78] and [MM82]. The interested reader may consult [Gol89] in this regard.

There is a somehow converse concept to unification called antiunification. For two given terms t_1 and t_2, one is searching for some term t and corresponding substitutions β_1 and β_2 satisfying both $t\beta_1 = t_1$ and $t\beta_2 = t_2$. Such a term t will be called an antiunifier, below. For antiunifiers, it is desirable to come up with most specific terms. (Trivially, if two terms are of the same sort, any variable of this sort works as an antiunifier.) For any two terms, some antiunifier t is most specific, if and only if there is no other antiunifier t' which results from t by some substitution β (i.e. $t' = t\beta$), where β contains at least one proper assignment $x \leftarrow t_0$ with $t_0 \in T(\Sigma, X) \backslash X$. Note that renaming substitutions are allowed. If one term t' is properly more specific than another term t, this is written as $t' \lhd t$.

Unification and antiunification are folklore in the free term algebra as considered here. The corresponding problems are very interesting, sometimes of a great complexity, and sometimes even unsolvable, if background knowledge is assumed. The most thoroughly studied form of background knowledge is equational theories. This, by the way, bridges the gap to term rewriting systems and techniques like KNUTH-BENDIX-completion. A deeper investigation exceeds the present paper. From the huge amount of literature concerning unification theory, the reader may consult [Sie89], for example. Concerning antiunification, there is considerably less topical literature. In the free term algebra, antiunification is just syntactical generalization. [Plo70] is a helpful reference in this regard.

3.2 Similarity of Terms

The key idea of a more flexible approach to similarity invoked here is to define similarity as some mapping into a highly structured partially ordered space. There are at least three origins of the ideas underlying the present approach.

First, some attempts to reflect the rich structures occuring in the application domain of the project FABEL already mentioned before (cf. [Hov93b], [Hov93a]) led to the idea to represent objects like supply-air net structures as terms over some appropriately tailored signature. This approach has been influenced by INDURKHYA's formalisms (cf. [Ind89], [Ind91]).

Second, there were some attempts to extend the inductive inference results sketched above to text patterns. This inevitably requires to reflect the similarity of two or more objects simultaneously. OHKAWA and his co-authors (cf. [OMBT92]) developed some deductive formalism for horn clauses. Although they considered only so-called "R-similar sets" for two atomic formulae, their approach may be easily generalized to any set of given atoms.

Third, SUMMERS' early work on inductive synthesis of LISP programs (cf. [Sum75], [Sum77]) is based on recognizing common term structures. The key idea is to process given input/output examples to express the outputs literally "in terms of the corresponding inputs". These techniques may be understood as guessing certain similarities of input/output examples. SUMMERS' approach applies to more than two objects in a quite natural way.

Compared to the deductive approach outlined in [OMBT92], the present approach is more inductive in spirit. This originates from its close connection to research in machine learning, esp. in inductive inference. There are several formalisms of a different strength.

Definition 1

The inductive similarity of a class C of terms is defined as a most specific antiunifier denoted by t_C.

This concept is very close to constructs in SUMMERS' work. For certain further work interpreting terms as cases (or, vice versa, formalizing cases by terms), it may be useful to have access to the corresponding substitutions reproducing terms from its antiunifier.

The importance of the following concept lies in its applicability to knowledge transfer. Compared to the concept before, the second one does almost completely reflect the knowledge contained in some collection of cases.

Definition 2

The strong inductive similarity of a class C of terms is defined as a pair $[t_C, \Gamma_C(t_C)]$, where t_C is a most specific antiunifier of C and $\Gamma_C(t_C)$ is defined by

$$\Gamma_C(t_C) = \{\gamma / \exists t \in C(t_C\gamma = t)\} \tag{9}$$

This easily allows to compute the similarity (formally: the inductive resp. strong inductive similarity) of a class of objects as well as the similarity between a class of objects and another individual object given. Furthermore, it allows to compare several similarities.

For instance, the terms $2 * sin(x + \pi)$ and $2 * sin(\psi - \varphi)$ yield the inductive similarity $2 * sin(x)$. Therefore, $sin(x + \pi)$ and $sin(\psi - \varphi)$ are more similar to each other than the first two terms. Furthermore, $sin(y_1 + y_2)$ and $sin(z)$ are of the same similarity as $sin(x + \pi)$ and $sin(\psi - \varphi)$. But from the viewpoint of strong inductive similarity, $sin(y_1 + y_2)$ and $sin(z)$ are properly more similar than $sin(x + \pi)$ and $sin(\psi - \varphi)$ are. There is not enough space here for further examples and discussions.

The reader may have recognized that one can imagine several refinements of the present approach, for instance, to make terms like $2 * sin(x)$ and $sin(x)$ more similar to each other. But this exceeds the space of the first approach presented here. In particular, it remains open how to embed the concepts introduced here into some framework for integrating standard and nonstandard approaches.

3.3 Computing Similarity

In [KSjSC85] (also adopted in [Alt93], for instance), there has been developed a scenario for case-based reasoning. According to this publication, the key steps are:

- Locate and retrieve potentially applicable cases from long term memory.
- Evaluate selected cases to determine the applicable ones.
- Transfer knowledge from the old case(s) to the current one.

From the viewpoint of classification, the last point is of less importance. The concept of strong inductive similarity seems to be of a particular value for knowledge transfer in case-based reasoning. To investigate this in depht is left to a forthcoming paper.

Instead, it will follow a brief discussion how to handle inductive similarity computationally. This is of some interest, as it is closely related to unification and antiunification. The importance of antiunification is obvious from the definitions above. Unification comes into play because of the necessity to find a most specific antiunifier. Algorithms known from the literature (cf. [PW78] and [MM82], for instance) may contribute to efficient procedures, in the future. Even ROBINSON's algorithm (cf. [Rob65]) may suggest algorithmic solutions in case-based reasoning.

The following algorithm is intended to illustrate the use of inductive similarity for case-based classification. Assume cases to be represented as labelled terms. There are two labels + and − indicating that some case does or does not represent some target concept. Fortunately, this does not require any additional formalism, as one can assume some additional operator, say "o", which connects terms and their corresponding values. For example, a case term t representing some concept may be either written as "$(t, +)$" or "$\circ(t, +)$". The latter notation allows to consider terms, only. One may assume any case base *CB* containing a finite number of positive and negative cases (examples and counter examples) collected in CB_+ and CB_-, respectively. These examples are labelled cases with the outermost operator o. In the worst case, the similarity of cases in CB_+, for instance, is expressed by the term $t_{CB_+} = \circ(x, +)$. Usually, there will be more structure in the left hand argument.

The functionality of the algorithm to be presented on the next page but one is very close to the behaviour of the seven algorithms published in [Aha91] and [AKA91], for instance. The key difference is that the similarites used are more complex in structure. In particular, there is the overall similarity of some collection of cases. Due to this richness in structure, there may occur some more interesting problems than under the standard approach.

The present algorithm should be understood as the prototype of some family of related algorithms. It is quite simple.

There is given *CB* consisting of CB_+ and CB_- as well as t_{CB_+} and t_{CB_-}. The current unlabelled term t has to be classified. z is any new variable occuring in none of the terms considered. One may assume any particular partial ordering $\prec$ to compare the complexity of terms. In many case, generality w.r.t. reducibility by substitutions (i.e. $\lhd$ introduced above) will do.

There is a particular procedure of lifting invoked in the algorithm. This should be briefly explained in some detail. For simplicity, one assumes that the variables in t_{CB_*} and t can be easily distinguished. Suppose all variables occuring in t_{CB_*} are x, with certain indices, whereas the variables of t are variants of y. If the two terms considered have a most general unifier β, this may be separated into two parts: $\beta = \beta_x \cup \beta_y$, where there are variables and terms meeting $\beta_x = \{x_{i_1} \leftarrow t'_1, \cdots, x_{i_k} \leftarrow t'_k\}$ and $\beta_y = \{y_{j_1} \leftarrow t''_1, \cdots, y_{j_l} \leftarrow t''_l\}$. t_{CB_*} is lifted as follows: Every variable y_{j_λ} has some occurence(s) p in t, i.e. $t_{|p} = y_{j_\lambda}$. To express this relationship, a variable occuring at some given position p is named y_p. Because of $t_{CB_*}\beta = t\beta$, it holds $t_{CB_*|p} = t_{j_\lambda}$. Lifting t_{CB_*} w.r.t. β means to replace at each such position p the corresponding subterm by its related variable. If these positions under consideration are $p_1, \cdots, p_m$, the meaning of lifting is formally expressed as

$$\uparrow_\beta (t_{CB_*}) = t_{CB_*}[p_1 \leftarrow y_{p_1}, \cdots, p_m \leftarrow y_{p_m}] \tag{10}$$

After lifting, it may be necessary to tune the notations of variables occuring in the lifted term.

Note that lifting is a standard technique for antiunification. In the algorithm below, it appears to be closely related to unification. For reasonable and efficient implementation, one should carefully shuffle the attempt to construct a most general unifier and the subsequently required lifting. Also the construction of an antiunifier, if necessary, may be started partially when trying unification. Thus, the algorithm introduced below is only describing the gist of a couple of procedures interacting with each other in a quite complex way.

Furthermore, if the algorithm turns out to be unable to classify a current term automatically, this may be managed in several ways. The algorithm describes a basic approach, only. It is out of the scope of the present paper to investigate heuristics or man-machine interaction for the type of case-based reasoning problems considered.

This completes the preparations for the following algorithm. It is intended to work as follows. Given any case base *CB* possibly containing both positive and negative examples to characterize some underlying concept, and getting fed in some current unclassified case, it tries to classify this case correctly and dates up the case base, if necessary.

Recall again, please, that the considerations on hand are purely syntactical. It is left to forthcoming investigations to take background knowledge into account. Thus, the following algorithm plays a role like unification does for theory unification in automated deduction. The more interesting problems are considerably more difficult, partially open, and may even turn out to be unsolvable, in general.

Algorithmus $\mathcal{A}_1$

begin
if t_{CB_+} and $\circ(t,z)$ are unifiable **then**
  begin
  return "+" to classify t ;
  if $\beta_y \neq \emptyset$ **then**
    begin
    $t_{CB_+} := \uparrow_\beta (t_{CB_+})$;
    $CB_+ := CB_+ \cup \{\circ(t,+)\}$
    end
  end
  else
  if t_{CB_-} and $\circ(t,z)$ are unifiable **then**
    begin
    return "−" to classify t ;
    $\beta_y \neq \emptyset$ **then**
      begin
      $t_{CB_-} := \uparrow_\beta (t_{CB_-})$;
      $CB_- := CB_- \cup \{\circ(t,-)\}$
      end
    end ;
if neither t_{CB_+} and $\circ(t,z)$ nor t_{CB_-} and $\circ(t,z)$ are unifiable **then**
  begin
  compute the antiunifier t_1 of t_{CB_+} and $\circ(t,z)$;
  compute the antiunifier t_2 of t_{CB_-} and $\circ(t,z)$; $\{t_i$ $(i = 1,2)$ is of the form $\circ(t'_i, z)$.$\}$
  if $t'_1 \prec t'_2$ **then**
    begin
    return "+" to classify t ;
    if $t_{CB_+} \lhd \circ(t_1,+)$ **then**
      begin
      $CB_+ := CB_+ \cup \{\circ(t,+)\}$;
      $t_{CB_+} := \circ(t_1,+)$
      end
    end ;
  if $t'_2 \prec t'_1$ **then**
    begin
    return "−" to classify t ;
    if $t_{CB_-} \lhd \circ(t_2,-)$ **then**
      begin
      $CB_- := CB_- \cup \{\circ(t,-)\}$;
      $t_{CB_-} := \circ(t_2,-)$
      end
    end ;
  if neither $t'_1 \prec t'_2$ nor $t'_2 \prec t'_1$ **then** { automatic classification impossible }
    begin
    ask the user to classify t ; get the user's input $\star \in \{+,-\}$;
    $CB_\star := CB_\star \cup \{\circ(t,\star)\}$;
    if $\star = +$ and $t_{CB_\star} \lhd \circ(t_1,+)$ **then** $t_{CB_+} := \circ(t_1,+)$;
    if $\star = -$ and $t_{CB_\star} \lhd \circ(t_2,-)$ **then** $t_{CB_-} := \circ(t_2,-)$
    end
  end
end.

An algorithm for strong inductive similarity is a little more complex, as it has to compute substitutions, additionally. But it may be designed as the algorithm before. Instead of going into these details, there will be discussed another version of the first algorithm which makes the relationship between unification and antiunification more explicit. For readability, the part dealt with user interaction will be skipped in the algorithm below. Moreover, to fit the text of the algorithm into the space of this page, certain computations are left out. But this is clearly indicated. Perhaps, some readers will even enjoy this.

Algorithmus $\mathcal{A}_2$

begin
compute the maximal subsets $D'_+ \subseteq D \cap D_+$ and $D'_- \subseteq D \cap D_-$ satisfying (for each corresponding index $\star \in \{+,-\}$)
  $D'_\star$ is a tree domain and
  $\forall p \in D'_\star(t_{CB_\star}(p) = t(p))$;
if neither $D'_+ \subset D'_-$ nor $D'_- \subset D'_+$ **then** user interaction ... **else**
  begin assume $\star \in \{+,-\}$ to be the index of the larger tree domain ;
  if $\forall p \in D_\star \backslash D'_\star(t_{CB_\star|p} \in X)$ and $\forall p,q \in D_\star \backslash D'_\star(t_{CB_\star|p} = t_{CB_\star|q} \rightarrow t_{|p} = t_{|q})$ **then**
    { there exists some substitution β satisfying $t_{CB_\star}\beta = t$ }
    return "$\star$" to classify t **else**
    if $\forall p \in D_\star \backslash D'_\star(t_{CB_\star|p} \in X)$, but $\exists p,q \in D_\star \backslash D'_\star(t_{CB_\star|p} = t_{CB_\star|q} \wedge t_{|p} \neq t_{|q})$ **then**
      begin
      while $\exists p,q \in D_\star \backslash D'_\star(t_{CB_\star|p} = t_{CB_\star|q} \wedge t_{|p} \neq t_{|q})$ **do**
        begin
        choose any such term $t_{|p}$;
        $D'' := \{q / t_{|p} = t_{|q}\} \cap (D_\star \backslash D'_\star)$
        choose any new variable x' not occuring in $t_{CB_\star}$;
        repeat
          choose $q \in D''$;
          $D'' := D'' \backslash \{q\}$;
          $t_{CB_\star} := t_{CB_\star}[t_{CB_\star|q} \leftarrow x']$
        until $D'' = \emptyset$
        end ;
      $CB_\star := CB_\star \cup \{o(t,\star)\}$
      end ;
  if $\exists p \in D_\star \backslash D'_\star(t_{CB_\star|p} \in T(\Sigma,X)\backslash X)$ **then**
    if $\forall p \in D_\star \backslash D'_\star(t_{CB_\star|p} \notin X \rightarrow t_{|p} \in X)$ **then**
      if $\forall p,q \in D_\star \backslash D'_\star(t_{CB_\star|p} \notin X \wedge t_{CB_\star|p} \notin X \wedge t_{|p} = t_{|q} \rightarrow t_{CB_\star|p} = t_{CB_\star|q})$ **then**
        { the terms $t_{CB_\star}$ and t are unifiable without confusion of variables }
        begin
        compute a most general unifier β of $t_{CB_\star}$ and t ;
        $t_{CB_\star} := \uparrow_\beta (t_{CB_\star})$;
        $CB_\star := CB_\star \cup \{o(t,\star)\}$
        end
  if $\exists p \in D_\star \backslash D'_\star(t_{CB_\star|p} \in T(\Sigma,X)\backslash X) \wedge$ ••• **then** •••
  if the terms $t_{CB_\star}$ and t are not unifiable **then** •••
  end
end.

The case base CB was assumed to be separated into CB_+ and CB_-. The corresponding inductive similarities are expressed by t_{CB_+} and t_{CB_-}, respectively. Any current term t has to be processed. The aim is to classify t and to update both the case base and its similarity expression, if necessary. For technical reasons, it is convenient to have notations for the tree domains of the three terms t_{CB_+}, t_{CB_-}, and t. These are D_+, D_-, and D, respectively. The explicit use of the underlying tree domains, which may be searched downwards starting at the root, indicates how to implement the algorithms investigated. To illustrate this, there have been presented three characteristic parts of the algorithm in some more detail. The two parts left out are those cases, where the algorithm can draw less advantage from the stepwise unification process before. This is the fourth case of standard unification and the fifth case, where the considered terms cannot be unified. In these cases, the upper part of the two terms (over the common tree domain $D'_\star$) has already been processed and may be kept unchanged. The remaining work is just unification resp. antiunification.

4. Further Approaches to Structural Similarity

A first natural generalization of the approaches above is to consider formulae. This does not need any further explanation, as in the theory-free case, unification and antiunification is unchanged. Therefore, it seems more reasonable to adopt the approaches above to sets of logical formulae where knowledge is explicitly available. This is an exciting area, and corresponding investigations will hopefully yield deeper insights into the relationship between deductive similarity of (sets of) formulae as introduced in [OMBT92] and inductive similarity as introduced here. However, in the author's opinion, one should develop the theory-based approach for the universal algebra framework before generalizing it to the logical framework.

The structures occuring in the application domain mentionend above (cf. [Hov93b], [Hov93a]) frequently suggest a graph-theoretic point of view. Hierarchical graph structures like pin graphs (cf. [HLW92], [AB92]) seem to be appropriate. This chapter is aimed at very briefly introducing an approach to structural similarity using pin graphs. It is more intended to widen the readers view at similarity than initiating an in depth discussion.

Pin graphs are finite collections of particular graphs which have certain distinguished properties. There are special nodes which may be replaced by complete individual graphs from the given collection. Some technical concepts guarantee non-conflicting replacements. One may drop technical details, here. The interested reader is directed to [HLW92], [AB92], for more details. In its elementary form, each pin graph may be flattened by doing all possible replacements. The result is an ordinary graph. In other words, pin graphs are structured presentations of ordinary graphs. They are offered here as an appropriate tool for case representation.

There are at least three reasons to favour pin graphs. One is that these structures seem naturally close to certain technical objects like supply-air systems to be represented. A second reason is that there are algorithms known to be more efficient on hierarchically structured graphs than on flat ones (cf. the references mentioned above). Finally, one may directly adopt the ideas developed in the preceding chapter. For graph structures, there are natural generalizations of unification and antiunification.

In the author's opinion, nonstandard approaches to similarity, which are intended to set the stage for case-based reasoning more flexible and expressive than today, may require combinations of several formal concepts as well as sophisticated heuristics and well-organized man-machine interaction. The paper on hand deals with particular problems of formalization, only. But already in this area, complex applications may require the combination of different concepts and techniques. For illustration, one may imagine term structures where particular leaves are graphs (perhaps even pin graphs) whereas others are tuples, as usual.

5. Summary

There are several approaches to similarity exhibiting a considerable number of deeper relations to other research areas. On the one hand, this points to the difficulty of the problem of formalizing similarity in a suitable way, of computing similarities, and of exploiting the similarities found. On the other hand, this indicates that there is a huge amount of knowledge and a remarkable number of results to be utilized for similarity-based reasoning. And last but not least, the approaches introduced here provide a large number of possible continuations and extensions as well as some exciting open problems.

In the author's opinion, there will be no single distinguished approach to similarity in case-based reasoning. Instead, powerful case-based reasoning approaches should allow to select from a couple of different concepts, and to integrate various approaches. It was the main intention of the present paper to propagate nonstandard ideas and to set the stage for some research in this direction.

Acknowledgement

The author's work presented here has been originally stimulated by MICHAEL M. RICHTER. In summer 1991 he asked some exciting questions about the relationship between inductive inference and case-based learning. Many colleagues contributed to a discussion about symmetry of similarity concepts. Most of them tended to defend the standard approach. They have been particularly helpful, as they excited the author to investigate other approaches in more detail. Within the BMFT joint venture FABEL, KATY BÖRNER has been particularly engaged for finding structural approachcs to similarity. Her considerations have been guided by BIPIN INDURKHYA's work on analogical reasoning (cf. [Ind89], [Ind91]). Those investigations have contributed to the author's work presented. Part of this work has been supported by the German Ministry for Research and Technology (BMFT) within the joint venture FABEL, grant no. 413-4001-01 IW 104 C, and within the joint venture GOSLER, grant no. 413-4001-01 IW 101 A.

References

[AB92] Helmut Alt and Johannes Blömer. Resemblance and symmetries of geometric patterns. In Burkhard Monien and Thomas Ottmann, editors, *Data Structures and Efficient Algorithms*, volume 594 of *Lecture Notes in Computer Science*, pages 1–24. Springer-Verlag, 1992.

[Aha91] David W. Aha. Case-based learning algorithms. In *DARPA Workshop on Case Based Reasoning*, pages 147–157. Morgan Kaufmann, 1991.

[AKA91] David W. Aha, Dennis Kibler, and Marc K. Albert. Instance-based learning algorithms. *Machine Learning*, 6:37–66, 1991.

[Alt93] Klaus-Dieter Althoff. *Eine fallbasierte Lernkomponente als integrierter Bestandteil der MOLTKE-Werkbank zur Diagnose technischer Systeme*, volume 23 of *DISKI, Dissertationen zur Künstlichen Intelligenz.* infix, 1993.

[Ang80] Dana Angluin. Finding patterns common to a set of strings. *Journal of Computer and System Sciences*, 21:46–62, 1980.

[AS83] Dana Angluin and Carl H. Smith. A survey of inductive inference: Theory and methods. *Computing Surveys*, 15:237–269, 1983.

[Gol67] E Mark Gold. Language identification in the limit. *Information and Control*, 14:447–474, 1967.

[Gol89] Hans-Joachim Goltz. Unifikationstheorie. In Jan Grabowski, Klaus P. Jantke, and Helmut Thiele, editors, *Grundlagen der Künstlichen Intelligenz*, pages 188–222. Akademie-Verlag Berlin, 1989.

[HLW92] Franz Höfting, Thomas Lengauer, and Egon Wanke. Processing of hierarchally defined graphs and graph families. In Burkhard Monien and Thomas Ottmann, editors, *Data Structures and Efficient Algorithms*, volume 594 of *Lecture Notes in Computer Science*, pages 44–69. Springer-Verlag, 1992.

[Hov93a] Ludger Hovestadt. A4 - digital building - extensive computer support for design, construction, and management of buildings. In *Proc. CAAD Futures'93, Pittsburgh, USA*, page to appear, 1993.

[Hov93b] Ludger Hovestadt. Armilla4 - an integrated building model based on visualisation. In *Proc. EuropIA '93, Delft, The Netherlands*, page to appear, 1993.

[Ind89] Bipin Indurkhya. Modes of analogy. In Klaus P. Jantke, editor, *Analogical and Inductive Inference, AII'89*, volume 397 of *Lecture Notes in Artificial Intelligence*, pages 217–230. Springer-Verlag, 1989.

[Ind91] Bipin Indurkhya. On the role of interpretative analogy in learning. *New Generation Computing*, 8(4):385–402, 1991.

[Jan89] Klaus P. Jantke. Algorithmic learning from incomplete information: Principles and problems. In J. Dassow and J. Kelemen, editors, *Machines, Languages, and Complexity*, Lecture Notes in Computer Science, pages 188–207. Springer-Verlag, 1989.

[Jan92a] Klaus P. Jantke. Case based learning and inductive inference. GOSLER Report 08/92, Technische Hochschule Leipzig, FB Mathematik & Informatik, February 1992.

[Jan92b] Klaus P. Jantke. Case based learning in inductive inference. In *Proc. of the 5th ACM Workshop on Computational Learning Theory, (COLT'92), July 27-29, 1992, Pittsburgh, PA, USA*, pages 218–223. ACM Press, 1992.

[JB81] Klaus P. Jantke and Hans-Rainer Beick. Combining postulates of naturalness in inductive inference. *EIK*, 17(8/9):465–484, 1981.

[Kod88] Yves Kodratoff. *Introduction to Machine Learning.* Pitman, 1988.

[KSjSC85] Janet L. Kolodner, Robert L. Simpson jr., and Katia Sycara-Cyranski. A process model of case-based reasoning in problem solving. In Aravind Joshi, editor, *Proc. IJCAI'85, Intern. Joint Conference on Artificial Intelligence, April 18-23, 1985, Los Angeles, CA, USA*, pages 284–290. Morgan Kaufmann Publ., 1985.

[KW80] Reinhard Klette and Rolf Wiehagen. Research in the theory of inductive inference by GDR mathematicians - a survey. *Information Sciences*, 22:149–169, 1980.

[Lak87] George Lakoff. *Women, Fire, and Dangerous Things*. The University of Chicago Press, 1987.

[MM82] A. Martelli and Ugo Montanari. An efficient unification algorithm. *ACM Transactions on Programming Languages and Systems*, 4:258–282, 1982.

[OMBT92] Takeano Ohkawa, Toshiaki Mori, Noboru Babaguchi, and Yoshikazu Tezuka. Analogical generalization. In *Intern. Conference on Fifth Generation Computer Systems, June 1-5, 1992*, volume 2, pages 497–513. Institute for New Generation Computer Technology (ICOT), Tokyo, Japan, 1992.

[Plo70] Gordon D. Plotkin. A note on inductive generalization. In B. Meltzer and D. Michie, editors, *Machine Intelligence 5*, pages 153–163. American Elsevier, 1970.

[PS92] Christian Posthoff and Michael Schlosser. Ähnlichkeitsbetrachtungen beim Problemlösen. In *Proc. Ähnlichkeit von Fällen beim fallbasierten Schließen, Workshop, June 25-26, 1992, Kaiserslautern*, pages 1–8. Universität Kaiserslautern, FB Informatik, SEKI Working Paper SWP-92-11, 1992.

[PW78] M.S. Paterson and M.N. Wegman. Linear unification. *Journal of Computer and Systems Science*, 16:158–167, 1978.

[Ric92] Michael M. Richter. Classification and learning of similarity measures. In *Proc. Ähnlichkeit von Fällen beim fallbasierten Schließen, Workshop, June 25-26, 1992, Kaiserslautern*, pages 1–8. Universität Kaiserslautern, FB Informatik, SEKI Working Paper SWP-92-11, 1992.

[Rob65] J.A. Robinson. A machine-oriented logic based on the resolution principle. *Journal of the ACM*, 12:23–41, 1965.

[Ros75] Eleanor Rosch. Cognitive reference points. *Cognitive Psychology*, 7:531–547, 1975.

[Sie89] Jörg H. Siekmann. Universal unification. In R.E. Shostak, editor, *Proc. CADE'84, 7th Conference on Automated Deduction*, volume 170 of *Lecture Notes in Computer Science*, pages 1–42. Springer–Verlag, 1989.

[Sum75] Philip D. Summers. *Program Construction from Examples*. PhD thesis, Yale University, Dept. Comp. Sci., 1975.

[Sum77] Philip D. Summers. A methodology for LISP program construction from examples. *Journal of the ACM*, 24(1):161–175, 1977.

[TG78] Amos Tverski and I. Gati. Studies of similarity. In Eleanor Rosch and B.B. Lloyd, editors, *Cognition and Categorization*, pages 79–98. Lawrence Erlbaum Assoc., 1978.

Learning from Cases for Classification Problem Solving

Frank Puppe

Universität Würzburg, Lehrstuhl für Informatik VI
Am Hubland, D-97074 Würzburg, Germany

Abstract: An overview of case-based learning techniques for classification problem solving from the research areas of expert systems, statistics, and neuronal nets is presented, together with some results from comparative evalutions. We broadly define a problem solver to be able to "learn from cases" if it usually performs better with every new case.

1. Introduction

Classification problem solving is a relatively well understood problem class. The task is to select diagnoses from a preenumerated set of alternatives on the basis of symptoms (data) observed. A set of different problem solving methods for classification has been described e.g. in [Benjamins 93] or [Puppe 93, Part III]. *Certain classification* methods like *decision trees* and *decision tables* comprise only categorical knowledge like a rule "if Symptom S then always diagnosis D". Only simple classification problems can be solved with these methods. Usually, the relations and/or the data must be qualified with uncertainty values. There are three main types for solving classification problems with uncertainty: Heuristic, causal and case-based classification. In *heuristic classification* [Clancey 85], the heuristic knowledge is coded in rules of the form "if S then D with some evidence value". For each diagnosis, the evidence values from the rules are comprised to an evidence score and (not taking into account special techniques for diagnosing multiple faults or differential diagnoses) the diagnosis with the highest accumulated score is selected. *Causal classification* comprises both *set-covering classification* (e.g. [Peng & Reggia 90]), where a typical piece of relational knowledge is represented by rules of the kind "diagnosis D may explain Symptom S" and that diagnosis is selected, which can explain best the observed symptoms, as well as *classification with functional or physical simulation models* (e.g. [Davis 84, de Kleer & Williams 89]) where such relations must be inferred from discrepancies with the model. In *case-based classification* [Kolodner 93], for a new case without solution the most similar case from a case base of correctly solved old cases is selected and its solution is taken over to the new case.

The key problem for classification problem solving is to acquire the detailed knowledge necessary for any method. While the standard way with domain experts and/or knowledge engineers building the knowledge bases seems to be feasible but quite expensive, there is now a great interest in learning techniques (e.g. [Kodratoff & Michalski 90]) which acquire the necessary knowledge automatically. However, since learning means knowledge transformation (from nothing comes nothing), the cruical question is what kind of knowledge is the starting point for learning. A very attractive form are cases, where a case contains a set of observed symptoms and the final solution. In its simplest form, a case consists of a set of attributes (A) and values (V) together with a diagnosis (D), e.g. $((A_1V_1)(A_2V_2)...(A_nV_n)D)$ or in vector representation with a fixed order of attributes: $(V_1V_2...V_nD)$. In this paper , we call the attribute-value pairs symptoms. Since most of the learning techniques require a single diagnosis as a solution, we do not discuss the problem of multiple diagnoses (e.g. [de Kleer & Williams 87]). Also, the problem of incremental data gathering is not considered in this overview.

Cases provide empirical knowledge about solutions to past problems, and these may hold valid for similar problems in the future. There are many different ways of generalizing the cases. A survey of these generalizations is the theme of this paper. More thorough treatments can be found in [Weiss 91, Winston 92, Michie 94] or in special literature about statistics (e.g. [James 85]), about neural nets (e.g. [Hinton 89] [Ritter, Martinez & Schulten 91]) or about machine learning (e.g. [Kodratoff & Michalski 90]. The last reference also contains an excellent survey on machine learning literature.

2. Categorization

An overview of the categorization of techniques for learning from cases is given in Fig. 1.

type of abstraction \ field	Expert systems	Statistics	Neuronal Nets
No abstractions	Case-based classification	Nearest Neighbor	
Simple abstractions		Bayes Theorem Linear decision functions	Perceptron Learning
Complex abstractions without background know	Inductive learning of decision trees or rules		Backpropagation Learning
Complex abstractions with background know	Explanation-based learn., adaptation of probabilities	Bayesian Networks	(pretrained neural nets)

Fig.1: Survey for case-based learning methods for classification problem solving. Explanation see text (know = knowledge).

The table contains two dimensions:

- The type of abstraction in the knowledge representation: It ranges from *no abstractions*, where cases remain unchanged, via *simple abstractions* with only simple symptom-diagnosis correlations via *complex abstractions* with complex rules or an intermediate diagnostic structure to *complex abstractions with background knowledge*, where the learning technique relies on additional knowledge which does not stem from the cases. Background knowledge may also be used in the category "no abstractions" for the similarity measure, which can be viewed as substitute for the generalizations of the other types.

- The research field: similar techniques have been developed in different research communities. We differentiate somehow arbitrarily between three communities: the *expert system* community, the *statistics* community and the *neuronal net* community.

We discuss the various techniques along a very simple, hypothetical example case base with 5 old cases with two symptoms and three diagnoses (fig. 2):

attributes \ cases	Case1	Case2	Case3	Case4	Case5	*New Case*
cough	yes	yes	no	yes	no	*yes*
fever	37	38,5	38	38	37	*37,8*
diagnosis	bronchitis	flu	other	flu	other	*??????*

Fig. 2: A small hypothetical example case base for illustration of the following techniques.

2.1 Simple abstractions

Theorem of Bayes

The main dimension for our categorization is the target problem solving type of the case-based learning procedure. A straightforward way is to compute simple, direct correlations between all symptoms and diagnoses, e.g. in 90% of all cases, where the diagnosis D were present, the symptom S was also present, i.e. $P(D|S) = 0.9$ (a symptom is assumed to be either present or absent). With such correlations between all symptoms and diagnoses and the a-priori-probabilities of the diagnoses $P(D)$, a variant of the Theorem of Bayes [Lusted 68] enables to calculate the most probable diagnosis from a set of observed symptoms (fig. 3).

$$P_r(D_i/S_1\&\ldots\&S_m) = \frac{P(D_i) * P(S_1/D_i) * \ldots * P(S_m/D_i)}{\sum_{j=1}^{n} P(D_j) * P(S_1/D_j) * \ldots * P(S_m/D_j)}$$

Fig. 3: Variant of Theorem of Bayes, for computing the relative probability P_r of a dig-nosis D from a set of diagnoses $D_1 \ldots D_n$, based on the observed symptoms $S_1 \ldots S_m$.

Example: With the case base from fig. 2, the following probabilities are calculated (note, that the numerical attribute "fever" had to be converted to a yes/no-attribute, for which we use the mean of 37 and 38):

P(flu) = 0,4 P(bronch) = 0,2 P(other) = 0,4
P(cough = yes / flu) = 1 P(cough = yes / bronch) = 1 P(cough = yes / other) = 0
P(fever > 37,5 / flu) = 1 P(fever > 37,5 / flu) = 0 P(fever > 37,5 / other) = 0,5

These probabilities determine that the diagnosis of the new case in fig. 2 must be "flu":
P_r(flu / cough = yes & fever = 37,8) = 0,4 * 1 * 1 / (0,4 + 0 + 0) = 0,4 / 0,4 = 1
P_r(bronch / cough = yes & fever = 37,8) = 0,2 * 1 * 0 / (0,4 + 0 + 0) = 0 / 0,4 = 0
P_r(other / cough = yes & fever = 37,8) = 0,4 * 0 * 0,5 / (0,4 + 0 + 0) = 0 / 0,4 = 0

Linear decision functions

A similar procedure uses linear decision functions [Fisher 36, O'Neil 92]. In this learning technique, for each diagnosis a set of weights is computed, one for each attribute. Given a new case, for each diagnosis its set of weights are multiplied with the observed values of the attributes and added up. The general form of linear classifiers is:

$$w_1 s_1 + w_2 s_2 + ... + w_n s_n - w_0$$

where $s_1 ... s_n$ represent the (normalized) attribute-values and $w_0 ... w_n$ the weights, which may be different for each diagnosis. The diagnosis with the highest score - based on the observed symptoms - will be selected.

There are many learning procedures ranging from computing the weights from the means and covariances of the attribute-values relative to the diagnoses to an incremental adaptation of the weights: (1) start with arbitrary weights and (2) for each misclassified case, (2a) lower the weights for a diagnosis, which was falsely selected and (2b) increase the weights for a diagnosis, which is the correct solution.

Example: In contrast to the Theorem of Bayes, linear decision functions can deal with continous values of attributes. However, it is very useful to normalize the raw values. In the example of fig. 2, a plausible normalization is to take the amount of the "fever"-value exceeding 36 degrees (i.e. subtracting 36 from the raw value) and to code cough = yes with the value of 2 and cough = no with the value of 0. A possible result of the lerning procedure yielding a set of weights $((w_1\ w_2\ w_0)$ where w_1 has to multipied with the normalized value of cough and w_2 with that of fever) for each diagnosis might be: flu = (1 1 3,5); bronchitis = (1 -1 0,5); other = (-1 0 1). The new case with the normalized symptoms cough = 2 and fever = 1,8 would result in the following scores for the diagnoses: flu: 2 + 1,8 - 3,5 = 0,3; bronchitis: 2 - 1,8 - 0,5 = -0,3; other: -2 + 0 - 1 = -3, so that the diagnosis "flu" will be selected. Fig. 4 show a graphical representation of linear decision functions.

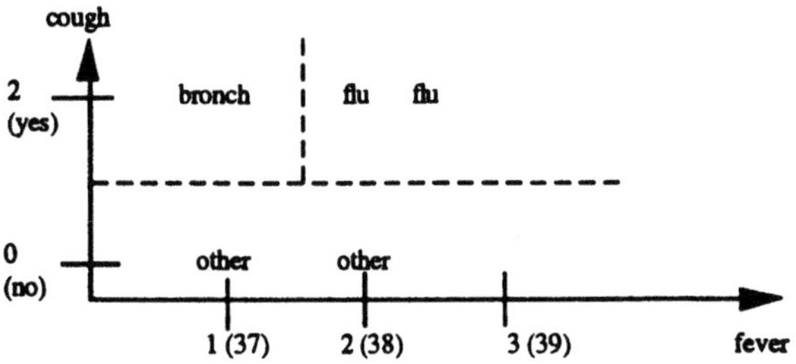

Fig. 4: The dotted lines represent the linear decision functions seperating the diagnoses in the cases of Fig.2.

A very similar learning procedure is used in *perceptrons*, i.e. a neural net with only two layers of neurons: one layer representing the symptoms (input) and the other layer representing the diagnoses (output). For every misclassified case, the weights between the symptom neurons and the diagnosis neurons are adapted in a similar way as described above.

Discussion

The strengths and the weaknesses of these approaches result from the restriction of the simple target knowledge representation. The learning procedures are straightforward and easy to implement. However, these approaches are unable to recognize multiple diagnoses or to deal with an incomplete set of diagnoses, because the relatively best diagnosis is always selected. Furthermore they cannot deal in a satisfactory manner with the XOR-problem, i.e. if a diagnosis has at least two disconnected symptom-regions like diagnosis $D1$ is established, if either $A1 = 0$ and $A2 = 0$ or if $A1 = 1$ and $A2 = 1$, while diagnosis $D2$ is established, if either $A1 = 0$ and $A2 = 1$ or $if A1 = 1$ and $A2 = 0$. Background knowledge is necessary to deal with such situations. In addition, the Theorem of Bayes requires independence of the symptoms, which in most problem areas is severely violated. However if these weaknesses are not too serious in a particular (usually not overly complex) domain, these techniques are quite attractive due to their simplicity.

2.2 Complex abstractions without background knowledge

A more powerful target knowledge represention may overcome the weaknesses of simple abstractions to a great deal, but makes the learning procedures more complicated. The basic idea is to introduce an intermediate diagnostic structure, i.e. objects which are neither observable symptoms nor final diagnosis, but intermediate states. A similar effect can be achieved with rules with complex preconditions. The difference is illustrated very well with ***backpropagation nets*** instead of perceptrons, where hidden layers are added between the symptoms and diagnoses. The additional power (e.g. the capability to solve the XOR-problem) must be payed for by a much more complicated learning procedure. We discuss two procedures for induction of decision trees and rules in more detail.

Induction of decision trees

The ID3 family of algorithms [Quinlan 86, 87]learns a decision tree which cannot only classify a new case correctly, but always determines the symptom which should be asked next, if only a partial description of the case is given initially. The learning algorithm starts with a set of cases and looks for the most informative attribute, whose values are used to partition the cases in as homogenous subsets as possible. The most informative attribute can be calculated with help of the entropy-formula:

$$-\sum_i p_i \log(p_i)$$

where p_i is the probability of the i-th diagnosis in the set of cases of a leave of the growing decision tree. The attribute with the highest difference between the entropy of the old set and the weighted sum of the entropies of the new partitions is selected. This procedure is repeated until in every set only cases with the same solution remain (i.e. with an entropy of 0). Since such exact decision trees tend to become rather large, afterwards a phase of tree pruning is added, which often includes also tradeoffs between diagnostic accuracy of the test cases and abstractions yielding smaller decision trees. Advanced examples of the ID3 family algorithms, which can also deal with numerical values of attributes and missing values, are e.g. the methods C4.5 [Quinlan 92] and NewId (see chapter 3 "evaluation").

Example: We first calculate the entropy of the total case set and then the entropy of all attributes and select the attribute with the highest gain as first choice for building the decision tree. The repetition of this procedure for the resulting subtrees is not shown, because it is trivial.
Entropy of total case set: -1/5 * $\log_3$ (1/5) - 2/5 * $\log_3$ (2/5) - 2/5 * $\log_3$ (2/5) = 0,293 + 0,334 +0,334 = 0,961
If we select the attribute cough as first decision in the decision tree, we get two sets: "cough = yes" with three cases with the diagnoses (bronch, flu, flu) and "cough = no" with two cases with the diagnoses (other, other) yielding the following entropies:
cough = yes: -1/3 * $\log_3$ (1/3) - 2/3 * $\log_3$ (2/3) = 0,333 + 0,246 = 0,579
cough = no: -2/2 * $\log_3$ (2/2) = 0
cough: 3/5 * 0,579 + 2/5 * 0 = 0,347,
i.e. we have an entropy gain of 0,961 - 0,347 = 0,611 if we know the value of cough.
If we select the attribute fever as first decision in the decision tree, we get two sets: "fever

> 37,5" with three cases with the diagnoses (flu, other, flu) and "fever ≤ 37,5" with two cases with the diagnoses (bronch, other) yielding the following entropies:
fever > 37,5: $-1/3 * \log_3 (1/3) - 2/3 * \log_3 (2/3) = 0{,}333 + 0{,}246 = 0{,}579$
fever ≤ 37,5: $-1/2 * \log_3 (1/2) - 1/2 * \log_3 (1/2) = 0{,}315 + 0{,}315 = 0{,}63$
fever: $3/5 * 0{,}579 + 2/5 * 0{,}63 = 0{,}347 + 0{,}252 = 0{,}597$
i.e we have an entropy gain of only 0,961 - 0,597 = 0,364 if we know the value of fever. Therefore, we first ask for cough. The complete decision tree is shown in Fig. 5

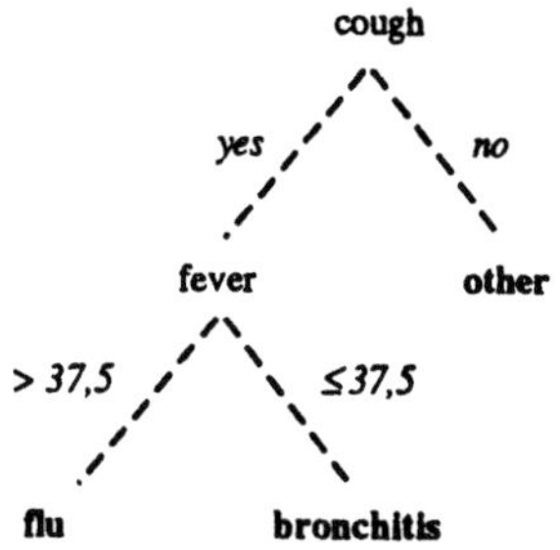

Fig. 5: Decision tree generated with an ID3 algorithm based on the entropy formula for the cases of fig.2.

Induction of rules

Decision trees have inherent difficulties when dealing with missing information, e.g. if the value of an attribute is unknown. This disadvantage can be compensated by rules, where each rule is syntactically independent of all other rules. It is possible to convert any decision tree to a set of usually rather large rules (i.e. rules with many preconditions), which can be pruned afterwards.

Example: The decision tree of Fig. 5 can be transformed in three rules:
If cough = yes *and* fever > 37,5 *then* flu
If cough = yes *and* fever ≤ 37,5 *then* bronchitis
If cough = no *then* other

An alternative way is to generate rules from the cases directly e.g. done by the AQ family of learning algorithms [Michalski 83]. Such an algorithm operates iteratively on each diagnosis, considering all cases containing the actual diagnosis as positive cases and all other as negative cases. The algorithm is very similar to the version space learning procedure [Mitchell 77], which tries to generalize the positive examples and to spezialize the "null-hypothesis" (that every case satisfies the diagnosis) with the negative examples. An advanced example of the AQ-family is the CN2 algorithm (see chapter 3 "evaluation").

Example: Learning the diagnosis "other", we have two positive cases (3 and 5) and three negative cases (1, 2, and 4). The most specific generalization of the positve cases yields the precondition: "Cough = no", while fever can take every value, because Case3 has fever > 37,5 and Case5 fever ≤ 37,5. Specialization of the "null-hypothes" with the negative examples yields: Casel → "cough = no or Fever > 37,5"; Case2 → "Cough = no"; Case4 → "cough = no". Since the most specific generalization and the most general specialization overlap with "cough = no", this rule-precondition is learned for the diagnosis "other".

2.3 Complex abstractions with background knowledge

Often, background knowledge of the domain is available, e.g. in most medical or technical domains. Incorporating it into the learning procedure may reduce the number of cases necessary for training and / or increase the quality of the output.

Bayesian networks

Bayesian networks [Pearl 88, Heckerman 91] overcome most of the limitations directly applying the Theorem of Bayes as described above. In addition to symptoms and diagnoses, Bayesian networks contain a net of intermediate states. The links between the states represent causal influence and are qualified by probabilities. While the probabilities may be computed from a case-set, the structure of the net represents causal background knowledge. If several links go into a state, each possible combination of predecessors must be given a probability on its own. Diagnosing a new case with such a network can be very time consuming if the network is large and neither a tree (i.e. singly-connected) nor at least nearly singly-connected.

Adaptation of probabilities in rules

A more general form of this technique is also popular in expert systems, where a backbone of a predefined rule structure including intermediate concepts is given and only the probabilities of the rules must be estimated from the cases. An example is the RX-Project [Blum 88]. When learning a new rule, RX checks for confounding variables (i.e. variables with a known influence on the postcondition or the pre- and the postcondition of the rule in question) in the case set. For identifying confounding variables RX uses background knowledge. Another class of algorithms compute the evidence values of the rules incrementally. They start with estimating the probabilities on the basis of a set of training cases and adapt them afterwards for every misclassified case in the training set. An example of this technique is BUBE [Bamberger et al. 93], which derives the rule structure from background knowledge and afterwards infers probabilities for the generated rules by incrementally adapting a rough estimation of the evidence values.

Explanation-based learning

Finally, we want to mention explanation-based learning [de Jong & Money 86], where a domain theory is compiled from an example case. Although the domain theory suffices for solving the case, it may be rather inefficient. If a solution to a new case has been found with the domain theory, it can be generalized into a rule, which is stored and should help to solve similar cases more efficiently. However, a problem is to find a trade-off between the overhead for checking maybe very specialized rules and their usefulness. In PRODIGY [Minton et al. 89], this trade-off is computed by an estimated utility for every rule and a counter how often the rule has been used in the past. If the ratio falls below a threshold, the rule will be "forgotten". At first sight, explanation-based learning differs from all other learning algorithms discussed so far, because the system learns nothing really "new", but only speeds up its interpreter. However the discussion of case-based reasoning techniques in the next sections questions this view.

2.4 No abstractions

The idea of case-based reasoning is to use the cases as they are without any transformations. For diagnosing a new case, just find a similar one in the case set and use its diagnoses as solution to the new case. The generalization is done by a similarity (distance) measure, which may vary from very simple to very sophisticated and knowledge-intensive ones.

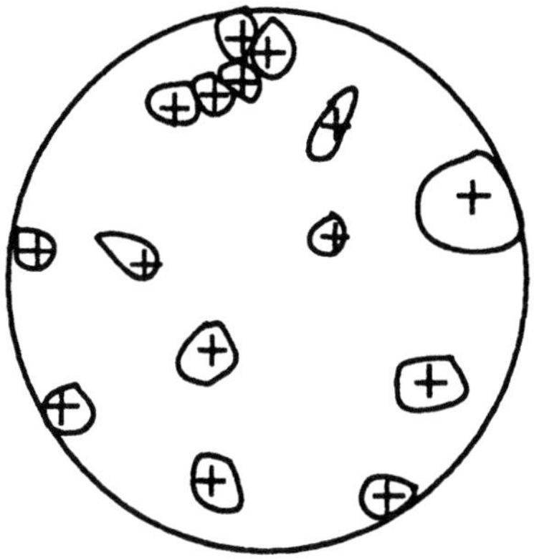

Fig. 6: Generalization in case-based reasoning. Each cross means a case, which should be imagined as a n-dimensional point. Each circle around a case represents the area determined by the similarity measure. If a new case lies within a circle, it may be viewed sufficiently similar to that case, so that its diagnoses may be used as solution of the new case. An alternative interpretation (typically used in the nearest neighbor method) is to select always that case to which the new case is closest and take over that solution.

Nearest-neighbor method

The basic form of case-based reasoning is the nearest-neighbor method [Hart 67]. In its simplest form, the distance between two cases is just the sum of the absolute differences of the values of each attribute-pair. More sophisticated distance measures may use functions on the raw-differences and take into account the different ranges of attributes and their different importances, which is done by a normalization of the values and by assigning a static weight to each attribute. However these extensions require knowledge.

Example: We code (similar as in the paragraph about linear decision functions) cough = yes with the value of 2 and cough = no with the value of 0 and use the euclidian distance measure (correponding to the graphical distances in fig. 6: take the root of the sum of the quadrats of the value differences). The distances of the New Case to the 5 old cases are:
Case1: $\sqrt{0^2 + 0,8^2} = 0,8$; Case2: $\sqrt{0^2 + 0,7^2} = 0,7$; Case3: $\sqrt{2^2 + 0,2^2} = 2,01$;
Case4: $\sqrt{0^2 + 0,2^2} = 2$; Case5: $\sqrt{2^2 + 0,8^2} = 2,15$.
The smallest distance is to Case4, therefore its diagnosis "flu" is assumed to be valid for the New Case too. The distances are illustrated by the dotted lines in fig. 7.

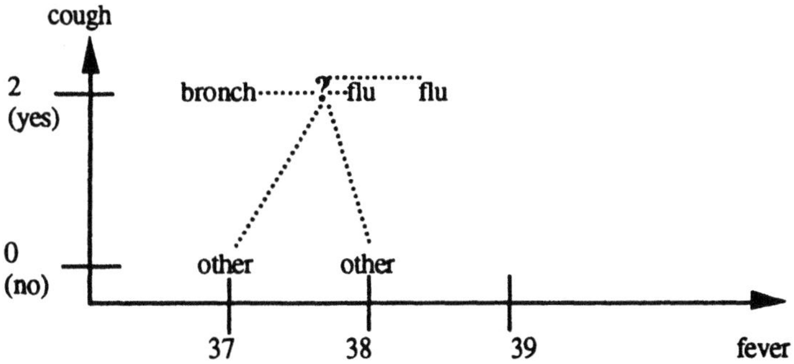

Fig. 7: Graphical illustration of the nearest neighbor method: The question mark represents the New Case, the old cases are represented by the names of their diagnoses and the dottet lines indicate the distance of the New Case to the 5 old cases.

Case-based classification

There are many opportunities for adding more knowledge to the similarity measure of case based reasoning [Kolodner 93]. The normalization mentioned above, which means computing the partial similarity between two corresponding attribute-values, can be supported by many predefined schemes. The static weight can be replaced by a dynamic weight, taking into account the normality or abnormality of the attribute value (the more abnormal the value, the higher the weight), the diagnoses of the old case under consideration (for reflecting the special importance of symptoms for particular diagnoses) and even preferences of the users. Another technique is to abstract the symptoms to more meaningful "symptom abstractions" before performing the case comparison. For example, it makes much more sense to compare the 'actual gas consumption' of two cars relative to their 'normal gas consumption' instead of comparing the raw numbers directly. Of course, various levels of symptom abstractions are possible, so that an optimal compromise between rule-based precomputation and case-based comparison must be found. This is also true for adding categorical rules for ruling out or establishing a diagnosis, if such rules exist. An example for a case-based system including all these techniques is CCC+ [Puppe & Goos 91]. In PATDEX [Weß 93], the similarity measure itself can be learned by an incremental adaptation procedure.

A great problem for case-based reasoning systems is the management of large case sets. In realistic applications, thousands, tens of thousands or more cases each with lots of attributes may occur. There is a storage and a search problem. In the worst case, the cases must be stored on secondary memory and all cases must be compared to find the most similar one. While a good solution to these problems might be a distribution of the cases to many processors and a parallel search, for sequential computers other techniques are required. The size of the case set being necessary in main memory can be decreased drastically, if very similar cases are grouped to clusters, which are represented by only one case of the cluster. A cluster can also be represented just by its key attributes. A procedure for finding clusters and its key attributes is presented in the UNIMEM learning algorithm [Lebowitz 90]. Searching through the case base may be speeded up by preselection techniques, e.g. indexing all cases by their main symptoms, which may be even more powerful, if symptom abstractions are available. Under certain circumstances precomputing a decision tree (K-D-Tree, see e.g. [Winston 92, 403–409]) for the case set reduces the time for finding the most similar case to a new case from linear ($O(n)$) to logarithmic ($O(\log n)$). However, currently there is little experience available for assessing the potential of case-based reasoning in large domains with large case bases.

3. Evaluation

The crucial question is the practical evaluation of the various learning techniques for classification problem solving, since it is impossible to predict their accuracy from their theoretical qualities only. Although there has been significant progress in recent time, realistic field tests are still lacking to my knowledge. "Realistic field tests" means, that a classification system has been built with a learning technique in the first place (not in retrospect), the domain is not small (lets say more than 100 observable symptoms and 50 final diagnoses) and it is in routine use (e.g. for more than two years).

However less ambitious evaluations exist. While in the first phase, only very simple data sets were used like the famous Iris data [Fisher 36] (150 cases, 4 attributes, 3 diagnoses), there are studies with more interesting data available now. A large comparative study with more than 20 learning algorithms of all above mentioned sorts (with the exception of learning procedures requiring background knowledge) and 7 data sets with 2 – 10 diagnoses, 6 – 40 attributes, and 270 – 18000 cases ranging from pattern recognition tasks to medical diagnostics has been performed in the Esprit-project "Statlog" [Michie 94]. One result is, that it is very difficult to describe the results in a few sentences, since the performances of the algorithms varies greatly: For some data sets, the nearest neighbor method performed best, for others, the linear or quadratic discriminant analyses won, and the C4.5 decision tree induction algorithm also performed best for one data set. However, winning one contest does not prevent these techniques to perform rather poorly in other data sets. This was also demonstrated in the neural net area: for example, in the Statlog-studies, back-propagation performed better than Kohonen nets, while in another experiment at Karlsruhe University, kohonen nets performed better [Rausseck 92]. The algorithms also vary greatly in many other dimensions like training and testing time, ability to deal with the value "unknown" for an attribute, incorperation of costs for misclassifying diagnoses, etc. (see [Michie 94]).

One of the participants of this study conducted another experiment [Huber & Nakhaeizadeh 93] and selected four learning algorithms (2 decision tree induction C4.5 and NewID, 1 rule induction CN2, and 1 linear disriminant analysis) to be tested in a realistic domain, where an expert system [Puppe, Legleitner & Huber 92] was in routine use. The results showed, that - if enough cases per diagnosis are available - the manually built expert system performed better, but the differences in the performance rates were not so big (97% to 93% for the best learning technique). However, for the rare diagnoses too few cases were available for the learning techniques, so that these diagnoses were not included in the study. This may indicate a principal limitation of the learning techniques. Another result of this study is, that the induction algorithms performed much better than the linear discriminant analysis technique (which may be a domain specific result however).

The comparison of learning techniques becomes extremely difficult when background knowledge is necessary, because the role of background knowledge is difficult to estimate for final performance. On the other hand, such techniques have the potential to require fewer cases, which is especially important for rare diagnoses. In the neuronal net area, pretraining a net with heuristic rules of the domain may have a beneficiary effect on backpropagation learning [Prem et al. 93]. In the area of rule-induction and case-based classification, we have performed a comparison with a knowledge base for book selection of 58 attributes, ca. 2000 cases and 15 diagnoses. The rule induction system BUBE (see above) uses background knowledge for infering the rule structure, while the case-based reasoner CCC+ (see above) uses some expert knowledge about the similarity measure. With both methods, little time was spent to enter the background knowledge and no attempt was made to optimize the knowledge afterwards. The results showed that both methods performed roughly equal at a 90% level, but the misclassified cases were largely disjunct for both methods. However, case-based classification seems not suitable for all domains. The evaluation of the case-based reasoner CASEY [Koton 88] in the heart failure domain with 240 cases showed a very unsatisfying performance, among other things due to the fact, that the patients usually had multiple diseases and therefore the probability of reusing a case was low [Aghassi 90].

4. Summary

Learning from cases for classification problem solving is a very active research area. There exist many approaches from different research communities (expert systems, statistics, neuronal nets), which sometimes follow similar ideas. So far the evaluations showed that the techniques perform respectable in non-trivial domains. However, they seem not to equal the performance of conventionally built expert systems currently. A crucial question is, how to deal with rare diagnoses, where only few cases are available. Another practical question is, which technique should be prefered under what circumstances.

References

AGHASSI, D. (1990): *Evaluating Case-Based Reasoning for Heart Failure Diagnosis*, Master Thesis, Massachusetts Institute of Technology, Computer Science Lab., MIT/LCS/TR-478

BAMBERGER, S., GAPPA, U., GOOS, K. UND POECK, K. (1993): Teilautomatische Wissenstransformationen zur Unterstützung der Wissensakquisition. In: *Proc. Expertensysteme-93*, Springer, Informatik Aktuell, 153–166

BENJAMINS, V. (1993): *Problem Solving Methods for Diagnosis*, PhD thesis, University of Amsterdam, Amsterdam, The Netherlands

BLUM, R. (1988): Modeling and Encoding Clinical Causal Relations in a Medical Knowledge Base, in : Miller, P. (ed.): *Selected Topics in Medical Artificial Intelligence*, Springer

CLANCEY, W. (1985): Heuristic Classification, *AI-Journal 20, 215–251*

DAVIS, R. (1984): Diagnostic Reasoning Based on Structure and Function, *AI Journal 24, 347–411*

DE JONG, G. AND MONEY, R. (1986): Explanation-Based Learning: An Alternative View, *Machine Learning 1, 145–176*

DE KLEER, J. AND WILLIAMS, B. (1987): Diagnosing Multiple Faults, *AI Journal 32, 97–130*

DE KLEER, J. AND WILLIAMS, B. (1989): Diagnosis with Behavioral Models, in: *IJCAI-89*, 1324–1330

FISHER, R. (1936): The Use of Multiple Measurements in Taxonomie Problems, *Annals of Eugenics 7, 179–188*

HART, P. (1967): The Condensed Nearest Neighbor Rule, *Transactions on Information Theory 14, 515–516*

HECKERMAN, D. (1991): *Probabilistic Similarity Networks*, MIT Press

HINTON, G. (1989): Connectionist Learning Procedures, *Ai-Journal 40, Special issue on Machine Learning, 185–234*

HUBER, K.-P. AND NAKHAEIZADEH, G. (1993): Maschinelle Lernverfahren zur Unterstützung beim Wissenserwerb von Diagnose-Expertensystemen, in: *Proc. Expertensysteme-93*, Springer, Informatik-Aktuell, 167–180

JAMES, M. (1985): *Classification Algorithms*, Wiley

KOTON, P. (1989): *Using Experience in Learning and Problem Solving*, PhD-Thesis, Massachusetts Institute of Technology, Computer Science Lab., MIT/LCS/TR-441

KODRATOFF, Y. AND MICHALSKI, R. (eds.) (1990): *Machine Learning Vol. III*, Morgan Kaufmann

KOLODNER, J. (1993): *Case-Based Reasoning*, Morgan Kaufmann

LEBOWITZ, M. (1990): The Utility of Similarity-Based Learning in a World Needing Explanantion, in: Kodratoff, Y. and Michalski, R. (eds.): *Machine Learning Vol. III, Chap. 15*, Morgan Kaufmann

LUSTED, L. (1968): *Introduction to Medical Decision Making*, Thomas Books

MICHALSKI, R. (1983): A Theory and Methodology of Inductive Learning, *AI-Journal 20, 111-161*

MINTON ET AL. (1989): Explanation-Based Learning: A Problems Solving Perspective, *AI-Journal 40, 63-118*

MICHIE, D., SPIEGELHALTER D., AND TAYLOR, C. (1994): *Book about the Esprit-Project Statlog*, Simon & Schuster, Campus 400, Maylands Avenue, Hemel Hepstead Herts HP2 7EZ, UK, to appear around February 1994

MITCHELL, T. (1977): Version Spaces: A Candidate Elimination Approach to Rule Learning, *IJCAI-77*, 305–310

O'NEIL, T. (1992): Error Rates of non-Bayes Classification Rules and the Robustness of Fishers Linear Discriminant Function, *Biometrika 79, 177-184*

PEARL, J. (1988): *Probabilistic Reasoning in Intelligent Systems*, Morgan Kaufman

PENG, Y. AND REGGIA, J. (1990): *Abductive Inference Models for Diagnostic Problem-Solving*, Springer

PREM, E., MACKINGER, M., AND DORFFNER, G. PORENTA, G. AND SOCHOR, H. (1993): Concept Support as a Method for Programming Neural Networks with Symbolic Knowledge, in: *Proc. GWAI-92*, Lecture Notes in Artificial Intelligence, Springer-Verlag, 166–175

PUPPE, F., AND GOOS, K. (1991): Improving Case-Based Classification with Expert Knowledge, in: *Proc. GWAI-91*, Springer, Informatik-Fachberichte 285, 196–205

PUPPE, F. (1993): *Systemtatic Introduction to Expert Systems*, Springer

QUINLAN, J. (1986): Induction of Decision Trees, *Machine Learning 1, 81-106*

QUINLAN, J. (1987): Simplifying Decision Trees, *International Journal of Man-Machine-Studies 27, 221-234*

QUINLAN, J. (1992): *C4.5: Programms for Machine Learning*, Morgan Kaufmann

RAUSSECK, M. (1992): *Diagnostik mit selbstorganisierenden Karten am Beispiel der Motordiagnose*, Diplomarbeit, Universität Karlsruhe, Institut für Logik, Komplexität und Deduktionssysteme

REGGIA, J. AND PENG, Y. (1990): *Abductive Inference Models for Diagnostic Problem-Solving*, Springer

RITTER, H. MARTINEZ, T. AND SCHULTEN, K. (1991): *Neuronale Netze*, Addison Wesley

WEISS, S. AND KULIKOWSKI, C. (1991): *Computer Systems that Learn*, Morgan Kaufmann

WESS, S. (1993): PATDEX - ein Ansatz zur wissensbasierten und inkrementellen Verbesserung von Ähnlichkeitsbewertungen in der fallbasierten Diagnostik, in: *Proc. Expertensysteme-93*, Springer, Informatik Aktuell, 42–55

WINSTON, P. (1992): *Artificial Intelligence*, Addison Wesley, 3rd edition

Methods of Phoneticizing in Regard to Spelling Variants of Medical Phrases

Rudolf-Josef Fischer

Institut für Medizinische Informatik und Biomathematik,
Universitätsklinikum Essen, Hufelandstr. 55, D-45122 Essen, Germany

Summary: Methods of automated indexing make use of indices with words to be sought. Mistakes made in the input text normally prevent the search to be successful. Therefore an investigation of a likely use of transformations was made in order to consider the most possible variants of spelling without allowing a great loss in the precision of indexing. For this purpose different groups of phonetic transformations were defined. Their effects were examined in regard to two different random samples of input texts: one of them consisting of texts which had to be successfully indexed, the other exclusively with examples to be rejected by the indexing algorithm. Some transformations were proven to be useful. Yet with them only the phonetical spelling errors (a small part of all occurring errors) are covered. Furthermore, significantly longer response times result. Therefore phonetic transformations benefit only when a method puts its main emphasis on a high degree of automation. On the contrary, if a semi-automated method is used, a well defined list of indexed alternatives which regards single errors in the input texts leads to an acceptable solution.

1. Introduction

Earlier publications report semi-automated methods of encoding medical phrases into a diagnosis table with standard texts. In doing so, the **input texts** (the medical phrases that were to be coded) were indexed by **descriptors** referring to the standard texts. These methods can be differentiated depending if word analysis occurs to find the descriptors, or if the descriptors are **pseudoroots** (shortened roots of words), merely searched as a character substring in the input texts.

The frequency of compound terms and alternative formulations of medical phrases in German ("Nierenrindenkarzinom" versus "Karzinom der Nierenrinde") makes a word analysis extraordinarily difficult. Brigl (1992) shows an example: even by application of the principle of longest match and backtracking, "Gehörgangsexostosen" is wrongly decomposed as "Gehörgang/sex/ost/os/en" because the corresponding word segments were found in the dictionaries of pseudoroots or endings, respectively.

For this reason, but also to save the amount of updating the dictionaries, Fischer (1993) preferred the substring search as a method of indexing during the semi-automated classification of input texts into **Weidtman's Diagnosis Table** (1989). Using pseudoroots as descriptors, the hereby results were extremely satisfactory.

2. Problem formulation

Every spelling error that falsifies even one descriptor almost always causes that the descriptor can no longer be found as a substring of the input text, thus the indexing fails. In order to allow at least the most usual spelling variants, in Fischer (1991) the input texts as well as the descriptors are submitted to **elementary transformations** (see table 1, first group). This leads to the question if there are further groups of transformations that take spelling errors into consideration without a great loss of precision or a huge increase of response time for the method.

3. Phonetic transformations already reported

Numerous methods are mentioned in literature, whereby groups of transformations are used in order to take spelling errors into account during the indexing of texts. A related task is to refind the falsified name of a client in a list of given names. The classical works hereto, done by Russell (1922), and further findings from Davidson (1962) and Greenfield (1977), already showed that the choice of transformation is dependent upon the method's needed degree of automation: the search for a name allows less automation and larger selection menues of alternatives than a diagnosis encoding which should save the working time of a terminology specialist.

Further specified transformations (of names) are the **Wiener Phonetik** and the **Kölner Phonetik** compared by Postel (1969). It seems clear that they cannot be readily applied to medical phrases, but can be used as directives to create transformation groups to be examined.

Knowlegde about error types and frequency found in medical texts was gained in studies about **automated correction of spelling errors**, for example by Joseph and Wong (1979), and Fischer (1980). They lead to the conclusion that the most frequent mistakes (deletion, insertion, or falsification of one character, transposition of two adjacent characters) can only be recognized with sufficient certainty using a word form dictionary and not considering compound terms. Both conditions don't meet the situation here.

A small part of spelling errors remains (according to Joseph and Wong about 10 %) that can be found by phonetic transformations.

4. Methods of solution

4.1 The groups of transformations used

Taking the semi-automated classification method described by Fischer (1993) as an example, different groups of transformations were examined in order to find out which of them lead to a still acceptable decrease of precision and response time.

After performing the "elementary transformations" (see table 1, first group), the second transformation dropped multiple adjacent occurences of the same character, leaving only one. This complies with the relatively frequent mistakes (5.65 % of all, not merely of phonetical errors, as Joseph and Wong counted), omitting one of two identical characters in succession, or duplicating a character, respectively. One should note that the effect of the chosen groups of transformations depends upon the sequence of its usage: if, for example, from the third group on a transformation results in further two adjacent identical characters, they are not summed up to one. Thereby the original order of characters in the input text is intended to be more exactly kept.

The third group of transformations (see table 1) takes into consideration that an "h" following a "t" or an "r" is mistakenly omitted or inserted, mainly in words with Grecian origin (for example in "Phthisis", "Rhythmus"); furthermore spelling variants varying "y" and "i" (for example "Oxyd" and "Oxid"), and mistakes writing "i" rather than "ie" and vice versa. The resulting transformation rule "y to ie" (or conversely) is also used by Brigl (1992), but probably due to the "ie"-to-"y" correspondence in pairs of English and German notions ("therapy" versus "Therapie").

The transformations of the fourth group (see table 2) resulted from common phonetic subgroups of Latin characters, used already by Russell in the SOUNDEX-method (1922), later by Herget in the "Wiener Phonetik" (1966) and by Postel in the "Kölner Phonetik" (1969).

first group of phonetic transformations ("elementary transformations"):		
all characters, except letters and digits, to just one blank as only word limiter; all letters to lower case		
additional character transformations	number of cases	number of descriptors involved
ä → ae	161	161
ö → oe	54	53
ü → ue	53	52
ß → ss	40	40
k → c	573	521
z → c	231	228
ph → f	155	149
no duplicate descriptors resulting		

second group of phonetic transformations:	
drop successive identical characters except one	
number of cases:	424
number of descriptors involved:	412
number of duplicate descriptors resulting:	4

third group of transformations:		
additional character transformations	number of cases	number of descriptors involved
th → t	90	89
rh → r	24	24
ie → i	94	94
y → i	323	306
no duplicate descriptors resulting		

Table 1: First three groups of transformations and their effects on the 2, 775 descriptors

fourth group of phonetic transformations:	
ch → c (but not in "sch") w → v, b → p, d → t g → c, j → i, m → n, qu → c	
number of duplicate descriptors resulting:	6

fifth group of transformations:	
drop all vowels	
number of duplicate descriptors resulting:	611

Table 2: Last two groups of transformations and number of duplicate descriptors resulting

The fifth group (see table 2) which omits all vowels is based on the principle often cited in literature, that vowels contain less information than consonants. It was yet expected, that character strings only consisting of transformated consonants would result in a no longer acceptable inaccuracy of the indexing algorithm. One notices that formal algorithms which are not able to decompose compound terms cannot observe transformation rules concerning initial sounds ("drop all but initial vowels").

4.2 Number of duplicate descriptors as a measure of declining precision

Transformations of the input texts and the descriptors may cause, that descriptors are no longer found in the input texts or, on the contrary, are found by substring search, though they were not in the original input text. Tables 1 and 2 show the effects of the transformations on 2,775 different descriptors. The numbers of affected characters ("cases") and descriptors presented allow to estimate the frequency and thus the importance of a transformation.

Furthermore the number of descriptors that coincide after applying a group of transformations indicates the degree of declining precision of the indexing algorithm part. Dropping of successive identical characters (see table 1, second group) resulted in the first 4 not unique descriptors: the roman digits "III" and "II" were projected on "I", furthermore "colli" and "Coli" as well as "XXY" and "XYY" coincided with another. The third group of transformations caused no double descriptors, the fourth 6 (see table 2), yet not significantly more. Finally the dropping of all vowels (fifth group) resulted in an explosion of descriptor matches, showing already, that the indexing does not work precisely enough any more.

4.3 Classifiability as a Measure of Precision

Weidtman's Diagnosis Table is a **thesaurus** of standard diagnosis texts in pediatrics. The classification algorithm (Fischer (1993)) using the first group of transformations tries to map an input text on items of the table. If there is only one semantically equivalent or closest ("correct") item, the input string is called **automatically classifiable**; if there is more than one correct item, it is (simply) **classifiable**; otherwise the input string is **not classifiable** and has to be rejected by the algorithm.

Thus, taking two random samples from the Children's Hospital of the University of Münster, 473 automatically classifiable and 463 not classifiable input texts respectively, the transformations beginning with the second group could affect the precision of the classification algorithm in two ways (see tables 3 to 6):

For an automatically classifiable input text there could be "only the correct item found" in the thesaurus (it remained automatically classifiable), or there could be "more than one item found" (it became only classifiable), or there could be "only one item found, but not the correct one" (it was misclassified), or there was "no item found" (it became not classifiable).

For a not classifiable input text there should be "no item found", thus it should remain not classifiable. But due to the transformations there could be one or more items found (mistakenly not rejected cases).

Tables 3 to 6 show the results. The classification algorithm is relatively stabile to the indexing error that a descriptor not originally contained in the input texts is found due to the transformations. The following should be noted here: when applying one group of transformations the problem cases found earlier are left out, so that in order to approximate the cumulated error rate the fractions caused by the individual transformations must be added.

5. Discussion of the tables

Table 3 shows 14 cases being no longer automatically classifiable after the second group of transformations. In 5 cases some descriptors became too short and unspecific and were therefore found too often as substring in the input text. For example the descriptor "cerr" from "Zerrung" was shortened to "cer", a common silable of many medical phrases. The remaining 9 cases of error were due to a word segment limit between the pair of identical characters, from which the transformation left only one. Thus, the string "hauttuberculose" changed to "hautuberculose". Here the descriptor "tuberculose" was found and deleted, so that the remaining "hau" no longer included the descriptor "haut". Therefore 2 input texts, which had to be classified automatically, were mapped only on an upper term ("Hauttuberkulose" on "Tuberkulose"); 7 others became not classifiable ("no item found").

Such mistakes can be counteracted by leaving the initial and the last character of a descriptor found in the input text and deleting only the rest.

	thesaurus texts				
input texts	only the correct item found	more than one item found	only one item found, but not the correct one	no item found	sum
formally classifiable	459 97.0 %	5 1.1 %	2 0.4 %	7 1.5 %	473
not formally classifiable	1 0.2 %	0	12 2.6 %	450 97.2 %	463

Table 3: Results achieved after the second group of transformations

For the 463 not classifiable input texts, normally no correct item can be found in the thesaurus, by definition. Thus, in 12 cases, where one item was found, this was erroneous and resulted from indexing errors caused by short descriptors. Nevertheless, there was one input text originally not classifiable due to a spelling error ("Wolf" instead of "Wolff"). Because this type of error is considered by the second group of transformations, the input text became (automatically) classifiable.

	thesaurus texts				
input texts	only the correct item found	more than one item found	only one item found, but not the correct one	no item found	sum
formally classifiable	458 99.8 %	0	1 0.2 %	0	459
not formally classifiable	0	0	3 0.7 %	447 99.3 %	450

Table 4: Results achieved after the third group of transformations

During the third group of transformations (see table 4) there arose the difficulties already mentioned, if word segment limits were touched, here by dropping "h" after "p" or "t" (for example in "Klumphand"). An examination of the descriptors showed, that the transformation "th" to "h" caused a phonetic error in 5 of 90 cases (examples are "Bluthochdruck" and "Gleithoden"), the transformation "rh" to "r" even in 7 of 24 cases ("Wasserhaushalt", "Verhalten" and others). Thus the input text "Körperhaltungsschwäche" was misclassified by "Schwäche", since the descriptor "haltung" could no longer be found in the transformated version "coerperaltungsschwaeche".

The 3 cases of misclassification in table 4 resulted from too shortened descriptors, the 10 cases in table 5, too.

	thesaurus texts				
input texts	only the correct item found	more than one item found	only one item found, but not the correct one	no item found	sum
formally classifiable	450 98.2 %	1 0.2 %	3 0.7 %	4 0.9 %	458
not formally classifiable	0	0	10 2.2 %	437 97.8 %	447

Table 5: Results achieved after the fourth group of transformations

Table 5 also shows 8 cases of input texts, which were no longer automatically classifiable. This is caused by the fact that subsequent transformations of the descriptors lead to less and less unspecific substrings, which are found too often in the (transformated) input texts. This development culminates in the situation shown by table 6, when the precision of the classification algorithm finally reached an unacceptable level.

	thesaurus texts				
input texts	only the correct item found	more than one item found	only one item found, but not the correct one	no item found	sum
formally classifiable	199 44.2 %	106 23.6 %	59 13.1 %	86 19.1 %	450
not formally classifiable	0	70 16.0 %	137 31.4 %	230 52.6 %	437

Table 6: Results achieved after the fifth group of transformations

With the number of substrings found also the response time is growing more and more. After the fourth group of transformations, classification of the two samples took twice the time needed for the original input texts.

It is nevertheless surprizing, how high the degree of precision remained, though the fourth group of transformations changed the descriptors so much, that human beings are hardly able to recognize the original forms. This must be attributed by all means to the classification algorithm and its very strict criterion for classifiability, rather than to the transformations considered here.

6. Results

Altogether the attained results can be summarized as follows. The most frequent spelling errors, namely any "typing errors", are unfortunately not captable by phonetic transformations. Methods of automatic correction of spelling errors fail due to the high portion of compound terms, typical for medical phrases in German.

With some improvements (to delete descriptors found in the input text not completely, and to use not too short descriptors), the tested transformations up to the fourth group yield acceptable results concerning precision and response time. Eventually one should do without dropping the "h" after "r", since this resulted too often in descriptors with phonetic errors.

The phonetic transformations are particularly advantageous, when the emphasis of the used classification method is put on a high degree of automation, that is when manual decision ought to be saved charging the response time.

Considering on the contrary (perhaps because of security reasons) the situation that for the most part a selection menue is presented for manual decision by the encoding staff, whereby short response times are important, one is inclined to dispense with the small amount of spelling errors recognized by phonetic transformations. (Eventually the second group and some of the third group of transformations are regarded as helpful.) Instead, the cases that one of the descriptors was not found due to a spelling error can be seized by the definition of the selection menue. The encoding staff can easily recognize these cases, even if they are not terminology specialists.

References:

BRIGL, B. (1992): Eine Methode zur automatischen lexikabasierten Indexierung von Diagnosen (LBI-Methode). *Report 1/1992 of Institut für Medizinische Biometrie und Informatik*, Heidelberg.

DAVIDSON, L. (1962): Retrieval of Misspelled Names in an Airline Passenger Record System. *Comm. Ass. comput. Machin., 5, 169–171.*

FISCHER, R.-J. (1980): *Automatische Schreibfehlerkorrektur in Texten.* Springer, Heidelberg.

FISCHER, R.-J. (1991): Vergleich verschiedener Methoden der Abbildung medizinischer Aussagen auf Texte einer standardisierten Terminologie. In: K. Überla, O. Rienhoff, and N. Victor (eds.): *Quantitative Methoden in der Epidemiologie,* Medizinische Informatik und Statistik, **72**. Springer, Heidelberg, 253–257.

FISCHER, R.-J. (1993): Pseudoroots as Descriptors for a Thesaurus Based on Weidtman's Diagnosis Table of Pediatrics. In: O. Opitz, B. Lausen, and R. Klar (eds.): *Information and Classification - Concepts, Methods and Applications.* Springer, Heidelberg (in print)

GREENFIELD, R.H. (1977): An Experiment to Measure the Performance of Phonetic Key Compression Retrieval Schemes. *Meth. Inform. Med., 16, 230–233.*

HERGET, M. (1966): Verwendung der Phonetik im Rahmen einer elektronischen Datenverarbeitungsanlage. *Kriminalistik, 20, 216f*

JOSEPH, D.M., and WONG, R.L. (1979): Correction of Misspellings and Typographical Errors in a Free-Text Medical English Information Storage and Retrieval System. *Meth. Inform. Med., 18, 228–234.*

POSTEL, H.J. (1969): Die Kölner Phonetik. *IBM-Nachrichten, 198, 925–931.*

RUSSELL, R.C. (1922): *U.S. Patent 1,435,663*

WEIDTMAN, V. (1989): *Diagnoseschlüssel für die Pädiatrie.* Springer, Heidelberg.

Disambiguating Lexical Meaning:
Conceptual meta–modelling as a means of controlling semantic language analysis

Josef Ingenerf

Institute of Medical Informatics and Biometry,
Medical Faculty of the Aachen University of Technology,
Pauwelsstraße 30, D–52057 Aachen, Germany

Summary: A formal terminology consists of a set of conceptual definitions for the semantical reconstruction of a vocabulary on an intensional level of description. The marking of comparatively abstract concepts as semantic categories and their relational positioning on a meta–level is shown to be instrumental in adapting the conceptual design to domain – specific characteristics. Such a meta–model implies that concepts subsumed by categories may share their compositional possibilities as regards the construction of complex structures. Our approach to language processing leads to an automatic derivation of contextual semantic information about the linguistic expressions under review. This information is encoded by means of values of certain attributes defined in a feature–based grammatical framework. A standard process controlling grammatical analysis, the unification of feature structures, is used for its evalution. One important example for the usefulness of this approach is the disambiguation of lexical meaning.

1. Introduction: The documentation of sublanguage statements

In the field of medical documentation, an automatic semantic analysis of language, as applied to medical sublanguage statements, get more and more important. As far as Germany is concerned, the latest health reform in 1993 has made this observation all the more pertinent. Diagnostic statements made in discharge summaries may be considered as an appropriate input material for this kind of linguistic analysis. Their clinical relevance is obvious, while their generally concise, nominal–syntagmatic style of writing allows an approach disregarding difficult text–pragmatic aspects of language. The most important syntactical constructions are shown in fig. 1.

Nominal lexemes	*word constructions*
Prefixation	"epi_cutaneous", "epi_gastrium", "post_infarct_syndrome"
Suffixation	"nephr_itis", "my_oma"
Composition	"inflammation of the gastric muscous membrane"
	in German: "Magenschleimhautentzündung"
Nominal syntagms	*sentences without verb phrases*
Nominal phrases	*without a function word*
Adjectival attribute	"acute nephritis"
Adverbial attribute	"one aspirin daily"
Nominal attribute	"morbus Hodgkin"
Nominal groups	*with one dominating nucleus*
Genetive attribute	"inflammation of the kidney", "ablatio retinae"
Prepositional attribute	"complaints after nephritis"
Nominal compositions	*with no dominating nucleus*
Conjunction	"swelling in knees and hands"

Fig. 1 "Relevant syntactical constructions in diagnostic statements"

A condensed, sober style of description is symptomatic for sub–languages [Schefe 1981]. Usually, the nominal form reflects a passive construction, while the subject of the active version of each reconstructed sentence will not be mentioned, as in the following examples:

The patient suffers from a
rupture of the front cross–ligament of the right knee–joint.
The cellular image (used for instance for a pathological analysis) shows an
adenocarcinoma ovarii with stenosis of the sigma and intraabdominal metastases.

These and other properties of an application–oriented sublanguage analysis lead to the name "language engineering" for the corresponding discipline ([Nagao 1988], [Hahn 1992]). Composition is the most important principle of constructing and condensing medical information. The semantic analysis of compound expressions poses a problem which still awaits a satisfactory solution. Compound expressions like "tracheobronchitis" contain no morpho–syntactical information concerning the relationships of the primitive concepts concerned. In paraphrases like "inflammation of trachea and bronchus", these relationships are marked by prepositions.

Modelling semantic phenomena is far more difficult than the grammatical representation and processing of language on a purely syntactical level. In medical practice, different kinds of taxonomic vocabularies are generally used for purposes of language–based controlled documentation. Their taxonomic structure seems to be similar; on the other hand, there are important differences. The central question is: "Why does concept A subsume concept B?" An answer can be found referring to the underlying logico–linguistic phenomenons, i.e. the dichotomy of "intension/extension" ([Ingenerf 1993], [Ingenerf 1994a]). It appears necessary to distinguish between an intensional and an extensional level of vocabulary (see chapter 2). The former is exemplified by formal terminologies, the latter by thesauri and, in a disjunct form, by systems of classification. Each of these systems has a specific practical application. A compositional semantic–oriented grammatical analysis of language, however, must be based on a formal terminology represented on an intensional level of vocabulary.

2. The compositional semantic analysis of language

This paper describes the automatized interpretation of sublanguage statements by means of a compositional semantic analysis of language. On the architectural level, one distinguishes between the following models of language processing:

1) Sublanguage expression → [syntax & semantic] → formalized reconstruction

2) Sublanguage expression → [syntax] → [semantic] → formalized reconstruction

3) Sublanguage expression → [syntax] ↑↓ [semantic] → formalized reconstruction

Typical representatives of the first architecture described are, in the first place, semantic grammars whose non–terminal categories have special semantic properties designed for a limited domain [Burton 1976]. The processing algorithms are generally based on various strategies of partial language analysis. Systems with an architecture of the second type treat semantic and syntactical analysis separately. On the basis of a completely developed parse tree, the semantic representation is generated in a second step [Sager et al. 1987]. In the present approach described in fig.2 (reflecting the third type of architecture), a link exists between syntactical and semantic analysis; both however, are functioning independently. As a result, contextual semantic information can already be considered while the syntactical analysis is in progress. This system can also be transferred to other semantic areas.

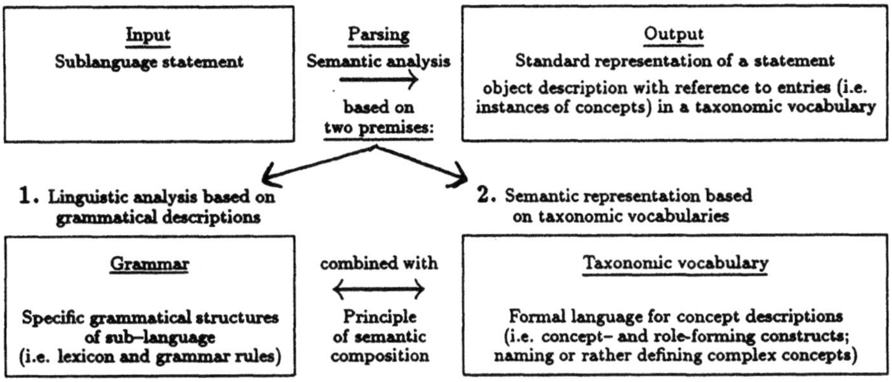

Fig. 2 "The compositional semantic analysis of language"

The semantic analysis of a sublanguage statement requires linguistic knowledge about the grammatical structures of the sublanguage and terminological knowledge about the semantic structures of concepts. Automatization means (first premise in fig.2) that entries of taxonomic vocabularies can be identified only through linguistic expressions. If these expressions are lexicalized, a reference to their lexical meaning in the vocabulary is sufficient (for instance, "myoma", "uterus"). All other cases require a semantic analysis of the linguistic expression (for instance, "myoma of the uterus"), complying with the compositional principle as no other information for the identification of concepts is available [Rahmstorf 1991].

The principle of semantic composition

The meaning of complex, non-lexicalized expressions can only be derived from the meaning of their parts and their syntactical composition.

The generation of semantic structures, as compared to the syntactical structures described in a 'Chomsky'-grammar, can and must be controlled through the introduction of appropriate categories to avoid effects of under- or over-generation. Semantic categories and relations in concept representation correspond to the syntactical categories and rules of a grammar.

G r a m m a r ("Generative Syntax")

Determination of syntactical categories,
for instance articles (Art), nouns (N), prepositions (Prep), nominal phrases (NP), etc.

Lexicon: for instance	definition of basic linguistic expressions (words or parts of words), "myoma" as N, "uterus" as N, "of_the" as Prep, etc.
Grammar: for instance	recursive generation of new linguistic expressions, NP ⟶ Art N, NP ⟶ N Prep N, etc.
from which follows:	"myoma of_the uterus" can be derived as an expression of the syntactical category 'NP'.

Fig. 3a "The generativity of syntactical structures"

Representation of concepts ("Generative Semantic")	
Determination of semantic categories as chosen concepts, for instance nosology (NOS), topography (TOPO), tissue (TISS), modifier (MOD), etc.	
Lexicon: for instance	definition of basic concepts and binary relations between concepts, "Rel_{Loc}" (localized on): relation between concepts from NOS and TOPO "Rel_{Tiss}" (tissue concerned): relation between concepts from NOS and TISS "Rel_{Mod}" (modified by): functional relation between concepts from TISS and MOD TUMOR *is_a* NOS UTERUS *is_a* TOPO, MUSCLE_TISSUE *is_a* TISS SMOOTH *is_a* MOD,
Formal language: for instance	recursive generation of new concepts (modified notation), SMOOTH_MUSCLE_TISSUE := MUSCLE_TISSUE *and* [partner_of_Rel_{Mod} *is* SMOOTH] MYOMA := TUMOR *and* [∀ partner_of_Rel_{Tiss}: partner *is_a* MUSCLE_TISSUE] LEIO_MYOMA := TUMOR *and* [∀ partner_of_Rel_{Tiss}: partner *is_a* SMOOTH_MUSCLE_TISSUE] MYOMA_OF_THE_UTERUS := MYOMA *and* [∃ partner_of_Rel_{Lok}: partner *is_a* UTERUS]
from which follows	"MYOMA_OF_THE_UTERUS *is_a* MYOMA", "LEIO_MYOMA *is_a* MYOMA" can be derived on a purely syntactical basis.

Fig. 3b "The generativity of semantic structures"

Feature–based grammar formalisms are used to describe the sub–language structures shown in fig.1 [Shieber 1986]. A process named chart parsing [Kilbury 1984] is used for an algorithmic realization of the grammatical analysis of language. The application of chart parsing and its extension towards an analysis based on feature grammars has been treated exhaustively by Gazdar und Mellish [1989].

The second premise in fig.2 requires a semantic representation based on taxonomic vocabularies. A serious approach to the establishment of such an underlying taxonomic relation (generalization/specialization) between concepts requires an exclusive use of terminological knowledge. In the philosophy of language, the term "intensional meaning" is used which indicates an exclusively intra–lingual or terminological meaning of a linguistic expression. This meaning is independent from the reality in the context of which the expression is interpreted. Through a semantic analysis based on the principle of composition the intensional meaning of a conceptual expression can be made fully explicit (see the deduction of the implicit knowledge "LEIO-MYOMA is_a MYOMA" in fig.3). The term "extensional meaning" is used when the meaning of a linguistic expression is defined by references to the domain. The success of such definitions and, respectively, the derivation of implicit knowledge about concepts can no longer be achieved by linguistic analysis alone (eg. "AIDS is_a LETHAL_ILLNES" as opposed to "AIDS is_an ACQUIRED_IMMUNE_DEFICIENCY_SYNDROME"). Instead, it is based on outside knowledge, that is, on real occurrences of Aids. A characterization of "AIDS" as "lethal" is based on empirical knowledge which tomorrow may no longer be up to date. Another example concerns region-dependent definitions. An illness like "sickle cell anemia" is terminologically defined as "hereditary anomaly of haemoglobin". In our Northern countries this illness is subjected to therapeutical treatment. In tropical countries, however, it serves as a protection against "malaria". The result is a different judgment about its clinical relevance. Most of the interesting knowledge concerning medical concepts is, of course, non–terminological knowledge, usually representing special clinical criteria. This must be distinguished from accepted intersubjective terminological knowledge.

Proposing a stringent view of this intensional way of defining concepts, Brachman (following Wood's criticism of semantic networks) developed the terminological knowledge representation system "KL–ONE" [Brachman, Schmolze 1985]. The model–theoretic semantic of this formalism permits a compositional calculation of the meaning of previously defined concepts. A second and most essential feature of this approach is especially concerned with the distinction between an intensional and an extensional level of description. Definitions of concepts (for instance, "MYOMA") are used to enrich the information available and to disambiguate lexical meaning. The meaning extracted from a diagnostic statement is represented on an extensional level. These so called object descriptions are derived, via formal statements on the basis of instances of defined concepts (for instance, "aMYOMA"), from a basic terminology (also called T–Box). They are stored in an object data base (also called A–Box).

Terminology (T-Box)	**Object-Data Base (A-Box)**
Box for terminological knowledge	Box for assertional knowledge
intensional knowledge	extensional knowledge (generally: basic facts)
concepts	objects (as instances of concepts)
sets	elements of sets
definitional, analytical knowledge	contingent, synthetic knowledge

Fig. 4 "Hybride formalism: terminology and object data base"

3. Reasons for the use of meta–modelling

A formalism of concept representation provides a formal language which allows the construction of more complex concepts on the basis of a set of primitive concepts as well as of roles linking different concepts. Such a language of concept representation is now introduced, linked to a related language of object description. The semiformal notation used so far is replaced by a formal notation, thus "^Concept" instead of "CONCEPT" and "$^{\vee}$Concept" instead of "aCONCEPT" (see Frost [1986, p. 385] for the prefixes "^" and "$^{\vee}$").

Formal terminology: Syntax

^ < concept name > $\sqsubseteq$ < concept term > "defines necessary conditions"
^ < concept name > = < concept term > "defines necessary and sufficient conditions"

Concept description language (interpreted as sets): Syntax

Given: - a set of concepts (concept names), incl. top 'T' and bottom '⊥'
- a set of roles (role names)

< concept term >	::=	^ < concept name >	'primitive concept'
	|	< concept term > $\sqcap$ < concept term >	'conjunction'
	|	$\forall$ < role > • < concept term >	'quantified value restr.'
	|	$\geq$ 1 < role >	'number restriction'
	|	$\leq$ 0 < role >	'number restriction'

Note: $(\forall^1 R \bullet {}^\wedge K) = (\forall R \bullet {}^\wedge K) \sqcap (\geq 1\ R)$

Object–Data–Base: Syntax

< TELL-DB > ::= (< object name >:< object term >) "introduces object terms"

Object description language (interpreted as truth values): Syntax

Given: - a set of concepts (concept names)
- a set of roles (role names)

< object term >	::=	$^{\vee}$ < concept name >	'primitive object term'
	|	$^{\vee}$ < concept name >	'relational attachment'
	|	**with** < role > • < object term >	
	|	< object term > **and** < object term >	'conjunction'

Further formal constructions and their model–theoretic interpretation with emphasis on the concept description language are discussed, among others, by von Luck [1991]. A large number of studies and results, which cannot be discussed here, concern the expressiveness and complexity of the inference algorithms based upon them (see for example Nebel [1990]).

A sublanguage, in its turn, provides the means to construct complex expressions from more simpler ones. The extraction of meaning from sublanguage structures is based on a morpho-syntactical level of description, in accordance with the principle of composition which

a) controls the extraction of meaning (assuring correct linguistic expressions) and

b) contributes to the extraction of meaning (helps to identify roles and arguments).

A non–lexicalized expression, for instance a diagnostic statement, may be given as input — for instance, "anemia after infection". Apart from the meaning of the components, the syntactical structure must be taken into account. The role "temp_-order" is explicitly given through the preposition "after", and the argument positions in the relation are determined by that syntactical construction as well. The interpretation of the resulting object term, "ˇInfection with temp_order • ˇAnemia", suggests that there exists an actual "infection" which is linked to an actual "anemia" by the directed relation "temp_order". The same object term is extracted from the statements "infection prior_to anemia" and "post-infectious anemia". It has already been mentioned above that compound expressions hide the inherent semantic roles linking parts of a concept.

- "Kopf–haut–entzündung", i.e.: (German) "ˇInflammation with role? • (ˇSkin with role? • ˇHead)"
- "tracheo–bronch–itis", i.e.: "ˇInflammation with role? • (ˇTrachea and? ˇBronchus)"

An appropriate determination of semantic categories leads to an implicit selection of the role or, alternatively, of the conjunction of components of an expression. What appears to be missing in this context is something like a case frame on which nearly all semantic approaches centered on verbs are based in computer linguistics [Fillmore 1968]. According to the valence of each verb, it is possible to determine which semantic partners it can be given in which syntactical position. The case frame based on the valence of "to throw" (for instance: "He throws the ball in(to) the box") expresses the semantic expectations inherent in the subject (instances of the concept ^Person), the direct object (^Object), the indirect accusative object (^Direction) and the indirect dative object (^Site). This idea of case frames is now transferred to the grammatical analysis of verb–less nominal syntagms discussed here. So–called context patterns are derived on the basis of the question: "Which semantic partners put into which relation allow the establishment of formal concept structures?"

Depending on the kind of application intended, a number of more abstract concepts of a given formal terminology (like in fig. 5) are designated as semantic categories.

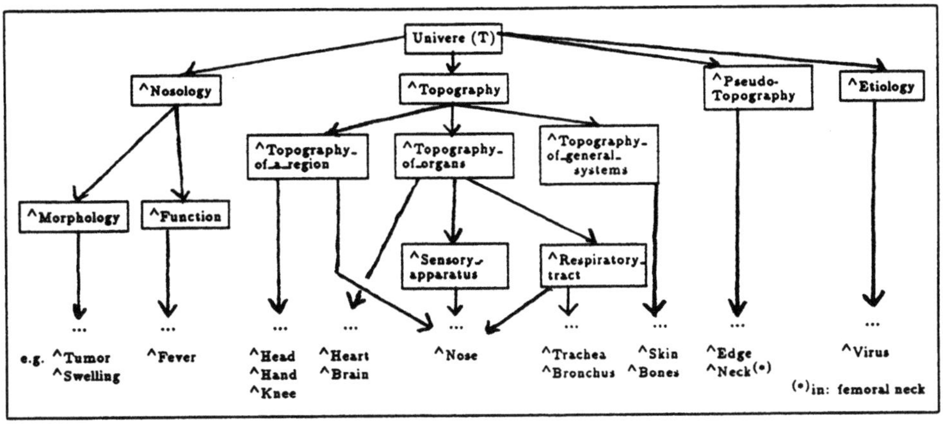

Fig. 5 "Upper portion incl. some primitive concepts of a typical medical taxonomy"

The categories chosen will, in their turn be positioned on a meta-level on a directed graph. By using, for instance, 'M1' in fig.6, the physician is enabled to refer to illnesses affecting certain regions or causes. The relation between the categories is implicitly contained within the model. He is also enabled further to differentiate such a model, for example ('M2'). To name this kind of implicitly defined roles encounters considerable difficulties. The use of the initials of the categories concerned has proved to be a suitable solution. The meta-models exemplified in fig.6 differ from such vocabularies as the multi-axial SNOMED-nomenclature [Wingert 1984] with its rigid categories (topography, morphology, function, etiology, etc.) in that the desired degree of conceptual generativity can be freely chosen.

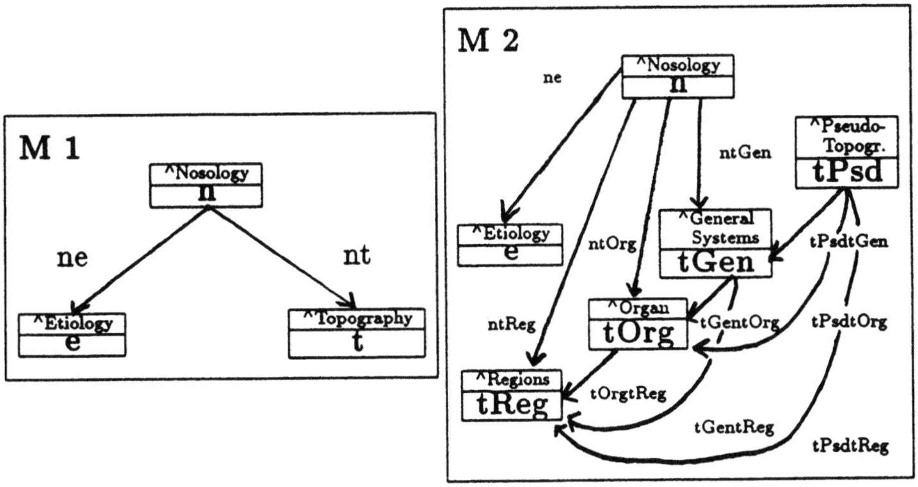

Fig. 6 "General versus special meta-model: directet graph with categories & roles"

Note: The following four levels of description have to be distinguished:

- based on concepts (concepts descriptions in a formal terminology, fig.5)
- based on categories as special concepts (meta-model, fig.6)
- based on instances of concepts (object descriptions in an object data base)

- based on medical sublanguage (here: diagnostic statements interpreted as object descriptions)

Each significant expression can be reconstructed, as indicated above, as an object term and thus as a formal statement based on instances of concepts. A semantic category can be related to the expressions through the subsumptive relation of the concepts with one of the designated concepts in fig. 6. The relations to dominant (father-) and dependant (son-) categories of such a model graph provide the contextual patterns, i.e. "With which partners in which relations a composition is possible". Two components within a compound expression are joined by conjunction if they belong to the same category. In the case of the example quoted above, "tracheo–bronch–itis", this leads to the desired effect.

— "tracheo[t]–bronch[t]–itis[n]" <u>i.e.</u>: "$^{\vee}$Inflammation **with** nt • ($^{\vee}$Trachea **and** $^{\vee}$Bronchus)"

"$^{\vee}$Trachea" and "$^{\vee}$Bronchus" are both localizations which are joined by conjunction. "inflammation" is a nosological term permitting, like the coordinated expression above, a relational composition via "nt" according to the meta-model **'M1'**. Additionally, the principle of composition requires a congruence test which implies that the dominant partner in terms of the semantic relation (in this case, "inflammation") has to be the dominant syntactical part of the expression as well. In this compound expression, the last element of the word (the suffix "–itis") contains its general meaning. Without such a congruence test, the different meanings of expressions like "knee joint" versus "joint knee" or "Nieren–becken" versus "Becken–niere" in german could not be distinguished.
Other compound expressions are treated accordingly, using the meta-model **'M1'**.

— "scalp inflammation" <u>i.e.</u>: "$^{\vee}$Inflammation **with** nt • ($^{\vee}$Skin **and** $^{\vee}$Head)"

This conjunctive interpretation, however, is misleading since the inflammation affects only the skin, not the head. As regards the German equivalent, i.e. "Kopf–haut–entzündung", the interpretation is made still more difficult by the ambiguity of the term "Haut" whose meaning is either "$^{\wedge}$Cutis", when applied to regions of the human body, or "$^{\vee}$Tunica", when applied to interior organs. The term "Kopf–haut" (scalp, with: "$^{\wedge}$Head $\sqsubseteq$ $^{\wedge}$Region") exemplifies the first, the term "Hirn–haut" (meninx, with: "$^{\wedge}$Brain $\sqsubseteq$ $^{\wedge}$Organ") the second possibility of semantic composition which this double meaning contains. Both "Kopf (head)" and "Hirn (brain)" belong to topographical categories. The meaning of the compound expression must remain ambiguous since the meta-model employed is not differentiated enough to disabiguate the german homonym "Haut". This differentiation, and the desired effect, can be achieved by employing instead the meta-model **'M2'** described in fig.6.

$$^{\wedge}Cutis \sqsubseteq {}^{\wedge}Haut \sqcap (\geq \mathbf{1} \ \text{tGentReg}) \sqcap (\leq \mathbf{0} \ \text{tGentOrg})$$

$$^{\wedge}Tunica \sqsubseteq {}^{\wedge}Haut \sqcap (\geq \mathbf{1} \ \text{tGentOrg}) \sqcap (\leq \mathbf{0} \ \text{tGentReg})$$

— "Kopf–haut–entzündung" <u>i.e.</u>: "$^{\vee}$Inflammation **with** ntGen• ($^{\vee}$Cutis **with** tGentReg• $^{\vee}$Head)"

— "Hirn–haut–entzündung" <u>i.e.</u>: "$^{\vee}$Inflammation **with** ntGen• ($^{\vee}$Tunica **with** tGentOrg• $^{\vee}$Brain)"

A last German example, "Rippenbruch" (rib fracture), shows once more that lexical ambiguity can be disambiguated by the instancing of defined concepts. It must be taken into account that "Bruch", in German medical terminology, may mean either a "fracture" or a "hernia". A grammatical derivation according to meta-model **'M2'** has the following results:

— the partial meaning1 = object term1 = $^{\vee}$Ribs
— the partial meaning2 = object term2 = $^{\vee}$Hernia <u>or</u> $^{\vee}$Fracture <u>with</u>:

$$\begin{array}{llllll}
^{\wedge}\text{Ribs} & \sqsubseteq & ^{\wedge}\text{Bones} & \sqcap & \forall^{1} & \text{tGentReg}\bullet\ ^{\wedge}\text{Thorax} \\
^{\wedge}\text{Fracture} & \sqsubseteq & ^{\wedge}\text{Bruch} & \sqcap & \forall^{1} & \text{ntGen}\bullet\ ^{\wedge}\text{Bones} \\
^{\wedge}\text{Hernia} & \sqsubseteq & ^{\wedge}\text{Bruch} & \sqcap & \forall^{1} & \text{ntGen}\bullet\ ^{\wedge}\text{Soft_parts}
\end{array}$$

The connecting relation "ntGen" is in congruence with the syntactical structure.

A wrong object term composition like "$^{\vee}$Hernia **with** ntGen• $^{\vee}$Ribs" is not successful. The concept definition of "^Hernia" demands that all partners referred to by the relation "ntGen" are instances of a concept subsumed under "^Soft_parts". The partner to which it is to be linked, however, is an instance of the concept "^Ribs". Thus, their composition must be a failure since it leads to a formal collision of disjunct, 'primitive' brother concepts.

— <u>disjoint</u> (^Soft_parts, ^Bones), and thus: $^{\wedge}\text{Soft\_parts} \sqcap {}^{\wedge}\text{Bones} = \perp$

As the desired representation for "Rippenbruch", the following object term emerges:

— $^{\vee}$Fracture **with** ntGen• $^{\vee}$Ribs

4. The semantic language processing system 'SMART'

Up to this point, due to their special relevance and to the problems connected with their application in medicine, a stress has been laid upon the semantic properties of compound expressions. The flexible principle of semantic analysis described here can be applied in an analogous way to the other syntactical constructions, to ensure, for instance, an adequate treatment of syntactical ambiguity.

— swelling (in knees and hands) <u>versus</u> (swelling in knees) and fever

Due to lack of space, it has not been possible to describe the actual transfer of semantic features (i.e. the semantic category and the contextual information) into a feature-based grammar formalism, and the modified method of chart parsing used for their interpretation. Such a combination of the three relevant formalisms - 'knowledge representation', 'grammar formalism', and 'parser' -, described in ([Ingenerf 1993], [Ingenerf 1994b]) leads to a user-friendly architecture of language analysis of the type described in fig.8. The reliability of this spectrum of methods has already been proved in the LILOG-project [Herzog, Rollinger 1991]. Here, it has been supplemented primarily by the idea of meta-modelling. Two related studies, each aiming at different results, must still be mentioned in passing. Pereira and Pollack [1991] want to "account for the influences of context on interpretation, while preserving compositionality to the extent possible", while Backofen et. al. [1991] describe the idea of "partial consistency checks during the unification of feature terms" and of "final consistency checks by classifying concept terms".
This approach has been implemented as a prototype named 'SMART' (System for processing Medical texts by the Analysis of concepts and their Representation on the basis of formal Terminologies) on a PC, using the tool ARITY-PROLOG. Together with a choice of parsing strategies and control mechanisms, the user disposes of a flexible and transparent environment for the development of a semantic language analysis based on formal terminologies as a suitable kind of taxonomic vocabulary.

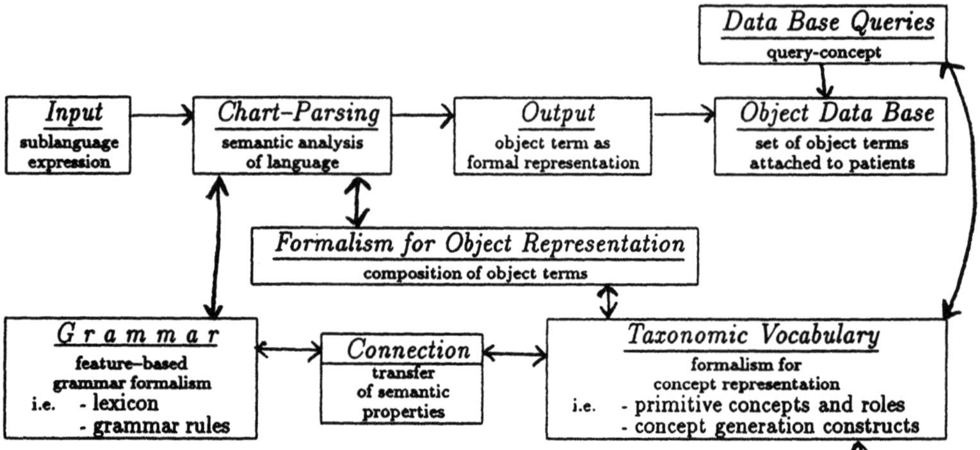

Fig. 7 "User–adaptable semantic language analysis"

5. Perspectives

The modelling of domain–specific concepts, using a terminological formalism of representation, determines the success of this approach. Doyle and Patil [1991], in their study "Two theses of knowledge representation", tend to criticize such a project of 'intensional concept representation'. They explicitly refer to a study by the American philosopher and logician Quine, "Two dogmas of empiricism", published in 1953, which has been very influential in that respect. They indicate the dilemma between, on the one hand, a precise semantic leading to correct and complete decision algorithms and, on the other hand, the question of how expressive and efficient the respective inference algorithms will turn out to be. The formal construction of 'role–subsumption', for instance, is of great interest to the approach described here (for instance, "ntGen $\sqsubseteq$ nt"), even though it leads to problems regarding the efficiency of the necessary inference algorithms. Despite these critical observations, the support for this approach to a formal reconstruction of terminological knowledge is increasing. This is especially true in the medical field where a great number of taxonomic vocabularies exists, providing a rich fund of terminological knowledge (cf. Rector et. al. [1992] or Schmiedel [1991]).

6. Literature

BACKOFEN, R., TROST, H., USZKOREIT, H. (1991). *Linking Typed Feature Formalisms and Terminilological Knowledge Representation Languages in Natural Language Front–Ends.* In: Brauer, W., Hernández, D. (Hrsg.). Verteilte Künstliche Intelligenz und kooperatives Arbeiten. Proc. 4. Int. GI-Kongreß 'Wissensbasierte Systeme', München, Okt. 1991, Informatik–Fachberichte 291. Berlin: Springer, 375–383.

BRACHMAN, R.J., SCHMOLZE, J.G. (1985). *An Overview of the KL–ONE Knowledge Representation System.* Cognitive Science 9, 171-216.

BURTON, R.R. (1976). *Semantic Grammar:* An Engineering Technique for Constructing Natural Language Understanding Systems, Rep. 3453, Bolt Beranek/Newman Inc., Dec.

DOYLE, J., PATIL, R.S. (1991). *Two theses of knowledge representation: language restrictions, taxonomic classification, and the utility of representation services.* Artificial Intelligence 48, 261-297.

FILLMORE, C. (1968). *The Case for Case.* In: Bach, E., Harms, R. (Eds.). Universals in Linguistic Theory. New York: Holt, 1-88.

FROST, R.A. (1986). *Introduction to Knowledge Base Systems.* London: Collins.

GAZDAR, G., MELLISH, C. (1989). *Natural Language Processing in PROLOG: An Introduction to Computional Linguistics.* Reading: Addison-Wesley.

HAHN, U. (1992). *Forschungsstrategien und Erkenntnisinteressen in der anwendungsorientierten automatischen Sprachverarbeitung — Überlegungen zu einer ingenieurorientierten Computerlinguistik.* Freiburg/im Breisgau: Technical Report 1/92.

HERZOG, O. ROLLINGER, D.-R. (EDS.) (1991). *Text Understanding in LILOG (Integrating Computational Linguistics and Artificial Intelligence: Final Report on the IBM Germany LILOG-Projekt),* Lecture Notes in Computer Sciences 546. Berlin: Springer.

INGENERF, J. (1993). *Benutzeranpaßbare semantische Sprachanalyse und Begriffsrepräsentation für die medizinische Dokumentation.* Dissertation an der Mathematisch-Naturwissenschaftlichen Fakultät, RWTH Aachen. Sankt Augustin: infix.

INGENERF, J. (1994A). *Die Notwendigkeit und das Dilemma monohierarchischer, nicht-kombinatorischer Klassifikationen.* Erscheint in: Ellsässer, K.H., Köhler, C.O. (Hrsg.). Grundlagen und Methoden der Klassifikation - Einsatzgebiete in der Medizin. Landsberg: ecomed, 30 Seiten.

INGENERF, J. (1994B). *Koppelung merkmalsbasierter Grammatikformalismen mit terminologischen Repräsentationsformalismen.* Eingereicht in: Künstliche Intelligenz, 16 Seiten.

Kilbury, J. (1985). Chart Parsing and the Early Algorithm. In: Klenk, U. (Hrsg.). Kontextfreie Syntaxen und verwandte Systeme. Tübingen: Niemeyer, 76-89.

LUCK, B. VON (1991). *Hybride logikbasierte Systeme.* Künstliche Intelligenz 2, 27-31.

NAGAO, M. (1988). *Panel: Language Engineering — The Real Bottle Neck of Natural Language Processing.* COLING Budapest - Proc. 12th Int. Conf. on Computional Linguistics, Vol.2. Budapest: John v. Neumann Society for Computing Sciences, 448-453.

NEBEL, B. (1990). *Terminological Reasoning is Inherently Intractable.* Artificial Intelligence 43, 235-249.

PEREIRA, F.C.N., POLLACK, M.E. (1991). *Incremental interpretation.* Artificial Intelligence 50, 37-82.

RAHMSTORF, G. (1991). *Compositional semantics and concept representation.* In: Bock, H.-H., Ihm, P. (Eds.). Classification, Data Analysis and Knowledge Organisation — Proc. of the 14th Annual Conf. of the GfKl in Marburg. Berlin: Springer, 163-172.

RECTOR, A.L., NOWLAN, W.A., KAY, S. (1992). *Conceptual knowledge: The core of medical information systems.* In: Lun, K.C., Degoulet, P., Piemme, T.E., Rienhoff, O. (Eds.) (1992). *MEDINFO-92 — Proceedings of the 7th World Congress on Medical Informatics in Geneva Palexpo, Switzerland.* Amsterdam: North-Holland.

SAGER, N., FRIEDMAN, C., LYMAN, M.S. (1987). *Medical Language Processing: Computer Management of Narrative Data.* Reading, Ma.: Addison Wesley.

SCHEFE, P. (1981). *Zur Funktionalität der Wissenschaftssprache — Am Beispiel der Medizin.* In: Bundgarten, T. (Hrsg.). Wissenschaftsssprache. München: Fink, 356-371.

SHIEBER, S.M. (1986). *An Introduction to Unification-based Approaches to Grammar.* CSLI Lecture Notes, no. 4.

SCHMIEDEL, A. (1991). *Modelling the Medical Data of the Heart Catheter Report.* Interner Arbeitsbericht, Projektgruppe Medizinische Informatik, Berlin.

WINGERT, F. (1984). *SNOMED — Systematische Nomenklatur der Medizin, 2 Bände.* New York: Springer.

Information Retrieval Using Conceptual Representations of Phrases

Gerhard Rahmstorf

IBM Institute for Knowledge Based Systems
Heidelberg

Summary: The information retrieval problem is described starting from an analysis of the concepts "user's information request" and "information offerings of texts". It is shown that natural language phrases are a more adequate medium for expressing information requests and information offerings than character string based query and indexing languages complemented by Boolean operators.. The phrases must be represented as concepts to reach a language invariant level for rule based relevance analysis. The special type of representation called advanced thesaurus is used for the semantic representation of natural language phrases and for relevance processing. The analysis of the retrieval problem leads to a symmetric system structure.

Introduction

The crucial problem of information retrieval (IR) is the analysis of the relevance of stored information. Given an information request, an IR system must check, whether stored information, mainly represented as texts, is relevant to the request.

The ultimate goal is 100% **precision**, i.e. each text presented as an answer to a query should contain relevant information. The search process should be exhaustive too. The **recall** would be optimal, when all texts containing relevant information are found and presented as an answer to the query.

A third goal is **specificity**: information requested by the user or offered by the texts stored in the data base should be expressible without limitations. It must be possible to specify any subject and any precise detail of the requested information the retrieval system has to search for.

The relevance problem of information retrieval systems must be solved by satisfying **conditions**:

Ease of use. The interaction with the system should be easy and effective. Complicated commands, difficult query languages and long-winded dialogue are not acceptable for information retrieval in professional environments.

Domain independence. Any work for customizing the data bases and components of the system to the application domains is expensive. The domain dependence of the IR system should be minimized. Domain dependence can originate from different factors such as the use of a domain dependent controlled vocabulary, domain specific indexing, domain specific "intelligent" dialogue with the user, etc.

Machine aided or automatic text indexing. Indexing should be supported by the IR system as much as possible. Each text of the data bases offered for users must be described by attributes, terms or phrases. Such index information stored in an index file is used for recognizing the relevance of a text for an actual query. Search results in terms of precision, recall, and specificity depend on the quality of the indexing process.

Response efficiency. The time the IR system needs to answer typical queries under defined hardware and load conditions must be minimized. In office applications, responses must be given within seconds. Longer response times are accepted for exhaustive searches for patents or very specialized scientific subjects.

Result presentation. The information retrieved by the system must be presented in a useful form and with necessary explanations. Hits can be ordered according to the relevance for the query. The selection of each text can be explained by highlighting terms, phrases or parts of the text which qualify the text as relevant.

User and data base services. Concurrent multi-user access and other features needed for professional data base services must be efficiently implemented. Information offered in the data bases of the system must be easily loaded and updated.

Classical Approach and new Trends

To satisfy these conditions, the relevance problem is solved in information retrieval systems currently marketed in a more or less similar way: the data base actually used for searching the requested information is organized as an inverted file of the significant character strings occurring in the texts. Significant character strings in a text are usually all words except determiners, conjunctions, prepositions etc. which are defined as so called stop words. The inverted file specifies also where the words occur in the text. Character strings (or words) combined by Boolean operators and some specific additional features such as contextual or positional operators are used as queries. These systems have been improved in the last decade by dialogue components, statistics and other service functions, but do not have much "intelligence".

Attempts have been made to improve the information retrieval systems by knowledge based techniques. Conventional information retrieval systems and expert systems differ in one essential point: information retrieval has been solved for many years by using cross products. This means the retrieval software is designed such that it can be used **independently of the domain** in which the customer will search. The semantics of the texts and the semantics of the query are not analyzed and represented. Precision and recall of the search is nearly totally independent of the application domain. Expert systems and all knowledge based techniques are quite different. They all use expert knowledge and are therefore strictly domain dependent. Knowledge about the domain must be represented. The knowledge base must be changed and extended permanently to reflect the evolution of the scientific world, technology, economy, society etc.

Knowledge based methods could be used as a technique to support and implement different functions of a conventional information retrieval system. [1]

An expert system could be complemented by a conventional information retrieval component to support the problem solving process, the interactive user interface or other components of the expert system. A symbiosis of the two types of systems is difficult, because the application and the philosophy of the systems are too different.

Proposal

It seems more adequate to start the development process for new retrieval techniques by a reanalysis of the application. The result of such an approach is summerized in a model (fig.1). The model shows the types of data and processes in information retrieval. The left side displays the processing of information requests. A similar

[1] Gauch (1992) describes some of the systems which assist during the query formulation and query reformulation stage. See also Morris (1991) and Smith et. al. (1989).

processing for the analysis of information offerings is shown on the right side. The results of both processes are conceptual representations. The representation of the information request and the information offerings by a common structure is the basis for the relevance analysis. A similar description of the basic processes of information retrieval has been given by W. B. Croft (1993).

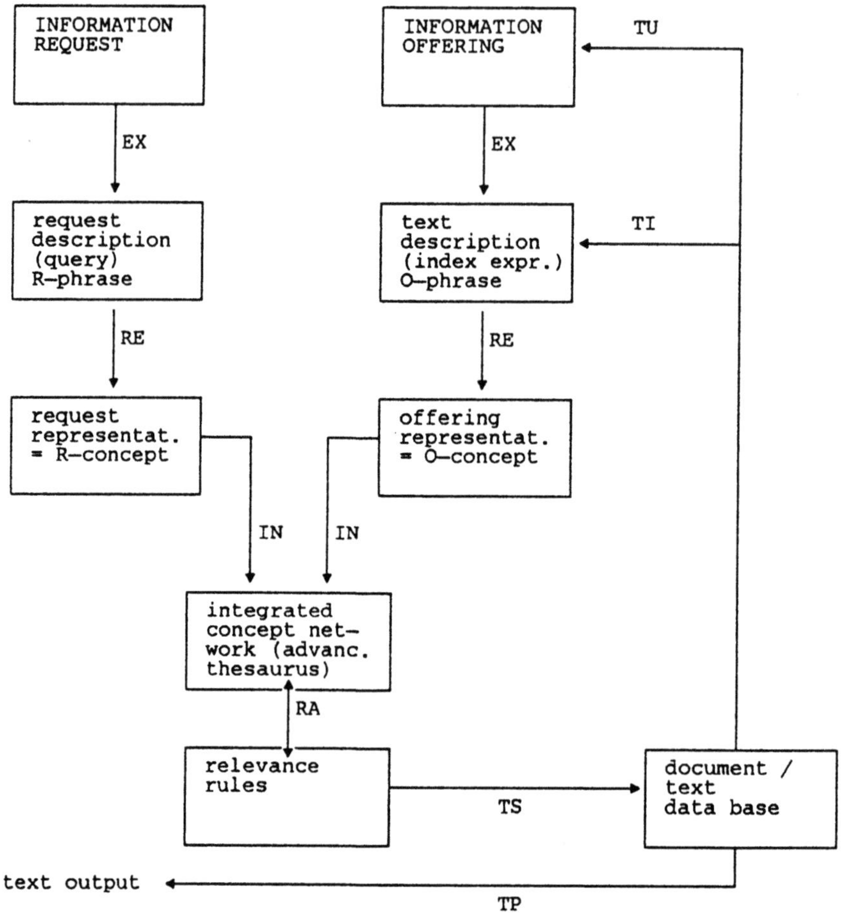

Figure 1. Architecture of an advanced retrieval system: EX = Expressing. RE = Representing. IN = Integration. RA = Relevance analysis. TU = Text understanding. TI = Text indexing. TS = Text selection. TP = Text presentation.

According to the structure of the model the following topics will be discussed here:

Basic concepts
(1) information request of the users
(2) information offerings of the texts
(3) relevance of information offerings for a given information request
Information request analysis
(4) expressing the information request using a query language
(5) representing the information request by conceptual structures
Analysis of information offerings

(6) expressing the information offering using an indexing language
(7) representing the information offering by conceptual structures
Relevance analysis
(8) integration of the representations in a conceptual system
(9) relevance based on conceptual relations

The problem of information retrieval is discussed here for information offerings represented as **texts**. Other information media such as images, tabular data, diagrams, image sequences, voice, etc. become more and more important for information retrieval systems. The term **document** refers to such publication units of the print media. But information retrieval must also support new styles of publishing and electronic communication. The basic information unit presented to users in a search report can be a complete or a partial document text, e.g. a paragraph of an article, or a textual unit in a hypertext system.

1. Information requests of the users

An information request can be understood as a knowledge state of a person. The person wants to change his state by getting information. The information request is a concept which determines the increase in knowledge the person wants to have from the information retrieval system.

Each information request should be expressible using a natural language. But not every query language used in information retrieval systems can express any possible information request.

Not every search is based on a real request for information on a subject. In the special cases (A) and (B) of the following list the requested information is restricted to specific documents or to formal attributes of texts:

(A) Searching for individual documents.
(B) Searching for any documents containing character strings.
(C) Searching for information on subjects.

(C) is the general case for information retrieval systems we will discuss here.

Searching for individual documents (A). The information request in the special case (A) can be characterized as an individual concept. The user searches for a single document or text. This document or text is an individual object which can be identified by a key. The key may be a document number or another unique character string. This key may be positioned anywhere in the text of the document. It is possible, that the user already knows the document, e.g. a clerk needs a copy of a business letter he has received and stored. The information contained in the document need not be new for the user. The appropriate expression for the request is a phrase like "document identified by xyz". An information request for a single object cannot be characterized as a generic concept. It is expressed by a proper name or phrase identifying an object. The user knows that there is only one single document which contains this key information. The search is complete when the single text is found which contains the key. Only this text is relevant according to the type A search.

Searching for documents containing character strings (B). The information request in the case (B) asks for documents or texts containing specified character strings. The information request may restrict the occurrence of the character strings by conditions, e. g. common occurrence of two or more strings. All documents in which the requested character strings occur as specified by the conditions are relevant to the user's information request according to the type B search. Typical users having this type of information requests are terminologists and linguists studying the use of the vocabulary of a language. They ask for the occurrence of given words, phrases or technical terms in the texts of a corpus. They are interested in the contexts of character strings, but not in the information which may be communicated by the retrieved texts.

Searching for information on subjects (C). The general information request in case (C) is characterized by information on a subject or special aspects of a subject. Subjects are topics of descriptions, argumentations, investigations, reports etc. The texts which communicate such descriptions, lines of reasoning etc. can transfer information to the reader by each sentence. The user asking for information on a subject does not want a unique document. He does not search texts satisfying formal conditions such as string occurrence. The information he wants must belong to the subject he selected. Usually, users select a subject in which they want to improve their knowledge by the process of searching. The motivation for the search can come from research work, technological development, problem solving, general interest in better education, examinations, etc. The information should be transported by whatever medium is available, e.g. text, images, diagrams, tables, video, etc. Texts are relevant to information request according to search typ C, when the conceptual representations of the information request and the conceptual representation of the information offering are in a specific conceptual relation. Some more terms have to be explained here to understand the idea of this relevance.

Clarification of the user's information request. Sometimes users are not sure what they need. It is supposed that at least some of those users need advice to understand and express their own information request. An initial dialogue of a user and an information retrieval system can support such users. The dialogue clarifies the type of the user's research interests and, in the case of type C, the requested subject.

2. Information offerings of the texts

The information offering of a text is a concept which classifies the text according to information which is communicated by reading the text.[2] Such a concept can also be called the subject of the text. Information request and information offering are both understood as "mental" concepts. Concepts can be expressed by different phrases of natural languages. The meanings of phrases can be approximately represented by structures which can be implemented in an information retrieval program. These structures are called "conceptual representations" or also "concepts" for short. The mental concepts information request and information offering can be compared on the basis of the conceptual representations which will be described later.

Subjects of textual descriptions, argumentations etc. classify text contents, not text forms. If a user knows the subject of a text, he knows only the class of the information which should be communicated by reading the text. The information of the text can only be communicated by reading and understanding the complete text. The information offering of a text must be determined and expressed.

The determination of the information offering for given texts is a well known prac-

[2]The information offering of a text is not the sequence of sentences which constitute the text. A text can be paraphrased or reformulated without a substantial change of the information mediated with the text. The text could also be translated from the original language to another language without much loss of information.

tical problem of scientific documentation and indexing. The theoretical analysis is difficult, because it includes other difficult problems such as semantics, text understanding, abstraction, concept generation etc. A reader which is competent in the language of the text and an expert of the domain of the text can usually classify the subject of a text by a concept expressed in a natural language. If a precise and complete classification is required, the indexing of texts can result in very special concepts.

The information offering of a text should be expressed by a natural language. The appropriate syntactical form of such an expression is a phrase. Text or paragraphs of texts are often precisely expressed by titles and headings. Such titles are the appropriate natural form for expressing an information offering of a short text. Longer texts usually deal with more subjects than shorter texts. The titles needed to express the information offering adequately become longer, the more pages the text has. Shorter texts are more effective for retrieval systems which shall provide specific information. Texts used as independent communication units should not be longer than a few pages. Very long documents must be decomposed into smaller text units which are then indexed separately. [3]

3. Relevance of information offerings for given requests

The term "relevance" is a well established term in information retrieval. But the meaning of this term reflects character string searching currently used. Relevance as described here is a relation of an information offering to an information request. Both arguments of this relations are concepts. If the concept "computer" is the information request, an information offering identified by the concept "personal computer" is relevant, although the concepts are not identical. The relevance relation is reflexive, transitive, but not symmetric.

When the relevance of two given concepts can be described, the relevance of the representation of the two concepts can be described using a formalism for the representation shortly described later. The relevance of natural language expressions describing information offerings and information requests is not as clear, because words and phrases may have more than one reading. Given the information request "band", "dance band" and "rubber band" cannot be relevant both.

4. Expressing the information request of a user

Subjects as possible information requests of users must be expressed by a language which can be recognized and interpreted by the information retrieval system. This process of formulating a query or an information offering is symbolized by **EX**. Natural language is the most flexible query language, because any kind of subject – general or specific, concrete or abstract – can be expressed.[4]

An information request can be expressed differently in a given natural language. If a user needs information on the subject discussed here, he can use a technical term or a phrase:

---

[3]Philosophical texts, historical documents and some other sorts of texts can be interpreted very differently depending on the intellectual interests, background and knowledge of the reader. Technical texts are usually not so much a subject of interpretation.

[4]Other ways of communication could be helpful for special applications. Non-verbal communication with pictures of objects, three-dimensional models of buildings, geographical maps, linear time scales etc. can also be used for specifying information requests in a man-machine dialogue. Non-verbal communication may support the users when the information retrieval systems offers no appropriate query language to express specific information requests.

(1) information retrieval
(2) procedures for recovering stored data to provide information on a given subject.

The second formulation is a noun phrase which defines "information retrieval" according to ISO 2382/1. More specific information requests can be expressed by phrases only, because no single words or terms are available for most of the subjects to be communicated in science and technology.

There are always two different sources for difficulties with query languages:

The query language is not correctly applied by the user
The query language cannot express the user's request

Users can make mistakes if the query language is difficult to understand. But inherent deficiencies of a query language cannot be removed by more learning efford. Information retrieval systems using Boolean expressions are unexcelled for information requests of the types A and B described above, but they do not have the same success with searching for information on subjects (case C). Boolean queries cannot be clearly interpreted as well defined concepts. A simple example with only two terms, ENERGY and TRANSPORTATION, connected by AND, will demonstrate the ambiguity. The concept behind this expression could be "transportation of energy" or "energy for transportation" or something else. Many texts dealing with the power demand of trucks, ships, planes, etc. would not be relevant if the first phrase describes what the user wants. Many other texts describing how electricity, oil, gas, etc. is transported using cable, pipelines, trucks, etc. would not be relevant if the second interpretation is the correct one. Current information retrieval systems ignore the important information of "stop words" such as prepositions. Grammatical constructions used to form precise and more specific expressions are not recognized.

The deficiency of the classical approach becomes more evident when the retrieval application is characterized by the following two conditions:

specific information request are given
no single technical term is available to express the request

The more terms or words are connected by Boolean operators, the more interpretations of the corresponding Boolean expressions are possible. The approach proposed here should overcome the limits of the current technology for information retrieval. It is characterized by the following features: The system supports natural language phrases instead of Boolean expressions as query statements. The system decides on the relevance of an information offering for a given information request on the base of conceptual representations of the phrases. The query phrases expressing information requests of users are called **R-phrases** in such a system (see fig. 1).

The syntax of every natural languages has many different constructions. The vocabulary includes common language words and technical terms. Most common words have more than one single reading. These features make natural languages powerful, but also complex. A balance of expressiveness and complexity guided by the state of the art in natural language processing must be found. One has to start with a restricted syntax. This can be done, because not all features of natural languages are needed for the expression of information requests. Types of phrases needed to express information requests are noun phrases, verbal phrases, prepositional phrases, adjective phrases, compound words, derivations and others.

5. Representing information requests by concepts

The user's query must be represented as a conceptual structure. This process is symbolized by **RE** in figure 1. The analysis is based on a lexicon, morphological rules, syntactical rules and semantical rules. A syntactical phrase structure diagram must be produced as an intermediary result. See figure 2 for an example. The phrase to be used as a query is "procedures for recovering stored data to provide information on a given subject". This phrase is conceptually equivalent with the noun compound "information retrieval". The result of parsing the phrase is the phrase structure diagram shown on the upper side of figure 2. This structure will be taken as input for the semantic analysis. Semantic rules will generate a conceptual structure shown on the lower side of figure 2. The concept representing the meaning of the whole query phrase has the label "information retrieval". We call this the **R-concept**, because it represents an information request. If the meaning of the noun compound "information retrieval" was already represented in the system, the semantic analysis of the query phrase will reproduce the conceptual structure for information retrieval and recognize the identity of the meanings.

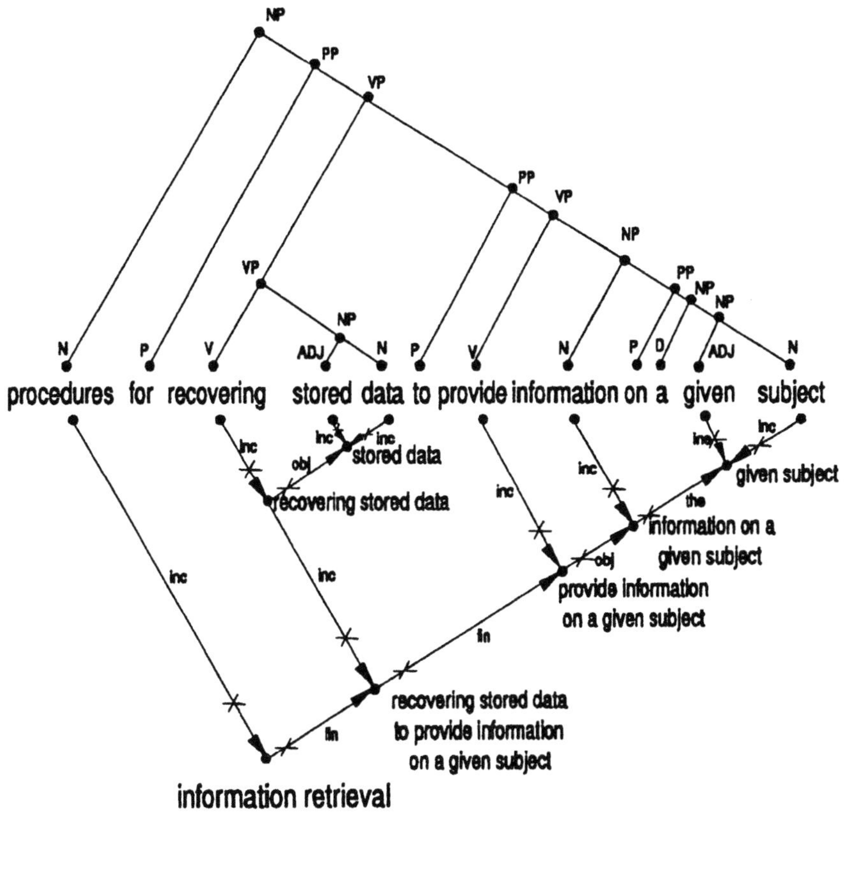

Figure 2. Representation of a phrase

The process of conceptual representation of a phrase is controlled by semantic rules. The words occurring in a phrase must be available in a lexicon. Such words of the phrase which are interpretable as concepts must be represented as nodes of the conceptual system from which the system starts the analysis of the given phrase. The conceptual representation of a complex phrase is a function of the representations of the phrase constituents and of the syntactic construction used for combining the constituents.

The conceptual structures can be described as directed labelled graphs with some additional restrictions. Nodes represent generic concepts. The label of a node represents a language expression for the concept in the given language. Concepts can be related by instances of conceptual relations. The type of the conceptual relation is symbolized by a label associated with the arcs. The generic relation INC(x,y) represents, that the concept x is a superordinate concept of the concept y. Each node is defined by only two binary relations. These defining relations of a generic concept are symbolized by the arcs with a cross close to the node they define. One of the two defining relations for a given generic concept must be a generic relation, where the argument y of the relation INC(x,y) refers to the defined concept.

Some types of phrases such as noun phrases NP and verb phrases VP are interpreted as concepts. The phrases "stored data", "recovering stored data" and others are expressions, interpreted as nodes of the conceptual structure. Prepositional phrases (PP) such as "on a given subject" are not interpreted as concepts, because they express components of more complex unit called concept relation. The preposition "on" is an indicator for a relation which links two concepts. The NP dominated by the PP is one of the concepts which are related. Most prepositions are ambiguous, e. g. "on" is an indicator for several relations. The relations which must be selected in the example is the "thematic relation" symbolized by THE(x,y). Selectional restrictions are used to limit the ambiguity.

The set of conceptual relations used for generating conceptual representations is determined by analyzing the morphological and syntactical constructions of natural language phrases. These relations are comparable with some of the binary relations used in thesauri.

This thesaurus should not be used as a conventional thesaurus for indexer and retrieval user. The advanced thesaurus has two main functions:

representation of natural language phrases
knowledge base for relevance deduction

The number of conceptual relations must be limited as in a thesaurus. Conceptual representations are generated for phrases expressing information requests and information offerings.

6. Expressing information offerings by an indexing language

An information offering characterizes the subject of a text. It is developed by reading and understanding the respective text. The process of text understanding is symbolized by **TU** in figure 1. When the text has been understood, the reader should be able to express the subject of the text (EX).

The process of **indexing** is well known in information science. Intellectual indexing is a labor intensive work. A specialist for indexing who should also be an expert in the domain must read the texts available for the system. Texts will be described using the text (or document) description language supported by the information retrieval system. The simplest form of indexing is done by selecting index terms from the text or from a list of a controlled vocabulary and use the set of the selected terms as a text description. It is not clear how to interpret such a set as a representation of the information offering of the text. This problem was already discussed for the corresponding problem of query formulation.

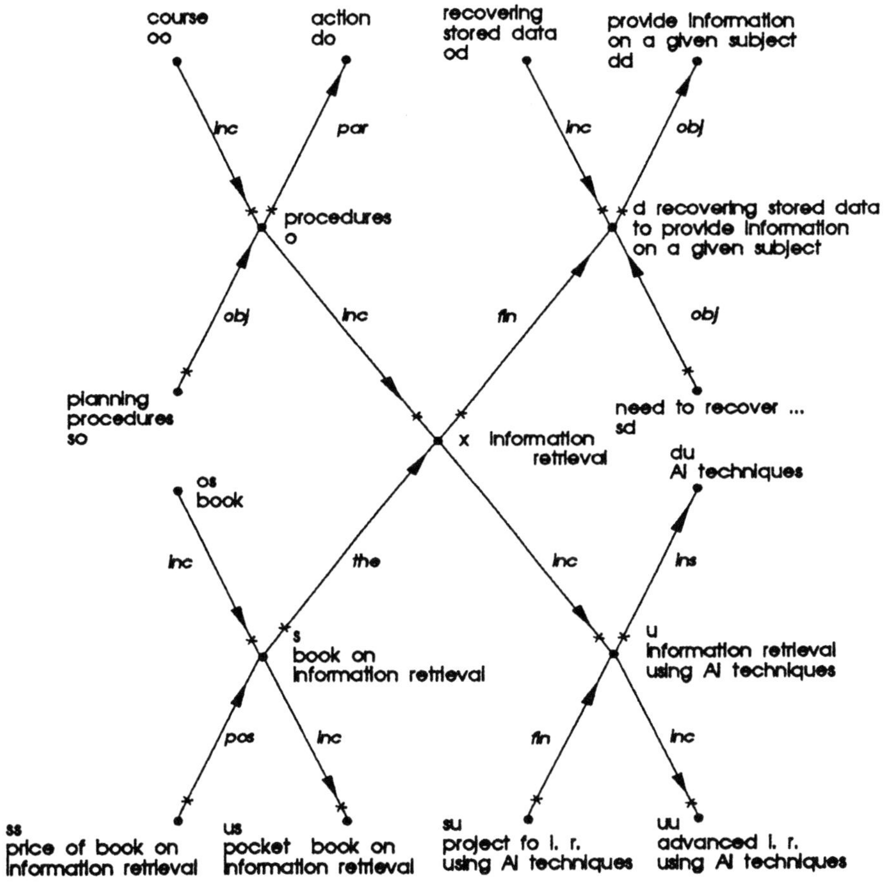

Figure 3. : The conceptual structure presented here shows the concept expressed by the term "information retrieval" and concepts in the environment.

A more precise text description would be a natural language phrase. The expression used to describe the information offering of a text is called an **O-phrase** for the text. The phrase can be the title of the text. Titles and headings can be recognized by their position, format or tags. The automatic generation of appropriate O-phrases for given texts, called **automatic indexing (TI)** by phrases, is an extremely difficult task. Such a task would involve natural language processing of free text, text interpretation, domain knowledge, evaluation, abstracting, and phrase generation. Salton et al. (1990) summarizes various linguistic approaches proposed for document analysis.

7. Representing information offerings

When an O-phrase for a given text is generated, this phrase can be represented conceptually by the same analysis used for queries. The result is an **O-concept**. The conceptual structure for the O-concept is generated by the same **representation language** used for queries.

8. Integration of representations in a common structure

Information requests (R-concepts) and information offerings (O-concepts) are stored in a common conceptual system. This advanced thesaurus is based on the conceptual representations for the words and technical terms usable in the phrases for queries and text descriptions. The concepts can be connected by **binary relations**. The relations are derived from the semantic analyses of natural languages. The relations such as causation, purpose, part, object etc. are identified by investigating natural language phrases and the syntactical constructions used in the phrases. The relations are described using paraphrases, selectional restrictions for the arguments and relational attributes such as transitivity etc. The relations used for conceptual representation are called the advanced thesaurus relations.

9. Relevance analysis based on conceptual relations

A relevance analyzer determines which information offerings have a certain minimal relevance for a given information request. The analysis is based on links between concepts represented in the advanced thesaurus. **Relevance** is a relation R(x,y), where x is a R-concept representing an information request and y is an O-concept representing an information offering. The shortest path which connects x with y is a sequence of links representing relations. This sequence of links describes qualitatively the relevance R(x,y). The path can also be characterized by a numeric weight. The process of calculating relevance starts from the R-concept. The relevance relation is not symmetric. The environment of the first degree of a concept x can be characterized by four relative positions of x (fig. 3). The symbol o(x) refers to the direct superordinate concept of x. The symbol d(x) refers to the difference concept of x. The representation formalism used here allows only one concept as o(x) and only one concept as d(x) for any given concept x which is defined by a phrase. The symbol u(x) identifies a subconcept position. The symbol s(x) is the label for a side concept of a concept x. The relative positions u(x) and s(x) are multiple positions, i.e. each concept can have many subconcepts and many side concepts. More than one concept can be a side concept of a given concept x, because the number of concepts to be defined by x as a difference position is not restricted.

The relevance R(x,y) depends on the relative position of y. Each relative position of a concept has a specific weight or relevance factor. Subconcepts have higher relevance than side concepts. The relevance of d-positions for a concept x depends on the type and direction of the relation which links x and d(x). This is also the case for s-positions. The factors which must be given to the different positions and different relations (for d- and s-positions) should be acquired by practical use of the system. The relevance(x,y) for the concepts x and y connected by any path can be calculated when these basic relevance factors are determined and a mode of factor summation for linking relations has been defined. This approach of relevance analysis does not use the terms or phrases of a given language associated with the concepts.

After having applied relevance rules to the R-concept, the relevant O-concepts are identified. The associated texts are selected from the text data base by the process **TS** (fig. 1) and can then presented as an answer to the user's query (process **TP** in fig. 1).

References:

CROFT, W. B. (1993): Knowledge-Based and Statistical Approaches to Text Retrieval. *IEEE Expert Intelligent Systems and their Applications 8, No. 2, 8–12.*

GAUCH, S. (1992): Intelligent Information Retrieval: An Introduction. *Journal of the American Society for Information Science 43, No. 2, 175–182.*

MORRIS, A. (1991): Experts Systems for Library and Information Services - A Review. *Information Processing and Management, 27, No. 6, 713–724.*

RAHMSTORF, G. (1990): Compositional Semantics and Concept Representation. In: H.-H. Bock and P. Ihm (eds.): *Classification, Data Analysis and Knowledge Representation. Models and Methods with Applications. Proceedings of the 14th Annual Conference of the Gesellschaft für Klassifikation e.V.*, Springer-Verlag Heidelberg, 163–172.

RAHMSTORF, G. (1983): Die semantischen Relationen in nominalen Ausdrücken des Deutschen. Diss. University of Mainz.

SALTON, G. et al. (1990): On the Application of Syntactic Methodologies in Automatic Text Analysis. *Information Processing and Management*, 26, No. 1, 73–92

SMITH, P. J. et al. (1989): Knowledge-Based Search Tactics for an Intelligent Intermediary System. *ACM Transactions on Information Systems*, 7, No. 3, 246–270.

Communication in Distributed Heterogenous Systems

Hans Czap

University of Trier, P.O.Box 3825, 54286 Trier, Germany
0651/201-2859, Fax 0651/201-3959,
E-mail: czap@uni-trier.dbp.de

Summary: Communication between different computing environments has to deal with heterogeneity. Specific problems of a more semantical nature which arise in a heterogenous distributed database environment are adressed here. Essentially, within an industrial environment, we consider two propositions represented by attributes A and B of different relations R_A resp. R_B belonging to heterogenous databases and the question under what conditions both propositions, i.e. A within R_A and B within R_B, have the same information content. This problem arises in the case of a dependency $B \leftarrow A$, and the need to maintain consistency. Typical applicational cases are presented in the introduction illustrating the following analysis. The insights won are applied to give first ideas how to establish a reference data base acting as intermediary for dependent attributes.

1. Communicational Problems with Business Data

Communication processes consist of statements regarding objects of reality. Allways, suited representations of such statements have to be considered in information processing.

Considering applications in business and administration, mainly simple statements regarding objects and their properties are of interest, as it is the subject of database theory. Since properties of objects are represented by attributes, communication processes involving distributed heterogenous systems, have to deal with representations of attributes and their assignments.

Example:
An US computer manufacturer operating on an international scale has a divisional organization oriented at national boundaries. This has been done in order to take account for national specialities. Consequently, the national divisions run their own local databases to support the national marketing activities. Figure 1 shows a typical situation of the German and British sales database.

revenue_Germany

Artikel	Preis	Menge
PC 286	900	100
PC 386	1700	270
Workst.	...	...

revenue_GB

article	price	quantity
PC 286	450	50
PC 386	800	200
...	...	...

Figure 1: Sales database of two different countries

(1) Different Names, Representations and Units of Scale

To control business succcess and as a bases for further planning processes the headquarter in US runs a summary database, consisting of aggregate national sales-figures (see table 2).

Report of product_groups:

product-group	country	factual revenue in US $	planned revenue in US $
Pers. Comp.	Germany	...	...
Pers. Comp.	Gr. Britain	...	...
Workstation	France	...	...

revenue_Germany

Artikel	Preis	Menge
PC 286	900	100
PC 386	1700	270
Workst.	...	...
...	...	...

revenue_GB

article	price	quantity
PC 286	450	50
PC 386	800	200
...	...	...
...	...	...

Figure 2: Combining heterogenous databases

Obviously, some problems occur:

- the attributes aggregated to the combined report of the summary database have different names in the national databases,
- they use different currency units
- they probably have different syntctical representations.

(2) Loss of Contextual Information
The phone system of the different sales representations has been connected to the local computer system. Thus establishing a phone connection to a customer or sending him a fax may be done automatically after pushing a button. Typically, the area code of a customer living in the same district, should not be stored in the database in order to support automatic dialing. The same holds for the national code of the international phone network. To supply high quality maintenance service (hot line) and enable additional marketing activities these adress-data are transfered to the business center in United States. The problem associated with this transfer consists in the loss of contextual information, since the area code and the national phone code are implicitly contained in the location of the computer system (respectively the location of the people responsible of the database).

(3) "Acquisition and Mergers" Problem
A year ago our computer manufacturer, let's call it firm A, has bought a different computer corporation, let's say firm B. Therefore, many business partners and quite a lot of OEM- articles (parts bought of other suppliers) still carry an identification of firm A as well as one of firm B. This redundancy causes unnecessary activities and many errors consuming thus a great deal of the expected synergy effects. But since organizational structures and the sequence of business events is very closely connected to a given numbering system, only step by step a change of these identification units will occur.

2. Statements and Reconstructability

2.1 Statements and Representations

The problems described above only constitute a subset of the problems one has to encounter in a multidatabase system (see Kim and Seo (1991) for a systematic classification). But these problems are related to the semantics of statements and their representations (for a more general treatment, see Wedekind (1992)).

Especially the cases (1) and (2) are the results of an incomplete representation of the intended statements. Take one of the lines of table 1, left side, for example

$$PC386, 1700, 270 \tag{1}$$

It has the interpretation that 270 items of the article 'PC 386' have been sold at a price of 1700 German Marks, whereas any line of the right table

$$PC386, 800, 200 \tag{2}$$

has the different interpretation 'quantity of 200 items sold at a price of 800 British pounds'. Because of the different interpretations, a simple transfer of the attributes of these databases to a single processing unit will not allow for the necessary calculations.

A projection π of the relation *revenue_GB* to the columns *'article'* and *'price'*

$$PC386, 800 \tag{3}$$

reduces the above combined statement to the most elementary form one may represent in a database: *"PC 386 has a price of 800 British pounds"*. This statement still combines different information units and thus may be reduced to more elementary propositions.

Especially one has to note the context-information in order to understand the explicitly represented statements, namely *"The relation considered has the attributes article and price"*, *"there is an article PC 386"*, *"the article PC 386 has a price with value 800"*, and *"price will be measured in British pounds"*.

Consequently, in the context of information interchange, one has to take care of:

- identification of the relation R_A considered, i.e. identification of the data bases DB_A and the name(R_A).
- possible the point in time.
  Since a data base is changing with time, updating procedures may change the statement considered, before an information transfer successfully has been completed.
- identification of the attributes involved in the statement.
- identification of the tuple the statement will rely on.
- possible the unit of measurement.

Additional information units are needed if classificational statements are considered.

Talking of attributes in the following, we generally mean attributes in first normal form (1NF), that is, attributes which have no syntactical structure. The above considerations may be formalized to:

Definition 1
Let DB_A be a database and A an 1NF-attribute of some relation R_A of DB_A. The name of A will be designed by **name(A)**, *its domain by* **dom(A)**.
A is called to measure a **physical quantity** *with dim(A) ('dimension') as the unit of mesurement, iff $A = (name(A), dom(A), dim(A))$.*

Definition 2
For $\mu \in R_A$ define $A(\mu)$ to be the value A takes on μ. Then the **predicate** $P(\mu, A(\mu))$ *reads as* **"μ has of property name(A) $A(\mu)$ units of measurement".**

Clearly, any tupel $\mu \in R_A$ must be identifyable to represent a statement, i.e. there is some *projection* **id** of R_A on a set of attributes which have key-property:

Definition 3
For any relation R of a database DB define $\mathbf{id}(R)$ *to be the projection of R to an* **identifying set of attributes of** *R. For* $\mu \in R_A$ *id(*μ*) designates a key of* μ. *Regarding propositions, there will be no distinction between* μ *and id(*μ*), i.e.* $P(\mu, A(\mu))$ *has the same interpretation as* $P(id(\mu), A(\mu))$.

2.2 Assignments of Elementary Predicates

In this chapter we ask, under what conditions two elementary predicates, $P_A(\mu, A(\mu))$, $\mu \in R_A$, and $P_B(\nu, B(\nu))$, $\nu \in R_B$, R_A and R_B relations of some databases DB_A and DB_B respectively, could be related each other without information loss.

Definition 4 (assignability)
Let R_A *and* R_B *be relations of the databases* DB_A *and* DB_B *and A and B attributes of* R_A *and* R_B *measuring physical quantities. Let* $id(R_A)$ *and* $id(R_B)$ *and the corresponding key-attributes be given.*
Let dom(A) respectively dom(B) designate the domain of A resp. B and $dom(id(R_A))$, *resp.* $dom(id(R_B))$ *the domain of the key-attributes of* R_A *resp.* R_B.

For $\mu \in R_A$ *a predicate* $P_A(\mu, A(\mu))$ *is called* **assignable without information loss** *to a predicate* $P_B(\nu, B(\nu))$, $\nu \in R_B$, *if and only if, there exists a function*

$$\begin{aligned} f = (f_R, f_a), f : (dom(id(R_A)), dom(A)) &\longrightarrow (dom(id(R_B)), dom(B)) \text{ with} \\ f_R : dom(id(R_A)) &\longrightarrow dom(id(R_B)), \text{ one-to-one and} \\ f_a : dom(A) &\longrightarrow dom(B), \text{ one-to-one.} \end{aligned}$$

Remarks:

(1) Considering statements $P_A(\mu, A(\mu))$ the tupel to which the statement is related to must be identifyable. Thus, in the definition 4 of assignability, it's not sufficient to have a function f_a of $dom(A)$ to $dom(B)$. In addition we need an assignment of tupels, i.e. the function f_R.

(2) In order to guarantee no information loss, one needs the property to regain the original predicate $P_A(\mu, A(\mu))$ out of $P_B(\nu, B(\nu))$, if $id(\nu) = f_R(id(\mu))$.
Clearly, having $\mu_1, \mu_2 \in R_A$, $\mu_1 \neq \mu_2$, and $id(\nu) = f_R(id(\mu_1)) = f_R(id(\mu_2))$, the different predicates $P_A(\mu_1, A(\mu_1))$ and $P_A(\mu_2, A(\mu_2))$ cannot be reconstructed. Thus f_R must be one-to-one (injective).
The same kind of argumentation can be applied to the case, where at two points in time t_1 and t_2, $t_1 < t_2$, $id(\mu_{t_1}) = id(\mu_{t_2})$, $A(\mu_{t_1}) \neq A(\mu_{t_2})$, and $B(\nu) = f_a(A(\mu_{t_1})) = f_a(A(\mu_{t_2}))$. This situation does not allow the reconstruction of the original statements. Clearly, with f_a to be injective, we avoid such situations.

2.3 Reconstruction of Elementary Predicates

For a **reconstruction** of the original statement $P_A(\mu, A(\mu))$ out of $P_B(\nu, B(\nu))$ the injectivity of the functions $f = (f_R, f_a)$ is not sufficient. Especially, in all those cases where f is not bijective, we need some criterium to decide for a given $P_B(\nu, B(\nu))$ whether $\nu \in R_B$ is an image of some $\mu \in R_A$ or not. This information can be stored in the range of values of B, in that case we will speak of a **complex attribute** B, or it can be stored in a separate attribute, that is B should be replaced by $[B_1, B_2]$. Semantically, there isn't much difference between both alternatives. As an illustration, take a local phone-number as attribute A of some relation and as attribute B the augmenation of A by the area code within a global database DB_B (table 3).

local database DB_A (1) $\Longrightarrow$ global database DB_B

phone_number
2012859

phone_number
0651-2012859

or (2) $\Longrightarrow$ global database DB_B

area code	local phone_number
0651	2012859

Figure 3: Alternatives for representing contextual information of phone numbers
(1): implicit storage (by value)
(2): explicit storage (by separate attribute)

In either case, implicit (by value) or explicit (by separate attribute) the origin of the tupel ν of R_B in DB_B may be derived by appliying some **selection mechanism** σ on R_B. To handle these cases instead of considering elementary predicates in R_B, i.e. just a single attribute B, rather a list of attributes $[B_0, B_1, \ldots, B_m]$, $m \geq 0$, of R_B should be considered. Reconstructability essentially involves a selection mechanism σ, i.e. some predicate involving $B = [B_0, B_1, \ldots, B_m]$ of R_B, to identify the origin of the tupels in R_B.

Theorem 5 (reconstructability)
Let R_A and R_B be relations of the databases DB_A and DB_B, A an attribute of R_A measuring a physical quantity, and $B = [B_0, B_1, \ldots, B_m]$, $m \geq 0$, a list of attributes of R_B.
To $\mu \in R_A$ assume $P_A(\mu, A(\mu))$ to be assignable by $f = (f_R, f_a)$ without information loss to the predicate $P_B(\nu, B_0(\nu))$, $\nu \in R_B$.
Based on $B = [B_0, B_1, \ldots, B_m]$ assume the existence of a selection $\sigma : dom(R_B) \longrightarrow dom(R_B)$ to identify the image of $f_R(dom(id(R_A)))$ within $dom(id(R_B))$.
Then for $\nu \in \sigma(R_B)$,

$$P_B(\nu, B_0(\nu)) \text{ is assignable to } P_A(\mu, A(\mu)) \text{ without information loss.}$$

In this case we call $P_A(\mu, A(\mu))$ to be **reconstructable** *out of $P_B(\nu, B(\nu))$, $\nu \in R_B$,*

$$P_B(\nu, B(\nu)) = P_B(\nu, B_0(\nu)) \wedge P_B(\nu, B_1(\nu)) \wedge \ldots \wedge P_B(\nu, B_m(\nu)),\ \nu \in R_B.$$

Proof: By definition, $\sigma(R_B)$ consists of those and only those tupels ν, having a projection π on $[id(R_B), B_0]$, which is the image of some $\mu \in R_A$ by $f = (f_R, f_a)$, i.e. $\sigma(dom(id(R_B))) = f_R(dom(id(R_A)))$ and $\sigma(dom(B_0)) = f_a(dom(A))$. Since f_R and f_a are injective on the sets $\sigma(dom(id(R_B)))$ resp. $\sigma(dom(B_0))$ they are invertible.

Remarks:

(1) Clearly, to know the origin of some statement does not automatically include all the contextual information of that statement. But at least, it involves the ability to find out these informations.

(2) Designing a database the applicational requirements define the **relevant context**. Especially for statements $P_A(\mu, A(\mu))$, A measuring a physical quantity, in most cases it will be reasonable to store the dimension $dim(A)$ of A, i.e. the unit of scale of A, explicitly. At least, we will need $dim(A)$ in order to define for the purpose of information interchange suited transfer units.

2.4 Structured Attributes and Classificational Statements

Besides attributes refering to physical quantites structured attributes occur very often. As example take any number of a numbering system like the telephone number, an article number or a customer number. Structured attributes carry different semantical units.

Example (phone number):

011496512012859	complete phone number
semantical units:	
011	international network (from USA)
49	country code
651	area code
201	organization number
2859	extension (within organization)

Definition 6 (structured attribute)
An 1NF-attribute A of some relation R_A is called structured, if there are some finite sets $Class_i$, $i = 1, \ldots, n$, $n \geq 0$, s.th.

$$A = (dom(A), name(A); c_1^A, c_2^A, \ldots, c_n^A),$$

where $c_i^A : dom(A) \longrightarrow Class_i$ (equivalence class i), $i = 1, \ldots, n$, $n \geq 0$.
[$n = 0$ denotes the case where $A = (dom(A), name(A))$].

Related to a structural attribute A of some relation R_A is a classificational statement of the kind:

Definition 7
For $\mu \in R_A$ the **classificational predicate** *$P_A(\mu, A(\mu))$ consists of the propositions*
μ has within the class (set) dom(A) the value $A(\mu)$ and
μ has within the class $c_1^A(dom(A))$ the value $c_1^A(\mu)$ and
...
μ has within the class $c_n^A(dom(A))$ the value $c_n^A(\mu)$.

Clearly, regarding assignments or reconstructions of such statements out of a statement $P_B(\nu, B(\nu))$, $\nu \in R_B$, B must carry a structure, comparable to that of A. That is, there must be some assignment of the equivalence classes as well.

Definition 8 (assignability of structured attributes)
Let R_A and R_B be relations of the databases DB_A and DB_B and
$A = (dom(A), name(A); c_1^A, c_2^A, \ldots, c_n^A)$, $n \geq 0$, and
$B = (dom(B), name(B); c_1^B, c_2^B, \ldots, c_m^B)$, $m \geq 0$, struct. attributes of R_A and R_B.
For $\mu \in R_A$ a classificational predicate $P_A(\mu, A(\mu))$ is called **assignable without information loss** *to a predicate $P_B(\nu, B(\nu))$, $\nu \in R_B$, if and only if,*
$m = n$ and there exists a function $f = (f_R, f_a, g_i)$, $i = 1, .., n$ with

$$\begin{aligned} f_R &: dom(id(R_A)) \longrightarrow dom(id(R_B)), \text{ one-to-one and} \\ f_a &: dom(A) \longrightarrow dom(B), \text{ one-to-one,} \\ g_i &: c_i^A(dom(A)) :\longrightarrow c_i^B(dom(B)), \text{ one-to-one, and} \\ g_i \circ c_i^A &= c_i^B \circ f_a, \; i = 1, .., n; \; n \geq 0. \end{aligned}$$

Remarks:

(1) This is a generalization of definition 4. Thus, as is shown there for f_R and f_a, injectivity is a necessary condition for a reconstruction of $P_A(\mu, A(\mu))$. As shown in chapter 2.3, theorem 5, some selection σ in addition is needed to have sufficency for an inverse assignment.

(2) The last condition, $g_i \circ c_i^A = c_i^B \circ f_a$, $i = 1, \ldots, n$; $n \geq 0$; states, that for any $\mu \in R_A$ the following diagramm commutes:

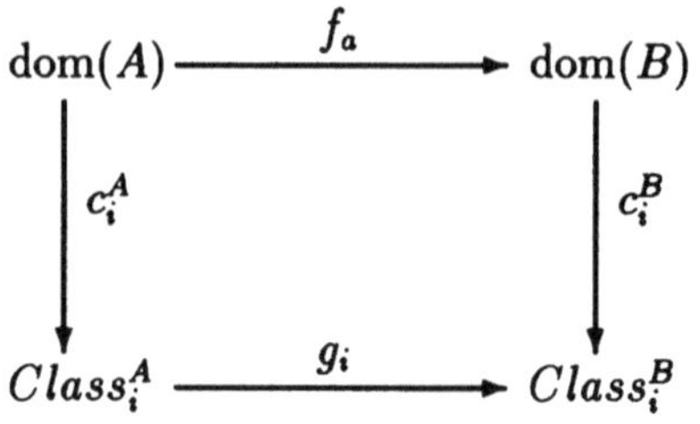

With respect to the reconstruction of $P_A(\mu, A(\mu))$ out of $P_B(\nu, B(\nu))$, $\nu \in R_B$, one may conclude:

Corollary 9 (reconstructability of structured attributes)
Let R_A and R_B be relations of the databases DB_A and DB_B,
$A = (dom(A), name(A); c_1^A, c_2^A, \ldots, c_n^A)$, $n \geq 0$, a structured attribute of R_A.
Let $B = [B_0, B_1, \ldots, B_k]$, $k \geq 0$, a list of 1NF-attributes of R_B with
$B = (dom(B), name(B); c_1^B, c_2^B, \ldots, c_n^B)$, $n \geq 0$.
To $\mu \in R_A$ assume $P_A(\mu, A(\mu))$ to be assignable by $f = (f_R, f_a, g_i)$, $i = 1, \ldots, n$, without information loss to the predicate $P_B(\nu, B_0(\nu))$, $\nu \in R_B$.
Assume the existence of a selection $\sigma : dom(R_B) \longrightarrow dom(R_B)$ to identify the image of $f_R(dom(id(R_A)))$ within $dom(id(R_B))$.
Then for $\nu \in R_B$,

$P_B(\nu, B_0(\nu))$ is assignable to $P_A(\mu, A(\mu))$ without information loss.

In this case $P_A(\mu, A(\mu))$ is called to be **reconstructable** *out of $P_B(\nu, B(\nu))$, $\nu \in R_B$,*

$$P_B(\nu, B(\nu)) = P_B(\nu, B_0(\nu)) \wedge P_B(\nu, B_1(\nu)) \wedge \ldots \wedge P_B(\nu, B_m(\nu)),\ \nu \in R_B.$$

Structured attributes, like a number system, as they occur in business and administration, usually cannot be assigned to or reconstructed from each other. This is, because organizational settings immediately are related to a number system. For example, parts of a customer number code the responsible sales representative, or a part number carries some assembly information. If these numbers are transferred to some different organization, as it is necessary in the case of acquisition and mergers, these classificational information items are not preserved.

3. Dependencies

In this chapter we consider a set of attributes $A_1, A_2, \ldots, A_n$ of some data bases $DB_{A_1}, \ldots, DB_{A_k}$, $k \geq 1$, and a dependent set of attributes $B_1, B_2, \ldots, B_m$, $B_j = f_j(A_1, A_2, \ldots, A_n)$, $j = 1, \ldots, m$, of a data base DB_B. If an update of one of the A_i is occuring, the question arises, how consistency of the dependent attributes B_j can be achieved and correspondingly, what kind of information must be transfered.

3.1 Strategies of Consistency

Consistency mechanisms can be distinguished by the event, which causes the "follow-up"-activities. Details may be found in Jablonski (1990), p.149ff.

(a) Active Attributes A_i
The independent attributes $A_1, A_2, \ldots, A_n$ can be active in the sense, that any update of one of the A_i, triggers the follow-up activities $B_j \leftarrow f_j(A_1, A_2, \ldots, A_n)$, $j = 1, .., m$. Thus, the knowledge which event causes the follow-up activity and the kind of activity has to be connected with the representation of the A_i.

(b) Passive Attributes A_i

The opposite strategy connects the knowledge of the necessary kind of activity with the B_j. The independent attributes $A_1, A_2, \ldots, A_n$ are passive in the sense, that any update of one of the A_i, causes no follow-up activity at all. Rather, the dependency condition $B_j \leftarrow f_j(A_1, A_2, \ldots, A_n), j = 1,.., m$ is evaluated every time a (read-)access to B_j occurs. Thus B_j actually is a virtual attribute, the evaluation mechanism $f_j(A_1, A_2, \ldots, A_n)$ matches the well known view-concept of relational databases.

Both strategies (a) and (b) are synchronous strategies, that is every update of an A_i in case (a) or any access of B_j in case (b) triggers an immediate action. As is well known (Jablonski (1990)), very soon these activities cause a system overload yielding lower performance.

(c) External event (trigger)

To avoid system overload an asynchronous strategy is highly recommended, that is, actions to achieve consistency between B_j and $A_1, A_2, \ldots, A_n$ are triggered only if some condition is met. The usual events for these triggers are:

- Meeting a time condition. A time-triggered update is the very classical form of data processing. In the bach-oriented environments of the 60s and early 70s during night hours the data stock had been updated by merging with the daily receipts and shipments.
- Change exceeding a definite limit. This is a value dependent condition, which has great importance in business applicatons. For example, any form of *'management by exception'* uses this trigger-mechanism to point to exceptional situations.
- Explicit request.

Which form of these asynchronous trigger mechanisms in a concrete situation is most suited, depends on the very situation and thus on the intended applicational environment.

In all the three cases there might be **conflicting updates**. Especially, as is usual in heterogenous environments, this will occur, if an independent update of the B_j is considered to be possible. Formally, we don't request that $\{A_1, A_2, \ldots, A_n\} \cap \{B_1, B_2, \ldots, B_m\} = \emptyset$. In these cases, some regulations of priority are necessary. These could consist of a time stamp connected to every update request, but also of a *'hierarchy of importance'* related to updates.

3.2 Reference Data Base

There is some additional argument against these three strategies (a), (b) and (c), which results from the need to adapt information systems to changing organizational structures. In the cases above, all the knowledge needed to augment the system, let's say to establish an additional attribute $B_{m+1} \leftarrow f_{m+1}(A_1, A_2, \ldots, A_n)$, is concentrated at the information providers, namely the attributes A_i and the databases DB_{A_i}, $i = 1, \ldots, n$. But the need of additional information and its use takes place at the attribute B_{m+1} at a presumable remote data base DB_B.

People at DB_B possibly don't know which kind of information would be available, and if they know, they don't know who is responsible for and what are the exact specifications of that information items. Thus, these strategies show a considerable inflexibility and therefore, sooner or later, the information supply of the supporting organization will be very poor.

To avoid these difficulties, the knowledge which information is available to others within an organization must be publicly available, that means available to definite user groups, independent of the actual information use. This idea leads to a

reference data base DB_R, serving as intermediary to organize the dependency $B_j \leftarrow f_j(A_1, A_2, \ldots, A_n)$. Clearly, the structure of the reference data base should allow to transfer statements without information loss.

Therefore, we need a relation R_R in DB_R and suited attributes $R_1, R_2, \ldots, R_r$ in R_R to be able to reconstruct the original predicates represented by the attributes A_i within some relation R_{A_i} of a database DB_{A_i}. Since DB_R is used by many different groups, any relation and attribute in DB_R needs some standardization but only to a degree to reconstruct the local information represented by the predicates $P_{A_i}(\mu, A_i(\mu))$, $\mu \in R_{A_i}$. It is not necessary to standardize all the local databases DB_{A_i}, as it's sometimes implied by the propagators of a corporate data model (Scheer(1988)).

Using a reference data base DB_R, the dependency $B_j \leftarrow f_j(A_1, A_2, \ldots, A_n)$ is subdivided into the dependencies
(1) $R_l \leftarrow g_l(A_1, A_2, \ldots, A_n)$, $l = 1, .., r$, and
(2) $B_j \leftarrow h_j(R_1, R_2, \ldots, R_r)$, $j = 1, .., m$.

Thus, the information providers A_i at the local sites have to know their *send-functions* g_l, which is a *get-funtion* for the intermediary R_l. The information requesters B_j, on the other hand, have to know their *get-strategy* h_j to extract the wished information. Clearly, depending on applicational needs, h_j equally well can be seen as a send strategy for the intermediary DB_R.

4. Exchange-Protocols and Middleware

The concept of a reference database as introduced in chapter 3.2 offers generalizations and modifications in different directions. Common to all these generalizations is a standardized intermediary between two communication partners A and B.

4.1 Peer to Peer Communication

Considering both communication-partners to be of the same rank (peer to peer) the intermediary functions as a protocoll for data interchange. For example, take the EDIFACT-norm (see Frank (1991)), which defines in the Data Message Directory the types of admissable messages. The above introduced *send-functions* g_l resp. *get-functions* h_j serve in the EDI-context as converters between the inhouse system and the standard. In addition, to support the pecularities and distinguished communication needs there are considerations to develop general data-models for specific branches (see Fischer (1993)).

The idea of a reference data base as a communication intermediary has been further elaborated by the author (see Czap (1991) and (1992)).

4.2 Middleware

If the communication partners A and B are considered on a vertical scale the reference data base DB_R introduced above generalizes in a natural way to the concept of a software layer situated between specific applicational requirements as well as front-end pecularities on one side and the hardware and systems software in use at the other side. This software layer is known as middleware (see DEC 93) and may be seen as a collection of transformation-services bridging the gap between applicational requirements and systems resources.

4.3 Trading

In a distributed computing environment, the concept of paragraphs 4.1 and 4.2 may be combined to a software layer, called TRADER (APM 1991) resp. BROKER (OMG

1992). It establishes a communication link between an information requester (client) and an information provider (server) stroing and administering the pecularities of A_i, g_l, B_j, h_j and the reference data base DB_R. In addition, depending on the specific context special policies for linking requesters and providers can be applied by the trading process.

These concepts introduced here are further evaluated within a research project at University of Trier, funded by Deutsche Forschungsgemeinschaft DFG, devoted to "Cooperation of Distributed Hospitals".

References:

APM (1991): *ARCHITECTURE PROJECTS MANAGEMENT Limited* (Hrsg), A Systems Designer's Introduction to the Architecture, Cambridge

CZAP, H. (1991): Kommunikation dezentraler betrieblicher Anwendungen. In: G. Schwichtenberg (Hrsg.): *Organisation und Betrieb von Informationssystemen*, Springer-Verlag, Berlin u.a.

CZAP, H. (1992): Informationsverarbeitung bei Homonymie und Synonymie der Unternehmensdaten. In: W. Gödert, P. Jaenecke, W. Schmitz-Esser (Hrsg.): *Fortschritte in der Wissensorganisation, Bd.2, Kognitive Ansätze zum Ordnen und Darstellen von Wissen*, Frankfurt/Main, 174-183.

DIGITAL EQUIPMENT COMPANY (DEC) (3-4/93): *Legacy-Applications: Common Object Request Broker Architecture (CORBA) und Application Control Architecture (ACA).*

FISCHER, J. (1993): Unternehmensübergreifende Datenmodellierung - der nächste folgerichtige Schritt der zwischenbetrieblichen Datenverarbeitung! *Wirtschaftsinformatik* 35 3, 241-254.

FRANK, U. (1991): Anwendungsnahe Standards der Datenverarbeitung: Anforderungen und Potentiale. - Illustriert am Beispiel von ODA/ODIF und EDIFACT. *Wirtschaftsinformatik*, 33/2, 100-111.

JABLONSKI, S. (1990): *Datenverwaltung in verteilten Systemen, Grundlagen und Lösungskonzepte*. Berlin, Heidelberg, New York.

KIM,Won, and SEO Jingyun (1991): Classifying Schematic and Data Heterogeneity in Multidatabase Systems. *IEEE, Computer*, Dezember 1991, 12-18.

OMG (1992): *The Common Object Request Broker Architecture (CORBA)*. Object Management Group (OMG) (ed.), Framingham Ma.

SCHEER, A.-W. (1988): Unternehmensdatenmodell als Grundlage integrierter Informationssysteme, *Zeitschrift für Betriebswirtschaft*, 58, 1091-1114.

WEDEKIND, H. (1992): *Objektorientierte Schema-Entwicklung. Ein kategorialer Ansatz für Datenbanken und Programmierung*. Reihe Informatik, Bd. 85, Hrsg. von Böhling, Kulisch, Maurer; BI Wissenschaftsverlag, Mannheim, Wien, Zürich.

Prerequisites and Development Perspectives for Information Processing in the Social Sciences

Matthias Herfurth

Informationszentrum Sozialwissenschaften
Lennéstr. 30, 5300 Bonn 1

Summary: As a result of the challenges of information processing as a "tool science of the mind", interesting prospects are opening up for the social sciences in the fields of increasing user orientation, discovery of semantics in data and potential uses of computer networks. Such methodological trends are therefore becoming increasingly important which on the one hand take account of the complexity of social processes - in science as well - and on the other hand promote the formulation of theories. This paper will describe the state of social sciences, in particular sociology, in concrete terms and will examine potential solutions to problems of epistemology and research practice by means of specific examples.

1. State description of sociology

The methodological world of the social sciences is divided. The Main Stream uses mathematically-oriented, quantitative empirical social research based on formal logic; however, qualitative, i.e. interpreting and comprehending "verstehende" approaches are still holding their ground. It would therefore be desirable to consolidate both lines of research in relation to the strength and extent of the theoretical foundations.

What then are the methodological conditions and decisive factors governing the contribution of social sciences, especially sociology, to the production of knowledge?

Irrespective of whether sociology is described as a key science of the 20th century or whether people complain - almost simultaneously - that it is socially irrelevant, it still needs to be explained why this discipline is always the object of alternating bouts of underestimation and excessive claims, overexpectations and harsh criticism (Zapf, 1987). A great many paradigms, schools of thought and research approaches intersect at the point which ranges from an understanding of sociology as a theory of society through to the science of individual phenomena and their conformity to natural laws. However, the wealth of information thus created only led to the further specialization and reinforcement of opinions of individual schools and, overall, to a diffuse state of knowledge. This shows that there are hefty discussions on occasions between quantitative and qualitative research in sociology, and also in psychology and economics, and that there appears to be a wide, often insurmountable gap between theory and empirics.

1.1 Quantitative methods

Empirical-analytical social research is now in an unfavorable position when it comes to the description and theory-derived explanation of social phenomena: its tools and methods are used and accepted throughout the world, but there are hardly any key findings on the most urgent problems in our society, for example the ecological crisis, mass migration of people seeking asylum, adolescents and violence, drug consumption, xenophobia, fundamentalism, etc. In this situation, the question must be asked whether the cognitive faculty of the actual repertoire of empirical tools is partly to blame. Empirical social research now works with sophisticated computer-assisted methods of data capture (for example Computer Assisted Telephone Interviewing,

CATI; Computer Assisted Personal Interviewing, CAPI) and, at times, with highly complex modeling and evaluation methods based on mathematical models (Andreß et al., 1992) like differential equation models, simulation methods, game theory and dynamic multi-level analysis. Factor, cluster and causal analyses have long been part of the permanent repertoire of the methods of empirical social research.

Therefore, empirical social research is neither an obscure secret science nor a simple opinion survey. Neither is it a small or illegitimate sister of actual statistics. It is a system of rules and experiences relating to the way in which theories are tested empirically in social sciences or socially significant data is captured reliably and accurately and analysed properly (Esser, 1988). The condition for the measurement of facts and processes is the formulation of variables. The standard methods of data analysis in the field of empirical social research typically relate to (causal) relationships between some variables, which sometimes are enriched through the introduction of spatial and time coordinates. Empirical data acquisition methods and statistical processing methods are regarded as objective, neutral and algorithmization methods which record reality in such a way that there is often no need to question decisions based on these methods. It may therefore be no accident that more and more people are talking about a relevance crisis in empirical social research as its data acquisition and processing methods become increasingly sophisticated, complicated and automated (Mayntz, 1985). It is recognized that the use of these methods alone does not represent an increase in knowledge. The more complex the formal algorithms, the more difficult it is for researchers to comprehend the contents of their decision-making sequence, document the associated parameters and reflect on suitable alternatives.

What is needed is a return to the formulation of theoretically derived perspectives from segments of social reality. Time and again it is asked whether analytical methodology can actually be applied in a practical manner to the field of sociological knowledge since this concerns social reality and not natural phenomena. The fact that humans are initially capable of concerted action (rational choice), but subjective interpretations have a strong effect on human behavior, still does not mean that behavioral regularities could not be identified. On the other hand, the methodological rules of practical analytical science have only partially promoted the scientific development of social sciences as a whole (Krotz, 1990; Lamnek, 1989; Garz and Kraimer, 1991).

Hardly any major social science variables can be measured directly, including seemingly objective parameters such as income or training not to mention social indicators such as satisfaction with life, political opinions etc. (Esser, 1988). In order to increase the meaning of results of empirical social research, individual action and process models are required as the hitherto missing link in the analysis of variable relationships (Faulbaum, 1991).

1.2 Qualitative methods

Qualitative research deals with the reconstruction of social quality and takes account of the original context of the examined social objects without depriving them of their independent character (Krotz, 1990).

The qualitative methodology bound by the interpretative paradigm is based primarily on individual case studies, qualitative (narrative) interviews, group discussions, participatory observation, content analysis and biographical methods for data acquisition. Evaluation methods are therefore concerned in particular with the explanation, structuring and categorization of the meaning of the basic data (texts), whereby the coincidence of data collection and analysis and theory formulation is emphasized (grounded theory) (Strauss, 1991).

The quality-oriented social researcher therefore uses throughout in individual steps the mathematics-based processing of data in order to provide answers to theoretical questions. They use methods for classification, reduction in information or for structuring relations in their data. The most important formal-structural methods in qualitative analysis are classification and type formation. They are regarded as abstraction methods which break up the data set according to different questions into classes. Different classifications with correct formal structures correspond to different abstractions and therefore to special questions and perspectives. They facilitate the building of different models. The object itself can then be reconstructed from the relationships between these classifications as a summary of its abstractions. In this respect, it is therefore not so much a question of a prove by statistics that the data belong to classes and are forming correlations and the like, but of judging the social objects as well as their immanent meanings, for example regarding the social structure, from these classifications (Krotz, 1990).

The characteristics of qualitative research are summarized as follows (Garz and Kraimer, 1991):

- Reality is constructed socially
- A coherent access point to reality is vital
- A case-related study with an associated type formation is a central element
- The researcher must become involved in practical work (going native).

2. Problems with sociological knowledge production and modeling from the viewpoints of epistemology and practice of scientific discovery

Knowledge progress in the social sciences is reduced significantly by the complexity of the object area. Social processes are based on natural characteristics, but require mental opinions about the processes as an explanatory factor. The reflexive and interactive nature of human awareness (i.e. the interpretation of an event changes) is an initial complexity dimension. The increase in information is therefore closely connected with progress in cognitive sciences. A second dimension is formed by context effects, i.e. the need to take account of environmental conditions. With abstraction processes, it is, for example, important to ask what aspects of the context should be considered or ignored. A third problem is posed by the social study objects themselves and their continuous change through mutual effects (Sallach, 1989).

According to Wilson (1981), the object area of conventional social sciences is made up of three dimensions:

- The objectivity of the social structure
- The transparency of the manifestations (displays) of social behavior
- The context dependency of the meaning of social types of behavior

Quantitative empirical social research stresses the first and second dimensions and ignores the context dependency of meaning or regards it as "confusing".

Qualitative research emphasizes the third dimension, but ignores the objectivity of the social structure and the obvious state of social conditions.

Ethnomethodologists explain the context dependency of meaning through the associated particulars which determine a social state as well as its indexicality and reflexivity. Indexicality means dependence on the surroundings in relation to identity and meaning. Reflexivity can be clarified by means of the tipping figure ("face/beaker") (Hebb, 1967). Wänke and Schwarz (1992) describe interesting contextual effects in standardized surveys: in comparative judgments, the direction of the comparison often determines the answer: tennis was judged as less exciting than soccer (35% more exciting versus 65% less exciting) when tennis (as a subject of comparison) was compared with soccer (as a referent of comparison), but as more exciting than soccer when soccer (subject) was compared with tennis (referent) (77% more exciting versus 15% less exciting).

The idea of context-free, but "context-sensitive" mechanisms of interaction is a great help in describing and explaining social interaction (Sacks et al., 1974). Context-free means that behavioral options are available for all participants in all interaction situations. Behavioral options are context-sensitive in the sense that an individual action is carried out in relation to its character and previous progress.

The social world is therefore composed of situation-related types of action of its participants. In other words, the social structure is a resource for and the result of social interactions alike.

The need to take account of both regularities and the context dependency of social behavior clearly shows that social phenomena cannot be merely explained by quantification and that a single explanation based on understanding is inappropriate. From a realistic point of view, quantitative and qualitative approaches are therefore not alternatives and objectivity is not a characteristic of knowledge which is derived from the use of certain types of scientific action. A convergence of methods (quantitative data may therefore be simply repeated qualitative observations) can thus be ascertained as well as the interdependence of methods; that is to say, the interpretation of quantitative data by the researcher is influenced by his qualitative understanding of the field of research in exactly the same way as his interpretation of qualitative data is derived from his knowledge of the regularities of the research area (Wilson, 1981).

A bridge will now be built between new approaches of information processing or specific information technologies and the improvement in the production of knowledge in social sciences in the sense of probing the depth of meaning of social phenomena.

Formulated as a heuristic leading idea the quantitative paradigm in social sciences at a theoretical analysis level can be compared with the representationalist, symbol-processing approach of the cognitive theorem of recent AI research - the qualitative paradigm with that of connectivism. At a methodical-technical level of information processing, relational databases have affinities with quantity-oriented variable sociology whereas object-oriented, semantic data models probably correspond more to qualitative social science research approaches.

The emergence theory premise that a cognitive world created by cognitive systems assists successful actions corresponds to the major themes of social science theory formulation by Marx and Weber through to Parsons. The fact that the social world is constructed and does not occur naturally, has long been accepted. However, the crucial questions are how this happens and what can be concluded from the circular nature of social reality construction for the organization of social systems and the prevention of their reproduction crises (Malsch, 1991). AI models of cognitive networks are potential sources of help in the problem horizon of the social science theory debate. Müller (1987), for example, offers approaches here with his attempt at real structure modeling for large social units which have more than 10^4 members. He sees the functioning of these social units in mechanisms which move in transformation processes between microphenomena and macrophenomena. In this case, there are intervening processes or filters at a meso-level. Müller uses the city of Osnabrück, which, according to his model, has 187 decision-making centers, as an example to show that the difficulty in describing large social units is due to complexity (the number of specific variables and their relationship is high), structural change (model parameters change continuously) and self-referentiality (intelligent decision-making processes take place).

Mayntz (1992) draws attention in this respect to possible transfer possibilities of recent, non-linear natural science models and theories - for example mathematical chaos theory or the concept of self-resembling structures - for social science problems such as the dynamics of social processes or the analysis of subcultures and communication processes.

3. Database technologies and knowledge-based systems

Current development trends in database software show parallels with the scientific problems of social sciences and can be used for new methodologies in model construction and the testing of theories:

3.1 Relational databases

Relational databases provide - at least in principle - a better variable representation for describing social systems, even in multi-level research designs. The disadvantages of relational technologies are the long-winded SQL statements which are required for (statistical) database analyses. For this reason, relational databases are frequently only used to store data and data is analysed via an interface in SPSS or similar statistical packages. The concept of the separation of data storage and data analysis into two software systems is extremely cumbersome. As a result of this, there are often loud calls to improve the database components in statistical analysis systems or conversely to extend database systems by means of suitable procedures for statistical analysis (Passenberger, 1989; Bertram, 1992). Relational databases depict the real world by first breaking it down into its individual components and then putting it back together again - as far as possible - within the framework of analyses. Relational databases are therefore based on the doctrine of atomism and correspond more to the empirical-analytical view of science (Bauer, 1992).

3.2 Object-oriented database systems

Object-oriented database systems, however, implement holistic principles. They attempt to describe complex entities by using a declarative language to enrich the relational model through special types of relations, thus creating a data model of high semantic expressiveness. Moreover, interactions between the objects, in other words an extensive network of characteristics and relationships, can be described. It is also possible to depict abstract data types and complex composed objects by generalization and to make specifications (Denert, 1992). This clearly shows that object-oriented data modeling is based on the same fundamental principle as action-oriented modeling in social sciences. As far as the field of social sciences is concerned, the positive interaction between object- oriented and graphical user interfaces as well the reusability of software due to high modularity in object orientation represent major advantages. Object orientation can also be used to describe and process complex data sets, to name one keyword from the current discussion on social science methods. However, object-oriented database systems are still in a research and development stage in the overwhelming majority of cases.

3.3 Knowledge-based systems

Knowledge-based systems represent a further stage in the field of database technologies. They can be used to define, describe, manipulate and call up analytical and synthetic knowledge, including structures. They also contain suitable inference methods for the derivation of new knowledge. This is the start of semantic information processing in which information is not only selected and thus ultimately reduced, but also information contents are interrelated and new information is derived from these or questions are answered intelligently. Knowledge-based information processing is possible, for example, using the declarative programming language PROLOG which describes variables, characteristics of variables and interrelationships, and makes logical derivations in the form of if-/then-rules. PROLOG can therefore be used to represent social science concepts (Brent, 1986; Herfurth et al., 1992).

Declarative programming languages are especially suitable for describing social science facts because they necessitate making meanings explicit, i.e. explaining social facts as precise as possible.

Various experiments and approaches relate knowledge-based description, modeling and programming to social science theories:

Historical materialism of Karl Marx (Klüver, 1991; Kiefer, 1991)
The elite theory of Vilfredo Pareto (Sallach, 1989)
The conflict theory of Georg Simmel (Sylvan and Glassner, 1985)
The group theory of George Homans (Gullahorn and Gullahorn, 1963); (Simon, 1952)
The social interaction theory of Irvin Goffmann (Brent, 1966) Program "N-ACT"
Heider's balance theory (Abelson, 1973) Program IDEOLOGY MACHINE
Small group theory (Doran, 1985) Program TEAMWORK
International relationships. Soviet Union, China (deB Mills, 1990)
Foreign policy, global model (Schrodt, 1988) Program PWORLD.

Thanks to their properties, knowledge-based systems have extended the scope of formal social science knowledge. They also formalize and thus maintain, refine and develop qualitative aspects of social science knowledge. Knowledge-based systems use the strengths of symbolic logic, but in a more diverse way than standard quantitative mathematical models. In other words, they offer a qualitative formalism. Garson (1990) provides an excellent overview of 26 computer programs which are, in his opinion, also suitable for social science applications. Knowledge-based language tools might become the appropriate formal instrument of social science and change methodology in human sciences thoroughly and positively (Manhart, 1991).

4. New word processing methods for knowledge production in the (qualitatively oriented) social sciences

It can often be difficult to reconcile abstract postulates and constructs in the analytical research approach with the prevailing empirical realities. Glaser and Strauss (1979) say that at the most only footnotes to aimless theories are produced, often at enormous formal and statistical expense.

What aids are now available within the qualitative paradigm in order to avoid pressing the mostly inhomogeneous, but rich sources of data into a rigid pattern of variables to obtain a reliable and valid evaluation of this data which is more or less economical, which can be subjected to intersubjective tests and which is sociologically relevant.

Salton (1987) states that good progress has been made over the course of 30 years in word processing and information retrieval in the areas of text editing, document generation, indexing, question formulation and dynamic searching in large databases. He says, however, that there have only been a few substantial developments in the field of text comprehension.

It is expensive to process alphanumeric data, i.e. text data, using the traditional social science-statistical software packages (SPSS, OSIRIS, SAS, PSTAT). However, the increasing use of scanners and suitable software will probably mean that this argument will become less important when it comes to data input. The same is true of the argument that only at least interval-scaled data could be analysed in a statistically discerning manner, i.e. multivariately, on account of the advance of non-metric, multivariate analysis methods (for example correspondence analysis, loglinear analysis).

In the field of text data processing, program packages for computer-assisted content analysis (CACA) such as Textpack 5 (Züll et al., 1991) or Intext (Klein, 1991) are much more efficient. In addition to diverse procedures relating to data input, data transfer, data processing and data exploration, the focal point of CACA is automatic text coding using a dictionary with the aim to elaborate the manifest statements in a text. However, CACA is not in widespread use at present, since there are only a very small number of up-to-date dictionaries which would be suitable for "linguistic surveying" of social object areas. In order to describe social conditions, it would be necessary to create parallels to the scale manual (ZUMA/IZ, 1983) which provides measuring instruments for content-related questions in standardized surveys.

Program packages, for example TACT (Bradley, 1990) are available for text-statistical analyses. These packages identify, localize, count, display in context the individual words and word patterns in the whole text or marked units, and can then be used as basic material for multivariate analyses (cluster analyses, factor analyses).

The question must now be asked whether there is an ideal research strategy for empirical social research in the combination of qualitative survey methods and computer-assisted quantifying evaluation methods. The problems which occur here are the time spent on surveys and evaluations, and ultimately the orientation towards the nomothetic paradigm.

What non-quantifying methods are available for the analysis of qualitative data? If structured texts are available, relational databases often correspond to the data material and the character of social science questions. Especially in the case of research designs, which interlink qualitative and quantitative approaches (questionnaires with open and closed answers), completely new prospects are opened up here for combined qualitative and quantitative data analysis, for example examination of the characteristic profiles of persons following a cluster analysis. However, the disadvantage of these database systems is the at times limitation of characters in text fields or - in many so-called free text systems - the unchangeable sequence of defined fields or limitation of the record length to the size of a screen. A new generation of software tools now offers resources especially for psychological and social science research.

The latest German software developments in the field of social science text data analysis, which satisfy the task of qualitative reasoning - to a varying degree - are:

- AQUAD* (Huber, 1990)
- ATLAS* (Muhr, 1991) and
- MAX* (Kuckartz, 1992).

Among other things, AQUAD can be used to carry out a qualitative hypothetical test, for example what logical links exist between codes (sequence, cluster, hierarchy, dimension or cause-effect relationship), as well as logical minimizing by analysing truth tables by means of Boolean algebra.

ATLAS, the result of an interdisciplinary research project at the Technical University of Berlin, currently has the following options: visualization and manipulation of semantic networks, a object-oriented user interface and knowledge-based method

*more information in the annex

advice for text interpretation.

The advantages of MAX include the extremely comfortable coding and search options (keyword in context; search in externally structured texts or coded data), the combination of qualitative and quantitative operations and the formation of typologies.

NUDIST* (Richards and Richards, 1987, 1992) is a sophisticated Australian program, especially for unstructured texts. The advantage of this program is its ability to run on different hardware platforms (Macintosh, PC, UNIX, VAX/VMS).

NUDIST can also be used to structure existing texts in different ways and to index them on an extremely fine basis. Moreover, it contains 14 different combinable retrieval commands. The design of NUDIST is based on the separation of the text database and the index database. The combination of document indexing and secondary indexing of already pre-processed text data means that qualitative analysis can be combined with aspects of variable analysis, i.e. not only word or topic frequencies can be counted, but also the proportions of file or text units (Tesch, 1990; Wang and Sidel, 1992).

Huber (1990) provides a summary of the elementary tasks of qualitative analysis:

- Reduction in data to semantic units (coding)
- Grouping of semantic units (data processing)
- (Re)construction of semantic contexts
- Qualitative meta-analysis and
- Cyclical evaluation procedures.

The representational methods and structural techniques of AI models in the framework of knowledge acquisition should also be examined to see whether they can be usefully applied to social science text processing (for example KADS, KRITON), that is to say as compensation for borrowing AI from empirical social research in relation to data acquisition methods. KEATS seems to be a suitable model (Motta et al., 1989). It is equipped with an editor for transcription analysis, graphic representation options as well as a hypertext-based tool for abstracting and structuring raw text.

5. Knowledge processing and communication

In future, computer networks will be a very useful tool in updating social science knowledge. Within the science network (WIN), Germany has an active social science user group which is headed by Professor Scheuch in Cologne.

These networks are not just used to replace electronic mail in all its variations or the "flying researcher", but also, for example, to link simulation programs on different computers, access extensive software catalogs, library inventories and general or specialized databases, as well as to stage computer conferences and to publish electronic magazines, etc. Special interest groups, which actively exchange information and solve problems, for example in the field of information retrieval (for example Lynch et al., IRLIST), [1] are being established in the networks.

If the computer was at first an increasingly efficient calculation machine and a data and information processing instrument, its communicative importance must also now be taken into account. The distinction between interpersonal communication and the movement of information inside a computer system becomes blurred, i.e. the subject/object asymmetry between a person and computers which transmit infor-

*more information in the annex

[1]Service Development: C. Lynch, N. Gusack, M. Engle; Distribution: University of California, Division of Library Automation, 3000 Lakeside Drive, Oakland, CA 94612-3550

mation becomes increasingly irrelevant when, for example, computers buy or sell shares. The new communicative situation shows, for example, that the demand for calculation services is linked to scientific cooperative relationships. The prominence of the network concept in recent computer semantics is also typical in this respect. Whereas the cognitive sciences are already living with the challenge of understanding the computer as a model of mental processes, work has only just begun on approaches to an interpretation of the use of computers in the communication process based on communication and action theories (Stichweh, 1989).

The increasing availability of resources required by the social scientist at a workstation represents other important development prospects for knowledge organization and distribution. The following tasks can be performed at a workstation: Use of an online library catalogue; keyword-controlled literature search; secondary analysis of data from other studies; explorative and final data analysis; simulations; comparisons between data and theories; communication with colleagues; writing research reports; sending literature to conference organizers, publishers and printing houses.

The integration of research activities and scientific communication means that the use of computers in knowledge processing is also a social activity with all the resulting consequences. A perhaps somewhat frightening scenario in this context is online publication and the astonishment of a researcher in the year 2030 over the fact that there were once libraries where books "did not talk to each other".

6. Summary and outlook

Schwartz and Ogilvy, two scientists from the Stanford Research Institute International, discovered a paradigm change in a comparative analysis of the scientific development of very different disciplines such as physics, chemistry, ecology, mathematics, philosophy, politics, psychology and linguistics more than ten years ago. The basic orientation of these sciences is changing as a result of this (Lincoln and Guba, 1985): From dominant paradigm towards emergent paradigm, it is from simple to complex; from hierarchy to heterarchy; from mechanical to holographic; from determinate to indeterminate; from linearly causal to mutually causal; from assembly to morphogenesis; from objective to perspective.

Even if a certain wishful thinking played a role in drafting this typology, these orientation principles would be well suited, for example, to a basic research or science policy program of a research ministry.

The objectives by which qualitative and quantitative social research can be better integrated fully correspond to these new perspectives. Variable relationships regarding social phenomena must also be understandable from preferences and perceptions of the agents in given action contexts. However, variable relationships and path models still do not constitute a theory. The complex statistical evaluation methods must not, however, be criticized - on the contrary, their further development is desirable - only their all too routinized application without reflecting empirical contexts and explaining contextual effects. The predicted warning that social science therefore has a future as a text science need not be alarming if processing mechanisms are available, are used and can be applied to a descriptive database which has priority concerning research logic.

Mohler (1992) imagines future empirical social research as an analysis of dynamic social systems in an international and intercultural context. In this respect, it must use a method mix which includes complex computer solutions to the formal and cognitive penetration of the object area. The institutes of the Association for Social

Science Infrastructure (GESIS)[2] have the basic task of preparing and making available a complex, i.e. integrated database for this purpose.

To this end a knowledge architecture, which dynamically supports, visualizes, locates and stores cognitive processes, must be attained in accordance with the concepts of hypertext (Idensen, 1993).

Heilig (1990) recommends a pragmatic, problem-oriented approach to these objectives: the development of integrated, but thematically limited multimedia applications (texts, data, graphics, movable images, sounds) - which can also be used for teaching purposes - based on at GESIS available and other data stocks in close cooperation with the scientific community. Heilig's concept calls for a knowledge structure analysis (Wersig, 1990) which combines a large number of methodical elements of both quantitative and qualitative social research. In addition to a continuous restructurization of known material for dealing more efficiently with poorly structured problems, this would also create interactive processes between knowledge producers, knowledge brokers and knowledge evaluators, which will lead to knowledge renewal and better distribution of knowledge (Bühl, 1984).

Computer models in the field of knowledge processing or recent information technologies are therefore faced with the following challenges (Tweney, 1990):

- Facility for considering heuristics at the widest range of levels
- Facility for processing random interactions between already fixed model parameters or variables
- Graphic display of study results
- Controllability of large knowledge bases
- Facility for "mistrusting", amending or making study results "softer" (fuzziness)

Schank (1987) hopes to obtain help from AI when he predicts that ten topics will be valid in any change in the basic definitions and orientations of artificial intelligence (Ryan, 1991): Decoding, representation, inference (extract meaning from input), control of combinatorial explosions, indexing, ability to predict and explain, dynamic modification, generalization and curiosity.

Computer programs:

Program:	**AQUAD 3.0** (Analyse qualitativer Daten mit Computerunterstützung)
Development:	Günter L. Huber, Arbeitsbereich für Pädagogische Psychologie, Institut für Erziehungswissenschaften I, Münzgasse 22-30, D-72070 Tübingen
Required hardware:	IBM-compatible PC, AT, XT (under MS-DOS from Version 2.0 onwards or DR-DOS)
Distribution:	Verlag Ingeborg Huber, Am Jürgenfels 29, D-87645 Schwangau
Price:	Demoversion from AQUAD 3.0 DM 25,- Software package AQUAD 3.0 (without manual) DM 198,- Software package AQUAD 3.0 with manual DM 258,-

[2]GESIS (Gesellschaft Sozialwissenschaftlicher Infrastruktureinrichtungen e.V., Association of Social Science Infrastructure Institutes Inc.) consists out of three institutions: a) Informationszentrum Sozialwissenschaften (Social Sciences Information Centre), Lennéstr. 30, 53113 Bonn, b) Zentralarchiv für empirische Sozialforschung for (Central Archive for Empirical Social Science Research), Bachemer Str. 40, 50931 Köln; c) Zentrum für Umfragen, Methoden und Analysen e.V. (Centre for Survey Research & Methodology), B 2,1, Postfach 122155, 68159 Mannheim

Program:	**ATLAS/ti** (Archiv für Technik, Lebenswelt und Alltagssprache)
Development:	H. Legewie, A. Böhm, E. Konrad, T. Muhr, J. Willenborg, F. Braun, H. Pishwa
Required hardware:	IBM-compatible PC, AT (with 4 MB working memory)
Distribution:	Technische Universität Berlin, Projekt ATLAS, Hardenbergstr. 28, D-10623 Berlin, Tel. +49-30-314-25400/+49-30-314-27 882
Price:	Price is not steady
Program:	**MAX**
Development:	U. Kuckartz, A. Maurer, Fachbereich Erziehungs- und Unterrichtswissenschaften, Freie Universität Berlin, Arnimallee 12, D-14195 Berlin (1989)
Required hardware:	IBM PS/2, PC, XT, AT and 100 % compatible tools
Distribution:	Büro für Softwareentwicklung und Sozialforschung Andreas Maurer, Schützallee 52, D-14169 Berlin
Price:	Demodisk for MAX text analysis system DM 25,- MAX text analysis system 3.3 DM 480,- plus packing and mailing costs
Program:	**NUDIST 2.2** (Non-numerical Unstructured Data Indexing Searching and Theorizing) (1991)
Development:	T. J. Richards, M. G. Richards, La Trobe University
Required hardware:	Macintosh, PC/386 (or compatible) under Windows, Unix Workstations, Unix and VAX/VMS Mainframes
Distribution:	Applied Computing Research Institute, NUDIST Pro-, ject La Trobe University, Bundoora, Vic 3083, Australia; 011-61-3-479-1311; FAX 011-61-3-470-4915; E-Mail NUDIST@LATAI.LAT.OZ.AU
Price:	stand-alone 250 $ (Australian) networked 1.000 $ (Australian)

References:

ABELSON, R.P. (1973): The Structure of a Belief System. In: R.C. Schank, and K.M. Colby (eds.): *Computer Models of Thought and Language.* Freeman, San Francisco

ANDREß, H.-J., HUININK, J., MEINKEN, H., RUMIANEK, D., SODEUR, W., and STURM, G. (eds.) (1992): *Theorie Daten Methoden.* R. Oldenbourg, München

BAUER, M. (1992): Schwerpunkt Relationale Datenbanken. In: *Computerwoche 34, 27 – 39*

BECKER, B., STEVEN, E., and STROHBACH, S. (1991): *Epistemologische und wissenssoziologische Aspekte maschineller Wissensverarbeitung.* Arbeitspapiere der GMD, Sankt Augustin

BERTRAM, H. (1992): *Persönliche Kommunikation im Rahmen der Gesellschaft Sozialwissenschaftlicher Infrastruktureinrichtungen e.V. (GESIS)*

BESSER, H., YAMASHITA, R., and DUSTER, T. (1989): New Computer Technologies and Social Science Research Methods. In: L. McCrank (ed.): *Databases in the Humanities and Social Sciences-4.* Learned Information, Medford, New Jersey, 61 – 69

BÖHM, A. (1989): *Bedarfserhebung für eine EDV-Unterstützung bei der Archivierung und Interpretation von Texten.* Forschungsbericht Nr. 89-1, Berlin

BRADLEY, J. (1990): *TACT 1.2.* TACT Distribution, Centre for Computing in the Humanities, University of Toronto, Ontario

BRENT, E. E., JR. (1986): Knowledge-Based Systems: A Qualitative Formalism. In: *Qualitative Sociology 9 (3), Fall, 256 - 282*

BRENT, E. E., JR., and ANDERSON, R. E. (1990): *Computer Applications in the Social Sciences.* McGraw Hill, New York, St. Louis u.a.

BÜHL, W. L. (1984): *Die Ordnung des Wissens.* Duncker & Humblot, Berlin

CICOUREL, A. V. (1970): *Methode und Messung in der Soziologie.* Suhrkamp, Frankfurt a.M. DENERT, E. (1992): Objektorientierung: Weg aus der Softwarekrise? In: *online 10/92, 50 - 52*

DIESING, P. (1979): *Patterns of Discovery in the Social Sciences.* Aldine, Hawthorne, N.Y.

DORAN, J. (1985): The Computational Approach to Knowledge, Communication and Structure in Multi-Actor Systems. In: G.N. Gilbert, and C. Heath: *Social Actor and Artificial Intelligence*, Gower, Aldershot

ESSER, H. (1987): Warum die Routine nicht weiterhilft - Überlegungen zur Kritik an der "Variablen-Soziologie". In: N. Müller, and H. Stachowiak (eds.): *Problemlösungsoperator Sozialwissenschaft*, I. Enke, Stuttgart, 230 - 245

ESSER, H. (1988): *Amtliche Statistik und empirische Sozialforschung: Bemerkungen zu einem (scheinbar) schwierigen Verhältnis.* Vortragsmanuskript für Statistische Woche Bremen

ESSER, H. (1990): Verfällt die "soziologische Methode"? In: W. Zapf (ed.): *Die Modernisierung moderner Gesellschaften*, Campus, Frankfurt/New York, 743 - 769

FAULBAUM, F. (1991): Von der Variablensoziologie zur empirischen Evaluation von Handlungs paradigmen. In: H. Esser, and K. Troitzsch (eds.): *Modellierung sozialer Prozesse.* Informationszentrum Sozialwissenschaften, Bonn, 111-138

FISCHER, K. (1989): Die kognitive Konstitution sozialer Strukturen. In: *Zeitschrift für Soziologie 18, 1, 16 - 34*

FLECK, C. (1992): Vom "Neuanfang" zur Disziplin? In: *Kölner Zeitschrift für Soziologie und Sozialpsychologie 4, 44, 747 - 765*

GANZHORN, K. (1991): Neue Herausforderungen der Informationsverarbeitung. In: *IBM-Nachrichten 41, 306, 16 - 25*

GARSON, D. G. (1990): Expert Systems: An Overview for Social Scientists. In: *Social Science Computer Review 8.3, 387 - 410*

GARZ, D., and KRAIMER, K. (eds.) (1991): *Qualitativ-empirische Sozialforschung. Konzepte, Methoden, Analysen.* Westdeutscher Verlag, Opladen

GIEGLER, H. (1992): Zur computerunterstützten Analyse sozialwissenschaftlicher Textdaten: Quantitative und qualitative Strategien. In: J. H. P. Hoffmeyer-Zlotnik (ed.): *Analyse verbaler Daten.* Westdeutscher Verlag, Opladen, 335 - 388

GLASER, B. G., and STRAUSS, A. L. (1967): *The Discovery of Grounded Theory. Strategies for Qualitative Research.* Aldine. Chicago

GULLAHORN, J. T., and GULLAHORN, J. E. (1963): A Computer Model of Elementary Social Behavior. In: E.A. Feigenbaum, and J. Feldman (eds.): *Computers and Thought*, McGraw Hill, New York

HABERMAS, J. (1967): Zur Logik der Sozialwissenschaften. In: H. G. Gadamer, and H.

Kuhn, (eds.): *Philosophische Rundschau, Beiheft 5*

HEBB, O. D. (1967): *Einführung in die moderne Psychologie.* J. Beltz. Weinheim and Berlin

HEILIG, G. K. (1990): *Expertise zur Entwicklung von computerbasierten Informations- und Instruktionssystemen.* Applied Systems Software, München

HENNINGS, R.-D. (1992): *Informations- und Wissensverarbeitung.* W. de Gruyter, Berlin, New York

HERFURTH, M. (1988): Notationen soziologischen Wissens. Ein Beitrag zur Architektur sozialwissenschaftlicher Wissensbasen. In: *Studien zur Klassifikation 18, 82 - 93*

HERFURTH, M., MUTSCHKE, P., and OHLY, H. P. (1992): Inference from bibliographic facts: A social network approach between front-ends and text comprehension. In: Zimmermann et al.: *Mensch und Maschine: Informationelle Schnittstellen der Kommunikation.* Universitätsverlag, Konstanz

HUBER, G. L. (1990): *AQUAD. Analyse qualitativer Daten mit Computerunterstützung.* I. Huber, Schwangau

IDENSEN, H. (1993): Hypertext als Utopie: Entwürfe postmoderner Schreibweisen und Kulturtechniken. In: *NfD 44, 37 - 42*

KIEFER, J. (1991): *Die Marxsche Theorie als Computersimulation: Versuch einer systemtheoretischen Explikation des Marxschen Forschungsprogramms.* P. Lang, Frankfurt a.M.

KLEIN, H. (1988): INTEXT – ein Programm zur computerunterstützten Inhaltsanalyse, in: F. Faulbaum, and H.-M. Uehlinger (eds.): *Fortschritte der Statistik-Software 1.* Stuttgart/ New York, 574 - 581

KLÜVER, J. (1991): Formale Rekonstruktion und vergleichende Rahmung soziologischer Theorien. In: *Zeitschrift für Soziologie 20, 3, 209 - 222*

KRIZ, J. (1985): Die Wirklichkeit empirischer Sozialforschung. In: W. Bonß, and H. Hartmann (eds.): *Soziale Welt. Entzauberte Wissenschaft.* O. Schwartz, Göttingen, 77 – 89

KROTZ, F. (1990): Lebenswelten in der Bundesrepublik Deutschland. In: *Sozialwissenschaftliche Studien, 24,* Opladen: Leske + Budrich

KUCKARTZ, U. (1992): *Textanalysesysteme für die Sozialwissenschaften. Einführung in MAX und TEXTBASE ALPHA.* G. Fischer, Stuttgart, Jena, New York.

KÜCHLER, M. (1980): Qualitative Sozialforschung: Modetrend oder Neuanfang? In: *Kölner Zeitschrift für Soziologie und Sozialpsychologie 32, 373 - 386*

LAMNEK, S. (1988, 1989): *Qualitative Sozialforschung, Bd. 1: Methodologie.* Weinheim and München: Psychologie Verlags Union; *Bd. 2: Methoden und Techniken.* München: Psychologie Verlags Union

LINCOLN, Y. S., and GUBA, E. G. (1985): *Naturalistic Inquiry.* Sage Publications, Beverly Hills.

LYNCH, C., GUSACK, N., and ENGLE, M. (eds.): *IRLIST Digest.* University of California, Division of Library Automation, Oakland

MALSCH, T. (1991): Rezension von F. Varela: Kognitionswissenschaft – Kognitionstechnik. Eine Skizze aktueller Perspektiven. In: *KI 4/91, 62 - 66*

MANHART, K. (1991): KI-Modellierung in den Sozialwissenschaften. In: *KI 2/1991, 32 - 40*

MANNHEIM, K. (1980): Eine soziologische Theorie der Kultur und ihrer Erkennbarkeit.

In: K. Mannheim: *Strukturen des Denkens*. Frankfurt a.M.: Suhrkamp Taschenbuch Wissenschaft 298, 155 - 322

MAYNTZ, R. (1992): Moderne Naturwissenschaft und Gesellschaftsverständnis: Was können die Sozialwissenschaften von den Naturwissenschaften lernen? In: *Jahrbuch Max-Planck-Gesellschaft, 52–58*

MAYNTZ, R. (1985): Über den begrenzten Nutzen methodologischer Regeln in der Sozialforschung. In: W. Bonß and H. Hartmann (eds.): *Entzauberte Wissenschaft. Soziale Welt*, O. Schwartz Göttingen, 65–76

MILLS, DEB W. (1990): Rule-Based Analysis of Sino-Soviet Negotiations. In: *Social Science Computer Review 8.2, 181 - 195*

MOHLER, P.H. (1992): *Ideenskizze eines Szenarios der Empirischen Sozialforschung*. Persönliche Kommunikation im Rahmen der Gesellschaft Sozialwissenschaftlicher Infrastruktureinrichtungen e.V. (GESIS)

MOTTA, E., RAJAN, T., and EISENSTADT, M. (1989): A Methodology and Tool For Knowledge Acquisition in Keats-2. In: G. Guida, and C. Tasso (eds.): *Topics in Expert System Design*. Methodologies and Tools, 297 - 322

MÜLLER, N. (1987): Überlegungen zu einer spezifischen sozialwissenschaftlich-planungstheoretischen Modelltheorie. In: N. Müller and H. Stachowiak (eds.): *Problemlösungsoperator Sozialwissenschaft*, I. Enke, Stuttgart, 323 - 344

MUHR, T. (1991): *ATLAS/ti. A Prototype for the Support of Text Interpretation*. Forschungsbericht Nr. 91-4, Berlin

OHLY, H. P. (in press): Knowledge-based Systems: Another Data Approach for Social Scientists? In: *Social Science Computer Review 93–11*

PASSENBERGER, J. (1989): *Computerintegrierte Informationsverarbeitung in der empirischen Sozialforschung*. Institut für Freie Berufe, Friedrich-Alexander-Universität Erlangen-Nürnberg

RICHARDS, L., and RICHARDS, T. (1987): Qualitative Data Analysis: Can Computers Do It? In: *Australian and New Zealand Journal of Sociology 23, 1, 23 - 35*

RICHARDS, L., and RICHARDS, T. (1992): Analyzing Unstructured Information: Can Computers Help? In: J. Weckert, and C. McDonald (eds.): *Library Hi Tech, Vol. 10*, 1-2, 95 - 109

RYAN, B. (1991): State of the Art. AI's Identity Crisis. In: *Byte, January, S. 237 - 246*

SACKS, H., SCHEGLOFF, E., and JEFFERSON, G. (1974): A Simplest Systematics for the Organization of Turn-Taking for Conversation. In: *Language 5, 696 - 735*

SALLACH, D. L. (1989): Toward an Expert Support System for Social Sciences. In: L. McCrank (ed.): *Databases in the Humanities and Social Sciences-4. Learned Information*, Medford, New Jersey, 571 - 577

SALTON, G. (1987): Historical Note: The Past Thirty Years in Information Retrieval. In: *Journal of the American Society for Information Science 38, 5, 375 - 380*

SCHANK, R. C., and CHILDERS, P. (1984): *The Cognitive Computer*. Addison-Wesley. Reading, Mass.

SCHANK, R. C. (1987): What Is AI Anyway? In: *The AI Magazine 8, 4*

SCHRODT, P. A. (1988): PWORLD: A Precedent-Based Global Simulation. In: *Social Science Computer Review 6.1, 27 - 42*

SIMON, H. A. (1952): Formal Theory of Interaction in Social Groups. In: *American Soci-*

ological Review 17, 202 - 211

STICHWEH, R. (1989): *Computer, Kommunikation und Wissenschaft: Telekommunikative Medien und Strukturen der Kommunikation im Wissenschaftssystem.* MPIFG Discussion Paper, Köln, 89/11

STRAUSS, A. L. (1991): *Grundlagen qualitativer Sozialforschung.* W. Fink, München

SYLVAN, D., and GLASSNER, B. (1985): *A Rationalist Methodology for the Social Sciences.* Basil Blackwell, Oxford

TESCH, R. (1990): *Qualitative Research: Analysis Types and Software Tools.* Falmer Press, Basingstoke

TWENEY, R. D. (1990): Five Questions for Computationalists. In: J. Shrager, and P. Langley (eds.): *Computational Models of Scientific Discovery and Theory Formation.* M. Kaufmann, San Mateo, 471–484

VARELA, F. (1990): *Kognitionswissenschaft - Kognitionstechnik. Eine Skizze aktueller Perspektiven.* Suhrkamp. Frankfurt a.M.

WÄNKE, M., and SCHWARZ, N. (1992): *Comparative Judgments: How the Direction of Comparison Determines the Answer.* ZUMA-Arbeitsbericht Nr. 92/15, Mannheim

WANG, C. and SIDEL, P.S. (1991): HyperResearch and NUDIST. Software Review, Researchware. Randolph, MA. In *Social Science Computer Review 10, 3*

WEISHAUPT, H. (1992): *Qualitative Forschung als Forschungstradition. Eine Analyse von Projektbeschreibungen der Forschungsdokumentation Sozialwissenschaften (FORIS).* Frankfurt a. M., Deutsches Institut für Internationale Pädagogische Forschung, Abteilung Bildungsökonomie (unpublished manuscript)

WERSIG, G. (1990): Informationswissenschaft in der Bundesrepublik Deutschland – gegenwärtiger Stand und Perspektiven. In: *Informatik. Berlin 37, 4, 126 - 129*

WESTLE, B. (1990): *Zur Verbindung qualitativer und quantitativer Forschungsverfahren in der empirischen Sozialforschung: Quantitative Analyse qualitativer Daten*, Arbeitspapier Nr. 310, J.-W.-Goethe-Universität Frankfurt and Universität Mannheim

WEYMANN, A. (1989): Soziologie – Schlüsselwissenschaft des 19. oder 20. Jahrhunderts? In: *Soziale Welt 1/2, 30, 133 - 141*

WILSON, T. P. (1981): Qualitative "versus" Quantitative Methods in Social Research: An Essay on Theory and Method. In: *ZUMA-Arbeitsbericht 19: Integration von qualitativen und quantitativen Forschungsansätzen, Mannheim, 37 - 69*

WINGENS, M., and FUCHS, S. (1989): Ist die Soziologie gesellschaftlich irrelevant? In: *Zeitschrift für Soziologie 18, 3, 208 - 219*

ZAPF, W. (1987): Rede anläßlich des Amtsantritts als WZB-Präsident. In: *WZB-Mitteilungen 37, 10–11* Zentrum für Umfragen, Methoden und Analysen (ZUMA)/Informationszentrum Sozialwissenschaften (IZ) (eds.) (1983): *ZUMA-Handbuch Sozialwissenschaftlicher Skalen, Teil 1 - 3*

ZÜLL, C., MOHLER, P. H., and GEIS, A. (1991): *Computerunterstützte Inhaltsanalyse mit TEXTPACK PC.* G. Fischer, Stuttgart, New York

Aspects of Coupling Logic Programming and Databases

Gregor Meyer[1], Christoph Beierle[1], Ralph Scheubrein[2]

[1] FernUniversität Hagen, Fachbereich Informatik
Bahnhofstr. 46-48, D-58084 Hagen, Germany
e-mail: {*Gregor.Meyer, Christoph.Beierle*}*@fernuni-hagen.de*

[2] Universität Hohenheim, Institut für Betriebswirtschaftlehre
Postfach 70 05 62, D-70593 Stuttgart, Germany

Summary: Emphasizing the pragmatic aspects of a close coupling of logic programming and databases, six coupling levels are distinguished, starting from a simple embedded SQL interface to using a deductive database component. As a concrete example, the PROTOS-L system is presented where several of these levels have been realized. A special database inference component extends the capabilities of a relational database to support DATALOG plus additional SQL functionality within PROTOS-L. Furthermore, there is an integrated access to the non-standard database system LILOG-DB which allows to use complex structured objects not only in the logic programming language but also in the database.*

1. Introduction

For more than 10 years the integration of standard database technology into logic programming and expert systems has been a major topic of research. This is well documented by many publications, special conferences (e.g. [Kerschberg (1986)]) and textbooks such as [Ullman (1988)], [Ceri et al. (1990)], [Naqvi and Tsur (1989)]. The major profit comes with the combination of reliable management of huge databases on one hand and flexible, easy to use programming languages and systems on the other. For the theoretical foundation of this integration known as deductive databases we refer to [Gallaire et al. (1984)]. In this paper we will emphasize the practical aspects.

Conventional database systems (DBMS) manage huge data sets and provide multi-user access to shared data, recovery, integrity checking, and more. On the other hand, knowledge based systems and logic programming systems feature powerful knowledge representation structures and flexible inference mechanisms. The practice shows that these 'intelligent' systems are limited in working with large permanent data volumes. In a company it is often required to manage the data with a DBMS in order to provide secure access for all users. Expert system shells and typical AI programming languages take this requirement into account by including some interface to an external DBMS.

However, there is a conflict between the two goals (1) persistent and efficient management of huge amounts of data, and (2) using expressive data structuring and query languages. Within various research areas this problem is attacked. AI systems try

***Acknowledgements:** Most of the work reported here was carried out within the EUREKA project PROTOS ("Logic Programming Tools for Building Expert Systems", EU56) while the authors were at the Institute for Knowledge Based Systems within the IBM Germany Scientific Center. We would like to thank all members of the PROTOS team at IBM for their enthusiastic support of the project.

to manage large data volumes, and database systems are extended with new data models; existing AI and DBMS systems are coupled or completely new Knowledge-Base Management Systems (KBMS) are designed. The various approaches to combine logic programming and databases can be divided into two classes. The *integration* of both areas aims to develop a single system providing the desired functionality, whereas in a *coupling* approach, two independently developed systems communicate with each other. Apart from the logic based approach there is much activity with similar goals in the object-oriented DBS scene. For an overview of current systems and a comparison of deductive and object-oriented DBS see [CACM (1991)] and [Ullman (1988,1991)].

The purpose of this paper is twofold: Concentrating on the coupling approach from the viewpoint of a logic programming system, we first distinguish six different coupling levels to a DBMS, starting from a simple embedded SQL interface to using to a deductive database. These levels are inspected taking a principal as well as a practical point of view (Section 2). Second, we present a concrete example of a system (the PROTOS-L system [Beierle (1992)], [Beierle et al. (1991a)]) where several of these coupling levels have been realized (Sections 3 and 4). In particular, we describe the access facilities to a relational DBMS as well as to a non-standard deductive database (LILOG-DB, [Ludwig (1990)]). Comparing both approaches (Section 5) shows that the coupling with the relational DBMS has several practical advantages, while the coupling with LILOG-DB provides a means to profit from higher level representation facilities.

2. Deductive Access to Relational Databases

For illustration purposes we will use a small path finding example. We assume a directed graph with weighted edges, e.g. 'arc(a,b,3)', where a path is a sequence of connected egdes in this graph.

2.1 Embedded SQL

By coupling two self-contained systems, the cost for re-implementing large parts of already existing components is saved. E.g., connecting a Prolog system with a DBMS will enhance the storage functionality of the Prolog system, which only provides sequential access to files, by allowing indexed operations on permanent relations. This connection can be realized by providing a kind of embedded-SQL within the Prolog system, e.g.:

```
arc(F,T,W) ← SELECT START, DEST, WEIGHT
             FROM 'DEMO2DATABASE $ TAB1'
             INTO F,T,W
```

Here, the edges of the graph are defined by the tuples in the TAB1 relation of an external SQL database. A path is defined by the standard Prolog definition

```
path(X,Y,W) ← arc(X,Y,W).
path(X,Z,W) ← arc(X,Y,U) ∧
              path(Y,Z,V) ∧
              W = U + V.
```

At this simple coupling level based on embedded SQL the full database language is available for defining, querying, and manipulating data. Major drawbacks are that the user has to know both the host language and SQL, including, e.g., system dependent naming rules, and furthermore there is no possiblity for any system to optimize the data transfers.

2.2 Predicate ∼ relation

A first improvement step is to replace a subset of the SQL query statements by logical notations, using the equivalence of facts of a predicate and tuples in a relation.

Assuming that all facts of the predicate 'arc' are stored in the database relation named 'TAB1', we only need a mapping between positional arguments in the Prolog predicate and named attributes in the database relation. This is defined by a simple sequence of relevant attribute names which are mapped one by one onto the respective argument positions of the predicate. Additionally, the name of the database is given as a symbolic constant. For instance, in

```
using DEMO2DATABASE.                        % which relational DB to use

rel arc: string x string x int.             % facts of 'arc' are stored
    dbrel arc is TAB1(START,DEST,WEIGHT).   % in the DB-table TAB1
```

the second line declares arc to be a three-place predicate with two string and one integer argument. The third line defines arc to correspond to the TAB1 relation where the START attribute corresponds to the first argument of arc, DEST to the second, and WEIGHT to the third argument.

Based on this link between predicates and external relations a simple coupled systems can be built. Each time a predicate linked to a relation is used, an appropriate query to the DBMS will be generated automatically taking the actual bindings into account. The results of that query are read into the Prolog system either implicitly by backtracking or by explicitly creating a handle (a cursor) for the result. Some commercial Prolog systems provide similar kinds of access to external databases. A basic problem with this simple linkage is that joins are computed in nested loops within the Prolog system. In contrast to the set-oriented query evaluation in the DBMS with special merge- or hash-joins, the Prolog evaluation is only term-oriented. This mismatch between 'broad data lines' in the DBMS and 'narrow data lines' in the Prolog system is commonly called *impedance mismatch* which is also already present in a coupling as described in 2.1.

2.3 DATALOG-rule ∼ relational algebra

More of the evaluation capabilities in the DBMS can be exploited by utilizing the similarity between SQL-queries (based on the relational algebra) and inference rules in logic. Every query which can be formulated in the relational algebra can also be written as a deduction rule in logic. The relevant subset of logic is called DATALOG, see [Ceri at al. (1990)] for more examples and explanations.

The body of a rule,

$$\mathsf{path(X,Z,W)} \leftarrow \mathsf{arc(X,Y,W1)} \wedge \mathsf{arc(Y,Z,W2)} \wedge \mathsf{W = W1 + W2}.$$

can be evaluated by a join over the relation linked to 'arc' instead of calling 'arc(Y,Z,W2)' each time 'arc(X,Y,W1)' gives a binding (after backtracking); especially for a large arc-relation this would result in a significant efficiency gain.

There are two problems which remain unsolved with this kind of coupling. First, many database query languages, most prominently SQL, extend the functionality of relational algebra by providing, e.g., aggregation functions. Furthermore, these extensions are difficult to express in logic. Second, logic programming languages inherently use recursion which is not part of standard database query languages.

2.4 Pragmatic requirements

Although the relational algebra is the foundation for relational DBMS, it does not include operations for updating data, transaction control, definition of access rights, etc. Practical database languages include special data types, such as DATE, and corresponding built-in functions. Usually it is also possible to build groups of records and sort them. E.g., given a relation with a company's cash-flow per day records, it is easy to partition this relation by months, compute the sums per month, and print the sums ordered by calendar. Simple data types and special functions, such as extended string matching, can be included into a logic programming language by defining appropriate built-ins. Tuple updates are similar to the assert/retract predicates in Prolog. More difficult is the integration of set-oriented updates, aggregations, and sorting. They require the introduction of the data type 'set' or 'relation', i.e., the arguments of a predicate are not atomic but can be sets of values. E.g., some languages use a notation like

$$\mathsf{longestpath(X,Z,L)} \leftarrow \mathsf{path(X,Y,L0)} \wedge \mathsf{L} = \mathsf{max(L0)}.$$

to compute the maximum of some values. Although this notation is very concise, it breaks with the clean logic style, because it is implicitly assumed that in a procedural manner all values for 'L0' are gathered before 'max' is applied to this set of values. Instead we prefer to make the set-grouping more explicit as in

$$\begin{aligned}\mathsf{longestpath(X,Z,L)} \leftarrow\ & \mathsf{LS} = \mathsf{dbsetof(L0,\ path(X,Y,L0))} \wedge \\ & \mathsf{L} = \mathsf{max(LS)}.\end{aligned}$$

where the built-in predicate '**dbsetof**' collects all values L0 given by path(X,Y,L0). This is very similar to the standard built-in '**setof**' in Prolog, except that **dbsetof** calls are specially optimized for DB access. Sets of values can also be used for set-oriented updates. The execution of the rule

$$\begin{aligned}\mathsf{upd(X,Y)} \leftarrow\ & \mathsf{WS} = \mathsf{dbsetof(W0,arc(X,Y,W0))} \wedge \\ & \mathsf{dbdeleteall(arc(A,B,W),W,WS)}.\end{aligned}$$

deletes all tuples in the relation corresponding to arc which have a weight W such that there is an edge from X to Y with weigth W. In this way we try to resolve the impedance mismatch in both directions, reading from and writing to the DBS.

2.5 Recursion

While it is almost trivial to formulate recursive queries in Prolog, recursion has enormous impacts on the query evaluation in DBMS and the coupling strategy because it is not supported by current query languages. Especially with large data sets it is highly important to have special and efficient algorithms (cf. the overview in [Bancilhon and Ramakrishnan (1986)] and the particular approach in the PROTOS-L system ([Meyer (1989)]). Our path-finding example will exhibit the problems if we use the Prolog top-down computation on top of database relations:

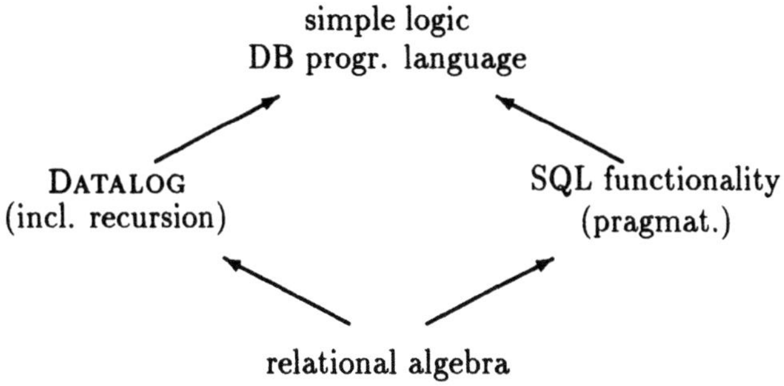

Figure 1: Extensions to the relational algebra

Set-oriented evaluation: Given the query path(start,dest,W) and the recursive path definition in 2.1, in the non-trivial case a node is searched, which is directly connected with 'start'. From there the searching process goes on recursively in order to find a path to 'dest'. If that path does not exist, another branch beginning at 'start' is tried. If the direct edges are stored in a DB-relation, then many 'small' queries are sent to the DBMS. In contrast, in a DBMS environment it is more useful to create fewer 'large' queries, e.g. selecting *all* nodes which are connected to 'start' directly with one query. Afterwards this set of 'one-step-nodes' can be used to determine the 'two-step-nodes' with the next query. This iteration is repeated until the 'dest'-node is reached (this is the computation of the transitive closure, see section 4.2).

Duplicates: Another problem with the SLD-resolution à la Prolog is the fact that it does not remember any resolution steps. If the depth-first search gets to a node which was already visited before, recursion goes on, because the (partial) solution from previous computations can not be reused. An extreme form of duplicate computation shows up when the data (the graph) contains cycles. Algorithms to compute the transitive closure of a DB-relation detect and prevent these replicated computations. By storing more information from the search process, similar to a breadth-first search, repetitions of queries and answers can be detected.

2.6 Deductive databases

While in the previous subsections we have only dealt with coupling a logic programming system and a DBMS , there are other *deductive DBMS* which integrate logic programming and database features. They differ from standard relational DBMS in that they usually have an extended data model and provide more powerful computation such as recursion. Although there is a trend for relational DBMS to include restricted forms of recursion as a built-in operation (e.g., [Shaw (1990)]) a major advantage of deductive DBMS besides recursion is their extended data representation facilities. E.g., in 2.4 we already mentioned the requirements for sets. They also play a central role in the deductive database $\mathcal{LDL}$ developed at MCC [Naqvi and Tsur (1989)], one of the currently most advanced deductive database systems. In this paper we will concentrate on another system, LILOG-DB, which provides an even richer data model.

Figure 1 briefly summarizes the previous subsections: Starting from relational algebra, the aim is to obtain a simple logic database programming language which includes DATALOG with recursion, but also satisfies pragmatic requirements as expressed by the SQL functionality.

3. PROTOS-L: Language and System

The logic programming language PROTOS-L ([Beierle (1992)], [Beierle et al. (1991a)]) which was developed within the European EUREKA project PROTOS ("Logic Programming Tools for Building Expert Sytems", EU 56), extends logic programming by providing various advanced features. Besides its database access facilities it has types, constraints, modules, and supports the development of high-level user interfaces. The typing concept allows for user-defined type hierarchies as well as for polymorphic types (c.f. [Smolka (1988)]), and the typing information is being used for consistency checks at compile time as well as for the reduction of the search space at run time [Beierle et al. (1991b)]. Variables can be constrained over finite domains in the sense of constraint logic programming. Via a small set of built-in predicates, PROTOS-L offers an easy-to-use object-oriented interface to OSF/Motif.

The module concept is similar to the one of Modula-2. It supports separate compilation, offers the definition of abstract data types and thus allows to hide implementation details in a module body. In PROTOS-L this abstraction even allows to change from one evaluation strategy to another one within a module without having to change the module interface which will be illustrated in the context of database access.

3.1 Language integration of database access

PROTOS-L offers *transparent* database access which is realized using modules. Each module has an *interface* where the exported predicates are declared, and a module *body* containig the predicate definitions. Since PROTOS-L is a typed language, every predicate declaration also specifies the types of its arguments. For instance, the interface definition

```
interface DEMO.

    rel          path: string x string x int.          % exported predicate 'path'

endinterface.
```

exports the predicate path having two string and one integer argument. An example of a corresponding so-called *program body* is

```
% standard version of path-finding example
module DEMO.

rel arc: string x string x int.                    % Facts (arcs) are given
    arc("a","b",3).                                % in the program explicitly
    arc("a","c",2).
    arc("b","c",4).

rel path: string x string x int.                   % recursive query
    path(X,Y,W) ← arc(X,Y,W).
    path(X,Z,W) ← arc(X,Y,U) ∧
                  path(Y,Z,V) ∧
                  W = U + V.
endmodule.
```

However, the DEMO interface can also be used for a *database body*

```
% second version with access to relational database
database_body DEMO using DEMO2DATABASE.             % which relational DB to use

rel arc: string x string x int.                     % facts of 'arc' are stored
    dbrel arc is TAB1(START,DEST,WEIGHT).           % in the DB-table TAB1

rel path: string x string x int.                    % recursive query
    path(X,Y,W) ← arc(X,Y,W).                       % (DATALOG + arithmetic)
    path(X,Z,W) ← arc(X,Y,U) ∧                      % formulation exactly as
                  path(Y,Z,V) ∧                     % in the original version!
                  W = U + V.
endmodule.
```

where instead of defining arc by program facts arc is defined by the tuples in the TAB1 relation of the database DEMO2DATABASE (c.f. 2.2). Note that the definition of path has not changed at all.

Thus, database access is transparent in the following sense: since queries to external databases are formulated in special module bodies via Horn clauses, the use of a DBS is not visible at the module interface. (In fact, in Section 4 we will introduce yet another kind of module body where the obove interface still fits.) Syntactically, there is essentially no distinction between the rules in a program body and in a database body. (In the latter, only function-free Horn logic and no extra-logical features like Cut are allowed.) The PROTOS-L system translates the rules automatically into the rules of the respective database language. The user does not have to know 'embedded SQL' or another query language of the particular DBS. In particular, it is very easy to move from a first prototype version of an application system containing just a few program facts to the full version actually accessing the database: All one has to do is to exchange and recompile the body of the respective module.

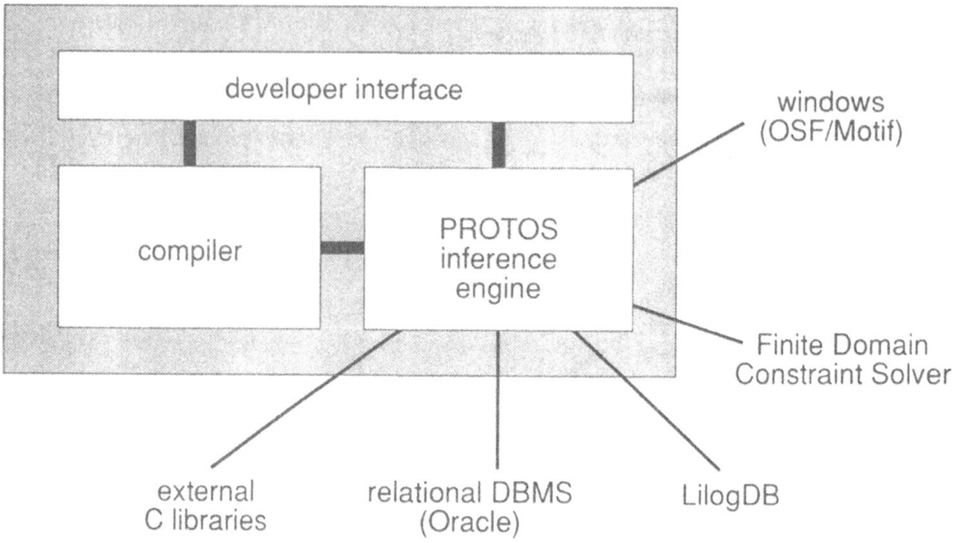

Figure 2: Architecture of the PROTOS-L system

3.2 Architecture of the PROTOS-L system

Figure 2 gives an overview of the architecture of the PROTOS-L system. It is a compiler-based system where the source code is translated into code for the PROTOS-L Inference Engine (PIE). At the heart of the PIE there is the PROTOS Abstract Machine (PAM) [Beierle et al. (1991b)], an extension of the Warren Abstract Machine [Warren (1983)] to the required polymorphic order-sorted unification. The PAM communicates with another specialized inference system, the Database Inference (DBI) [Meyer (1989)] which realizes a simple deductive DBS over a *relational DBMS* (see 3.3). There is also an interface to LILOG-DB ([Scheubrein (1991)], see Section 4). Following the idea to couple independently developed subsystems in order to obtain greatest possible flexibility, also the finite domain constraint solver communicates with the PAM over a well-defined interface. The PROTOS Window Manager (PWM) realizes the object-oriented view of OSF/Motif and thus supports the easy development of high-level user interfaces.

The compiler of the PROTOS-L system is implemented in TEL [Smolka (1988)] and a new version is currently being implemented in PROTOS-L. The inference engine is implemented in C. It is available under AIX on IBM RS/6000 (with OSF/Motif and the database systems ORACLE and LILOG-DB), IBM RT/PC 6150 (with SQL/RT), and PS/2.

3.3 Relational database inference in PROTOS-L

From **database_body** modules the PROTOS-L compiler generates code for the DBI, and the DBI transforms the compiled rules of a database body into corresponding queries to the database. The language allowed in such database bodies incorporates DATALOG (full recursion, no function symbols) and arithmetic operations. Thus, the DBI implementation fully covers the database access levels described in 2.2, 2.3, and

2.5. The evaluation of the rules is done by the DBI in cooperation with the external DBS. The evaluation strategy is a mixed bottom-up/top-down approach which could be described as set-oriented SLD resolution. The principal idea is to translate non-recursive rules directly into database views, whereas recursive rules are translated to a sequence of database queries. Where it is possible, several single subqueries are put together into a single DBS query. The internal interface to the database system is 'embedded SQL', in contrast to other deductive database systems where for efficiency reasons a lower database access level is used, e.g. in the ECRC systems EDUCE and EKS [Bocca (1986), Vieille et al. (1990)]. This approach was not followed in PROTOS-L in order to simplify, having a standardized interface, the use of different DB systems, and to be able to integrate, e.g., 'remote database access' easily.

The PROTOS-L database access facilities have already been used successfully in various applications like route finding and production planning. An important experience was that the pragmatic requirements as described in 2.4 play a significant role in such applications. Therefore, update operations for the insertion and deletion of elements as well as transactions have already been built into the language, and also a direct 'embedded SQL' interface as sketched in 2.1.

However, on several occasions, it appeared that for instance set-valued expressions were missing as well as set-oriented update facilities. Whereas a PROTOS-L **database_body** supports DATALOG plus arithmetic, the deductive database system $\mathcal{LDL}$ also has, besides sets and set-oriented updates, functional symbols and other constructs such as 'if-then-else'. The PROTOS-L coupling with the non-standard DB system LILOG-DB goes in a similar direction.

4. Coupling with LILOG-DB

LILOG-DB [Ludwig (1990)] is a database system for the support of knowledge representation systems, applications in computational linguisitics, and logic programming environments. It was developed in the context of the IBM Germany LILOG project [Herzog and Rollinger (1991)] at the universities of Aachen and Trier. The LILOG-DB data model (feature term data model, FTDM) extends the relational model, since for instance attributes do not have to be atomic any more, thus allowing for structured objects, and even extends the $\mathcal{LDL}$ data model by including structures with attributes. The LILOG-DB kernel directly supports complex objects and provides specialized storing and accessing facilities [Ley and Walter (1990)], and the query language also includes recursion.

This extended DBS permits another representation of the path finding example using structured terms. We can attach to each node a list of all outgoing edges and their weights; e.g. 'arcs(a,[arcto(b,3),arcto(c,2)])'. Of course, this represenation can not be expresssed directly in the relational model, but both in PROTOS-L program bodies and in particular in the LILOG-DB data model. Before describing the LILOG-DB data model in more detail, we give an example for the language integration into PROTOS-L. Analogously to the relational database access (3.1) there is a special type of module body (**lilog_db_body**) which contains essentially ordinary PROTOS-L syntax, but which is evaluated by queries to LILOG-DB. Thus the third module body version of our example could look like:

```
% third version with access to LILOG-DB
lilog_db_body DEMO using DEMO3DATABASE.      % which LILOG-DB to use

type connection := {arcto: string x int}.    % type definitions

rel arc: string x list(connection).          % 'arc'-facts stored in
    lilog_db_rel arc is_set SET1.            % LILOG-DB as SET1

rel path: string x string x int.             % recursive query
    path(X,Y,W) ← arc(X,L) ∧                 % (no simple DATALOG,
                  member(arcto(Y,W),L).      %  uses
    path(X,Z,W) ← arc(X,L) ∧                 %  function symbols)
                  member(arcto(Y,U),L) ∧
                  path(Y,Z,V) ∧
                  W = U + V.

rel member: T x list(T).                     % standard predicate for
    member(X,[X|Tail]).                      % membership in a list
    member(X,[Y|Tail]) ← member(X,Tail).

endmodule.
```

In some systems sophisticated mappings are put on top of a standard (relational) DBS. Complex objects and operations available to the user are automatically mapped onto the flat structure of the underlying DBS. This approach has some practical advantages, but leads to inefficiencies, due to the conversion overhead and limited optimization possibilities.

4.1 Extended data model

Since LILOG-DB supports structured terms, principally all data structures known from Prolog are available in LILOG-DB. In addition, instead of the ordinary positional notation, one can use structures with named arguments, the so-called feature terms. These are of the general form

$$\langle\ label_1 : term_1, \ldots, label_n : term_n\ \rangle$$

For instance, an edge between two nodes could be represented by the feature term

⟨ start: a, dest: b, weight: 3 ⟩

The properties of feature terms are determined in particular by the unification operation which is based on an *open-world semantics*. Two feature terms are unifiable if all common attributes are compatible. Thus, for the unification of two feature terms the sequential order of the attributes must not be identical. For instance, unification of the two terms

⟨ start: a, dest: b, ⟩ and ⟨ start: X, weight: 3 ⟩

yields the term given above, where the variable X has been bound to 'a'. [1]

As argued in 3.3 it is desirable to also have sets as terms (analogously to lists). This possibility is also given in LILOG-DB's data model. Additionaly, sets play the role of relations in LILOG-DB. All operations are applied to sets, resulting again in a set. In contrast to tuples in a relation, the individual terms in a set do not have to have a fixed schema; instead, all kinds of terms are allowed in a LILOG-DB set. The decision to use general sets instead of more restrictive relations rests on the observation that in AI application knowledge is often complex and yet incomplete.

4.2 Operations

DB queries, for instance in the form of deduction rules, are translated into the 'extended feature term algebra' (EFTA) [Ludwig and Walter (1990)] which extends the relational algebra. Besides the ordinary relational algebra operations which are also available in EFTA, two important extensions are the *conditional constructors* and the *closure operator*.

Whereas relational algebra tuples are always at least in first normal form, EFTA allows arbitrary complex terms. Therefore, the concept of attributes was generalized to *paths*, consisting of a sequence of feature names and position numbers followed by the □ symbol, e.g. 1.weight.□ . Applying this path to a term, one obtains the value of the attribute "weight" in the first argument of the term; thus, for the term f(⟨ dest:b, weight:20 ⟩,d) one obtains the value '20'.

For the construction of complex terms, "cons(c,t)" applies a constructor 'c' to a term 't'. For instance, cons(f(⟨ weight:□ ⟩),20) generates the term f(⟨ weight:20 ⟩).

The application of a conditional constructor depends on the evaluation of a condition. For instance, applying

if 2.□ < 5 then arcto(1.□,short) else arcto(1.□,long)

to the term arcto(b,3) yields the term arcto(b,short).

There are additional operations available, like term inspecting operations. W.r.t. the relational algebra operators like selection, join, difference etc., one of the most important additions in EFTA is the closure operator Φ which allows to express recursive queries. The rules

path(X,Y) ← arc(X,Y).
path(X,Z) ← arc(X,Y) ∧ path(Y,Z).

can be translated into the EFTA expression

$$\Phi_{<if(1.2.\Box=2.1.\Box)\ then\ arc(1.1.\Box,2.2.\Box)\ (1\times 1)>} <\mathrm{ARCS}>$$

where ARCS denotes a set of binary 'arc' terms.

The closure operator Φ currently emulates the so-called "naive machine" (cf. [Bancilhon and Ramakrishnan (1986)]): When evaluating $\Phi_{\langle e1,\ldots,en\rangle}\langle I1,\ldots,In\rangle$ the (syntactically as a tuple) given vector <e1,...,en> of n EFTA expressions is applied iteratively

[1]The type concept of LILOG-DB also allows for restrictions on terms so that unification works under a 'Closed World Assumption'.

to the vector <I1,...,In> of the n initial arguments until no more terms are generated. For instance, applying the Φ expression given above to the set

ARCS = {arc(a,b),arc(a,c),arc(b,c),arc(c,d)}

yields 'result' by the union of intermediate results 'res_i' of the individual iteration steps:

res_0 = ARCS = {arc(a,b),arc(a,c),arc(b,c),arc(c,d)}
res_1 = {arc(a,c),arc(b,d)} $\cup$ res_0
res_2 = {arc(a,d)} $\cup$ res_1
result = res_2

4.3 LILOG-DB access in PROTOS-L

As with relational DB access, access to LILOG-DB is hidden in special module bodies as illustrated above. A **lilog_db_body** is compiled 'horizontally' (i.e. on the same level of abstraction) into an equivalent EFTA query term in two phases (cf. [Ludwig and Walter (1990)], [Scheubrein (1991)]):

1. At compile time, a general query term is generated.
2. At run time, the actual variable bindings are propagated into the general query. These enriched queries can be evaluated directly by the LILOG-DB query processor.

The data model required for this and the corresponding powerful query language are supported by LILOG-DB. The efficiency of LILOG-DB results from the fact that the complex data objects are directly represented in the lower system levels, and that the most important operations on these data objects are already provided by the database kernel. At the PROTOS-L interface to LILOG-DB no extensive computations are necessary as in the DBI component, since they are carried out by the LILOG-DB query processor.

5. Comparison

The coupling between the PROTOS-L inference engine and an *external relational DBMS* presented in Section 3 (cf. Figure 3) is characterized by:

- The external DBMS is utilized as far as possible. Permanent storage of data, multi user access, data security, etc. can be used without any re-implementation.
- The logic program is independent from the DBMS and its query language.
- Deduction rules are translated into appropriate database queries automatically.
- The coupling should provide a proper superset of the SQL-functionality.
- There are several levels of database access, from which the programmer can choose the most appropriate.

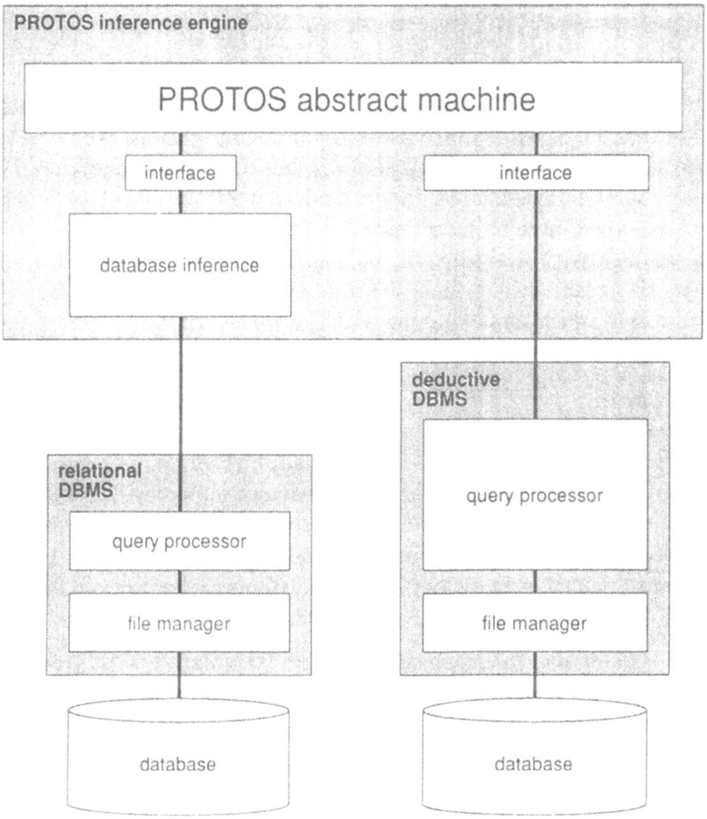

Figure 3: PROTOS-L Inference Engine with its two database interfaces

From a practical viewpoint there is an important consequence from using a relational DBMS: The standardized data model can be (and is) used by other applications. Hence, the DBS can not only be used to store data permanently and safe, but is also a communication medium to exchange data with other programs. This is especially useful for production planning applications, as they are developed in the PROTOS project.

Disadvantages of the coupling with external DBS, and at the same time differences from the LILOG-DB interface are:

- Without function sysmbols and sets, several problems are difficult to express.
- Repeated conversion of queries and data results is a big overhead.
- Possibilities for optimization are restricted because the database inference in the PROTOS-L system and the DBMS are not integrated.

The DBI is integrated into the PIE and can only use local information for building evaluation plans. The common data model merely includes integers and strings. Internal informations of the DBMS (e.g., which relations are buffered in main memory, which access paths exist) are hardly availabe. But such information is essential for an efficient query evaluation.

Instead, the interface to LILOG-DB is "broader", allowing to store all possible ground facts in the database in a compact manner. Several decuction rules can be combined into one DB-query. The integrated DB query processor is able to handle recursion.

The data model of LILOG-DB is very rich, including lists, structures, feature terms, and sets, and is combined with appropriate extended operations, such as the closure operator. The rich data model of LILOG-DB has the advantage that there is no separation between the Prolog data model and that of the database. With regard to efficiency reasons we have to note that there may be an overhead in applications which can be realized with a simpler relational DBMS. The extended system shows its power when the applications have are more complex structure, e.g. in natural language understanding systems [Herzog and Rollinger (1991)].

6. Conclusions

In this paper, we have investigated a close coupling of logic programming with databases, where we have identified six different coupling levels. As a particular example of a concrete approach we have presented the PROTOS-L system. By coupling two database systems to the programming language PROTOS-L, this system is adapted for implementing knowledge based systems using both methods of logic programming and capabilities of database systems.

Queries and rules referring to database facts are formulated with Horn clauses, in the same way as other rules. The programmer does not need to know the concrete query languages of the particular DBMS. For software engineering purpose it is very advantageous to have a powerful and coherent formalism for all systems. This goal is reached by compiling logical queries into the equivalent database queries automatically. The module concept of PROTOS-L serves as a means to abstract from the operational differences: The concrete implementation and evaluation mechanism is hidden to users of a module.

The tight coupling of the PAM to the relational DBS, implemented by the DBI, offers access to approved features of the DBS and includes the possibility to formulate recursive queries in the database modules. On the other hand, the link to LILOG-DB provides access to a new extended DBS, which has comfortable features for defining and manipulating complex objects. Current work on PROTOS-L includes a type-safe object-oriented language extension providing hierarchically ordered attributed structures similar to feature terms found in LILOG-DB.

References

BANCILHON, F., AND RAMAKRISHNAN, R. (1986): An amateur's introduction to recursive query processing strategies. In *ACM SIGMOD International Conference on Management of Data, Washington D.C.*, pp. 16 – 52.

BEIERLE, C. (1992): Logic programming with typed unification and its realization on an abstract machine. *IBM Journal of Research and Development 36*, 3, 375–390.

BEIERLE, C., BÖTTCHER, S., AND MEYER, G. (1991a): Draft report of the logic programming language PROTOS-L. IWBS Report 175, IBM Scientific Center, Institute for Knowledge Based Systems, Stuttgart.

BEIERLE, C., MEYER, G., AND SEMLE, H. (1991b): Extending the Warren Abstract Machine to polymorphic order-sorted resolution. In *Logic Programming: Proceedings of the 1991 International Symposium, San Diego, CA* (Cambridge, MA, USA), V. Saraswat and K. Ueda, Eds., The MIT Press, pp. 272–286.

BOCCA, J. (1986): EDUCE: A marriage of convenience: Prolog and a relational DBMS. In *Proceedings of the Symposium on Logic Programming, Salt Lake City, Utah*, IEEE Computer Society Press.

CACM (1991) Special section: Next-generation database systems. *Communications of the ACM 34*, 10.

CERI, S., GOTTLOB, G., AND TANCA, L. (1990): *Logic Programming and Databases.* Surveys in Computer Science. Springer Verlag.

GALLAIRE, H., MINKER, J., AND NICOLAS, J.-M. (1984): Logic and databases: A deductive approach. *Computing Surveys 16*, 2.

HERZOG, O., AND ROLLINGER, C.-R., Eds. (1991): *Text Understanding in LILOG - Integrating Computational Linguisitcs and Artificial Intelligence - Final Report on the IBM Germany LILOG-Project*, vol. 546 of *Lecture Notes in Computer Science.* Springer Verlag.

KERSCHBERG, L., Ed. (1986) *Proceedings of the First International Conference on Expert Database Systems, Charleston, South Carolina*, Benjamin/Cummings. published as: Expert Database Systems, 1987.

LEY, M., AND WALTER, B. (1990): Der LILOG-DB-Fact-Manager: Ein Datenbankkern zur Speicherung variabel strukturierte Objekte. *Inf. Forschung und Entwicklung 5.*

LUDWIG, T. (1990): A brief overview of LILOG-DB. In *Proceedings of the 1990 Data Engineering Conference, Los Angeles, CA*, IEEE Computer Society Press, pp. 420–427.

LUDWIG, T., AND WALTER, B. EFTA: A database algebra for deductive retrieval of feature terms. *Data & Knowledge Engineering 6* (1990), 125–149.

MEYER, G. (1989): Evaluation of database inference rules in the PROTOS-L system. Diplomarbeit Nr. 630, Universität Stuttgart, Inst. für Informatik, Stuttgart. (in German).

NAQVI, S., AND TSUR, S. (1989): *A Logical Language for Data and Knowledge Bases.* Principles of Computer Science. Computer Science Press, Inc., New York.

SCHEUBREIN, R. (1991): Integration of a deductive database system into a logic programming language. IWBS Report 176, IBM Scientific Center, Institute for Knowledge Based Systems, Stuttgart. Diplomarbeit, Univ. Kaiserslautern, (in German).

SHAW, P. (1990): Database language standards: Past, present, and future. In *Proceedings of the International Symposium on Database Systems of the 90s; Müggelsee, Berlin*, A. Blaser, Ed., vol. 466 of *Lecture Notes in Computer Science*, Springer Verlag, pp. 55–80.

SMOLKA, G. (1988): TEL (Version 0.9) Report and User Manual. SEKI Report SR-87-11, University of Kaiserslautern.

ULLMAN, J. D. (1988): *Principles of Database and Knowledgebase Systems.* Computer Science Press, Inc., Rockville, Maryland.

ULLMAN, J. D. (1991): A comparison of deductive and object-oriented database systems. In *Proceedings of the Second International Conference on Deductive and Object-Oriented Databases, DOOD'91*, C. Delobel, M. Kifer, and Y. Masunaga, Eds., vol. 566 of *Lecture Notes in Computer Science*, Springer Verlag, pp. 263–277.

VIEILLE, L., BAYER, P., KÜCHENHOFF, V., AND LEFEBVRE, A. (1990): EKS-V1, a short overview. In *AAAI-90 Workshop on Knowledge Base Management System, Boston, USA.*

WARREN, D. H. D. (1983): An abstract Prolog instruction set. Technnical Note 309, SRI International, DEC, Menlo Park.

Processing Scientific Networks in Bibliographic Databases

Peter Mutschke

Informationszentrum Sozialwissenschaften
Lennéstr. 30, D–53113 Bonn, Germany
e-mail: mutschke@iz–bonn.gesis.dbp.de

Summary: The paper presents a knowledge–based model for processing network–like relationships between scientists in bibliographic databases. The need to reason about scientific structures has turned out to be a remaining problem in the field of person–related Information Retrieval (IR) considering research contexts. The method, which is introduced here, determines science–structural relationships between all scientists in sample domains of a database including their semantic proximity. As an expressive description of the "world" studied at an instantaneous point of time, such networks of personal relations can be used to draw inferences concerning the importance of scientists for a certain topic in regard to their structural position within a complete "research landscape".

1. Introduction

Scientific literature, project and lesson databases basicly offer an extensive foundation for a description and analysis of "research landscapes". These facts relate, for example, to the main topics of scientists, their publication and research behavior, theme distribution, scientific relationships between persons (or institutions) as well as the change of these characteristics in the course of time and within spatial conditions. The key problem in the use of bibliographic databases as regards these analysis perspectives is the low level of exploitation of the database contents because of the limited functionality of the relational algebra and, in particular, the document orientation of the classical IR process (Ohly (1991)). Since conventional IR–systems only permit the selection of precisely that information which is explicitly represented in the internal database scheme, they cannot deal with knowledge about thematic and scientific contexts in which documents, authors or researchers are embedded in.

The major aim of the AKCESS[1] project at the Social Science Information Centre is to process science–structure knowledge for identifying research networks from literature, project and academic lesson data, in order to derive the structural positions of scientists within these contexts and, finally, their relevance to a specific user–defined topic. From a formal point of view, this approach involves forming and processing virtual relations by extending the poor inference capabilities of Boolean algebra by the deductive power of the predicate logic (Appelrath (1985)).

The so far implemented prototype of AKCESS supports a person–related retrieval by processing scientific context knowledge. The method represents a pattern matching which evaluates the semantic proximity between content interests of the user and thematic concepts of scientists in the database. Furthermore, basic properties of scientists in relation to their individual publications, projects or university lessons, e.g. 'author of a book', 'editor of a proceedings volume', 'head of a thesis project' and so on, as well as structural relationships between persons via these characteristics are taken into account with a view to identify "expert neighbourhoods". The goal of this approach is to raise the relevance of a person under review by the relevance of the

[1] Assistance by Knowledge-based Context Evaluation in Social Science Information Retrieval

research context in which one of his documents appears. For instance, the relevance of an editor of a proceedings volume increases with the weight of its individual articles. These modules are described more detailed in (Herfurth et al. (1993)).

The need to reason about network-like relationships has proved out to be a recurring problem in the field of evaluating the relevance of scientists for a certain domain. The above mentioned approach interprets the local environment of a document in respect to a specific query, i.e. its relationships to other directly connected documents. However, what about the environment of the environment? To deal with this problem, a method has been formulated which recognizes the complete "research landscape" of a domain, irrespective of a special query, in order to consider questions such as "Who is the most central scientist in sociology of family?" or "What about scientific schools in system theory?". The method, which is introduced here, identifies scientific relationships between all persons in the database and thus also permits the identification of transitive and hierarchical relations between these scientists including their semantic proximity.

2. Processing science–structural networks

2.1 Scientific relationships

The problem–solving task is to derive structural relationships between all scientists in a bibliographic database. The reasoning method building up the personal networks refers to a set of inference rules representing the formal relationships between two scientists which should be considered during the problem–solving process. In this respect, seven elementary *connected* relations, usually existing between two persons p_i and p_j in a scientific community, are perceived as most relevant to our work:

connected(Pi, Pj, above):- editor_of(Pi, Pj).
connected(Pi, Pj, equal):- co_author_of(Pi, Pj).
connected(Pi, Pj, equal):- co_editor_of(Pi, Pj).
connected(Pi, Pj, above):- projecthead_of(Pi, Pj).
connected(Pi, Pj, equal):- co_projecthead_of(Pi, Pj).
connected(Pi, Pj, equal):- co_projectmember_of(Pi, Pj).
connected(Pi, Pj, equal):- co_teacher_of(Pi, Pj).

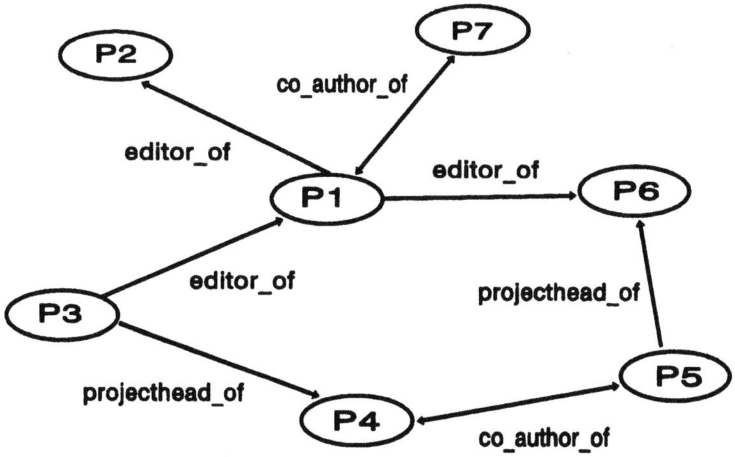

Fig. 1: Scientific Relationships Graph

These relations can be used to express other relationships that can be found between two persons, e.g. the relationship between persons who are authors in the same proceedings volume can be derived by using the *editor_of* rule representing an editor–author–relationship. In respect to the aim of AKCESS (evaluating relevant persons) the hierarchical relationships between scientists are on special interest, such as p_i is *above* p_j if p_i is, for instance, projecthead of p_j.

The relationships between persons are maintained in a network where the nodes represent individual scientists. Each arc in the graph represents two characteristics of how two scientists are related to one another: the label $arc(p_i, p_j)$ of a link, for instance *editor_of*, describes the structural relation between p_i and p_j, its direction indicates the hierarchical relationship between two persons (see Fig. 1).

To allow uncertainty, the weight of each relationship, i.e. an edge in the graph, is calculated in two ways:

(1) Firstly by means of a statistical Equivalence Index $E(p_i, p_j)$ (Grivel and Lamirel (1993)) for the measurement of co–occurrences of two persons p_i and p_j in the complete data set, given by the following equation:

$$E(p_i, p_j) = \frac{CO(p_i, p_j)^2}{F(p_i) \cdot F(p_j)}, \tag{1}$$

where $CO(p_i, p_j)$ is the number of co–occurrences of p_i and p_j; $F(p_i)$ is the frequency of p_i, and $F(p_j)$ is the total number of p_j in the data set. This allows a production of binary equivalence relations between scientists in the data corpus, e.g. CO(P1,P2) = 1.0 or CO(P1,P3) = 0.2, saying that P1 always occurs with P2, but not very frequently with P3. Thus, it can be concluded that P1 has a very strong (statistical) relationship to P2, but a relatively weak one to P3.

(2) Secondly an arc is valued by specifying a user–defined weight $W(arc(p_i, p_j))$ for each *connected* relationship, for example:

weight(editor_of, 0.1).
weight(co_editor_of, 0.4).
weight(co_projectmember_of, 1.0).
weight(co_projecthead_of, 1.0).
weight(projecthead_of, 0.6).
weight(co_author_of, 1.0).
weight(co_teacher_of, 0.8).

In order to express the semantic strength $S(p_i, p_j)$ of each binary relationship in the graph, the average of $E(p_i, p_j)$ and $W(arc(p_i, p_j))$ is used.

2.2 Maintaining knowledge about scientific relationships

The main reasoning done by the system is to derive transitive relations between all persons studied including their hierarchical and semantic relationships. The algorithm furthermore determines whether a new relationship between a pair of scientists is consistent with the network of relations which already exists. The problem–solving process works in a similar fashion to Allen's temporal reasoning (Allen (1984), Richter (1992)), but is based on deductive inference. In order to obtain this functionality, a set of transitivity rules defining a hierarchical relationship between two scientists p_i and p_j over a third person p_k is referenced (see Fig. 2), where $r1$ is the hierarchical relation between p_i and p_k, and $r2$ the relation between p_k and p_j.

$p_k\ r2\ p_j$ $p_i\ r1\ p_k$	$<$	$>$	$=$
$<$	$<$	$=$[2]	$<$
$>$	$=$[3]	$>$	$>$
$=$	$<$	$>$	$=$

Fig. 2: The Transitivity Table for Hierarchical Relationships

When a personal relation is retrieved from the database by firing one of the *connected* rules listed above, all consequences through the network are computed, i.e. the relationships between all scientists in the network are calculated new on account of the new fact added to the system. In doing so it is provided that a network always maintains complete and correct information about how its nodes are related to one another, i.e. the relationship, in particular the path, between each pair of scientists is definite. Derived relationships are obtained by examining the transitive closure of the personal relations defining the network and by applying the transitivity rules determining hierarchical relationships. These new facts are then asserted to the network in the same way, which may in turn produce further relationships between other scientists.

Consider a simple example of this method on the base of the binary relationships from Fig. 1. In the following $\rightarrow$ symbolizes an *above*– and $\leftrightarrow$ an *equal*-relationship (the arcs are labeled with the path between two persons):

Assume we have the facts

$$P1 \longrightarrow P2$$
$$P3 \longrightarrow P1$$

the system inferes

$$P3 \xrightarrow{[P1]} P2.$$

If we enter

$$P3 \longrightarrow P4$$

the algorithm computes

$$P4 \xrightarrow{[P3,P1]} P2$$
$$P4 \xleftrightarrow{[P3]} P1.$$

Propagating with all binary relations from Fig. 1 gives the following furtheron derived relationships defining the network:

$$P3 \xrightarrow{[P1]} P7$$
$$P7 \xrightarrow{[P1]} P2$$
$$P4 \xleftrightarrow{[P3,P1]} P7$$
$$P3 \xrightarrow{[P1]} P6$$
$$P4 \xrightarrow{[P5]} P6$$

[2]Exception: p_k is directly connected to p_i (resp. p_j), p_k is *above* a person q, and q is *above* p_j (resp. p_i)

[3]Exception: p_i (resp. p_j) is directly connected to p_k, p_j (resp. p_i) is *above* a person q, and q is *above* p_k

$$\text{P7} \xrightarrow{[\text{P1}]} \text{P6}$$
$$\text{P6} \xleftrightarrow{[\text{P1}]} \text{P2}$$
$$\text{P5} \xrightarrow{[\text{P6,P1}]} \text{P2}$$
$$\text{P3} \xrightarrow{[\text{P1,P6}]} \text{P5}$$
$$\text{P7} \xleftrightarrow{[\text{P1,P6}]} \text{P5}$$
$$\text{P1} \xleftrightarrow{[\text{P6}]} \text{P5}.$$

While building up the networks of personal scientific relationships, the shortest transitive relationships between persons are identified and stored as derived facts to reinforce the database. Furthermore the semantic proximity $SP(p,q)$ of each relationship in a network is calculated on account of the shortness of the distance between the two scientists p and q in the graph at the one hand side and the statistical and semantic strength $S(p_i,p_j)$ of each binary relation on the path at the other hand side. This is done by multiplicating all $S(p_i,p_j)$ on the path between p and q.

A basic property of the reasoning method is that it rejects relationships which do not consist with existing relations in a network. Anyway, the paths between nodes and the semantic proximity $SP(p_i,p_j)$ of each relationship are updated while propagating the network. But it should be noted that this algorithm only guarantees consistency between three binary nodes, i.e. local consistency. In principle, a new or derived relationship is rejected from the network if its semantic proximity index is weaker than that of the relation to which it does not consist. If a relationship entered to the system causes a local inconsistency, then the weakest relation will be rejected from the network. If this fails, i.e. if the contradictionary relationships have the same strength, a conflict exists, and a decision rule is referenced governing a conflict solution. In the actual state of the system all participating individuals are set to *equal*, e.g. if $p_i \rightarrow p_j$ and $p_j \rightarrow p_k$ and $p_k \rightarrow p_i$ then $p_i \leftrightarrow p_j \leftrightarrow p_k$. In both cases the network will be propagated new starting with consistent relationships. In order to ensure global consistency, a pair of scientists is set to *equal* if a contradiction occurs.

2.3 Analysing scientific networks

A network of scientific relationships can be used to analyse the structure of a certain "research landscape". For instance, the relative position of scientists to other persons in the same domain can be retrieved. Before considering some global analysing methods currently used in AKCESS, it should be noted that networks containing less than three nodes are ignored.

If a network has been built up completely, its hierarchical relationships can be used to define a tree-like structure of personal relations. Thus, the "highest" position in a network can be derived, i.e. the person which is on the top of the structure. Concerning the graph presented in Fig. 1, for example, P3 is "on the top" from a logical point of view since there is no person in the network *above* P3. In order to obtain a measure to express the hierarchical level $H(p)$ of a scientist p, the following equation is used:

$$H(p) = \frac{m(p)}{n}. \tag{2}$$

Here, n is the number of scientists in the network, and $m(p)$ is the number of all relationships where p is *above* or *equal* to a person. Thus, considering our sample network from Fig. 1, P3 is "on the top" since $H(\text{P3}) = 1.0$.

Concerning the main goal of AKCESS (evaluating relevant persons) the most central position in a network of scientific relationships is on special interest. In order to compute a centrality index for a scientist in a given network, the following equations are used, where $\{p_1, ..., p_n\}$ is the set of all persons in the network different from p and n is the number of scientists in the network, i.e. nodes in the graph, different from p:

(1) centrality-by-distance $C_d(p)$ which expresses the relation between the size of the network of p and the distance of p to each other scientist (Scott (1991)):

$$C_d(p) = \frac{\sum_{i=1}^{n} D(p, p_i)}{n}, \; p \neq p_i, \tag{3}$$

where $D(p, p_i)$ is the number of links between p and p_i;

(2) centrality-by-semantic-proximity $C_s(p)$ which is the average of all semantic nearness indices associated to p:

$$C_s(p) = \frac{1}{n} \cdot \sum_{i=1}^{n} SP(p, p_i), \; p \neq p_i, \tag{4}$$

where $SP(p, p_i)$ is the semantic proximity index of the relationship between two scientists p and p_i in a network;

(3) total-centrality $C(p)$ given by the average of $C_d(p)$ and $C_s(p)$.

Applying (1), for instance, on the basis of the sample network from Fig. 1 the system concludes that P1 is the most central person in the network since $C_d(\text{P1}) = 1.33$ is the strongest centrality-by-distance index we could obtain.

In respect to the main task of AKCESS (evaluating experts), scientists with the most strongest and, at the same time, most superordinated relationships to all other members of a network are considered. This is obtained by applying a relevance index $R(p)$ which results from the average of $C(p)$ and $H(p)$.

3. Conclusion

The system described represents a method for deductive reasoning about structural relationships between scientists in sample domains of bibliographic databases, in order to propagate a network of scientific structures. Processing of network-like relationships enables inferences to be drawn concerning the importance of scientists in consideration of their structural position within a certain research field. As an expressive description of the "world" studied at an instantaneous point of time, such networks of personal relations can also be used to carry out a formal analysis of complete "research landscapes" which is basicly non-topic-related, i.e. irrespective of a specific question of a user. This could be done, for instance, with regard to a typology of scientists by considering the structural relationships between persons in a database.

Currently, a representation which deals with different types of *above–* and *equal–* relationships and a differentiated belief revision model for treating conflict decisions are under development. Since the representation captures formal connections between scientists and implicit hierarchies of science-structural relationships, the system will be extended by a representation which is able to maintain thematic relationships between scientists. Furthermore, a reasoner based on constraint propagation should be examined, in order to deal with a "Possible Worlds" semantics. The system presented has been implemented in Prolog running under SCO–Unix; a Windows version written in C++ is planned.

References:

ALLEN, J.F. (1984): Towards a general theory of action and time. *Artificial Intelligence, 23, 123-154.*

APPELRATH, H.-J. (1985): *Von Datenbanken zu Expertensystemen.* Springer-Verlag, Berlin, Heidelberg.

GRIVEL, L., and LAMIREL, J.-C. (1993): SDOC. An analysis tool for scientometric studies integrated in an hypermedia environment. *Communicating Knowledge in Organizations. Proceedings of the 4th International Conference on Cognitive and Computer Science for Organizations.* Montreal, 146–154.

HERFURTH, M., and MUTSCHKE, P., and OHLY, H.P. (1993): AKCESS: Konzeptorientiertes Retrieval mit bibliographischem Kontextwissen. In: G. Knorz et al. (eds.): *Information Retrieval '93. Von der Modellierung zur Anwendung. Proceedings der 1. Tagung der Fachgruppe Information Retrieval der Gesellschaft für Informatik.* Universitätsverlag, Konstanz, 199–207.

OHLY, H.P. (1991): Conceptual information retrieval by knowledge-based programming techniques. *International Classification, 18-3, 148-152.*

RICHTER, M.M. (1992): *Prinzipien der Künstlichen Intelligenz.* B.G. Teubner, Stuttgart.

SCOTT, John (1991): *Social Network Analysis. A Handbook.* Sage Publications, London.

Object-Oriented Systems Analysis Applied to the Method Component of a Knowledge-Based System for Data Analysis

Michael Rundshagen, Stefan Marx

Lehrstuhl für Wirtschaftsinformatik III, Universität Mannheim,
Schloß, 68131 Mannheim

Summary: The application of object-oriented modeling concepts to the early stages of the software life cycle promises a smooth transition from analysis to design, and from design to implementation. In this article we present the development of an analysis model for the method component of WIMDAS using Coad and Yourdon's well documented object-oriented analysis approach. Emphasis is laid on the order of necessary activities during systems analysis.

1. Introduction

The emergence of object-oriented programming languages in the last years led to a demand of new (object-oriented) approaches in the early stages of the software engineering process. Traditional top-down functional design approaches require a mapping between inhomogeneous modeling concepts in transition from system design to implementation with an object-oriented programming language (see e.g. Henderson-Sellers and Edwards (1990)).

Methods proposed by Booch (1991), Coad and Yourdon (1991), or Shlaer and Mellor (1988, 1992) are meeting the requirements of an object-oriented analysis method and allow the use of homogeneous modeling concepts. Thus, the results of each stage in the software development process are directly mapped into the following stages (see, e.g., Schader and Rundshagen (1993)).

Difficulties in applying object-oriented analysis methods occur from the vague descriptions of necessary activities. The vagueness of these descriptions is caused by the lack of experience in applying object-oriented approaches in systems analysis.

Coad and Yourdon's approach of object-oriented analysis (OOA) contains detailed descriptions of major activities which are necessary to apply their method. The order of analysis activities depends on the experience of the system analyst.

We chose this method for modeling a part of the development of the knowledge-based system WIMDAS[1] because it is well-known. This article emphasizes the succession and repetition of activities during the development of an analysis model.

2. Object-Oriented Analysis (OOA)

In this section the Coad and Yourdon (1991) method for object-oriented analysis is introduced.

Their OOA model is structured in five layers. Within the Class layer classes and objects are modeled. Potential classes are structured by generalization-specialization

[1] wissensbasiertes Marketingdaten Analysesystem

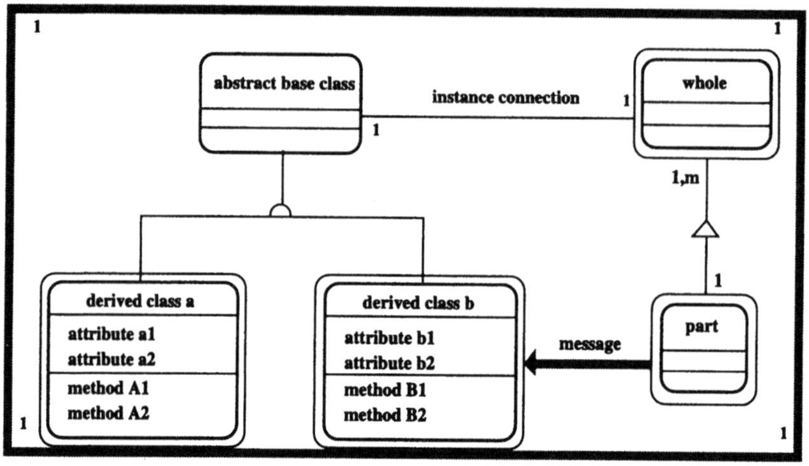

Fig. 1: OOA notation

and whole-part structures within the Structure layer. The properties which characterize each object are defined by attributes and associations between objects (instance connections). The definition of attributes and associations is done within the Attribute layer. Within the Method layer the operations which each class provides are defined. How the objects communicate with messages is described within this layer using message connections. The Method layer also contains detailed specifications of each class, attribute and instance connection. For more complex analysis models a Subject layer is introduced to divide the model into smaller groups of classes which are closely related to each other (e.g. by structures).

The five-layer model may be interpreted like overlapping transparencies which gradually present more and more detail. Coad and Yourdon propose the following order of the activities in OOA: 1. Finding Classes, 2. Identifying Structures, 3. Defining Subjects, 4. Defining Attributes, 5. Defining Services.

Figure 1 shows the notation of the modeling constructs used in OOA. Each class of the analysis model has to be specified with a detailed class description. Attributes are specified by their unit of measure, range, precision, default value etc.

Further, the class description contains information about external inputs or outputs. An object state diagram shows the potential states and corresponding transitions of each object within a class. Finally, services are specified by structured english or by graphical notation similar to a flow chart.

3. The Problem Domain

The WIMDAS system, jointly developed by groups at the Universities of Karlsruhe, Mannheim, and Federal Armed Forces Hamburg, has evolved into a distributed system. It consists of five main components: The user interface, the knowledge-based component, the data management component, the method component, and the data display system which are running on various machines in a network. Design and implementation issues concerning the first four components have already been discussed in a variety of publications (see for example Baier and Marx (1991), Hantelmann (1992), Kiel and Schader (1991)). The method component is currently simulated by shell scripts generated by the user interface. It is the topic of the following object-oriented analysis.

Starting point of any kind of analysis is to investigate the task the software has to perform. This exploration of the problem domain gives insight into the problem and the system's responsibilities.

The method component is responsible of the evaluation of complex *proposals* for data analysis the user wishes to be computed. He or she selects from a bulk of conceptual proposals, generated by the knowledge-based component and based upon the description of available input matrices and the answers to the system's questions. A proposal arranges matrices and methods in a bipartite directed acyclic graph where the available (input) matrices are the source nodes, the inner nodes are methods and intermediate results and the sinks are the resulting matrices (for a comprehensive discussion see Marx and Rundshagen (1992)).

The evaluation of a proposal is performed by the machines of the network. The analysis component is responsible for the scheduling, distribution, and monitoring of the methods of the proposals to be evaluated. In this task, we have to consider the following points:

1. The machines are of different quality and have variable load.
2. Methods may be bound to particular machines. Programs from statistical software packages should be usable.
3. Machines may become available or may become inaccessible. (Variability of the network.)
4. The quality of any computation may be assessed according to certain criteria.
5. The knowledge-based component usually generates many proposals which are fairly similar, i.e. the proposals contain common subgraphs and therefore common intermediate results.
6. During the work of the analysis component new proposals may arrive from other consultations of the system.
7. Information about the state of the evaluation of a proposal and a specific computation should be provided to the user.
8. The evaluation should be "as fast as possible".

In spite of the fact that the last goal needs a more thorough specification we can draw the following conclusions.

- Every proposal must be treated with consideration of the whole set of proposals to be evaluated. This set must be assessed whenever new proposals arrive.
- Methods must be scheduled dynamically, i.e., whenever a method can be started and not statically, for instance at the time the proposal arrives.
- The analysis component must be able to get necessary information at every point in time, e.g., considering the properties of machines and methods, the load of machines etc.

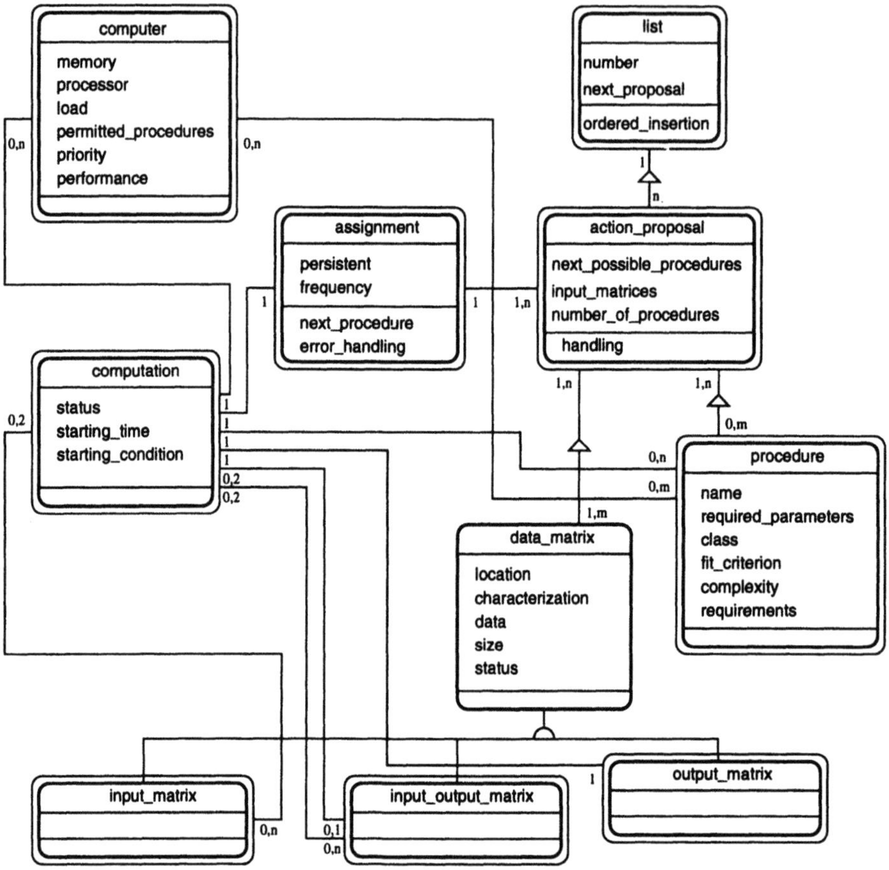

Fig. 2: First analysis model of WIMDAS analysis component.

4. Modeling the System

The system requirements outlined in the last section lead to a first possible analysis model depicted in fig. 2, which can be interpreted as follows. Each proposal contains procedures and data matrices which may be input, output, or input-output matrices, respectively. An assignment initiates a computation of a procedure with the appropriate matrices on a specific computer. The set of proposals to be evaluated is ordered in a list.

Classes were chosen according to criteria mentioned by Coad and Yourdon. The class *computer*, for example, corresponds to a device in the problem domain (the physically existing computers of the network) with which the analysis component has to interact. Therefore, it is a prime candidate for a class in the analysis model.

The classes are naturally structured. *data_matrix* is the abstract base class (there is no data_matrix as such) for the three specializations *input_matrix, output_matrix* and *input_output_matrix*. A *proposal* consists of data matrices and procedures. The proposals are collected in a *list*.

For reasons of space we only discuss the attribute layer concerning the class *proposal*. Each *proposal* relates to as many *assignments* as it contains *procedures*, since every *procedure* results in one *computation* on one specific machine. Because every *proposal* contains at least one *procedure* the cardinality of this relationship is $(1, n)$ on the side of the *proposal*. On the contrary, each *assignment* is bound to exactly one *proposal* (and to one *procedure* of one *proposal*). Therefore, the appropriate cardinality is 1 for the *assignment* side. Every *proposal* records which of its *procedures* are ready to be computed next, which are its *input matrices* (stored in WIMDAS's database) and how many different *procedures* it contains.

When developing the method layer the algorithmically simple services are not included in the model. Such services are for instance the generation of objects, setting and reading of attributes, and so on. Thus, we find only a few services in the model. Class *list* handles the ordered insertion of arriving *proposals*, class *assignment* determines the procedure to be computed and has to cope with any errors which might arise during processing of any computation.

We omit the message connections between objects in the first analysis model, since at the beginning of the analysis the identification of data structures within the problem domain is of major concern. At this stage of the analysis, the classes found so far are further explained in accompanying object specifications, which may for example include object state diagrams, dependencies between services or instance connections, and ranges of attributes.

Coad and Yourdon strictly distinguish between the two phases analysis and design (despite the fact, that the smooth transition from analysis to design is often quoted as one advantage of the object-oriented paradigm (Rundshagen (1993)). Therefore every class (structure, attribute, service ...) of the first analysis model has to be checked whether it is really necessary in order to precisely describe the problem domain or if it merely reflects a premature decision regarding aspects of design or implementation. As an aid for reviewing an analysis model Coad and Yourdon provide a catalog of criteria possible classes should fulfill. Classes should

- store information, which is needed in the system.
- provide operations, which are needed in the system.
- have more than one attribute.
- have more than one object.
- contain attributes, which are valid for every object.
- contain services, which are applicable to every object.
- only be based upon problem domain requirements (as opposed to reasons of technologies for example).
- not contain derivable information.

The revision of the first analysis model with respect to this catalog reveals several weak points and finally yields the model depicted in fig. 3. The class *list* for instance handles how proposals are organized in the system and should at earliest be introduced in the design phase, if ever. In fig. 3 the *list* is dropped, instead we are using an instance connection between proposals to model that every proposal is connected to others. The specializations of *data_matrix* in the first model are excessive and can be better modeled as one distinguishing attribute *type_of_matrix*. Further attributes are removed, because they are derivable (*number_of_procedures* of class *proposal*) or because they resemble referential attributes (*permitted_procedures* or *input_matrices*).

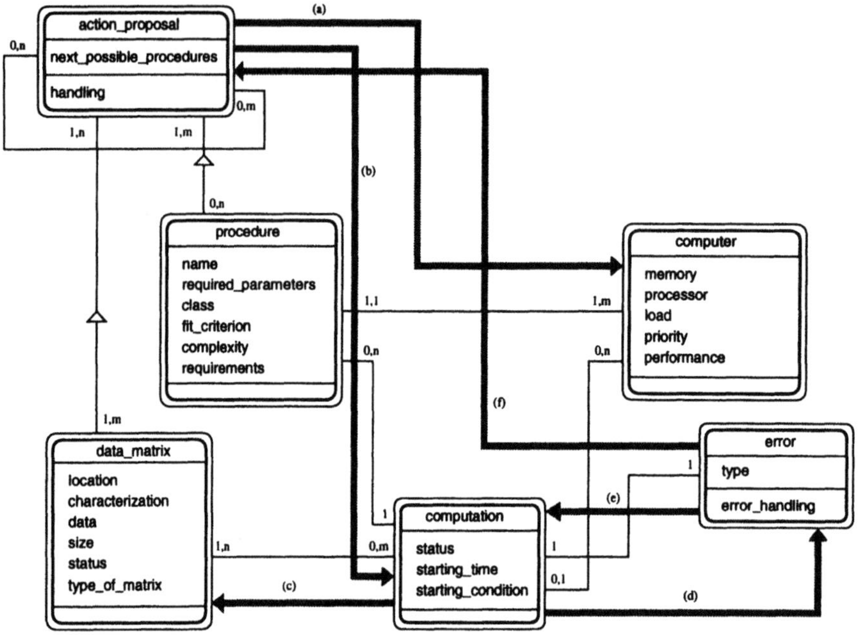

Fig. 3: Final analysis model.

The class *assignment* contains information which is better moved to class *proposal*. Attributes *persistence* and *frequency* fit nicely into the service *handling* of *proposal*. The same holds for service *next_procedure*. Therefore, *assignment* finally changes to *error*, which models only errors and their treatment.

The final model in fig. 3 now shows message connections which are labeled for the sake of the following description. Whenever an object of class *proposal* is created (external input) service *handling* first determines all computable procedures then questions all *computers* to get their current state (message (a)). Based upon this information a *computation* is created (message (b)), which creates a *data_matrix*, if it performs without error (message (c)) but creates an *error* otherwise (message (d)). An *error* object tries to resolve the error condition to continue the computation (message (e)) or has to notify the *proposal* about the failure (message (f)).

5. Conclusion

Passing the development of the final analysis model in review, we see that a few of the requirements mentioned in section 3 have not had any particular influence on the model. Some points, for example the information of the user, become only important in the design and implementation phase of the system developed. The analysis model has only to provide the necessary prerequisites to fulfill the requirements. In our example it is sufficient to incorporate state information for every computation.

The requirements cited in section 3 have different realizations in the analysis model. Requirement 2) for example is modeled as an instance connection between class *procedure* and *computer* but point 3) is expressed by the message of service *handling* to the objects of class *computer*.

It is the task of the system analyst to consider and incorporate the requirements in his model of the system. During this process he or she heavily depends on his or her experience and competence. Our example shows that the analyst, even when using object-oriented methods, cannot achieve a complete model in one single step. On the contrary, a stepwise approach with repeated reconsidering of the model is recommended to avoid typical errors like, for example, the inflation of the model by unnecessary specialization.

Our experience shows, that the analyst is normally tempted to anticipate decisions which should be made in the design or implementation phase (referential attributes, derivable attributes, ...). At least in object-oriented analysis according to Coad and Yourdon this mixing of phases should be avoided. The first analysis model should therefore be checked and expurgated, if necessary, of any such influences.

In conclusion, the method of Coad and Yourdon for object-oriented analysis is a useful aid for object-oriented system development. Objects and structures of the problem domain are modeled by means of an easily understandable notation and many hints are given for the development of an analysis model. But to assess the method more thoroughly the transformation and realization of an analysis model in the later phases design and implementation needs further investigation.

References:

BAIER, D., and MARX, S. (1991): Data management in a knowledge-based system for marketing research. In: Schader (ed.): *Analyzing and Modeling Data and Knowledge.* Springer-Verlag, Berlin Heidelberg New York, 189–197.

BOOCH, G. (1991): *Object Oriented Design with Applications.* Benjamin-Cummings, Redwood City, California.

COAD, P., and YOURDON, E. (1991): *Object-Oriented Analysis.* second edn., Prentice-Hall, Englewood Cliffs, New Jersey.

HANTELMANN, F. (1992): Application-interfaces for ai-environments in example of the wimdas-project. In: W. Gaul et.al. (eds.): *Operations Research Proceedings 1991.* Springer Verlag, 660–667.

HENDERSON-SELLERS, B., and EDWARDS, J. M. (1990): The Object-Oriented Systems Life Cycle. *Communications of the ACM, 33, 142–159.*

KIEL, R., and SCHADER, M. (1991): Knowledge representation in a system for marketing research. In: M. Schader (ed.): *Analyzing and Modeling Data and Knowledge.* Springer-Verlag, Berlin Heidelberg New York, 179–187.

MARX, S., and RUNDSHAGEN, M. (1992): *Systemanalyse nach Coad und Yourdon am Beispiel der Methodenkomponente von* WIMDAS. Discussion Paper no. 2-92, Lehrstuhl für Wirtschaftsinformatik III, Universität Mannheim.

RUNDSHAGEN, M. (1993): *Phasenmodelle zur Beschreibung von objektorientierten Softwarelebenszyklen.* Discussion Paper no. 1-93, Lehrstuhl für Wirtschaftsinformatik III, Universität Mannheim.

SCHADER, M., and RUNDSHAGEN, M. (1993): Ein Vergleich objektorientierter Analysemethoden. In: Niedereichholz, J., and Schuhmann, W. (eds.): *Wirtschaftsinformatik – Beiträge zur modernen Unternehmensführung.* Campus, Frankfurt a. M., to appear.

SHLAER, S., and MELLOR, S. M. (1988): *Object-Oriented Systems Analysis, Modeling the World in Data.* Yourdon Press, Englewood Cliffs, New Jersey.

SHLAER, S., and MELLOR, S. M. (1992): *Object Lifecycles, Modeling the World in States.* Yourdon Press, Englewood Cliffs, New Jersey.

Classification Based Query Evaluation in Knowledge Base Systems

Martin Staudt, Hans W. Nissen

RWTH Aachen, Informatik V, Ahornstr. 55, D-52056 Aachen, Germany

Summary[1]: Classification as structuring principle for data is a common basis of many knowledge base systems. A newer idea is to employ classification for query evaluation purposes as recently done in concept language implementations and in the advanced deductive database system *ConceptBase*. The question is: Can we amalgamate the advantages of the techniques used in these approaches? This paper first proposes the notion of query classes with a deductive interpretation and then describes an orthogonal integration of a concept language for types of classes. Classification on the type lattice leads to an optimized query evaluation process.

1. Introduction

The rather spongy term 'knowledge base systems' in our understanding covers database systems that go beyond the standard relational, hierarchical and network databases concerning structuring capabilities of data as well as the representation of inference knowledge. In the meantime several knowledge representation systems have grown out of the AI area which are intended for storing and managing larger amounts of data and therefore also fall under this category.

In the database field relational databases were extended to deductive databases which offer rules (mostly horn clauses with limited negation) for specifying intensional relations. Object-oriented databases took over principles from object-oriented programming such as objects, classes and methods. Since general optimization algorithms for object-oriented systems must fail due to the high expressive power of the language usually allowed for programming methods, there have been several approaches to reconcile the deductive and object-oriented paradigms within integrated systems. These systems have a clear theoretical semantics which enables the development of suited implementation and optimization techniques. In addition they offer the rich object-oriented structuring facilities for representing data. Most of these so called *deductive object bases* are based on extensions of the basic deductive, relational model, e.g., with sets, lists, or nested relations.

The deductive object base system *ConceptBase*[2] [Jarke (1992)] with its data model O-Telos pursues a more conservative approach which leads to a semantics that is exactly a perfect model semantics as in deductive databases with stratified negation, together with the three object-oriented abstraction principles *classification/instantiation*, *generalization/specialization* and *aggregation/decomposition*. The system allows the specification of methods in the form of deductive rules and integrity constraints. Queries in ConceptBase are represented as special classes, the answers to a query are those objects in an object base that can be classified in the respective *query class* since they fulfill its (necessary and sufficient) membership constraints. Query evaluation is done using the conventional deductive database algorithms.

---

[1]This work was supported by the Commission of the European Communities under ESPRIT BRA 6810 (Compulog 2)

[2]A scaled-down version is available via anonymous ftp from ftp.informatik.rwth-aachen.de directory pub/CB.

Whereas the principle of classification is common as structuring principle for knowledge in object-oriented databases as well as in many knowledge representation (KR) systems, query evaluation by classification is a newer approach. In the KR area this idea is especially found in *concept language* based systems which have their origin in the system KL-ONE [Brachman and Schmolze (1985)].

Concept languages allow the definition of so called concepts which in fact correspond to special structured first order logical formulas over unary (concept membership) and binary relations (role membership). Concept definitions in systems as, e.g., CLASSIC [Brachman et al. (1991)] or CANDIDE [Beck et al. (1989)], serve as schema definitions for objects that are asserted as instances of concepts and fillers of roles for other objects. In the last years there was made significant progress in examining the complexity of subsumption between concepts [Donini et al.(1991)], i.e., in the determination wether the set of instances necessarily (forced by concept definitions) include each other. Based on this relationship concepts can be organized (classified) in a hierarchy. Queries in concept languages are just concepts which are defined as any other schema constituting concept. As a consequence queries can be automatically positioned within the concept hierarchy. All instances of subsumed concepts are answer instances of the query concept [Lenzerini and Schaerf (1991)].

A very important advantage of concept languages is the availability of subsumption deciding algorithms with known complexity, whereas for the case of Datalog$^\neg$ the subsumption of rule sets resulting from deduction rules and especially query classes in systems as ConceptBase, is not decidable in general.

In the following section we introduce the basic properties of the O-Telos data model implemented in ConceptBase together with its embedded query language CBQL. Then we propose in section 3 an integration of a subsumption checking component into the deductive query evaluator of ConceptBase for optimization purposes, especially the reuse of materialized views. The subsumption checker works on a type system orthogonally defined on the object base. In section 4 we give a short summary and conclude with some remarks about related and future work.

2. Queries as Classes

The O-Telos data model [Jeusfeld (1992)] provides a deductive formalization of an object-oriented data model and is a variant of the language Telos [Mylopoulos et al. (1990)]. We introduce the basic properties of O-Telos and the query language CBQL which represents queries as classes. The syntax of query classes is a class definition with superclasses, attributes, and a membership condition. The interpretation is based on a deductive rule containing predicates for each of the clauses in the class definition. Answers are those objects of the knowledge base that can be classified as instances of the query class.

2.1 Objects in O-Telos

Deductive object bases can be defined as deductive databases with an always present set of axioms formalizing the three object-oriented abstraction dimensions of classification, aggregation and generalization.

We define a deductive object base as a triple $DOB = (OB, R, IC)$ where OB is the extensional object base, R and IC contain deduction rules and integrity constraints. The set of axioms is represented as predefined formulas in R and IC. In addition we require that the perfect model of $OB \cup R$ entails IC.

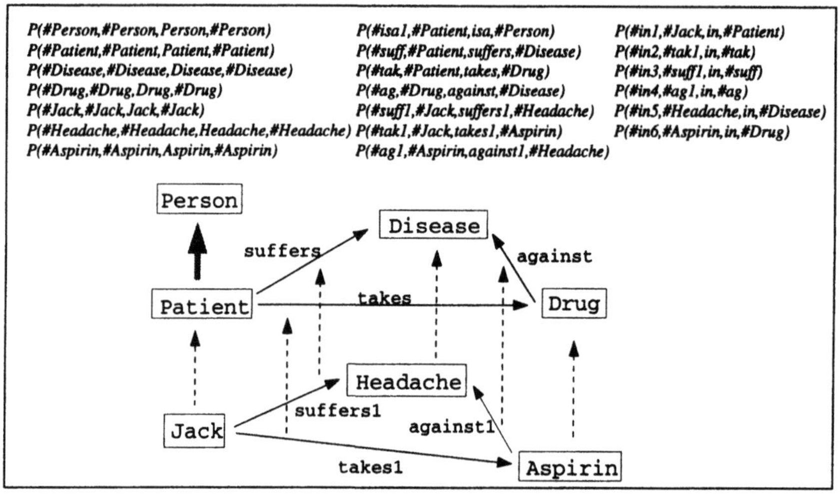

Figure 1: Tuple and semantic net view of deductive object bases

Let ID and LAB be sets of identifiers, and labels respectively. An **extensional O-Telos object base** is a finite subset

$$OB \subseteq \{P(o,x,l,y) \mid o,x,y \in ID, l \in LAB\}.$$

The elements of OB are called **objects** with identifier o, source and destination components x and y and name or label l.

A different view on such an extensional object base is a graphical semantic net style view. Objects of the form $P(o,o,l,o)$ (called *individuals*) are represented as nodes with label l. Instantiations of the form $P(o,x,in,c)$ (= "x is in class c") and specializations of the form $P(o,c,isa,d)$ (= "c is subclass of d") are represented as dotted resp. shaded directed links. All other objects $P(o,x,l,y)$ are called attributes and are drawn as l labeled directed links between x and y. The examples in this paper can be thought of as an excerpt from a medical database and are based on the following simple scenario:

Patients are persons, suffer from diseases and take drugs. which have an effect on diseases.

Fig. 1 shows the example graph derived from an object base OB representing this scenario. OB contains one concrete patient `Jack` who is instance of the object `Patient` and takes the drug `Aspirin` and suffers from `Headache`. In the following the object identifier of an individual with name i is written as `#i`.

A third view on deductive object bases results in a frame-based notation of objects which only relies on object labels. Around the label l of an object o we group the labels of all other objects which have o as source component. The frame notation of the objects `Patient`, `Drug` and `Jack` is, e.g.

```
Patient isA Person with
  attribute
    suffers:Disease;
    takes:Drugend
end
```

```
Drug with
  attribute
    against:Disease
end
```

```
Jack in Patient with
  suffers
    suffers1:Headache
  takes
    takes1:Aspirin
end
```

[Jeusfeld (1992)] specifies the complete set of axioms for a deductive object base in the form of certain facts, rules, and integrity constraints that must be part of any

O-Telos object base. Here, we only refer to a few important ones. From the P-tuple representation three new predicates In, Isa and A can be derived. We say that $In(x,y)$ resp. $Isa(c,d)$ holds in DOB if OB contains the object $P(o,x,in,y)$ resp. $P(o,c,isa,d)$. Similarly, $A(x,m,y)$ is defined as follows:

$$\forall o,x,l,y,p,c,m,d\; P(o,x,l,y) \wedge P(p,c,m,d) \wedge In(o,p) \Rightarrow A(x,m,y)$$

These literals are used when defining the complete axiomatization but also when specifying deductive rules and integrity constraints. For our example above we get for instance

$$In(\#Jack, \#Patient),\; Isa(\#Patient, \#Person) \text{ and}$$
$$A(\#Jack, suffers, \#Headache).$$

Instances of an object in O-Telos are also instances of its superclasses. In the example object base `Jack` is also an instance of `Person`. The instantiation link between the `drug1` attribute of `Jack` and the `takes` attribute of `Patient` satisfies another condition, namely that instantiation relationships between objects imply instantiation relationships between their source respectively destination components. A similar restriction holds for specialization links and allows a consistent way of refining attributes of classes by their subclasses. Another important axiom states the conditions for multiple generalization by ensuring that for all attributes one unique attribute class can be found.

In addition to the O-Telos axioms the sets R and IC contain application specific deduction rules and integrity constraints which are specified in a many-sorted first order language interpreted under closed-world assumption. Variable quantifications range over classes and are interpreted as instantiation relationships. The admissible literals of the language are in principle the predicates In, Isa and A. [Jeusfeld and Jarke (1991), Jeusfeld and Staudt (1993)] describe the language for rules in more detail and discuss applicable implementation techniques as e.g., a compilation into Datalog$^{\neg}$, a language utilized in many deductive database approaches.

2.2 Query Classes

The query language CBQL [Staudt 1990] embedded in O-Telos represents queries as classes whose instances are the answer objects to the query. A specific predefined object `QueryClass` may have as instances so-called query classes which contain necessary and sufficient membership conditions for their instances (answers). These conditions can be used to check whether a given object is an instance of a query class or not. On the other hand, they can be used to compute the set of answer objects.

Query classes have superclasses to which they are connected by isa-links. These superclasses restrict the set of possible answers to their common instances. Therefore, query class definitions are object-preserving. Two different kinds of query class attributes can be distinguished.

Retrieved attributes are already defined for one of the superclasses of the query class. An explicit specification of such an attribute in a query class description means that answer instances are given back with values for this attribute, similar to relational projection. In addition a necessary condition for the instantiation by an admissible value is included. The attributes of superclasses can also be refined, i.e., the target class is substituted by a subclass which results in an additional value restriction. If we assume a subclass `Antibiotics` of `Drug` the query class

```
QueryClass MaleOldAntibioticsPatient isA MalePatient,OldPatient with
  attribute
    takes:Antibiotics
end
```

has those male and old patients as instances who take antibiotics. In addition these drugs are included in the answer description.

Computed attributes have values derived in the query evaluation process. Neither the extensional object base contains this relationship between the answer instance and the attribute value, nor is it inferable by deduction rules. For prescribing how to deduce these new relationships by analogy to deduction rules and integrity constraints many-sorted first order logic expressions are admissible as building elements for query classes.

```
QueryClass WrongDrugPatient isA Patient with
  attribute
    wrong:Drug
  constraint
    wrongconstr:$ (this takes wrong) and
                  not exists d/Disease
                      (this suffers d) and (wrong against d) $
end
```

The predicates used here (e.g., `(this suffers d)`) are variants of the predicates A, In, Isa of the previous subsection (e.g., $A(this, suffers, d)$) where we use labels (external names) of objects here instead of object identifiers. The constant `this` refers to the answer instances of `WrongDrugPatient`, namely all patients who take a drug which is against a disease they don't suffer from. The computed attribute `wrong` is identified with the variable of the same name within the formula. All deduced wrong drugs are part of the answer.

The logical expression in query classes descriptions can also contain arbitrary other constraints for the answer instances which would then work like the selection operation in relational databases.

In order to avoid the frequent reformulation of similar more specialized queries, attributes of query classes can be parameterized using an attribute category *parameter*. Substitution of a concrete value for such an attribute or specialization of its target class by a subclass leads to a subclass of the original query which implies a subset relationship of the answer sets.

In deductive databases queries are represented as intensional relations derived by a set of rules. Starting from O-Telos object bases as just special cases of deductive databases it should be possible to extend this view to query classes. The definition of a query class `Q` induces a so called *query literal* $Q(x, x_1, ..., x_n)$ whose arity depends on the number of attributes and parameters of `Q`. The first argument x of Q stands for the answer object identifiers. Query classes are transformed into rules concluding their corresponding query literals. Our example query class `WrongDrugPatient` can be transformed into the following rule:

$$\begin{aligned} \forall\, i, v \quad & In(i, \#Patient) \wedge In(v, \#Drug) \wedge \\ & \wedge\, A(i, takes, v) \wedge \neg\exists\, d\, In(d, \#Disease) \wedge \\ & \wedge\, A(i, suffers, d) \wedge A(v, against, d) \\ & \Rightarrow WrongDrugPatient(i, v) \end{aligned}$$

The deduction rule form of queries allows the application of the whole array of implementation techniques known from the deductive database area [Ceri et al.(1990)]. Since query classes define views on a given object base we can, e.g., adopt view maintenance algorithms in order to keep stored answers correct with the changing rest of the database, and view update algorithms which allow to realize given updates on intensional defined knowledge (deducible attributes or again query classes) by suited base data changes [Staudt et al.(1993)].

3. Queries and Concepts

In programming languages objects (variables, functions, etc.) usually have types. Type constructors for tuples ([]), sets ({ }), lists etc. allow to handle objects of arbitrary complex structure. In the same way most object-oriented databases distinguish between objects and values on the one hand and classes and types on the other hand. Objects have an unique identity and an assigned value of a certain type. Classes are object containers which collect objects of the same type (sometimes called member type).

O-Telos does not incorporate a clear distinction between classes and types. Nevertheless we can 'extract' type information about objects from class definitions. This divides a class definition into a 'clean' part to be understood as a type definition, and a 'dirty' part in which all those aspects are defined for which the chosen type system is not powerful enough. Our example class

```
Patient isA Person with
  attribute
    suffers:Disease;
    takes:Drug
end
```

carries the information that patients have attributes `takes` and `suffers`, each with certain value restrictions. In addition, patients have attributes (e.g. `name` and `address`) defined for the superclass `Person`. Under consideration that `takes` and `suffers` can have multiple fillers, a (member) type of `Patient` could be the following:

$$type(\texttt{Patient}) = [suffers : \{type(\texttt{Disease})\}, takes : \{type(\texttt{Drug})\}, \\ name : string, address : type(\texttt{Address})]$$

[Borgida (1992)] shows the connection between the plain type languages in object-oriented databases and the family of concept-based description languages which allow additional more powerful restrictions on objects of a certain type. Following the syntax used in the concept language literature for our patient example, we get a member type *PatientT* for `Patient`.

$$\forall suffers.DiseaseT \sqcap \forall takes.DrugT \sqcap PersonT$$

where $\forall r.D$ denotes a type where the values for an attribute r are restricted to instances of the type D and $\sqcap$ denotes an intersection (conjunction) of types. As a shorthand notation for $type(\texttt{A})$ we write AT for any class `A`. The type information included in the frame-like O-Telos view can be thought of as a layer orthogonal to the object base definition presented in section 2. As in concept languages, this type layer is organized in a lattice structure.

The 'type view' of deductive object bases introduced so far only used the $\forall$- and $\sqcap$- constructor for types and represented necessary conditions in the sense that e.g. every patient is also a person and that all attribute fillers for the `takes` attribute are drugs. Both conditions are guaranteed by O-Telos axioms. In a next step we allow other concept language constructors; for example, the entity relationship approach employs atleast ($\geq n\, r$) and atmost ($\leq n\, r$) cardinality restrictions for attributes r. Syntactically, this can be represented by introducing new built-in attribute categories `atleastn` and `atmostn`. Note, that such cardinality restrictions can be expressed by logical formulas too, but not in a very elegant way.

Type restrictions may be interpreted as sufficient membership conditions for their instances which in concept languages leads to the distinction between (defined) *concepts* and *primitive concepts* (the latter with only necessary conditions). The simple class definitions we had until now have primitive concepts as their types. Primitive concepts (as e.g., *PatientT*) are directly subordinated under their corresponding type expression in the lattice which exactly represents the interpretation as necessary conditions. In fact we have for `Patient`:

$$PatientT \sqsubseteq \forall suffers.DiseaseT \sqcap \forall takes.DrugT \sqcap PersonT$$

Defined concepts are related to their defining expression by equality. At the object layer we introduce a class `Concept` that includes all those classes that have a (defined) concept as their type. These classes are called *concept classes* in the following. The class `DrugPatient` contains all those patients that take atleast one drug. Since we want this to be interpreted as a sufficient condition for being instance of `DrugPatient` it is defined as instance of `Concept`.

```
Concept DrugPatient isA Patient with
  atleast1
    takes:Drug
end
```

At the type layer this corresponds to

$$DrugPatientT = PatientT \sqcap \geq 1\, takes$$

The definition is nearly[3] the same as defining `DrugPatient` as a query class:

```
QueryClass DrugPatient isA Patient with
  attribute
    takes:Drug
end
```

In so far we can say that query classes have a concept type. The implicit necessary ($\cong$ `atleast1`) condition of attributes and the interpretation of all conditions in query class descriptions as sufficient results in the same deducible set of instances. With the introduction of additional type constructs and their corresponding object counterparts we can apply the well understood inference algorithms for concept languages at the type level of a deductive object base to infer new information at the object level. E.g., if we define a class `AntibioticsPatient` just as a specialization of `Patient` with a value restriction for `takes` to drugs of the class `Antibiotics` with the additional requirement of at least one filler for `takes`, we can nevertheless place `AntibioticsPatient` automatically with respect to the isa-hierarchy below `DrugPatient` in the object base applying the rule

$$type(\texttt{A}) \sqsubseteq type(\texttt{B}) \Rightarrow (A\, isa\, B)$$

for each instance `B` of `Concept` and each object `A` where $\sqsubseteq$ denotes the subsumption relationship on the type lattice.

In the example an isa-relationship between `AntibioticsPatient` and `DrugPatient` can be concluded from

$$\begin{aligned} type(\texttt{AntibioticsPatient}) &\sqsubseteq PatientT \sqcap \forall takes.AntibioticsT \sqcap \geq 1\, takes \\ &\sqsubseteq PatientT \sqcap \geq 1\, takes = type(\texttt{DrugPatient}) \end{aligned}$$

[3]apart from the projection on the `takes` attribute included in the query class description.

provided that $AntibioticsT \sqsubseteq DrugT$. It follows from these considerations that the isa-hierarchy in the object base reflects the subsumption hierarchy at the type layer by deduced isa-relationships for concept classes. Since an O-Telos object base may have an arbitrary number of layers we reach different type layers for each isa-hierarchy in the object base and hence a multi-level simulation of a concept language based knowledge base system. In the following we will concentrate on the relationship between concepts at the type layer and query classes.

With a more comfortable type language than presented here many parts of a query class definition `Q` can be expressed in a structural format which directly can be mapped to a concept type CT. With these (clean) parts we can perform a classification along isa-hierarchies. For the exact type QT of `Q` which cannot be described at whole within the type language in general (e.g., in the presence of rules) we get the relationship $QT \sqsubseteq CT$. The classification process starts with all directly mentioned supertypes S_iT in CT and then tries to determine wether CT is subsumed by any defined concept DT below S_iT (i.e. $CT \sqsubseteq DT$). From this we can especially conclude that $QT \sqsubseteq DT$ holds at the type layer and subordinate `Q` below `D` in the isa-hierarchy at the object layer. This subsumption test is repeated for all subconcepts of DT etc. until the most specific subsumer(s) T_iT of CT and hence of QT are found. A subsequent deductive query evaluation process would then work on a restricted searchspace (i.e., consider the instances of $\mathtt{T}_i$ instead of $\mathtt{S}_i$). Concepts that are subsumed by CT do not help very much in that they denote sets of answer candidates, since their specific further restrictions can contradict the 'dirty' parts of `Q` not covered by CT.

If `Q` however doesn't contain any 'dirty' parts ($QT = CT$) we can gain a significantly improved evaluation since then we can respect that QT itself eventually subsumes other defined or primitive (i.e., types of regular classes) concepts. If there is any defined concept DT below the S_iT and $DT \sqsubseteq QT$ then every instance of DT is definitely an instance of QT resp. `Q`. The same holds if there is a primitive concept subsumed by QT. In order to get an answer to `Q` one can just take the union of all instances for these subsumed concepts and only check all additional instances of the direct supertypes of QT by the deductive query evaluator for membership in `Q`.

Besides the search space restriction during the query evaluation process an important optimization concerns the reuse of precomputed resp. stored answers for other queries also called *materialized views*. In this case we classify the type of a query against the types of these materialized queries. It should be obvious that we can only detect subsumption relationships between two queries where at least one has a completely defined concept as type. We can distinguish the same possible situations for the incoming query as above. The most common situation is the following: A materialized query has a (defined) concept type that subsumes the type of the query to be evaluated. The set of objects relevant for the answer can then be restricted to the answers of the materialized queries and only some (hopefully inexpensive) selections will be necessary to get the final answer set.

4. Conclusions

In this paper we discussed two different classification techniques for query evaluation in knowledge-base systems and proposed an orthogonal integration into one hybrid query evaluator. The representation of queries as special classes in the deductive object base system ConceptBase is combined with a type system based on a concept language which manages the hierarchy on object and especially query types. Fig. 2 presents the graphical usage environment of ConceptBase with the example query classes of section 2. The deductive query evaluation component of ConceptBase

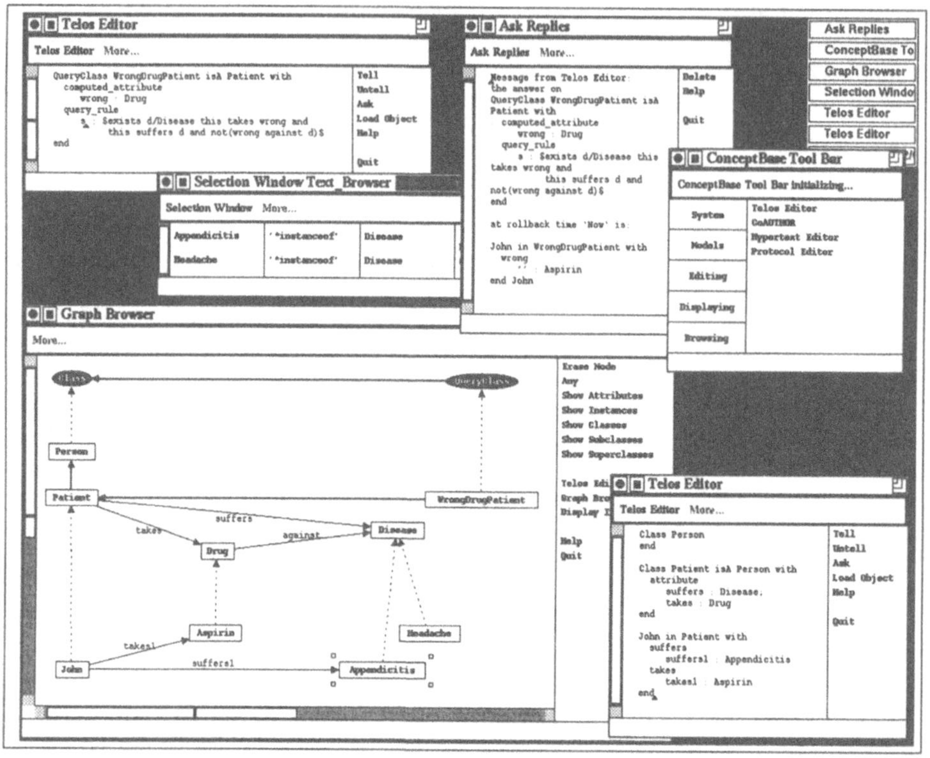

Figure 2: Query classes and objects in ConceptBase

utilizes the classification mechanism on the type system for optimization purposes, namely search space restriction and reuse of materialized views. Whereas in the general case we cannot decide subsumption between queries and/or constraints and rules in a deductive database system, decidable reasoning in a more or less powerful type language seems to be a good formal basis for a further examination of optimization opportunities. The next steps following our proposal will be a careful formal description of the integration between the object base and the type system, especially of the way how to extract type information from object bases. We are currently experimenting with concept languages of different power and corresponding syntactic O-Telos constructs, namely new predefined attribute categories, and check their relevance both for representing queries and for modelling with concept classes in general. An issue not addressed in this paper is the effect of computed subsumption relationships on the structure of explicitly stored information. One important advantage concerning the implementation of a subsumption algorithm is that in our case we don't need to rely on a complete algorithm. Opposed to pure concept language implementations this means only less optimization but not less inference esp. incomplete answers to queries. In comparison with other work done in the area of integrating the notions of concepts, classes, and deductive rules into one unique framework for representing knowledge as well as queries, we can state that previous efforts concentrate on only two of the three aspects at a time. An approach combining Datalog with a concept language into one hybrid language (similar to typed logic programming) is reported in [Donini et al.(1991)]. Several data models and query languages amalgamating deductive and object-oriented features were proposed in [Kim et al. (1990), Delobel et al. (1991)]. An example for integrating object-oriented and concept language technology is presented in [Scholl et al. (1991)].

References:

BECK, H.W., GALA, S.K., and NAVATHE, S.B. (1989): Classification as a query processing technique in the CANDIDE semantic data model. In: *Proc. 5th Int. Conf. on Data Engineering, 572–581.*

BRACHMAN, R.J., MCGUINESS, D.L., PATEL-SCHNEIDER, P.F., RESNICK, L.A., and BORGIDA, A. (1991): Living with CLASSIC: When and how to use a KL-ONE-like language. In: J. Sowa (ed.): *Principles of Semantic Networks*, Morgan Kaufmann.

BORGIDA, A. (1992): From type systems to knowledge representation: natural semantics specifications for description logics. *Int. Journal of Intelligent and Cooperative Information Systems, 1(1), 93–126.*

BRACHMAN, R.J., and SCHMOLZE, J.G.(1985): An overview of the KL-ONE knowledge representation system. *Cognitive Science, 9 (2), 171–216.*

CERI, S., GOTTLOB, G., and TANCA L. (1990): *Logic programming and databases.* Springer-Verlag.

DELOBEL, C., KIFER, M., MASUNAGA, Y., (eds.) (1991): *Proc. 2nd Int. Conf. on Deductive and Object-Oriented Databases, LNCS 566*, Springer-Verlag.

DONINI, F.M., LENZERINI, M., NARDI, D., and NUTT, W. (1991): The complexity of concept languages. In: *Proc. 3rd Int. Conf. Principles of Knowledge Representation and Reasoning.*

DONINI, F.M., LENZERINI, M., NARDI, D., and SCHAERF, A. (1991): A hybrid system with datalog and concept languages. In: Ardizzone E., Gaglio S., Sorbello F., (eds.): *Trends in Artificial Intelligence*, LNAI 549, Springer Verlag, 88-97.

JARKE, M., (ed.) (1992): *ConceptBase V3.1 user manual.* Report Aachener Informatik-Berichte Nr. 92-17, RWTH Aachen, Germany.

JEUSFELD, M. (1992): *Update control in deductive object bases* (in German). Infix-Verlag, St.Augustin, Germany.

JEUSFELD, M., and JARKE, M. (1991): From relational to object-oriented integrity simplification. In [Delobel et al. (1991)], *460–477.*

JEUSFELD, M., and STAUDT, M. (1993): Query optimization in deductive object bases. In: Freytag et al. (eds.): *Query Processing for Advanced Database Systems*, Morgan-Kaufmann.

KIM, W., NICOLAS, J.-M., Nishio, S., (eds.) (1990): *Proc. 1st Int. Conf. on Deductive and Object-Oriented Databases*, North-Holland.

LENZERINI, M., and SCHAERF, A. (1991): Concept languages as query languages. In: *Proc. AAAI, 471–476.*

MYLOPOULOS, J., BORGIDA, A., JARKE M., and KOUBARAKIS, M. (1990): Telos: a language for representing knowledge about information systems. *ACM Trans. Information Systems, 8 (4), 325–362.*

STAUDT, M. (1990): *Query representation and evaluation in deductive object bases* (in German). Diploma thesis, Universität Passau, Germany.

STAUDT, M., NISSEN, H., JEUSFELD, M. (1993): Query by Class, Rule and Concept. To appear in *Applied Intelligence*, Oct. 1993.

SCHOLL, M.H., LAASCH C., TRESCH M. (1991): "Updatable views in object oriented databases." In [Delobel et al. (1991)], *189–207.*

Neural Networks:
Architectures, Learning and Performance

Reiner Kree

Institut für Theoretische Physik,
Bunsenstrasse 9, W-3400 Göttingen, Germany

Summary: We review recent work on some principles of a theory of neural networks. Contributions about network architectures, performance criteria, learning objectives and learning strategies from statistical mechanics, computer sciences, statistics and approximation theory are presented.

1. Introduction

This article is about artificial neural networks without presenting a single application. Even the word "backpropagation" is rarely used. It mainly reviews recent developments from disjoint fields of research which all seem to converge to something that perhaps will deserve the name "theory of neural networks" in the future. Do we need such a theory? Successes of neural networks in real world applications can hardly be overlooked by now. Failures are not so easily detected because they are neither published nor sold. But researchers and engineers know that there have been such failures. The successful applications have been developed without a sophisticated theory. So have the failures. Most of the mathematical procedures which are performed by neural networks are quite well known in approximation theory and statistics.

The exciting thing for many people (including the author) about these devices is that they sometimes work surprisingly well and sometimes, -even when they fail, they do so in a surprising manner. It is surprising because the way it fails is often vaguely reminiscent of human failure in cognitive or intelligent problem solving. It tends to produce erroneous answers which are more like "blurrbible " in style than like "qwxc35fvm&d", which might well be the outcome of an optimized but very slightly deranged computer program. This gives them an air of being made for the "real world", rather than for some high precision task in numerics. The engineering successes seem to support this view up to now.

Thus if there is something special about neural networks which makes them work well in real applications, this also tells us something about the structure of the real world problems. Do not forget that the models of artificial neural networks, i.e. the biological ones have been optimized over a long period to cope with such problems.

If we are lucky, a theory of artificial neural networks may have impacts in three directions: similar to ,- let us say-, Maxwell's equations of electrodynamics it may provide an exact framework for engineering; it may tell us something about the environment of problems we all live in and finally it may help to understand the strategies nature has equipped us with to survive in such an environment. Why not take an optimistic view and try?

2. Artificial Wilderness

2.1 Computational Environment

The stage on which neural networks are supposed to act is a "real world". They are not especially designed to be universal machines like a conventional computer. Rather, their raison d'etre is to behave as fit as possible in a "computational environment" of specific problems. The **environment** provides questions or inputs **x** which are drawn from some **input space I** according to some law μ. Commonly, the inputs **x** are N_{in}-dimensional real or binary vectors, i.e. $\mathbf{I} = \mathbf{R}^{N_{in}}$ or $\{+1,-1\}^{N_{in}}$. μ may be a fixed probability distribution on **I** or it may denote some **I**-valued (stochastic) process with realizations $\{\mathbf{x(t)} : t \geq 0\}$. In exceptional cases it may also be possible that μ becomes a deterministic law, prescribing an exactly defined sequence of inputs. Generically, however, real world data are corrupted by noise, incomplete and only partially controllable.

The neural networks have to perform useful computations in a given input environment $\{\mu, \mathbf{I}\}$ in order to survive, i.e. not been thrown away by the designing engineer. The usefulness of a computation is inferred from the networks outputs **y** which are elements of the **output space O** using some **evaluation function** $E(\mathbf{y}\cdots)$. ($\mathbf{O}=\{+1,-1\}^{N_{out}}$ or $\mathbf{R}^{N_{out}}$ is sufficient for most of the present discussion.) The evaluation function measures some error,loss (or gain) connected with the network's output and may be regarded as the environment's response to the network's computations. The special feature of neural networks which makes them fit for survival is their ability to use the information contained in E in order to improve their performance by adaptively changing their structure. This process of **learning** is discussed in detail in sect. 4. and 5.

Now we may sum up the above and define a computational environment $\mathcal{E}$ as consisting of an input space **I**, a law μ to generate inputs, an output space **O** and an evaluation function E:

$$\mathcal{E} = \{\mathbf{I}, \mu, \mathbf{O}, E\} \tag{1}$$

According to the type of information that is provided by the evaluation function one distinguishes between three different situations which we will subsequently discuss: supervised, reinforcing and unsupervised evaluation.

2.2 Supervised Evaluation

In this case, answers cannot only be obtained from the network but also -via an experiment- from the real world. Thus we can construct sets $\mathcal{T}= \{(\mathbf{x}_i, \bar{\mathbf{y}}_i) : i = 1 \ldots p\}$ of input questions and corresponding experimental results $\bar{\mathbf{y}}_i$. For example, we may use a network to predict tomorrow's mean outdoor temperature on the basis of some of today's meteorological data. We can compare the network prediction with the actually measured value (one day later) as many times as we want. For supervised evaluation, E is chosen to measure the distance between the network output **y** and the actually observed $\bar{\mathbf{y}}$. The most frequently encountered distance functions in the current literature are the squared error

$$E_{SE}(\mathbf{y}, \bar{\mathbf{y}}) = \sum_{k=1}^{N_{out}} (y_k - \bar{y}_k)^2 \tag{2}$$

and the cross entropy for outputs $0 \leq y_k \leq 1$

$$E_{CE}(\mathbf{y}, \bar{\mathbf{y}}) = - \sum_{k=1}^{N_{out}} [\bar{y}_k ln(y_k) + (1 - \bar{y}_k) ln(1 - y_k)] \tag{3}$$

Generally, each specific computational task may require its own specially designed distance measure.

The network will only have a chance to infer useful outputs from the inputs, if there are relations between inputs and experimental results in the real world. For our present purposes the relation may be a function (**rule** V:$\mathbf{I} \rightarrow \mathbf{O}$ or a set of consistent conditional probabilities $\Pi(A \subseteq \mathbf{O} \mid \mathbf{x})$ (i.e. the probabilities that the result is contained in (measurable) subsets A of output space under the condition that $\mathbf{x}$ is the input). The corresponding probability density will be denoted by $\pi(\mathbf{y} \mid \mathbf{x})$.

Remark 2.2.1: If the inputs are drawn from $\mathbf{I}$ according to a fixed probability distribution μ, the joint probability $\omega(A \subseteq \mathbf{O}, B \subseteq \mathbf{I})$ of inputs and results is related to Π by

$$\omega(A \subseteq \mathbf{O}, B \subseteq \mathbf{I}) = \int_{\mathbf{x} \in B} d\mu(\mathbf{x}) \Pi(A \mid \mathbf{x}) \tag{4}$$

where $d\mu$ denotes the measure on $\mathbf{I}$ corresponding to the distribution μ.

Remark 2.2.2: It may also be useful to describe the stochastic relationships between inputs and results in a slightly different way. If we have the a priori information, that there is a set $\mathcal{V}$ of functions g: $\mathbf{I} \rightarrow \mathbf{O}$ (which can be equipped with probability measures) such that one of these functions is realized in every single observation, i.e. $\bar{\mathbf{y}}_i = g_i(\mathbf{x}_i)$ and the g_i are drawn according to a distribution ν from $\mathcal{V}$ the conditional probabilities Π can be expressed as

$$\Pi(A \mid \mathbf{x}) = \int_{\mathcal{V}_{A,x}} d\nu(g) \tag{5}$$

The integral extends over the subset $\mathcal{V}_{A,x}$ which contains all functions g with g($\mathbf{x}$)=$\bar{\mathbf{y}}$ $\in$ A. $\mathcal{V}$ is called the **rule space** and the parts of rule space which are compatible with a given set $\mathcal{T}$ of inputs and results is known in AI–literature as **version space** $\mathcal{V}(\mathcal{T})$.

Remark 2.2.3: In standard applications of neural networks with supervised evaluation, the information which is provided by comparing the network outputs with finitely many experiments (collected in a **trainig set** $\mathcal{T}$) via the evaluation function E is used to adapt the network in such a way that it performs well for all the further inputs. The process of **programming** known from conventional algorithmic machines is thus replaced by **learning** from examples and **generalizing**; i.e. extracting those aspects of the real world relation which lead to a good evaluation (via E) of the answers to all further questions . The important ability to generalize will be discussed in section 4.

2.3 Reinforcing Evaluation

In a number of applications a direct comparison between the network output and real world data is not possible, but it is still known, whether the output is "correct" or "wrong" (other types of classification, including fuzzy ones are also possible). As an example imagine a decision performed on the basis of the output of a network which predicts inaccessible data. If the decision turned out to be the correct one you may like to "reward" the network and otherwise "punish" it. This corresponds to a binary valued evaluation function: E($\mathbf{y}$)=+1 if "correct" and it is -1 if "wrong". Note, however, that due to the stochastic relationships between data in the real world, the same input–output pair may sometimes be classified as correct and sometimes as wrong! Obviously, reinforcement is a weaker form of supervision. Therefore the disussion of section 2.2 applies to this situation, too.

2.4 Unsupervised Evaluation

There are also situations in which no real world information about the output is available or needed for the evaluation. As an example, consider a network that shall perform a data compression by **vector quantization**, i.e. it has to find a set of prototypes $\{\mathbf{y}_a \mid a = 1 \ldots M\}$ to a given input data set $\{\mathbf{x}_i \mid i = 1 \ldots N\}$ with $N_{in} = N_{out}$ and $M \ll N$ which provide a good, coarse grained approximation of the input data. The output can be evaluated from the distances between data points $\mathbf{x}_i$ and their nearest prototype $\mathbf{y}_{n(i)}$. A commonly used evaluation function is given by:

$$E(\mathbf{y}) = \sum_{i=1}^{N} \| \mathbf{x}_i - \mathbf{y}_{n(i)} \|^2 \tag{6}$$

Loosely speaking, networks with unsupervised evaluation act as **feature detectors** (principle components, novelty etc) . It has been pointed out by Barlow (1989) that such networks operate on the (information theoretic) redundancy of the input data, because in a completely random data set there are no features to be discovered.

3. Artificial Agents

3.1 Definition of Architecture

The previous section offered a preliminary glimpse on the wildernes in which neural networks have to survive. Now it is time to introduce some of the main characters of this species of computational devices.

For the purposes of the current review we define a **neural network** as a set $\{\mathcal{N}_i \mid i = 1 \ldots N\}$ of elementary processing units $\mathcal{N}_i$ called **neurons** which are placed on nodes $i = 1 \ldots N$ of a directed graph $\mathcal{G}$ (see fig.1). The remaining $N_{in} + N_{out}$ nodes of $\mathcal{G}$ are either (N_{in}) pure sources or (N_{out}) pure sinks. They constitute the locations of input and output of the network. The direction of the edges represent the flow of information. Thus each neuron $\mathcal{N}_i$ receives input from n_i(=number of incoming edges at node i) different nodes which are either occupied by other neurons or are network input nodes. The input and output spaces of single neurons are denoted by $\mathbf{I}_i$ and $\mathbf{O}_i$, respectively. The computations performed by neuron $\mathcal{N}_i$ are described by a function $f_i : \mathbf{I}_i \to \mathbf{O}_i$ for **deterministic** neurons and conditional probabilities $f_i(\mathbf{y} \mid \mathbf{x})$ for **stochastic** neurons. To condense notation we choose the same symbol f_i for the two different objects. They are both called **updating rules**.

The most interesting aspect about a neuron is the adaptability of its updating rule. The f_i can be varied within a certain class F_i of functions and most commonly these classes are parametrized by a finite dimensional vector of parameters so that $F_i = \{f(\cdot; \mathbf{w_i}) \mid \mathbf{w_i} \in \mathbf{W_i}\}$. The W_i dimensional vector $\mathbf{w_i}$ is called **weight vector** or **synaptic weights** of neuron $\mathcal{N}_i$. Correspondingly $\mathbf{W_i}$ is referred to as **weight space**.

Remark 3.1.1: To specify the operation of the complete network we do not only need the updating functions of single neurons and the graph $\mathcal{G}$ (**structure graph**) but also the flow of information in time. If input $\mathbf{x}$ is presented to the net, the output will in general depend on the sequence of updatings of the neurons. Most applications use either a **synchronous updating** of all neurons during a single time step (**parallel dynamics P**) or the updating of a single neuron at one time step (**sequential dynamics S**). In the latter case one has to specify the sequence of updatings which of course offers many possibilities. An interesting case is **random sequential updating RS** where at every time step one of the N neurons is chosen randomly to be updated. One should keep in mind that due to the nontrivial time structure of information flow, a neural network cannot be completely described as a function from input space to output space. Rather **it maps temporal sequences**

of inputs to temporal sequences of outputs.

Now we can give a general definition of a **neural network architecture**$\mathcal{A}$: it consists of a structure graph $\mathcal{G}$, the classes F_i of updating rules and the choice of an updating dynamics $\mathbf{D}$

$$\mathcal{A} = \{\mathcal{G}, \{F_i \mid i = 1 \dots N\}, \mathbf{D}\} \tag{7}$$

In the following subsections we briefly describe some special but widely used structure graphs and updating rules.

3.2 Special Structure Graphs

3.2.1 Feed Forward Networks: They are characterized by acyclic structure graphs $\mathcal{G} = \mathcal{L}_k$, with nodes belonging to one of k+2 layers . A node of layer l only has incoming edges from nodes of layers $m < l$. There are k layers of deterministic neurons, one input and one output layer. With **D=P** a **static input** $\mathbf{x_{in}}$ generates a **static output y** after k time steps which is considered as the result of the computation.

3.2.2 Non-Recurrent Networks: As in feed forward networks the information flow is unidirectional between layers and the graph is acyclic. However, there are edges within the layers. Again a static input produces a static output after finitely many steps (equal to the number of nodes visited by the longest path from input to output layer) if the neurons are deterministic.

3.2.3 Recurrent Networks: The structure graph contains closed loops and consequently the network's output to a static input will generally be a temporal sequence which may not settle down to some static **y** in finite time. Thus one has to specify more precisely what is to be considered as a result of the computation. An important subclass of recurrent networks which have been studied and have found a variety of applications are the **attractor neural networks**. For these networks, the output sequence ends up on low dimensional attractors (fixed points, cycles, low dimensional strange attractors) and the input space **I** is partitioned into subsets $\mathbf{I}^\mu$ (**domains or basins of attraction**) such that inputs $x \in \mathbf{I}^\mu$ lead to output sequences approaching the μ'th attractor. Static outputs in finite time can be obtained if the attractors are **terminal fixed point** attractors which have been proposed by Zak (1988) as useful memory devices. Let us finally mention two special types of recurrent structure graphs which underly some well studied architectures: the graph $\mathcal{G}_{max}$ where every neuron is connected to every other neuron and **sparse random** graphs which are constructed from $\mathcal{G}_{max}$ by removing edges at random with probability 1-p, $p \ll 1$. The dynamics for large nets (i.e. for $N \to \infty$) of this type can be solved exactly if $p = \mathcal{O}(K/N)$ with $K = o(lnN)$ (see Derrida, Gardner et al (1987) and Kree, Zippelius (1987).

3.3 Special Neurons

3.3.1 Deterministic Threshold Neurons: The prototype of threshold neurons is the McCulloch–Pitts neuron (1943). It is characterized by a binary output space $\mathbf{O}_i = \{+1, -1\}$,a linear summation of inputs and a threshold. $F_i = F^{MP}$ is defined as the set of all functions

$$f_i(\mathbf{x}) = \text{sign}(\mathbf{w}_i \cdot \mathbf{x} - \theta_i) \tag{8}$$

with threshold $\theta_i \in \mathbf{R}$ and real weight vector $\mathbf{w}_i \in \mathbf{R}^{n_i}$ which can be normalized, $\| \mathbf{w}_i \| = 1$. Note that the elements of the weight vector are in one-to-one correspondence with the directed edges of the structure graph. These neurons can be considered as a very crude caricature of physiological neurons where a weight w_{ij} corresponds to a synaptic connection between presynaptic neuron j and postsynaptic neuron i and the scalar product $h_i = \mathbf{w}_i \cdot \mathbf{x}$ models the postsynaptic potential. If the sign function is replaced by some other function, several variants of McCulloch–Pitts neurons are obtained:

Linear threshold neuron F^L: sign(z) replaced by $z\theta(z)$

Sigmoidal threshold neuron F^σ: sign(z) replaced by a sigmoidal function, e.g. $\tanh(\beta z)$. Note that for $\beta \to \infty$ the McCulloch-Pitts neuron reappears.

3.3.2 Stochastic Threshold Neurons F^θ : They are defined by conditional probabilities for the binary outputs

$$f_i(y = +1 \mid \mathbf{x}) = g_\beta(\mathbf{w}_i \cdot \mathbf{x} - \theta_i) \tag{9}$$

with $g_\beta(z) = [1 + \exp(-2\beta z)]^{-1}$. Note that for $\beta \to \infty$ we again recover the McCulloch–Pitts neuron. In fact the stochastic threshold neuron is closely related to the sigmoidal threshold neuron because the latter is the average of the stochastic updating functions defined by equ.(9).

3.3.3 Boolean nets: If the input spaces $\mathbf{I}_i$ are binary, i.e. $\{+1, -1\}^{n_i}$ and the output space is binary, too, the update functions are Boolean functions of n_i variables and we may choose F_i as the set of all such Boolean functions. Architectures built from such elements are called Boolean nets (Wong and Sherrington (1988)).

We will not enlargen this list further. There is a plethora of different types of neurons, both for computational purposes and for modelling aspects of physiological neurons.

3.4 Special Architectures for Some Special Tasks

3.4.1 Perceptrons as classifiers: The simple perceptron is a feed forward network with N input nodes, one output node and only one neuron $\mathcal{A}^P = \{\mathcal{L}_1, F^{MP}, \mathbf{P}\}$ (Rosenblatt (1962). Note that in an $\mathcal{L}_1$-network all the neurons perform their computations independently. Thus it is sufficient to consider just one neuron. The inputs may be either real numbers or binaries. This architecture can be used to classify inputs according to their position with respect to the hyperplane $\mathbf{w} \cdot \mathbf{x} = \theta$ in $\mathbf{R}^N$. Input vectors $\mathbf{x}_+$ classified as $+1$ are seperated from vectors $\mathbf{x}_-$ classified as -1 by this hyperplane.

A standard problem for a perceptron is to learn a given classification (the so called **target**) from examples. Obviously this is impossible for a simple perceptron if the target is not a linearly seperable one. Therefore this architecture is unable to perform such simple Boolean classifications as the XOR (Minsky, Papert (1969)). However, a straightforward generalization of the simple perceptron to feed forward networks with more than one layer of neurons is able to solve **any** given classification task. In fact it can be shown that an architecture with N inputs, one layer of threshold neurons (usually called **hidden units** because they are not connected to input or output nodes) and one McCulloch-Pitts neuron in a second layer is able to realize every Boolean function of N binary inputs. (The network is rather easy to construct explicitely by choosing weights such that exactly one of the hidden units is activated ($\sim$+1) for each of the 2^N possible inputs . Unfortunately this construction requires exponentially many hidden units.)

If the McCulloch-Pitts neuron is replaced by a simple weighted sum of the outputs of the hidden units, every continuous function $\mathbf{R}^N \to \mathbf{R}$ can be approximated with arbitrary accuracy (Cybenko(1989), Hornik, Stinchcombet al (1989),). In this sense Multi Layer Networks (MLN) are universal classifiers and universal approximations.

3.4.2 Perceptrons for Time Series Prediction and Sequence Recognition: A frequently encountered task in practical applications is to predict the next point in a time series from N preceding data points or to recognize the N data points as a specific sequence and classify it accordingly (for example, think of recognizing a string of phonemes as a particular word). The N preceding data points may be taken as input for a perceptron. Technically this can be realized as a **tapped delay line** (storing N values and presenting them simultaneously to the network). This scheme is widely used in speech recognition (see Lippmann(1989) and references therein).

3.4.3 Feed Forward Networks for Feature Extraction: We have already encountered the example of data compression by vector quantization in 2.4. There is an extensive literature about this subject. The architectures are all fairly simple, consisting mostly of one or two layers of neurons. The interesting aspects are connected with the learning algorithms. We refer the reader to (Kohonen (1989), Ritter, Schulten et al (1992), Hertz, Krogh et al (1991), Linsker (1988))

3.4.4 Feed Forward Networks for Optimization Problems: Unsupervised feed forward networks have also been applied to some optimization problems like, e.g., the **travelling salesman problem**. See Durbin, Willshaw (1987) and Peterson (1990) for benchmark tests which compare different algorithms.

3.4.5 Attractor Neural Networks for Association and Information Storage: To every input $\mathbf{x}$ the network produces a sequence $\mathbf{y}(t)$ which approaches attractor number k if the input was taken from the corresponding basin of attraction. This process may be interpreted as an **associative recall** of stored data. These data may be either time independent (fixed point attractors), periodic sequences (limit cycles) or aperiodic sequences. The possibility of **temporal association** is a unique feature of recurrent nets. Whereas for fixed points the information which can be stored and associatively recalled is connected to the volume of the domains of attraction, it can become much larger in the case of temporal association. Note that the information is stored in a **content addressable** way in these networks.

If input and output nodes are connected to the same set of neurons, the network is called **auto-associative** (otherwise **hetero-associative**). Hetero-association with hidden units can alternatively be considered as information processing rather than storage(see 3.4.6). Neural networks with **symmetric weights**, i.e. $w_{ij} = w_{ji}$ and **D=P, RS** and without hidden units have been proposed by Little (1974) and Hopfield (1982) as associative memories for static patterns. Hopfield pointed out the analogy of these systems with disordered magnetic systems called spin glasses. This analogy revived the interest of many physicists in neural networks (for a recent review see Sherrington (1993))and the application of techniques from statistical mechanics has turned out to be quite successful ever since.(see e.g. Hertz, Krogh et al (1992), Domany, vanHemmen et al (ed) (1992) and the review of Watkin, Rau, et al (preprint 1992) on learning)

3.4.6 Boltzmann Machines for Problems with Reinforcing Evaluation: Boltzmann machines are recurrent networks which operate with stochastic threshold neurons and possess hidden units. The standard architecture has symmetric synaptic weights. In this case the stationary probability $p_{st}(s_1, \ldots, s_N)$ of all the binary neuronal outputs has the form of a Boltzmann distribution $p_{st} = \exp(-\beta H)/Z$ with $H = -(1/2)\sum w_{ij}s_is_j$ and is reached asymptotically for $t \to \infty$.

Obviously, Boltzmann machines maybe used to model stochastic rules of the real world. But they are also useful for other tasks which require a stochastic search. In reinforcing evaluation, e.g., the network gets no hint from the environment in which direction it should change if its output had been wrong. In this situation it has to undertake some random search which can be conducted by a Boltzmann machine.

(Hinton and Sejnowski (1986)).

Recurrent networks with deterministic and stochastic threshold units have been applied to a large variety of other problems. We do not have space enough even to mention most of them. Let us merely give some examples: time series prediction (Wulff(1992)); graph recognition (Kree, Zippelius (1988)); recursive filtering (Nerrand, Roussel-Ragot et al (1993); nonlinear optimization (Tsirukis, Reklaitis et al (1989); Hopfield, Tank (1986))).

4. Artificial Learning

Now that we have got an impression about the anatomy of neural networks as it is encoded in their architectures we can turn to the most important question: How can they survive in a computational environment? As the only forces of "natural selection" are the performance criteria set up by the designer, we will first discuss these and then consider some survival (i.e. learning) strategies.

Let us consider an architecture $\mathcal{A} = \{\mathcal{G}, \{F_i(\mathbf{w_i} \mid i = 1 \dots N\}, \mathbf{D}\}$ in an environment $\mathcal{E}$ with a probability distribution μ for input questions and evaluation function E. Currently, two approaches to the objectives of learning are discussed in the literature. The **information theoretic** or **statistical** approach and the **probably almost correct PAC** learning approach from computer sciences. As they differ both in the structure of the environment and the required performance criteria, we will introduce them sequentially. We will restrict the discussion to cases where the result of the computation consists in a single element of the output space. The more general case of temporal sequences as output can be treated similarly but the topic is much more technical. We describe the network by a conditional probability density $p_{NN}(\mathbf{y} \mid \mathbf{x}, \mathbf{w})$. A deterministic net corresponding to a fixed map $\mathbf{y} = \mathbf{\Phi}(\mathbf{x}, \mathbf{w})$ is included as a limiting case for which p_{NN} becomes a Dirac distribution $\delta(\mathbf{y} - \mathbf{\Phi}(\mathbf{x}, \mathbf{w}))$. The evaluation function $E(\mathbf{y}, \bar{\mathbf{y}}; \mathbf{x}, \mathbf{w})$ will be denoted by $E(\mathbf{\Phi}, \bar{\mathbf{y}}; \mathbf{x}, \mathbf{w})$ for a deterministic net, by $E(\mathbf{\Phi}, V; \mathbf{x}, \mathbf{w})$ for a deterministic net and a deterministic target rule V and by $E(\mathbf{x}, \mathbf{w})$ for unsupervised evaluation.

4.1 Statistical Approach

(Carnevali and Paternello (1987), Tishby, Levin et al (1989); Levin, Tishby et al. (1990)) In this approach we require that the network performs well **on the average**. More precisely, suppose that a specific learning algorithm **L** has produced a distribution in weight space with the density $\lambda_{\mathcal{T}}(\mathbf{w}; \mathbf{L})$ from a training set $\mathcal{T}$ with p example questions drawn from **I** according to the distribution μ. We may pick one particular solution $\hat{\mathbf{w}}$ produced by **L** and ask further questions to this specific network. The performance of $\hat{\mathbf{w}}$ is measured by the **generalization error** :

$$\epsilon(\hat{\mathbf{w}}; \mathbf{L}; \mathcal{T}) = \int E(\mathbf{y}, \bar{\mathbf{y}}, \mathbf{x}, \hat{\mathbf{w}}) p_{NN}(\mathbf{y} \mid \mathbf{x}, \hat{\mathbf{w}}) \pi(\bar{\mathbf{y}} \mid \mathbf{x}) d\mu(\mathbf{x}) \tag{10}$$

The performance of the algorithm is evaluated by

$$\epsilon(\mathbf{L}, \mathcal{T}) = \int \epsilon(\mathbf{w}; \mathbf{L}; \mathcal{T}) \lambda_{\mathcal{T}}(\mathbf{w}) d\mathbf{w} = < \epsilon >_{\lambda_{\mathcal{T}}} \tag{11}$$

This performance measure still depends on the specific training set. In statistical learning we only require that the **averagely trained network performs well on the average**. The corresponding performance criterion is the **average generalization error**

$$\epsilon(p, \mathbf{L}) = \int \epsilon(\mathbf{L}, \mathcal{T}) \prod_{\mu=1}^{p} d\omega(\mathbf{y}_\mu, \mathbf{x}_\mu) = << \epsilon(\mathbf{L}, \mathcal{T}) >>_{\mathcal{T}} \tag{12}$$

which only depends on the size of the training set and the learning algorithm.

Minimizing the generalization error $\min_{\mathbf{w}\in\mathbf{W}} \epsilon(\mathbf{w}) = \epsilon(\mathbf{w}^{opt})$ cannot be accomplished by any algorithm a priori because we do not know the distributions ω, μ and the conditional probabilities Π which encode the real world relation between inputs and results. Instead we can collect a sample of p measurements into a **trainig set** $\mathcal{T}$ and use it for estimates of the distributions and of $\mathbf{w}^{opt}$. It is precisely this information which is available for learning algorithms **L**. We can explicitely measure the performance of the network on the training set (**training error**):

$$e_{\mathcal{T}}(\mathbf{w}) = \frac{1}{p}\sum_{i=1}^{p} E(\bar{\mathbf{y}}_i, \mathbf{y}_i, \mathbf{x}_i, \mathbf{w}) \tag{13}$$

4.1.1 Bayesian Learning: Learning could be realized as a Bayesian estimate. All the networks $\mathbf{w} \in \mathbf{W}$ are considered as hypotheses with a prior distribution $\lambda_0(\mathbf{w})$. The posterior distribution after we have received the additional information contained in the training set $\mathcal{T}$ is denoted by $\lambda_{\mathcal{T}}^{Bayes}(\mathbf{w})$. The corresponding predictive conditional probability then becomes an average over the posterior distribution:

$$p_{\mathcal{T}}(\mathbf{y} \mid \mathbf{x}) = \int p(\mathbf{y} \mid \mathbf{x}, \mathbf{w}) \lambda_{\mathcal{T}}^{Bayes}(\mathbf{w}) d\mathbf{w} \tag{14}$$

The posterior distribution itself is explicitly given by

$$\lambda_{\mathcal{T}}^{Bayes}(\mathbf{w}) = \frac{\lambda_0(\mathbf{w}) \prod_{i=1}^{p} p_{NN}(\bar{\mathbf{y}}_i \mid \mathbf{x}_i, \mathbf{w})}{Z_p(\mathcal{T})} \tag{15}$$

where the normalization constant $Z_p(\mathcal{T})$ is a measure of those weights which are compatible with $\mathcal{T}$. In particular, $Z_{p+1}/Z_p = \text{prob}(\bar{\mathbf{y}}_{p+1} \mid \mathbf{x}_{p+1}, \mathcal{T})$ is the **generalization probability**. Using the predictive distribution as an estimator for $\pi(\bar{\mathbf{y}}, \mathbf{x})$ is called **Bayes algorithm**. Note, however, that $p_{\mathcal{T}}$ may not be realizable within the given architecture. Opper and Haussler (1991) constructed a neural network which performs the Bayesian algorithm for linearly seperable rules. Even for this simple class of problems a two layer perceptron with an infinite number of hidden units is necessary. Instead, we can try to minimize the posterior generalization error within the architecture. This is called **optimal learning** (Watkin (1992)). Picking a specific network at random according to the posterior distribution is called the **Gibbs learning algorithm**.

4.1.2 Statistical Physics and Learning: Now let us assume that the only information we have about the learning algorithm is that it produces an average training error $e(p, \mathbf{L}) = < e_{\mathcal{T}} >_{\lambda_{\mathcal{T}}}$, i.e. we may fix

$$\int \left(\sum_{\mu=1}^{p} E(\mathbf{y}_\mu, \mathbf{x}_\mu, \mathbf{w}) \right) \lambda_{\mathcal{T}}(\mathbf{w}) d\mathbf{w} = e(p, \mathbf{L}) \tag{16}$$

The resulting distributions $\lambda_{\mathcal{T}}$ in weight space may be estimated by the extrema of the entropy

$$S(\lambda_{\mathcal{T}}; \lambda_0) = \int \ln \left[\frac{\lambda_{\mathcal{T}}(\mathbf{w})}{\lambda_0(\mathbf{w})} \right] \lambda_{\mathcal{T}}(\mathbf{w}) d\mathbf{w} \tag{17}$$

with respect to normalized densities $\lambda_{\mathcal{T}}$ under the additional constraint equ. (16). The result is the Boltzmann distribution (see sect. 3.4.3)

$$\lambda_{\mathcal{T}}(\mathbf{w}) = \frac{\lambda_0(\mathbf{w})}{Z} \exp[-\beta \sum_{\mu=1}^{p} E(\mathbf{y}_\mu, \mathbf{x}_{mu}, \mathbf{w})] \tag{18}$$

It is precisely at this point, where techniques from statistical physics can be used with advantage.(Seung, Sompolinsky et al (1992); Györgyi (1990)) Inserting equ. (18) into equ. (11) and equ. (12) we arrive at averages which appear in the statistical physics of disordered materials. Several powerful computational techniques and approximation schemes have been developed. For example Tishby, Levin et al (1989) propose to apply the so called **annealed approximation** which independently averages the denominator and the numerator in equ (18). Bilbro and Van den Bout (1992) demonstrate that efficient learning algorithms can be constructed from this approximation. Seung, Sompolinsky et al. (1992) use the method of **replicas** well known from spin glass physics (see Mezard, Parisi et al (ed) (1986)) to investigate the average in several special cases.

As a result of the statistical learning theory we get the so called **learning curves**, i.e. the dependence of $\epsilon(p)$ and e(p) on the size of the trainig set. The statistical theory becomes particularly powerful for large neural networks. In this case it is possible to show that the errors become **self-averaging**. This means that for a net with N neurons and a training with $p = \alpha N$ examples the quantities $\epsilon(p)/p$ and $e(p)/p$ converge to limits for $N \to \infty$ either almost sure or in probability. (White (1989), Seung, Sompolinsky et al. (1992)). Several remarkable results have been obtained in this way (see Watkin, Rau et al (1992) for a recent review). According to this line of work there are certain **universality classes** of learning machines which are distinguished by the asymptotic $\alpha \to \infty$ behaviour of their learning curves or the number of networks $\mathcal{N}(\epsilon)$ with fixed ϵ within a given architecture for small ϵ. $\epsilon(\alpha)$ either approaches its limit value as a power law $\mid \epsilon(\alpha) - \epsilon(\infty) \mid \sim \alpha^{-\gamma}$ (if $d(\ln \mathcal{N})/d\epsilon$ diverges for $\epsilon \to 0$) or there is a discontinuous transition from finite to zero generalization error at some finite α_c (If $d(\ln \mathcal{N})/d\epsilon \to 0$ for $\epsilon \to 0$).(see Seung, Sompolinsky et al. (1992), Amari, Fujita et al (1992))

4.1.3 Entropy and Intelligence of Architectures: The learning theory scetched above has much more implications than just the learning curves. Let us consider deterministic networks $y = \Phi(\mathbf{x}, \mathbf{w})$ and deterministic target rules $\bar{y} = V(\mathbf{x})$. To keep the discussion as simple as possible let us assume that $y, \bar{y} \in \{0, 1\}$; $\mathbf{x} \in \{+1, -1\}^N$ and $\mathbf{w} \in \{+1, -1\}^W$. The generalization function $\epsilon(V, \mathbf{w}) = \sum_{\mathbf{x}} (V(\mathbf{x}) - \Phi(\mathbf{x}))^2$ counts the number of outputs which disagree with the target rule. The version space of rule V occupies a fraction of the weight space $\mathbf{W}$ (which contains 2^W elements) given by $g(V, \Phi) = 2^{-W} \sum_{\mathbf{w}} \delta[E(V, \mathbf{w})]$. ($\delta(x) = 1$ if $x = 0$ and zero otherwise). The **entropy of the architecture** is defined by

$$S(\Phi) = -\sum_V g(V, \Phi) \ln[g(V, \Phi)] \tag{19}$$

This number varies between zero (only one rule realizable) and $\ln(2)\min(2^N, W)$ (a RAM for $W \geq 2^N$) and is a measure of the diversity of rules which are implementable within the architecture. The Kullback-Leibler entropy $S(\Lambda \mid \Phi) = -\sum_V g(V, \Phi) \ln[g(V, \Lambda)/g(V, \Phi)]$ provides a natural **similarity measure between architectures**. It has been suggested by Parisi(1992) that the **intelligence** of an architecture may be defined by the similarity of that architecture to a Turing machine T with W bits of code. The output of this Turing machine is 1 if it stops and 0 otherwise. The version spaces of such a machine occupy fractions $g(V, T)$ which can be estimated as $\approx 2^{-\mathcal{C}(V)}$ where $\mathcal{C}(V)$ is the **algorithmic complexity** of rule V if $\mathcal{C}(V) \leq W$. This definition is a quantitative measure of the ability of a network to find the simplest rule which fits the training data (see Parisi (1992)).

4.2 PAC Learning

In probably almost correct learning networks lead harder lives than in statistical learning. Instead of requiring that the network predictions should be good on average, it is required that they are good enough in the **worst case**. More precisely one fixes two performance parameters $0 \leq \delta, \eta \leq 1$ and requires that a learning algorithm be able to produce a network with probability $1 - \delta$ that gives correct answers to future examples with probability at least $1 - \eta$ on the basis of a training set with p examples.(Valiant (1984))

Learning is considered as a process which picks a special hypotheses $\mathbf{w}$ from the available weight space. The probability that net $\mathbf{w}$ gives correct predictions i.e. the **generalization probability** is denoted by $g(\mathbf{w})$. The frequencies of correct predictions of $\mathbf{w}$ on the training set are denoted by $\nu(\mathbf{w}, p)$. PAC learning theories ask whether the following bound can be satisfied:

$$\text{Prob}\left\{\sup_{\mathbf{w}\in\mathbf{W}} \mid \nu(\mathbf{w},p) - g(\mathbf{w}) \mid > \eta\right\} \leq \delta \qquad (20)$$

Let us discuss the relevance of this bound to learning a bit further. In sect.3.4.1 we have stated that multi-layered perceptrons are universal classifiers. Thus we know that within this class we can make the generalization probability as close to one as we like if we have a deterministic environment. In this case we could simply ask how the $\nu(p)$ approaches 1 with increasing training set provided we would know the learning algorithm which leads to the optimal solution. But, alas, we do not! We circumvent our ignorance by requiring that the difference between training and generalization probability is **uniformly bounded** no matter whether the net has good or poor generalization abilities. Thus there is additional "selection pressure" even on the more "stupid" representatives of the network species.

The central result about the uniform convergence with increasing training set is due to Vapnik and Chervonenkis (1971). They proved the following inequality valid for a general classifier with binary output:

$$\text{Prob}\left\{\sup_{\mathbf{w}\in\mathbf{W}} \mid \nu(\mathbf{w},p) - g(\mathbf{w}) \mid > \eta\right\} \leq 4m(2p;\mathbf{W})\exp[-\eta^2 p/8] \qquad (21)$$

m(p) measures the maximum number of binary functions that can be realized by all $\mathbf{w} \in \mathbf{W}$ on the p example inputs. It can be shown that it is either equal to 2^p or is bounded from above by a polynomial $p^{d_{VC}} + 1$. Only the latter case allows for generalization, otherwise, the network is essentially a look-up table. d_{VC} is called **Vapnik-Chervonenkis dimension**. It should be obvious now that d_{VC} is decisive for determining the size of a net and a training set which can give a desired valid generalization. Baum and Haussler (1989) proved that in a feed forward net with N threshold neurons and W weights one has confidence at least $1 - 8\exp(-3W/2)$ that the net will correctly classify future examples with probability $1 - \eta$ if it is possible to load $p \geq (32W/\eta)\ln(32N/\eta)$ randomly chosen training examples such that they are correctly classified with probability $1 - \eta/2$.

5. Strategies of Artificial Learning

5.1 Learning by Adapting Weights

Up to now we have said nothing about explicit learning algorithms. From the above discussion we may extract that learning algorithms are mainly concerned with **estimation** of unknown probability distributions and **optimization** in weight space.

There is a variety of algorithms around for such purposes which may be used and some have been reinvented for network purposes.

Minimizing the training error$E = \sum_{\mu=1}^{p} E(\mathbf{y}_\mu, \mathbf{x}_\mu, \mathbf{w})$ may be tried with local optimization algorithms as, e.g., steepest descent. The particular form (2) (squared error) leads to the simplest version of **backpropagation learning**, which can be efficiently implemented for multi-layer perceptrons. In fact most of the practical applications use variants of this type of algorithm. Local optimization may fail if there are several suboptimal solutions or the "training error landscape" develops flat portions or the solution space consists of many disjoint parts in weight space. From statistical mechanics calculations the latter situation has actually been found for perceptrons with hidden units (Barkai, Hansel et al. (1990), Engel, Köhler et al (1992)). Nevertheless, recent numerical studies by Eisenstein and Kanter (1993) indicate that backpropagation learning is capable to load a special perceptron architectures with one hidden layer (the parity machine) up to its theoretical limit $\alpha_c(N) = p_c/N$ in **learning times** $t_l \sim (\alpha_c(N) - \alpha)^{-2}$ which seems acceptable if this result will not be spoiled with increasing size of the net. (That remains to be tested.)

Remark 5.1.1: We like to mention just one special global optimization method because it provides a natural link to the statistical learning theory. This is **simulated annealing** which combines gradient descent with random search. For continuously varying weights, it can be defined by a stochastic differential equation

$$\begin{aligned} \frac{d\mathbf{w}}{dt} &= -\nabla_{\mathbf{w}} E + \boldsymbol{\xi}(t) \\ < \boldsymbol{\xi}(t) > &= 0 \\ < \xi(t)_r \xi(t')_s > &= (2/\beta)\delta_{rs}\delta(t-t') \end{aligned} \tag{22}$$

with a white noise source $\boldsymbol{\xi}$. The noise parameter β is tuned up during a run of the algorithm. For fixed β this algorithm produces a distribution $\lambda(\mathbf{w})$ which is exactly the Boltzmann distribution equ. (18) we found in statistical learning theory and therefore it is an explicit formulation of the Gibbs learning algorithm we encountered in sect. 4.1.1. Learning times can been studied for this algorithm by standard methods of statistical mechanics (see, e.g. Horner (1992) for interesting work on the perceptron with binary weights).

5.2 Learning by Adapting the Network Structures. Multi-Agents

The learning time is a performance criterion which is of much practical importance but is very hard to analyze theoretically. In the context of PAC learning it is explicitely required, that an acceptable algorithm shall only use a time which is bounded above by a polynomial of the net size N: $t_l \leq N^s$. There are some general results in this direction which seem quite discouraging (see Judd (1988), Blum und Rivest (1988)): the problem of deciding wether a given set of examples can be reproduced by a net with appropriately chosen weights is NP-complete. Though implications of this result for today's practical applications are hard to judge it cannot be denied, that learning time is one of the very weak spots of neural networks. The engineering response to this challenge is a growing industry of clever tricks to speed up backpropagation.

Another way out of this principle dilemma is to enlargen the adaptability of the network architectures by **adaptively varying the structure graph**. There are two classes of such algorithms:**constructive algorithms** which try to build a net by adding neurons or layers until the training set is learned without errors (see e.g. Marchand,Golea et al (1990), most of the existing algorithms are described in Hertz,

Krogh et al. (1991)). Unfortunately, the task of finding a minimal size architecture for a given training set is also NP-hard (Blum and Rivest (1988)). Therefore there is the danger that the constructed net is much too complex for the underlying task and consequently shows poor generalization.

The other class of algorithms has up to now rarely been recognized as belonging to one class at all. Let us call them **multi-agent** neural network models. In these models, one considers a collection of neural networks which may vary in architecture and trains them together, i.e. with a common evaluation function. In contrast to other methods which also use ensembles of nets (e.g. genetic algorithms) the crucial feature is that the networks shall perform as an **organization** of communicating individual nets in the end. One well studied simple example of this type of models is the **query-by-committee** algorithm, analyzed recently by Seung, Opper et al. (1992). This is a collection of simple perceptrons (samplers) with weight vectors which are randomly trained in version space. They **select** as a new question one on whose answer as many samplers as possible disagree. In this way, the machine can get a finite amount of information out of questions even for $\alpha \to \infty$ in contrast to training on a random set. As a consequence the generalization error decreases **exponentially**. Unfortunately, the time that the committee needs to decide upon a question diverges as $\sim 1/\epsilon$. In this example, the communication between the agents is prefixed.

The case of mixtures of "local experts" has also been considered in a number of examples (Baxt (1992), Jacobs, Jordan et al (1991), Kree and Müller (1991)). In these approaches, the nets are pretrained on different data and in a second stage of unsupervised (competetive or cooperative) learning the communication structure is organized. A communication structure differs from a choice of weights because it allows for changes of the architecture in response to new questions, thus collecting the computational power which the net believes is necessary to produce a correct answer. These models are not only interesting as learning algorithms but also as models for cognitive psychology and organization theory (Kree and Müller (1993)).

References:

AMARI, S., FUJITA, N., and SHINOMOTO, S. (1992): Four Types of Learning Curves. *Neural Computation, 4, 605–618.*

ABU-MOSTAFA, Y.S. (1989): The Vapnik-Chervonenkis Dimension: Information versus Complexity in Learning. *Neural Computation, 1, 312-317.*

BAUM, E.B., and HAUSSLER, D. (1988): What Size Net Gives Valid Generalization? *Neural Computation, 1, 151–160.*

BAXT, W.G. (1992): Improving the Accuracy of an Artificial Neural Network Using Multiple Differently Trained Networks. *Neural Computation, 4, 772–780*

BARLOW, H.B. (1989): Unsupervised Learning *Neural Computation, 1, 295–311*

BILBRO, G.L., and VAN DEN BOUT, D.E. (1992): Maximum Entropy and Learning Theory. *Neural Computation, 4, 839–853.*

BLUM, A., and RIVEST, R.L. (1988): Training a 3-node neural network is NP-complete In:*Proc. Workshop on Computational Learning Theory.*, 9–18.

CARNEVALI, P., and PATARNELLO, S. (1987): Exhaustive Thermodynamical Analysis of Boolean Learning Networks. *Europhys. Lett. 4, 1199–1204.*

CYBENKO, G. (1988): Continuous Valued Neural Networks with Two Hidden Layers are Sufficient. *Technical Report, Department of Computer Science, Tufts Univ., Medford, MA.*

DERRIDA, B., GARDNER, E., and ZIPPELIUS, A. (1987): An Exactly Solvable Model of an Asymmetric Neural Network *Europhys. Lett., 4, 167.*

DOMANY, E., VAN HEMMEN, J.L., SCHULTEN, K. (ed.) (1992): *Models of Neural Networks*, Springer, Berlin.

DURBIN, R., and WIILSHAW, D. (1987): An Analogue Approach to the Travelling Salesman Problem Using an Elastic Net Method. *Nature, 326, 689–691.*

GYÖRGYI, G. (1990): Inference of a Rule by a Neural Network with Thermal Noise. *Phys. Rev. Lett., 64, 2957–2960.*

HERTZ, J., KROGH, A., and PALMER, R.G. (1991):*Introduction to the Theory of Neural Computation*. Addison-Wesley, Redwood City.

HINTON, G.E., and SEJNOWSKI, T.J. (1986): Learning and Relearning in Boltzmann machines. In: Rumelhart, McClelland (ed.) *Parallel Distributed Processing.*, Vol.1, Ch. 7.

HOPFIELD, J.J. (1982): Neural Networks and Physical Systems with Emergent Collective Computational Properties. Proc. Nat. Acad. Sc. USA, 81, 3088–3092

HORNER, H. (1992):Dynamics of Learning in a Binary Perceptron. *Zeitschrift f. Phys. B, 86, 291–308*

HORNIK, K., STINCHCOMB, M., and WHITE, H. (1989): Multilayer Feedforward Networks are Universal Approximators. *Neural Networks, 2, 359–368*

JACOBS, R.A., JORDAN, M.I., NOWLAN, S.J., and HINTON, G.E. (1991): Adaptive Mixtures of Local Experts *Neural Computation, 3, 79–87.*

JUDD, S. (1988): On the Complexity of Loading Shallow Neural Networks. *Journal of Complexity, 4.*

KOHONEN, T. (1989): *Self-Organization and Associative Memory.* Springer, Berlin.

KREE, R. and ZIPPELIUS, A. (1987): Continuous–Time Dynamics of Asymmetrically Diluted Networks. *Phys. Rev. A, 36, 4421–4427*

KREE, R., and ZIPPELIUS, A. (1988): Recognition of Topological Features of Graphs and Images in Neural Networks. *Journ. Phys. A, 21, L813-L818*

KREE, R., and MÜLLER, A. (1991): Classification Properties of Communicating Neural Networks. In: *Proceedings of the 16th Annual Meeting of the GfKl, Dortmund.*

KREE, R., and MÜLLER, A. (1993): Multi-Agent Neural Network Models In: *Proc.III. European Congress of Psychology, Tampere, Finland*

KULLBACK, S., and LEIBLER, R.A. (1951): On Information and Sufficiency. *Ann. Math. Stat., 22, 79–86*

LEVIN, E., TISHBY, N., and SOLLA, S. (1990): A Statistical Approach to Learning and Generalization in Layered Neural Networks. *Proc. IEEE, 78, 1568–1574.*

LINSKER, R. (1988): Self-Organization in a Perceptual Network. *Computer, 105–117*

LIPPMANN, R.P.(1989): Review of Neural Networks for Speech Recognition. *Neural Computation, 1,1–38.*

LITTLE, W.A. (1974): The Existence of Persistent States in the Brain. *Math. Biosc., 19, 101–120.*

MARCHAND, M., GOLEA, M., and RUJAN, P. (1990): A Convergence Theorem for Sequential Learning in Two-Layer Perceptrons. *Europhys. Lett, 11, 487–492.*

MCCULLOCH, W.S., PITTS, W. (1943): A Logical Calculus of Ideas Immanent in Nervous Activity. *Bull. Math. Biophys., 5, 115–133.*

MEZARD, M., PARISI, G., and VIRASORO, M.A. (ed.) (1986): *Spin Glass Theory and Beyond*, World Scientific, Singapure

MINSKY, M.L., and PAPERT, S.A. (1969): *Perceptrons*. MIT Press, Cambridge, MA.

NERRAND, O., ROUSSEL-RAGOT, P., PERSONNAZ, L., DREYFUS, G., and MARCOS, S. (1993): Neural Networks and Non-Linear Adaptive Filtering:Unifying Concepts and New Algorithms. *preprint*

OPPER, M., and HAUSSLER, D. (1991): Generalization Performance of Bayes Optimal Classification Algorithm for Learning a Perceptron. *Phys. Rev. Lett., 66, 2677–2680*

PARISI, G. (1992): On the Classification of Learning Machines. *Network, 3, 259–265.*

PETERSON, C. (1990): Parallel Distributed Approaches to Combinatorial Optimization: Benchmark Studies on Travelling Salesman Problem. *Neural Computation, 3, 261–269.*

RITTER, H., MARTINETZ, T., and SCHULTEN, K (1991): *Neural Computation and Self-Organizing Maps*. Addison-Wesley, Reading, MA.

ROSENBLATT, F. (1962): *Principles of Neurodynamics*. Spartan, New York

SEUNG, H.S., OPPER, M., and SOMPOLINSKY, H. (1992). In: *Proc. Annual ACM Workshop on Computation and Learning Theory*, 287–294.

SEUNG, H.S., SOMPOLINSKY, H., and TISHBY, N. (1992): Statistical Mechanics of Learning from Examples. *Phys. Rev. A, 45, 6065-6091*

SHERRINGTON, D. (1993): Neural Networks: The Spin Glass Approach. *preprint*

TANK, D.W., and HOPFIELD, J.J. (1986): Simple Neural Optimization Networks *IEEE Trans. Circuits and Systems, 33, 533-541.*

TISHBY, N., LEVIN, E., and SOLLA, S. (1989): Consistent Inference of Probabilities in Layered Networks. *IJCNN, pp.II, 403-410*

TSIRUKIS, G., REKLAITIS, V., and TENORIO, M.F. (1989): Nonlinear Optimization Using Generalized Hopfield Networks. *Neural Computation, 1, 511–521.*

VALIANT, L.G. (1984): A Theory of the Learnable. *Commun. ACM., 27, 1134–1142.*

VAPNIK, V.N., and CHERVONENKIS, A. (1971): On the Uniform Convergence of Relative Frequencies of Events to Their Probabilities. *Theory Prob. Appl., 16, 264–280.*

WATKIN, T.L.H., (1992): Optimal Learning with a Neural Network *preprint*

WATKIN, T.L.H., RAU, A., and BIEHL, M. (1992): The Statistical Mechanics of Learning a Rule (to be published in *Rev. Mod. Phys.*)

WHITE, H. (1989): Learning in Artificial Neural Networks: A Statistical Perspective. *Neural Computation, 1, 425–464.*

WONG, K.Y.M., SHERRINGTON, D. (1988): Storage Properties of Randomly Connected Boolean Neural Networks for Associative Memory. *Europhys. Lett., 7, 197–201.*

WULFF, N.H. (1992): Learning Dynamics with Recurrent Networks. *NORDITA preprint*

ZAK, M. (1988): Terminal Attractors for Addressable Memory in Neural Networks. *Phys. Lett. 133A, 18–22.*

Reasoning with Uncertainty in Diagnostic Systems

Sigrid Pöhlmann
University of Munich, Seminar für Spezialgebiete der Statistik
Ludwigstr. 33, D-80539 München, Germany

Summary: This paper discusses aspects of uncertainty in diagnostic systems and shows in particular what opportunities arise in the use of interval probabilities. It also demonstrates that this method has its own problems which have to be taken into consideration.

I. Methods for describing and combining uncertain knowledge

In literature many proposals for describing uncertainty in diagnostic systems are found. Probably the most well-known is the expert system MYCIN. It was developed 1976 by Edward Hance Shortliffe and Bruce G. Buchanan for use in medical diagnosis (Shortliffe (1976)). It was based on probability theory, but it uses ad hoc methods to handle these probabilities. Let us present briefly its main properties.

In the simplest case there is given a hypothesis E (it may be a disease or a certain state of nature), an observation Z (for example the result of a test or a certain symptom) and the probability $\mathrm{P}(E|Z)$ - the probability of E given the observation Z - and let $\mathrm{P}(E)$ be the prior probability of E. For example a physician knows the probability $\mathrm{P}(E)$ for a certain disease in the whole population. The probability of this disease is $\mathrm{P}(E|Z)$ if a certain symptom Z was observed. In such a situation MYCIN uses two measures, the measure of increased belief MB and the measure of increased disbelief MD:

$$\mathrm{MB}(E,Z) = \begin{cases} 1 & \text{if } \mathrm{P}(E)=1 \\ \dfrac{\max\left[\mathrm{P}(E|Z),\mathrm{P}(E)\right]-\mathrm{P}(E)}{1-\mathrm{P}(E)} & \text{otherwise} \end{cases}$$

$$\mathrm{MD}(E,Z) = \begin{cases} 1 & \text{if } \mathrm{P}(E)=0 \\ \dfrac{\mathrm{P}(E)-\min\left[\mathrm{P}(E|Z),\mathrm{P}(E)\right]}{\mathrm{P}(E)} & \text{otherwise} \end{cases}$$

These are measures of a change in belief but not measures of absolute belief.

One of the major problems in diagnostic systems is the combination of evidence resulting from several sources of information. The problem can be described as follows: There are two sources of information, Z_1 and Z_2, and one diagnosis E (or $\neg E$) is asked for. It is known how to conclude from Z_1 to E and how to conclude from Z_2 to E. Our aim is the conclusion from Z_1 and Z_2 to E. Examples are two alarm units in a power plant, or medical diagnoses on the basis of two symptoms or the coincidence of two hints.

In MYCIN for each symptom we have a measure of increased belief $\mathrm{MB}(E,Z_j)$ and a measure of increased disbelief $\mathrm{MD}(E,Z_j)$ $(j=1,2)$. The goal is the probability — respectively the MB and the MD — for a certain disease E if symptom Z_1 and symptom Z_2 are observed. In an earlier version the expert system MYCIN used separate formulas for combining the MBs and the MDs. In a later version MBs and MDs are combined to a single number which is called the certainty factor CF:

$$\mathrm{CF}(E,Z) = \mathrm{MB}(E,Z) - \mathrm{MD}(E,Z)$$

The combination rule for this certainty factor is given by:

$$\mathrm{CF} = \begin{cases} \mathrm{CF}_1 + \mathrm{CF}_2 - \mathrm{CF}_1 \cdot \mathrm{CF}_2 & \text{for } \mathrm{CF}_1, \mathrm{CF}_2 > 0 \\ \mathrm{CF}_1 + \mathrm{CF}_2 + \mathrm{CF}_1 \cdot \mathrm{CF}_2 & \text{for } \mathrm{CF}_1, \mathrm{CF}_2 < 0 \\ \text{undefined} & \text{for } \mathrm{CF}_1 \cdot \mathrm{CF}_2 = -1 \\ \dfrac{\mathrm{CF}_1 + \mathrm{CF}_2}{1 - \min[|\mathrm{CF}_1|, |\mathrm{CF}_2|]} & \text{otherwise} \end{cases}$$

where CF_1 is the certainty factor for the observation Z_1 and CF_2 is the certainty factor for the observation Z_2. In MYCIN the certainty factors of several hypotheses are compared and therapy is based on the hypothesis with the highest CF-value (respectively the hypotheses with high CF-values).

Both combination rules are only ad hoc methods and no theoretical foundation can be given.

For a discussion of the advantages respectively disadvantages of the two combination rules see for example Shortliffe/Buchanan (1985) and Weichselberger/Pöhlmann (1990). Let us mention only one point: the MB and MD are relative measures characterizing a change of belief. If we consider an example: The prior probability of the first disease is very large and after the observation Z it has not changed very much; the prior probability of a second disease is very small and after the observation Z it has increased relatively strongly, but still by far does not reach the probability of the first one. Then MYCIN states that the second disease is given, not the first. (For detailed examples see Weichselberger/Pöhlmann (1990).)

It is very often cited that in many cases available knowledge is not sufficient to identify probabilities with a precise number, even for an expert. It is only possible to specify a lower limit L and an upper limit U. For example, an expert (or a physician) would not be able to describe the probability for a certain disease as exactly 38%; he could only say that this probability is between 35% and 40%. Classical probability theory could not deal with such an interval probability.

One theory, which is said to handle this kind of uncertain knowledge is the Dempster-Shafer theory of belief-functions. The starting point of this theory is the basic probability assignment $m(E)$ of a certain state of nature E. The features of this m are the following:

$$0 \le m(E) \le 1$$
$$m(\emptyset) = 0$$
$$\sum_{E \subset \Theta} m(E) = 1$$

where Θ is the frame of discernment (the union of all states of nature).

From the last formula it is easily seen that m need not be a probability in general (except in special cases). In 1975 Glenn Shafer, a student of Arthur P. Dempster's, had introduced these basic probability assignments (Shafer (1975)) . But he did not define what m has to be. If the basic probability assignments for all hypotheses are given, the belief function for a certain hypothesis E may be calculated:

$$\mathrm{Bel}(E) = \sum_{E' \subset E} m(E') \quad .$$

With this notation Shafer defines the so-called upper probability of E:

$$P^*(E) = 1 - \mathrm{Bel}(\neg E) \quad .$$

In the case of two independent sources of information yielding $m_1(E_i)$ and $m_2(E_i)$ the basic probability assignments produced by the combined information are given by the Dempster-Shafer combination rule:

$$m(E) = \frac{\sum\limits_{\substack{i,j \\ E_i \cap E_j = E}} m_1(E_i) m_2(E_j)}{1 - \sum\limits_{\substack{i,j \\ E_i \cap E_j = \emptyset}} m_1(E_i) m_2(E_j)}$$

With this "combined" m the combined belief function of E can be calculated.

For this combination rule it is assumed that the two sources of information are independent. Dempster describes his ideas about this assumption in the following way (Dempster (1967)):

> "The mechanism adopted here assumes independence of the sources, a concept whose real world meaning is not so easily described as its mathematical definition. Opinions of different people based on overlapping experiences could not be regarded as independent sources. Different measurements by different observers on different equipment would often be regarded as independent, but so would different measurements by one observer on one piece of equipment: here the question concerns independence of errors."

There are many objections against the use of Dempster-Shafer theory for expert systems (for example see Weichselberger/Pöhlmann (1990), Zadeh (1979)). Therefore the question arises whether it is possible to use interval probability in diagnostic systems.

II. Combining uncertain knowledge with probability theory

Firstly we consider the case that probabilities are given precisely, i.e. they are point probabilities. It is assumed that probabilities of hypotheses given certain observations are used, as is done in MYCIN (the "direct" method). With an other method (the "analytic" method) the expert estimates the probability of observations given certain hypotheses, as in the expert system HUGIN (Lauritzen/Spiegelhalter (1988)). Which method is used depends on the type of information the expert can give[1].

In this paper we only deal with the direct method, i.e. with probabilities of a disease E given a certain symptom Z: $P(E|Z)$. The general model is that there are k states of nature E_i:

$$E_1, \ldots, E_k$$

and l units Z_j (alarm units or tests or symptoms):

$$Z_1, \ldots, Z_l$$

[1]The only way, by now, to combine the analytic method with weaker information is found in Shaefer, Shenoy, Mellouli (1987), where belief functions are used for this purpose.

and each unit Z_j produces exactly one of n_j signs:

$$Z_j = \left(Z_{j1}, \ldots, Z_{jn_j}\right) \quad .$$

To derive a simple combination formula we have to demand the mutual k-independence (see Weichselberger/Pöhlmann 1990):

$$\mathrm{P}(Z_{1r_1} \cap \cdots \cap Z_{lr_l} | E_i) = \prod_{j=1}^{l} \mathrm{P}(Z_{jr_j} | E_i)$$

for all $r_j = 1, \ldots, n_j; j = 1, \ldots, l$ and for all $i = 1, \ldots, k$.

This is a relatively strong assumption, but it is justifiable in practical circumstances, because it may be taken that whatever the state of nature, the alarm units or the symptoms do not influence each other[2].

If the conditional probabilities $\mathrm{P}(E_i|Z_{jr_j})$ are given, it follows that the total probability:

$$\mathrm{P}(E_i) = \sum_{r_1=1}^{n_1} \mathrm{P}(E_i|Z_{1r_1})\mathrm{P}(Z_{1r_1}) = \cdots = \sum_{r_l=1}^{n_l} \mathrm{P}(E_i|Z_{lr_l})\mathrm{P}(Z_{lr_l}) \qquad (i = 1, \ldots, k)$$

where $\mathrm{P}(Z_{jr_j})$ is the probability that unit Z_j shows sign Z_{jr_j}. This system of equations represents restrictions for the conditional probabilities $\mathrm{P}(E_i|Z_{jr_j})$, which have to be taken into account whenever these probabilities are estimated: the system of the conditional probabilities $\mathrm{P}(E_i|Z_{jr_j})$ and the system of the total probabilities $\mathrm{P}(E_i)$ have to fit together. If conditional probabilities $\mathrm{P}(E_i|Z_{jr_j})$ and total probabilities $\mathrm{P}(E_i)$ are given, probabilities $\mathrm{P}(Z_{jr_j})$ must exist, so that the formula given above is valid. If these conditions are fulfilled, the following combination rule can be derived:

$$\begin{aligned}
\mathrm{P}(E_{i^*}|Z_{1r_1} \cap \cdots \cap Z_{lr_l}) &= \frac{\dfrac{1}{\mathrm{P}(E_{i^*})^{l-1}} \prod_{j=1}^{l} \mathrm{P}(E_{i^*}|Z_{jr_j})}{\sum_{i=1}^{k} \dfrac{1}{\mathrm{P}(E_i)^{l-1}} \prod_{j=1}^{l} \mathrm{P}(E_i|Z_{jr_j})} \\
&= \frac{\mathrm{P}(E_{i^*}) \prod_{j=1}^{l} \dfrac{\mathrm{P}(E_{i^*}|Z_{jr_j})}{\mathrm{P}(E_{i^*})}}{\sum_{i=1}^{k} \mathrm{P}(E_i) \prod_{j=1}^{l} \dfrac{\mathrm{P}(E_i|Z_{jr_j})}{\mathrm{P}(E_i)}}
\end{aligned}$$

$$(i^* = 1, \ldots, k; r_j = 1, \ldots, n_j, j = 1, \ldots, l)$$

which is based on Bayes' theorem. (For example compare Weichselberger/Pöhlmann (1990).)

We now extend this formula to interval probabilities. At first we consider the simplest case with only two states of nature $\{E_S, \neg E_S\}$ and two alarm units with only two

[2]This assumption is discussed by Dempster (1967). It should be noted that by some authors a much stronger independence assumption is used, which leads to contradictions (for example see Johnson (1986) and Pednault, Zucker and Muresan (1981)), since independence may be assumed separately for each of several distributions, but not at the same time for a mixture of such distributions.

signs (for simplicity let us call them + and −). We are interested in the combined probability of the state of nature E_S under the condition that unit Z_1 shows sign + and unit Z_2 shows sign +. The conditional probabilities $P(E_S|Z_{1+})$, $P(E_S|Z_{2+})$ and the total probability $P(E_S)$ have to be given. In the case of interval probabilities, the basic information for our problem may be described by:

$$\begin{aligned} L_1 &\le \omega_1 = P(E_S|Z_{1+}) \le U_1 \\ L_2 &\le \omega_2 = P(E_S|Z_{2+}) \le U_2 \\ L &\le \omega = \quad P(E_S) \quad \le U \end{aligned}$$

For abbreviation we use x_{++} for the probability $P(E_S|Z_{1+} \wedge Z_{2+})$. Then the formula for point probabilities reads as follows:

$$x_{++} = \frac{\dfrac{\omega_1\omega_2}{\omega}}{\dfrac{\omega_1\omega_2}{\omega} + \dfrac{(1-\omega_1)(1-\omega_2)}{1-\omega}} .$$

Since the system of conditional probabilities and the total probabilities have to fit together (theorem on total probability), we get:

$$P(E_S) = P(Z_{j+}) \cdot P(E_S|Z_{j+}) + P(Z_{j-}) \cdot P(E_S|Z_j-) \qquad j = 1,2$$

or

$$\omega = p_j\omega_j + (1-p_j)\overline{\omega}_j \qquad j = 1,2$$

where $\overline{\omega}_j = P(E_S|Z_{j-})$ and $p_j = P(Z_{j+})$.

With the upper limits U and U_j and the lower limits L and L_j we derive a first solution (without additional information) of our problem:

$$\frac{\dfrac{L_1L_2}{U}}{\dfrac{L_1L_2}{U} + \dfrac{(1-L_1)(1-L_2)}{1-U}} \le x_{++} \le \frac{\dfrac{U_1U_2}{L}}{\dfrac{U_1U_2}{L} + \dfrac{(1-U_1)(1-U_2)}{1-L}}$$

(If one denominator is zero, for example $L = 0$, we get for the upper limit of x_{++}: $U(x_{++}) = 1$. If $U = 1$ we get for the lower limit of x_{++}: $L(x_{++}) = 0$. In the following it is always possible to consider the case of a zero denominator separately. Therefore it will not be stated explicitly.)

Now the question arises whether it is possible that the conditional probability, for example $P(E_S|Z_{1+})$, is equal to its upper limit U_1 and simultaneously the total probability is equal to lower limit L. Normally this cannot be the case. The same is true for the second unit. Therefore we have to take into account the additional information for the conditional probability $P(E_S|Z_{j-})$. Let us suppose that also in this case an interval probability is given:

$$\overline{L}_j \le \overline{\omega}_j = P(E_S|Z_{j-}) \le \overline{U}_j, \qquad j = 1,2$$

The term "additional information" is not quite correct. It only means that this information is not directly used in the combination rule for x_{++}. But it is implicitly used in the formula for the total probability. Since the expert system has to take into account all possible outcomes, both types of information, ω_j and $\overline{\omega}_j$, are of the same importance.

Let us now investigate the connection between conditional probabilities and total probabilities. At first we consider one system and the total probability:

$$L_1 \le \omega_1 \le U_1$$
$$\overline{L}_1 \le \overline{\omega}_1 \le \overline{U}_1$$
$$L \le \omega \le U$$

The prior probability ω is a linear combination of the conditional probabilities ω_1 and $\overline{\omega}_1$:

$$\omega = p_1\omega_1 + (1 - p_1)\overline{\omega}_1$$

Therefore a probability p_1 must exist:

$$p_1 = \frac{\omega - \overline{\omega}_1}{\omega_1 - \overline{\omega}_1} \in [0;1]$$

This is true if and only if the values for ω_1 and $\overline{\omega}_1$ fit together, i.e. they can occur simultaneously and produce a common ω.

The value for p_1 depends on the values of ω_1 and $\overline{\omega}_1$. Therefore there exist many values of p_1 and the demand is, that these values constitute a probability interval $\left[p_1^L; p_1^U\right]$. This leads to the following definition (see Weichselberger/Pöhlmann (1990)).

Definition:

A given system

$$L_1 \le \omega_1 \le U_1$$
$$\overline{L}_1 \le \overline{\omega}_1 \le \overline{U}_1$$
$$L \le \omega \le U$$

is called **interval-admissible**, iff:

$$[L, U] = \left\{\omega \,|\, \omega = p_1\omega_1 + (1 - p_1)\overline{\omega}_1; L_1 \le \omega_1 \le U_1, \overline{L}_1 \le \overline{\omega}_1 \le \overline{U}_1, \; 0 \le p_1^L \le p_1 \le p_1^U \le 1\right\}$$

This definition means that for all values of ω_1 and $\overline{\omega}_1$ in the given intervals and for all p_1 in the interval $\left[p_1^L; p_1^U\right]$ the resulting ω lies within its limits L or U. Another demand (implicitly given in this definition) is that the limits for the prior probability ω (L and U) have to be reached.

Not every given system is interval-admissible from the beginning. If a system is not interval-admissible, it can be made interval-admissible by modification of some limits. Which of these modifications are used, must be decided by the expert. (For further detail see Weichselberger/Pöhlmann (1990).)

In the following it is supposed that all systems are interval-admissible. Then the combination rule for two systems can be derived in the following way:

Our basic information is that:

$$L_1 \le \omega_1 \le U_1 \qquad\qquad L_2 \le \omega_2 \le U_2$$
$$\overline{L}_1 \le \overline{\omega}_1 \le \overline{U}_1 \qquad\qquad \overline{L}_2 \le \overline{\omega}_2 \le \overline{U}_2$$
$$L \le \omega \le U$$
$$p_1^L \le p_1 \le p_1^U \qquad\qquad p_2^L \le p_2 \le p_2^U$$

is an interval-admissible system.

Therefore

$$\min \omega(\omega_1,\omega_2) \le \omega \le \max \omega(\omega_1,\omega_2)$$

$$\text{with} \quad \begin{aligned} \min\omega(\omega_1,\omega_2) &= \max_{j=1,2}\left\{\min\left[p_j^L\omega_j + (1-p_j^L)\overline{L}_j; p_j^U\omega_j + (1-p_j^U)\overline{L}_j\right]\right\} \\ \max\omega(\omega_1,\omega_2) &= \min_{j=1,2}\left\{\max\left[p_j^L\omega_j + (1-p_j^L)\overline{U}_j; p_j^U\omega_j + (1-p_j^U)\overline{U}_j\right]\right\} \end{aligned}$$

has to be valid.

This leads to the

Combination Rule:

If $0 < \max\omega(L_1, L_2) < 1$ and $0 < \min\omega(U_1, U_2) < 1$:

$$\frac{\frac{L_1L_2}{\max\omega(L_1,L_2)}}{\frac{L_1L_2}{\max\omega(L_1,L_2)} + \frac{(1-L_1)(1-L_2)}{1-\max\omega(L_1,L_2)}} \le x_{++} \le \frac{\frac{U_1U_2}{\min\omega(U_1,U_2)}}{\frac{U_1U_2}{\min\omega(U_1,U_2)} + \frac{(1-U_1)(1-U_2)}{1-\min\omega(U_1,U_2)}}$$

The following example compares the solution without additional information and the combination rule:

Example:

$$\begin{aligned} 0.1 &\le \omega_j \le 0.2 \\ 0.01 &\le \overline{\omega}_j \le 0.02 \qquad (j=1,2) \qquad p_j^L = 0.5,\ p_j^U = 0.6 \\ 0.055 &\le \omega \le 0.128 \end{aligned}$$

$0.1447 \le x_{++} \le 0.3476$ (combination rule)

$0.0776 \le x_{++} \le 0.5178$ (solution without additional information)

It is evident that analogous formulas for the other signs, namely for x_{+-}, x_{-+} and x_{--} can be derived. Also the formulas can be extended to the case of more than two units (for detail see Weichselberger/Pöhlmann (1990)). We use the following notation:

$$\mathrm{P}(E_s|Z_{11} \wedge \cdots \wedge Z_{l1}) =: x_{l+}$$

The combination rule for point probabilities is :

$$x_{l+} = \frac{\frac{\omega_1 \cdot \ldots \cdot \omega_l}{\omega^{l-1}}}{\frac{\omega_1 \cdot \ldots \cdot \omega_l}{\omega^{l-1}} + \frac{(1-\omega_1)\cdot \ldots \cdot (1-\omega_l)}{(1-\omega)^{l-1}}}$$

In case of interval probabilities the following interval-admissible systems have to be given:

$$\begin{aligned} L_1 \le \omega_1 \le U_1, &\quad \overline{L}_1 \le \overline{\omega}_1 \le \overline{U}_1 \\ \vdots \qquad & \qquad \vdots \\ L_l \le \omega_l \le U_l, &\quad \overline{L}_l \le \overline{\omega}_l \le \overline{U}_l \end{aligned}$$

For the solution without additional information we get (if $U < 1$ and $L > 0$):

$$\frac{\dfrac{L_1 \cdot \ldots \cdot L_l}{U^{l-1}}}{\dfrac{L_1 \cdot \ldots \cdot L_l}{U^{l-1}} + \dfrac{(1-L_1)\cdot \ldots \cdot (1-L_l)}{(1-U)^{l-1}}} \le x_{l+} \le \frac{\dfrac{U_1 \cdot \ldots \cdot U_l}{L^{l-1}}}{\dfrac{U_1 \cdot \ldots \cdot U_l}{L^{l-1}} + \dfrac{(1-U_1)\cdot \ldots \cdot (1-U_l)}{(1-L)^{l-1}}}$$

If $L = 0$: $U(x_{1+}) = 1$
If $U = 1$: $L(x_{1+}) = 1$

and for the combination rule we obtain:

Theorem:

If $p_j^L \le p_j \le p_l^U,\ j = 1, \ldots, l$
and $0 < \max\omega(L_1, \ldots, L_l) < 0$, $0 < \min\omega(U_1, \ldots, U_l) < 1$ then:

$$\frac{\dfrac{L_1 \cdot \ldots \cdot L_l}{[\max\omega(L_1,\ldots,L_l)]^{l-1}}}{\dfrac{L_1 \cdot \ldots \cdot L_l}{[\max\omega(L_1,\ldots,L_l)]^{l-1}} + \dfrac{(1-L_1)\cdot \ldots \cdot (1-L_l)}{[1-\max\omega(L_1,\ldots,L_l)]^{l-1}}} \le x_{++} \le$$

$$\le \frac{\dfrac{U_1 \cdot \ldots \cdot U_l}{[\min\omega(U_1,\ldots,U_l)]^{l-1}}}{\dfrac{U_1 \cdot \ldots \cdot U_l}{[\min\omega(U_1,\ldots,U_l)]^{l-1}} + \dfrac{(1-U_1)\cdot \ldots \cdot (1-U_l)}{[1-\min\omega(U_1,\ldots,U_l)]^{l-1}}}$$

where

$$\min\omega(\omega_1, \ldots, \omega_l) = \max_{j=1,\ldots,l}\left\{\min\left[p_j^L\omega_j + (1-p_j^L)\overline{L}_j; p_j^U\omega_j + (1-p_j^U)\overline{L}_j\right]\right\}$$
$$\max\omega(\omega_1, \ldots, \omega_l) = \min_{j=1,\ldots,l}\left\{\max\left[p_j^L\omega_j + (1-p_j^L)\overline{U}_j; p_j^U\omega_j + (1-p_j^U)\overline{U}_j\right]\right\}$$

Now let us consider some problems arising with the use of interval probabilities.

III. Problems with interval probabilities

The first problem is the very important one of interval-admissiblity. All interval probabilities have a significant property: the inaccuracy of the probability estimates, resulting in the length of the probability interval. The estimates of the conditional probabilities should not be of significantly lower accuracy than the estimates of the total probability. Otherwise the total probability can never be the result of a weighted average of the conditional probabilities. Therefore modifications of the conditional probabilities — which are indeed improvements — are possible if the information about the total probability is taken as reliable, but the system is not interval-admissible. The problem with these modifications is that there is a choice between different modifications and only the expert can decide which of the possible modifications is justifiable.

The next problem appears in the combination of knowledge. Let us consider the following example. The prior limits are:

$$L = 0.4 \qquad U = 0.6$$

If the first symptom Q_1 is observed these limits increase to 0.7 and 0.8; if this symptom is not observed these limits decrease to 0.1 and 0.2:

Q_1:	$L_1 = 0.7$	$U_1 = 0.8$	$0.5 \leq p_1 \leq 0.667$
	$\overline{L}_1 = 0.1$	$\overline{U}_1 = 0.2$	

We also have symptoms Q_2, Q_3 and Q_4 with the following limits:

Q_2:	$L_2 = 0.8$	$U_2 = 0.9$	$0.2 \leq p_2 \leq 0.4$
	$\overline{L}_2 = 0.3$	$\overline{U}_2 = 0.4$	
Q_3:	$L_3 = 0.7$	$U_3 = 0.9$	$p_3 = 0.4$
	$\overline{L}_3 = 0.2$	$\overline{U}_3 = 0.4$	
Q_4:	$L_4 = 0.5$	$U_4 = 0.55$	$0.667 \leq p_4 \leq 1.0$
	$\overline{L}_4 = 0.2$	$\overline{U}_4 = 0.7$	

All these four systems are interval-admissible with the corresponding intervals for p_j.

If the physician searches for the first symptom and finds that this symptom appears, the probability of having the disease increases to 70% – 80%. If he then asks for the second symptom, and if this symptom is also given, the probability of having this disease increases to:

$$0.891 \leq x_{++} \leq 0.978 \quad .$$

If he combines a third symptom which is also at the sign +, we get

$$0.949 \leq x_{+++} \leq 0.997 \quad .$$

We find not only an increase in the probability for having the disease but also a shortening of the interval (our information has gained precision). Now let us ask for the fourth symptom. If it turns out that this symptom has the sign +, we get

$$0.945 \leq x_{++++} \leq 0.998$$

and if this is not the case (the sign -) we get

$$0.811 \leq x_{+++-} \leq 0.999 \quad .$$

In both cases the interval becomes larger; the accuracy is reduced.

This fact can be demonstrated in a simpler way: We only use symptom Q_1 and symptom Q_4. If symptom Q_1 is at sign +, the probability of having the disease is between 0.7 and 0.8 and if symptom Q_4 is at sign +, too, we obtain:

$$0.671 \leq x_{++} \leq 0.857 \quad .$$

If symptom Q_1 is at sign + and symptom Q_4 is at sign –, we arrive at:

$$0.338 \leq x_{+-} \leq 0.903 \quad .$$

The observation of the first symptom leads to an improvement of our information, but the observation of symptom Q_4 leads to a deterioration since the limits become wider. The same is true if symptom Q_1 is at sign $-$ and symptom Q_4 is at $+$:

$$0.078 \leq x_{-+} \leq 0.272$$

and if both Q_1 and Q_4 are at sign $-$:

$$0.022 \leq x_{--} \leq 0.368 \quad .$$

If we ask how this result is created, we have to go back to the original limits for symptom Q_4. If symptom Q_4 is true $(+)$, the prior limits (0.4 and 0.6) become restricted, while they are broadened if Q_4 is at sign $-$. This means that symptom Q_4 may "spoil" the available information, since the probability of a negative outcome $(-)$ at Q_4 is between 0 and 0.333. This possibility must be taken into consideration. Therefore it is better to eliminate symptom Q_4 from the stock of our information basis.

In the case of using interval probabilities, the process of gathering information has to be planned very carefully. Problems of this type never occur if point probabilities are used. Therefore statisticians using interval probabilities should become accustomed to these new aspects.

References:

BUCHANAN, B.G., and SHORTLIFFE, E.H. (1985) (Eds.): *Rule-based expert systems: the MYCIN experiments of the Stanford heuristic programming project*, Addison-Wesley.

DEMPSTER, A.P. (1967): Upper and lower probabilities induced by a multivalued mapping, *Ann. Math. Stat. 38, 325–339.*

JOHNSON, R.W. (1986): Independence and Baysian Updating Methods. *Artificial Intelligence 29, 217–222.*

LAURITZEN, S.L, and SPIEGELHALTER, D.J. (1988): Local computations with probabilities on graphical structures and their application to expert systems, *JRSS, B, Vol 50, 2, 187–224.*

PEDNAULT, E.P.D., ZUCKER, S.W. and MURESAN, L.V. (1981): On the independece assumption underlying subjective bayesian updating, *Artificial Intelligence 16, 213–222.*

SHAFER, G. (1975): *A mathematical theory of evidence*, Princeton University Press.

SHAFER, G., SHENOY, P., and MELLOULI, K. (1987): Propagating Belief Functions in Qualitative Markov Trees, *Inter. J. of Approx. Reas. 1, 349–400.*

SHORTLIFFE, E.H. (1976): *Computer-based medical consultations: MYCIN* Elsevier Computer Science Library.

WEICHSELBERGER, K., and PÖHLMANN, S. (1990): *A Methodology for Uncertainty in Knowledge-Based Systems*, Springer.

ZADEH, L.A. (1979): On the validity of Dempster's rule of combination of evidence, Memorandum No. UCB/ERL M 79/24, Electronic Research Laboratory, University of California, Berkeley.

Der klassische und der moderne Begriff des Begriffs.

Gedanken zur Geschichte der Begriffsbildung in den exakten Wissenschaften

Christian Thiel

Institut für Philosophie der Universität Erlangen-Nürnberg,
Bismarckstr. 1, D-91054 Erlangen, Germany

Summary: Up to the present day, difficulties have confronted all attempts at establishing a theory of concepts that would comprise the various kinds of concept-formation in the disciplines of the spectrum of sciences. Not a few philosophical dictionaries, under the entry "concept", still offer doctrines which were current far back in the history of philosophy and have little in common with concept-formations in the sciences of today.

The paper aims at an improvement in this situation. After a sketch of the "classical" notion of *concept*, already developed in antiquity (essentially a logic of "classification", although "class-formation" in its present understanding had not yet been conceived), the canonical modern doctrine of concepts is outlined. With an eye to application in the exact sciences, it is shown how in the nineteenth century the view of *concept* as an additive complex of characteristics yields to a functional approach systematized, in the last quarter of the century, by classical quantificational logic. Almost simultaneously, Mach, Frege, Peano, Weyl and others set out to shape the modern theory of abstraction. It is these two theories that today permit philosophers of science not only to deal with functional processes of concept-formation but also to represent in a formally correct manner metalinguistic propositions about concepts and their properties. Thus it seems that the fundamental tasks of a modern theory of *concept* have finally been taken care of.

1. „Begriff" in der Antike und in der traditionellen Logik

Begriffe stehen nach der alltäglichen Auffassung im Gegensatz zum Anschaulichen, sie sind etwas Allgemeines im Unterschied zum Individuellen oder Speziellen, etwas Abstraktes im Unterschied zum Konkreten. Dabei sieht man sie sei es als Mittel, sei es als Produkte unseres „Begreifens", das dabei nicht als eine logische, sondern als eine psychische Operation verstanden wird. Tatsächlich teilt uns noch 1971 der Artikel „Begriffsbildung" im *Historischen Wörterbuch der Philosophie* mit, unter „Begriffsbildung" werde „der *psychologische* Vorgang verstanden, der zur Kategorisierung von Objekten oder Ereignissen führt".[1] Das englische „concept" und seine Entsprechungen in den romanischen Sprachen scheinen ebenfalls auf eine psychologische oder zumindest erkenntnistheoretische Urbedeutung hinzuweisen.

Dieser Eindruck täuscht jedoch. Als Boethius durch die lateinischen Wörter „conceptus", „conceptio" und „terminus" das Aristotelische „horos" (ὅρος) übersetzte, traf er dieses bereits in metaphorischer Verwendung an. Ursprünglich bezeichnet nämlich „horos" ebenso wie „terminus" einen Steinhaufen oder Grenzstein zum Zwecke der Abgrenzung eines Ackers oder anderen Grundbesitzes von benachbartem Gebiet; und die Vornahme solcher Markierungen ist, wenngleich von psychischen Akten begleitet,

[1]Foppa (1971), 787. Hervorhebung im Original.

sicherlich selbst kein solcher. Überraschenderweise gilt das auch von unserem deutschen „Begriff": die mittelhochdeutsche Bedeutung von „begrif" als „Umfang, Bezirk" hält sich noch bis zum Ende des 17. Jahrhunderts, wo uns ein Wörterbuch der deutschen Sprache den „Begriff einer Stadt" als den „ambitus urbis" erklärt.[2] Das Bilden eines Begriffs, technisch gesprochen seine Definition, ist immer und vor allem ein Begrenzen.

Möglicherweise hat diese Stammbedeutung das in der Logikgeschichte nachweisbare Überwiegen der extensionalen Auffassung des Begriffs, also den Vorrang des Begriffsumfangs gegenüber dem Begriffsinhalt, gefördert. Wir finden diese Grundlage, obwohl die Lehre von den Begriffsverhältnissen und der Definition beim Platonischen Sokrates ihren Ursprung hat und von Aristoteles theoretisch ausgearbeitet wurde, erstmals in der Einleitungsschrift (Εἰσαγωγή, Isagoge) des Porphyrius (232–304) zu den *Kategorien* des Aristoteles. Hier werden die Grundbegriffe der traditionellen Begriffslogik eingeführt — in Boethius' lateinischer Übersetzung *genus*, *species*, *differentia* [*specifica*], *proprium* und *accidens*. Mit ihrer Hilfe werden dann die Über- und Unterordnungsverhältnisse zwischen Begriffen beschrieben, und ihre Veranschaulichung führt (wohl wenig später) zu der sog. „Begriffspyramide", die in verschiedenen isomorphen Formen auftritt, und zu den später oft liebevoll ausgeführten „Porphyrschen Bäumen", die man wohl als der lebendigen Naturwirklichkeit näherstehend empfand.[3]

Die Verhältnisse und Beziehungen innerhalb des Geästs dieser Bäume sind das Thema der klassischen Lehren von Prädikation und Subsumption, von Inhalt und Umfang der Begriffe, von der Definition, der Division und Klassifikation. Logikhistoriker und Logikhistorikerinnen, die nicht bloß antiquarisch arbeiten, finden es atemberaubend, welchen Umfang und welche Komplexität die durch die heutigen Probleme der Dokumentation und Datenverwaltung wiederbelebte und ausgebaute Lehre von der Klassifikation angenommen hat.[4] Und doch täuschen diese Anwendungen traditioneller Denkformen leicht darüber hinweg, daß deren Reichweite immanente Grenzen gesetzt sind, die sich bei der Metamorphose der Logik in der zweiten Hälfte des 19. Jahrhunderts in so gravierender Weise zeigten, daß ich die Rede von einer damals zutage getretenen „Krise der Merkmalslogik" für gerechtfertigt halte (weshalb ich diese Wendung auch in den Titel meines Vortrags aufgenommen habe). Worum geht es dabei?

2. Opposition gegen die „Merkmalslogik" im 19. Jahrhundert

Die traditionelle Rede von Merkmalen wurde vor allem durch zwei Mängel beeinträchtigt, deren ersten Rainer Stuhlmann-Laeisz im Kapitel „Merkmalslehre" seines Buches über Kants Logik klar beschrieben hat. Ein Merkmal dient danach, wie wohl auch der Name andeuten soll, als „Erkenntnisgrund"; es soll „eine Eigenschaft anzeigen, die hinreichend ist dafür, daß ein Gegenstand unter einen bestimmten Begriff

[2]Stieler, Kaspar: *Der Teutschen Sprache Stammbaum und Fortwachs* (Nürnberg 1691), zitiert nach dem Artikel „Begriff" in Kluge (1960), 60b.

[3]Die von Platon im *Sophistes* (219a ff.) gegebene dihairetische Definition der Angelfischerei setze ich als bekannt voraus. Die beiden ersten der vier Abbildungen auf den unserem Beitrag unmittelbar folgenden Seiten finden sich als Tafeln am Schluß von Purchotius (1730); sie zeigen eine „arbor Porphyriana" und eine „arbor Purchotiana". Die dann folgende „naturalistische" Arbor Porphyriana stammt aus Reviczky (1757). Modern und schmucklos präsentiert sich als vierte Abbildung die Klassifikation der Schalter in Warner (1966), wiedergegeben nach Heyde (1973), 16.

[4]Eine kritische Darstellung der traditionellen Lehre von Division und Klassifikation findet man in Joseph (1906), ch. V: "Of the Rules of Definition and Division: Classification and Dichotomy".

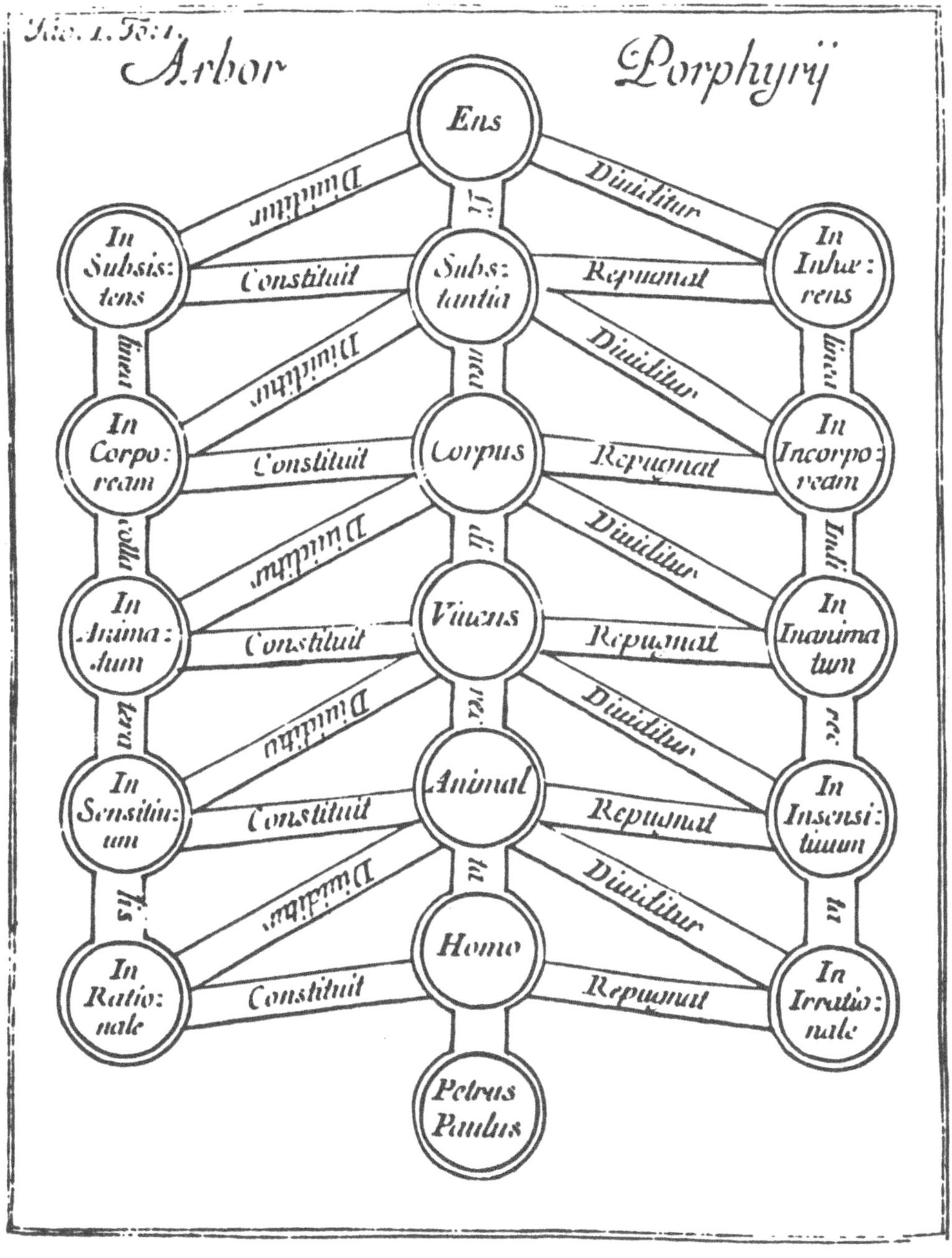
Arbor Porphyrij
Ens
Diuiditur
Diuiditur
In Subsis: tens
Constituit
Subs: tantia
Repugnat
In Inhæ: rens
In Corpo: ream
Constituit
Corpus
Repugnat
In Incorpo: ream
In Anima: tum
Constituit
Viuens
Repugnat
In Inanima tum
In Sensitiu: um
Constituit
Animal
Repugnat
In Insensi: tiuum
In Ratio: nale
Constituit
Homo
Repugnat
In Irratio: nale
Petrus Paulus

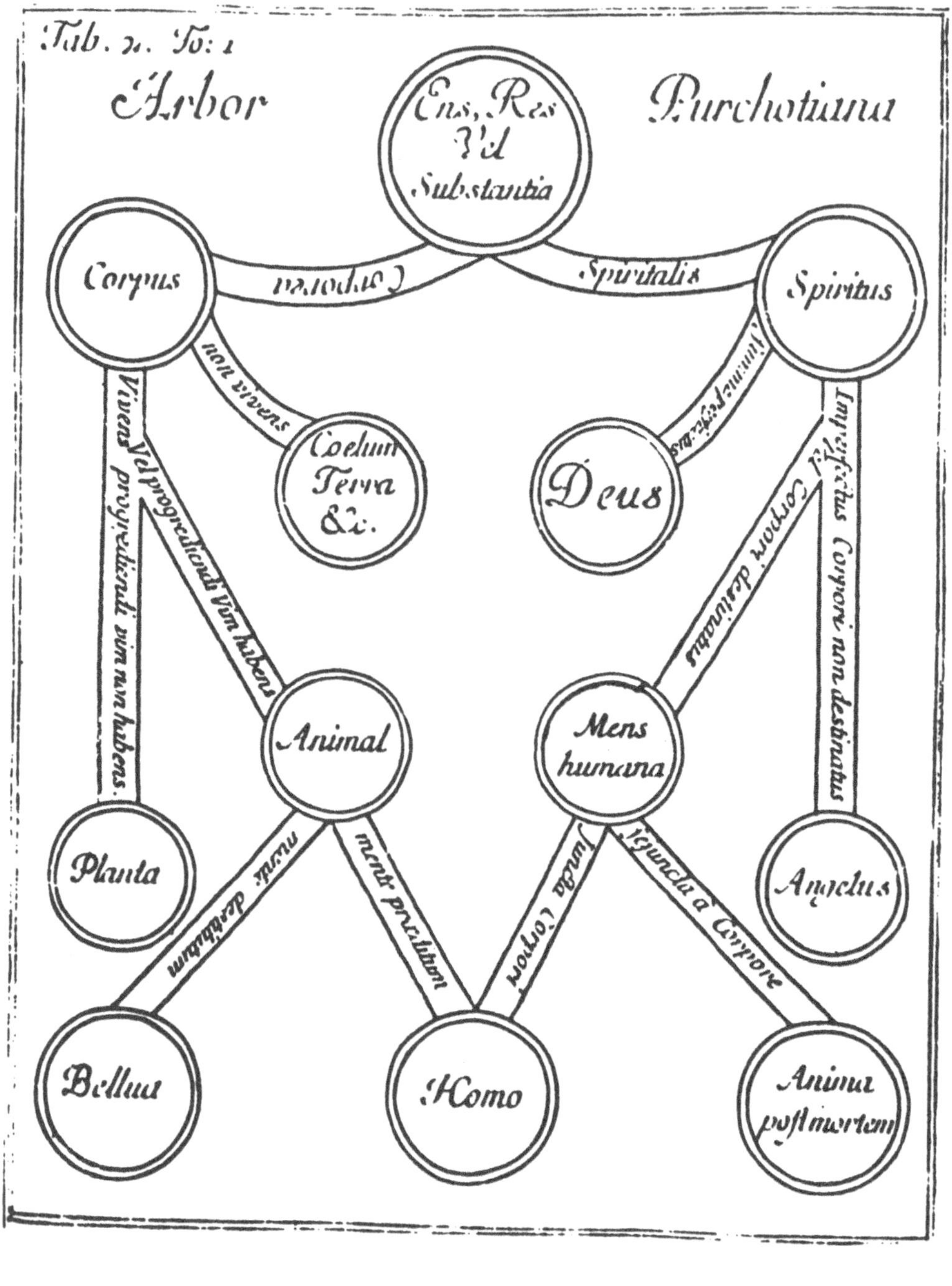
Tab. 2. To: 1
Arbor
Purchotiana
Ens, Res Vel Substantia
Corpus
Corporea
Spiritalis
Spiritus
non vivens
Coelum Terra &c.
summe perfectus
Deus
Vivens Vel progrediendi Vim habens
Vivens progrediendi vim non habens.
Imperfectus Vel Corpori destinatus
Imperfectus Corpori non destinatus
Animal
Mens humana
Planta
mente destitutum
mente praeditum
Juncta Corpori
Sejuncta a Corpore
Angelus
Bellua
Homo
Anima post mortem

ARBOR PORPHYRIANA
Ens
genus generalissimum
per se
in alio
Substantia
Accidens
SUBSTANTIA
CORPOREA
INCORPOREA
Coelum
Spiritus
CORPUS
ANIMATUM
INANIMATUM
planta
lapis
VIVENS
SENSITIVUM
INSENSITIVUM
animal
planta
ANIMAL
RATIONALE
IRRATIONALE
homo
brutum
HOMO GALLUS OVIS
SPECIES
Martin Prokes: sculp. Tyrnaviae

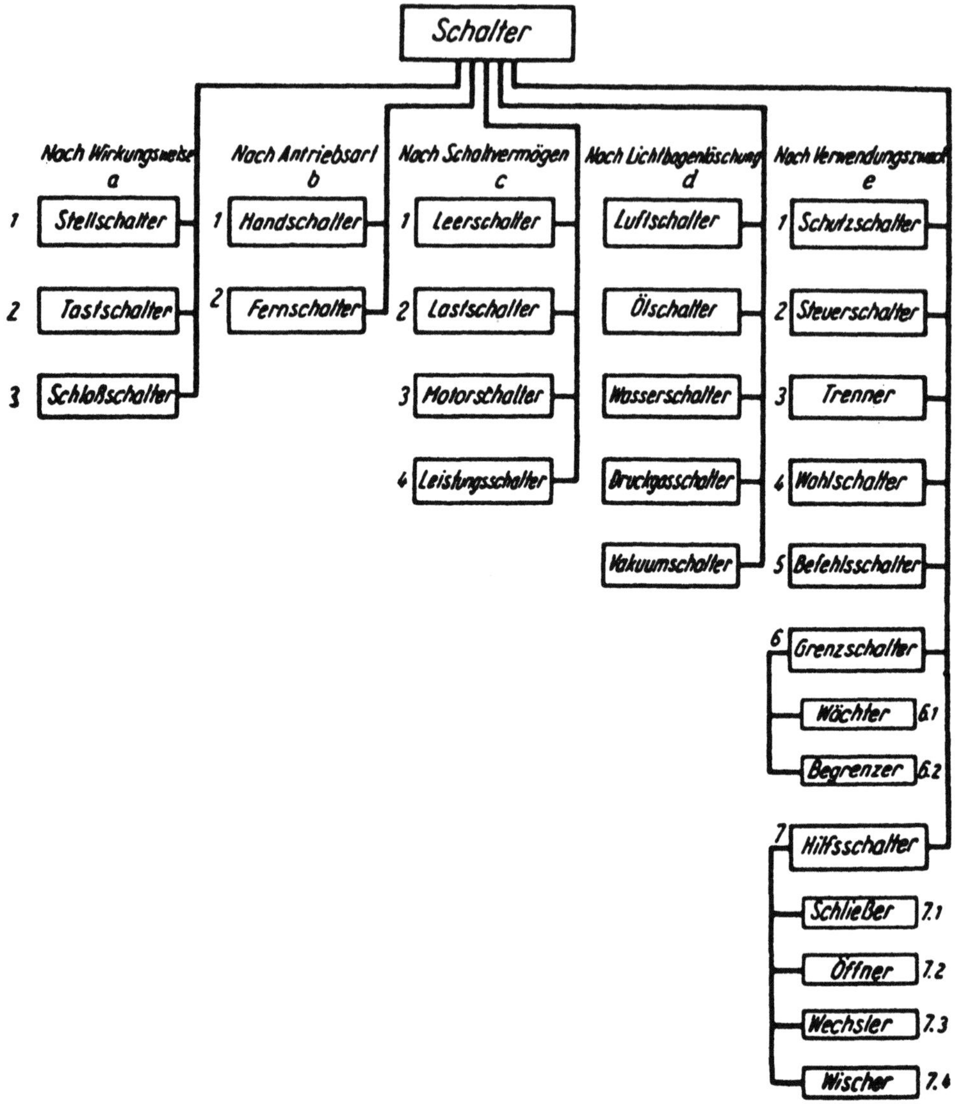
Schalter
Nach Wirkungsweise
a
1 Stellschalter
2 Tastschalter
3 Schloßschalter
Nach Antriebsart
b
1 Handschalter
2 Fernschalter
Nach Schaltvermögen
c
1 Leerschalter
2 Lastschalter
3 Motorschalter
4 Leistungsschalter
Nach Lichtbogenlöschung
d
Luftschalter
Ölschalter
Wasserschalter
Druckgasschalter
Vakuumschalter
Nach Verwendungszweck
e
1 Schutzschalter
2 Steuerschalter
3 Trenner
4 Wahlschalter
5 Befehlsschalter
6 Grenzschalter
Wächter 6.1
Begrenzer 6.2
7 Hilfsschalter
Schließer 7.1
Öffner 7.2
Wechsler 7.3
Wischer 7.4

fällt"[5] — haben nur Menschen Hände, so ist, eine Hand zu besitzen, Merkmal des Menschen, Kennzeichen des Menschseins. Stuhlmann-Laeisz hat a.a.O. überzeugend gezeigt, daß diese Auffassung, die sich so bei Kant und dem von ihm benutzten Lehrbuchautor Georg Friedrich Meier findet,[6] unverträglich ist mit der zugleich vertretenen Auffassung, Merkmale eines Begriffs seien dessen Teilvorstellungen. Diese Inkonsistenz bildet den ersten Mangel der traditionellen Merkmalslehre, auch wenn ich nicht überprüft habe, wie verbreitet die simultane Vertretung der beiden genannten Positionen war. Den zweiten Mangel sehe ich darin, daß die Auffassung der Merkmale als Teile von Begriffen in dem Normalfall, daß Gegenstände unter diese Begriffe fallen, nicht sonderlich damit harmoniert, daß man die Merkmale diesen Gegenständen zuschreibt, für die sie dann wie soeben skizziert „Kennzeichen" sein sollen — und Drobisch bedient sich 1851 noch dieser Redeweise. Zu den wenigen Autoren, die sich für eine eindeutige Position und Terminologie entscheiden, gehören Frege und Twardowski. Frege schreibt 1884 in *Die Grundlagen der Arithmetik*: „Unter Eigenschaften, die von einem Begriffe ausgesagt werden, verstehe ich natürlich nicht die Merkmale, die den Begriff zusammensetzen. Diese sind Eigenschaften der Dinge, die unter den Begriff fallen, nicht des Begriffes".[7] Twardowski kommt 1894 in seiner Habilitationsschrift nach umsichtigen Überlegungen über „Die dem Terminus ‚Merkmal' anhaftende Zweideutigkeit" (so die Überschrift des § 8) zu dem Ergebnis, „dass als Merkmale immer nur Teile des Gegenstandes einer Vorstellung, niemals jedoch Teile des Vorstellungsinhaltes zu bezeichnen seien".[8] Trotz eines Reprints 1982 scheint diese wichtige Arbeit Twardowskis (die auch noch einen § 13 „Das Merkmal" enthält) heute sowohl in der Klassifikationstheorie als auch in der philosophischen Logik fast unbekannt zu sein. John Findlay hat sie als "one of the most interesting treatises in the whole range of modern philosophy" bezeichnet.[9]

Die meisten Autoren entscheiden sich überhaupt nicht und verwenden „Merkmal" nebeneinander in beiden Bedeutungen; das entstandene terminologische Chaos führt noch 1929 drastisch — wenn auch ungewollt — der Artikel „Merkmal" in Rudolf Eislers *Wörterbuch der philosophischen Begriffe* vor Augen.[10] Drobisch spricht 1851 im § 16 der zweiten Auflage seiner Logik mehrfach davon, daß Objekte Merkmale „haben", und die Beispiele für disjunkte Merkmale („Roth und gelb z.B. oder süss und bitter", a.a.O., 48) legen eine Identifikation mit *Eigenschaften* der Objekte geradezu nahe. In den folgenden Auflagen (^{3}1863, ^{4}1875, ^{5}1887) ist diese Redeweise verschwunden.

Aber zurück zu den Merkmalen als Teilbegriffen! Nach Drobischs Formulierung schon in der ersten Auflage seiner Logik[11] „zerfällt [...] der Inhalt eines Begriffs als ein *Ganzes in Theile*; diese heißen *Merkmale*" (S. 10), und der § 18 beginnt (S. 14) mit den Worten: „Da Merkmale selbst Begriffe sind [...]". Im Zentrum der jetzt aufkommenden Kritik an der Merkmalsvorstellung steht jedoch nicht diese störende Doppeldeutigkeit, sondern eine Inkonsistenz anderer Art. Drobisch hatte schon 1836 die übliche Darstellung des Begriffsinhalts als *Summe seiner Merkmale* zu verbessern versucht, als er schrieb:[12]

Die Art der Verbindung der Merkmale im Begriffe ist keineswegs ein bloßes Ne-

[5] Stuhlmann-Laeisz (1976), 90.
[6] Meier (1752a), Meier (1752b/1760). Zu den Nachweisen bei Kant vgl. Stuhlmann-Laeisz (1976).
[7] Frege (1884), 64.
[8] Twardowski (1894), 46.
[9] Findlay (1963), 8.
[10] Eisler (1929), 123 f.
[11] Drobisch (1836).
[12] Drobisch (1836), 13.

beneinanderstellen, sondern eine *Bestimmung* des einen Merkmals oder des bereits gebildeten Complexes von Merkmalen durch das noch hinzukommende (also nicht analog der Addition, sondern der Multiplication). Sie führt daher sehr passend den Namen der *Determination.*

Er hatte sich diesen recht diskutablen Ansatz aber gleich selbst wieder verdorben durch die Auffassung der Abstraktion und der Determination als einander „entgegengesetzte logische Operationen“ (ibid.). Eine Determination eines Begriffs aufheben heißt dann (trotz der bekundeten Einsicht in die Verschiedenartigkeit der möglichen Determinationen), Merkmale, also Teile des Begriffs wegnehmen — eben „abstrahieren“. In einer ansonsten sehr wohlwollenden Rezension des Werkes äußerte sich Herbart skeptisch, ob es der Logik je gelingen werde, die verschiedenen möglichen Arten der Merkmalsverbindungen zu erfassen.[13] Trendelenburg hob in seinen *Logischen Untersuchungen* hervor, daß bei Drobisch wie in der Tradition die Merkmale „summirt oder multiplicirt“ seien und „daher auch wie Summanden oder Factoren nach einer beliebigen Reihenfolge getrennt werden“ könnten.[14] In der Tat erläutert Drobisch 1836 die Abstraktion durch ein reichlich unglückliches Beispiel. Er schreibt (a.a.O., 12 f.):

> Bewegung ist stetige Veränderung des Ortes. Abstrahire ich nun zunächst das Merkmal der Stetigkeit, sodann das des Ortes, so entsteht die Begriffsreihe: Bewegung, Ortsveränderung, Veränderung; geht man nun umgekehrt von diesem letzten Begriffe aus und fügt ihm zuerst das Merkmal der Stetigkeit bei, so kommt die von der vorigen im Mittelgliede verschiedene Reihe: Veränderung, stetige Veränderung, Bewegung.

Da keine Reihenfolge des Vorgehens vor einer anderen ausgezeichnet wird, scheint Drobisch tatsächlich die Merkmale selbst als gleichwertig zu behandeln. Das problematische Beispiel ist schon in der zweiten Auflage wieder verschwunden, die „egalitäre“ Merkmalsauffassung aber keineswegs. Sie wird lediglich eingehender erläutert:[15]

> Man kann den Inhalt eines Begriffs nicht als die *Summe* seiner Merkmale bezeichnen, auch wenn man darunter etwas Allgemeineres versteht als eine Summe von Zahlen und Grössen. In der Summe haben nämlich die Summanden nur eine ganz *äusserliche* Gemeinschaft, sie bestehen neben und an einander. Qualitativ Verschiedenes nach Art der Summe zusammengefasst kann nur ein *Aggregat* geben (vgl. § 29). Der Inhalt des Begriffs ist aber kein blosses Aggregat seiner Merkmale, sondern eine *innerliche* Verbindung, eine Durchdringung, deren Ergebniss man eher ein *Product* als eine Summe nennen kann.

Trendelenburg kommentierte dies in der zweiten Auflage seiner *Logischen Untersuchungen* trocken mit den Worten, hier sei „doch nur ein arithmetisches Bild in ein anderes verwandelt worden“,[16] das, wie Gegenbeispiele sofort zeigten, ebenso danebentreffe wie das erste.

Die Summenauffassung des Begriffs war wohl zu dieser Zeit bereits „out“, mit Sicherheit jedenfalls bei den Herbartianern. „Es ist neuerlich mit Nachdruck und grossem

---

[13]Herbart (1836), auch in Herbart (1852), 750–754.

[14]Trendelenburg (1840), II 159. Zur Diskussion der Frage vgl. auch Band I der zweiten, ergänzten Auflage Trendelenburg (1862), 21 f.

[15]Drobisch (1851), 28.

[16]Trendelenburg (1862), 22.

Rechte darauf aufmerksam gemacht worden", schreibt beispielsweise 1843 Franz Exner, „wie sehr man fehle, wenn man die Bestandtheile der Begriffe als blosse Summanden denke"[17] und er zitiert dafür Bolzanos *Wissenschaftslehre* und Trendelenburgs *Logische Untersuchungen*, merkwürdigerweise nicht Drobisch, den er doch bei Trendelenburg zitiert gefunden hat. Für Exner ist die Sache erledigt: Die „Summenform [...] hat kein einziger brauchbarer Begriff" (Exner (1843), 20); „Summen sind die Begriffe nun einmal nicht" (ibid., 24). Mit einiger Skepsis, aber doch als bedenkenswerte Anregung betrachte ich Exners Hinweis auf S. 21 seiner Abhandlung, schon Leibniz habe in seinem Fragment *Fundamenta calculi ratiocinatoris*[18] auch den Fall der Ungleichartigkeit von Begriffskomponenten gesehen und berücksichtigt, eine Einsicht, die dann erst im 19. Jahrhundert — wohl unabhängig von Leibniz — neu entdeckt worden sei.[19]

Den für die folgende Entwicklung entscheidenden Schritt scheint Lotze getan zu haben. In seiner Logik[20] von 1874 kritisiert er „die geläufige falsche Meinung, als seien ganz allgemein die Bestandtheile des Begriffs gleichwerthig, jeder mit dem Ganzen des Inhalts ebenso verbunden wie jeder andere, und jeder erste mit dem zweiten ebenso wie dieser mit dem dritten." Man darf vermuten, daß er sich auf das unglückliche Beispiel Drobischs bezieht, wenn er (S. 46 f.) schreibt:

> Der Begriff der Bewegung enthält nicht blos überhaupt die Theilvorstellungen Ort, Veränderung, Richtung, Geschwindigkeit; sondern Richtung und Geschwindigkeit sind, beide in verschiedenem Sinne, Bestimmungen der Veränderung; der Ort, da er ja verlassen wird, kann am wenigsten ein Merkmal des Begriffs heißen, er ist ein Beziehungspunkt für die Vorstellung der Veränderung [...] Die Verfolgung dieser Mannigfaltigkeit ist zu weitläufig; zu der Ueberzeugung aber würde sie ersichtlich führen, daß im Allgemeinen die Merkmale eines Begriffs nicht gleichwerthig einander coordinirt sind, daß sie vielmehr in den mannigfaltigsten Stellungen sich auf einander beziehen, einander verschiedenartige Anlagerungen vorschreiben und so sich wechselseitig *determiniren*; daß ein zutreffendes Symbol für den Bau eines Begriffs nicht die Gleichung $S = a + b + c + d \ldots$ sondern höchstens die Bezeichnung $S = F(a, b, c, \ldots)$ ist, welcher mathematische Ausdruck eben nur andeutet, daß a, b, c,... auf eine im Einzelfall genau angebbare, im Allgemeinen höchst vielförmige Weise verknüpft werden müssen, um den Werth von S zu ergeben.

Weshalb habe ich die wohl 1836 bei Drobisch beginnende und fast vier Jahrzehnte später in die zitierten Worte Lotzes mündende Entwicklung so ausführlich referiert? Der Grund ist, daß nach heutiger Auffassung Lotze und Frege die „herausragenden Repräsentanten"[21] der traditionellen bzw. der modernen Logik im 19. Jahrhundert sind, daß Frege als erster Begriffe als spezielle Funktionen aufgefaßt hat, und daß ein Teil der Sekundärliteratur diesen Schritt als von Lotze beeinflußt ansieht. Ist er das wirklich?

Für eine bejahende Antwort scheint eine zweite Übereinstimmung beider Denker zu sprechen. Frege hat 1879 im Vorwort zu seiner *Begriffsschrift* versichert, ihm hätten die „Bestrebungen, durch Auffassung des Begriffs als Summe seiner Merkmale eine

[17] Exner (1843), 20.

[18] Leibniz (1840), 92–94.

[19] Meine Skepsis gründet sich darauf, daß Leibniz an der gemeinten Stelle am Schluß des Fragments vom ungleichartigen Eingehen von *Buchstaben* („characteres") in *Formeln* spricht, wobei ich eine Analogie zwischen *formulae* und *notiones* zwar angedeutet finde, aber eine Aussage über *Begriffs*komponenten allenfalls aus der übergreifenden Feststellung entnehmen könnte: „*Compositio omnis* est *aequiformis* vel *disquiformis*" (Leibniz (1840), 94a; die erste Hervorhebung von mir, C.T.).

[20] Lotze (1874), 46 f.

[21] Vgl. Gabriel (1989), XI–XXXV, vor allem XIII und XXV.

künstliche Aehnlichkeit [sc. zur Arithmetik] herzustellen, [...] durchaus fern gelegen".[22] Andererseits hat Lotze, ohne Frege zu erwähnen und wohl auch ohne dessen Schrift zu kennen, ein Jahr später in einer neu in die zweite Auflage seiner Logik aufgenommenen „Anmerkung über logischen Calcül"[23] die zeitgenössische Algebra der Logik abgelehnt, weil sie erneut den verfehlten Versuch der Merkmalslogik unternehme, die qualitativ höchst unterschiedlichen Determinationsweisen der Merkmale eines Begriffs durch kombinatorische Rechnung mit gleichartigen Elementen zu erfassen. Daß Lotzes Einfall, die mögliche Vielfalt jener Determinationsweisen durch Verwendung eines allgemeinen Funktionssymbols „$F(\ldots)$" zu signalisieren, noch keine Antizipation von Freges Auffassung des Begriffs als einstellige Funktion erster Stufe mit ausschließlich Wahrheitswerten als Werten bedeutet, darauf hat Gottfried Gabriel (a.a.O.) überzeugend hingewiesen: der Begriff ist bei Lotze eine Funktion seiner (als Argumente aufgefaßten) Merkmale.

An der funktionalen Betrachtungsweise als solcher fanden die Logiker und Methodologen rasch Gefallen. Wilhelm Wundt etwa spricht schon 1883 in seiner *Logik* von „der *analytischen Definition im engeren Sinne*, die symbolisch immer in der Form einer Functionsgleichung

$$M = F(a, b \ldots u, v)$$

ausgedrückt werden kann, wo M den zu definirenden Begriff, $a, b \ldots$ die constanten, $u, v \ldots$ die variablen Elemente, in die derselbe zerlegt wird, und endlich das Zeichen F die Functionsbeziehung bezeichnet, die zwischen allen diesen Elementen stattfindet".[24]

Doch allzu wenig sollte man aus den Berührungspunkten zwischen Lotze und Frege auch nicht machen. Frege ist im gleichen Vorwort zur *Begriffsschrift* stolz darauf, gezeigt zu haben, „wie die Auffassung eines Inhalts als Function eines Argumentes begriffbildend wirkt".[25] Dies ist nunmehr *Freges* Alternative zur Merkmalslogik, und es antizipiert ersichtlich seine spätere, in systematischer Form erstmals in *Function und Begriff* 1891 vorgetragene Lehre von den Begriffen als speziellen Funktionen. Aber Frege sieht vor allem, daß die Art der funktionalen Verknüpfung der Argumente nicht gänzlich unbestimmt bleiben muß. Freges für heutige Ohren ungewöhnlich klingende Erklärung der Begriffe als einstellige Funktionen erster Stufe, deren Werte ausschließlich Wahrheitswerte sind, heißt ja nichts anderes, als daß er als Symbole für Begriffe *Aussageformen* mit einer Leerstelle verwendet, die bei Ausfüllung der Leerstelle mit einem Gegenstandsnamen in eine *Aussage* übergehen, die innerhalb der klassischen Logik dann entweder wahr oder falsch ist. Läßt man für solche Aussageformen und Aussagen junktorenlogische Verknüpfungen und für Aussageformen Quantifikationen zu, so beschreibt der Bereich des dann mit Begriffssymbolen Ausdrückbaren die „möglichen Determinationen" von Begriffen.

Ihre Vielfalt ist nicht in jeder Hinsicht unbegrenzt, sondern eine überschaubare Unendlichkeit; doch was sich mit ihr ausdrücken läßt, übertrifft gewaltig alles, was die alte Merkmalslogik erfassen konnte. Dies wird schon an ganz einfachen Beispielen klar, etwa an der *gleichmäßigen Stetigkeit* einer auf der reellen Zahlengeraden definierten Funktion f durch

$$\bigwedge_\varepsilon \bigvee_\delta \bigwedge_x \bigwedge_y (| x - y |< \delta \rightarrow | fx - fy |< \varepsilon)$$

[22] Frege (1879), IV.

[23] Lotze (1880), 256–269.

[24] Wundt (1883), 39. Man beachte, daß der Begriff M bei Wundt wie bei Lotze dabei zwar durch einen mehrstelligen Funktionsausdruck definiert wird, selbst aber im Fregeschen Sinne einstellig ist, insofern er stets nur von *einem* Gegenstand ausgesagt wird.

[25] Frege (1879), VII.

(mit allen positiven reellen — oder auch rationalen — Zahlen als Variabilitätsbereich von δ und ε und allen reellen Zahlen als Variabilitätsbereich von x und y). „Stetigkeit" läßt sich auf keine mir bekannte vernünftige Weise als „Teilbegriff" oder „Merkmal" des so dargestellten Begriffs auffassen, erst recht nicht „Gleichmäßigkeit" als Determination von „Stetigkeit". Und doch ist die gewöhnliche Stetigkeit, schon von Frege begriffsschriftlich äquivalent zu unserer Symbolisierung

$$\bigwedge_x \bigwedge_\varepsilon \bigvee_\delta \bigwedge_y (| x - y | < \delta \rightarrow | fx - fy | < \varepsilon)$$

definiert, in der gleichmäßigen Stetigkeit „enthalten". Aber natürlich nicht wie ein Summand in einer Summe oder ein Faktor in einem Produkt, sondern jetzt in dem präzisen Sinne, daß für jede bestimmte Funktion f der Ausdruck ihrer Stetigkeit aus dem Ausdruck ihrer gleichmäßigen Stetigkeit *quantorenlogisch folgt*.

So eindrucksvoll dieses Beispiel sein mag: ob man den Leistungen und Intentionen der traditionellen Logik gerecht wird, wenn man die von Frege geschaffene klassische Quantorenlogik als Erweiterung der traditionellen Lehre vom Begriffsinhalt auffaßt, und die durch Hinzunahme eines Komprehensionsaxioms und ggf. weiterer Axiome entstehende elementare Mengenlehre als Erweiterung der traditionellen Lehre vom Begriffsumfang, ist eine ganz andere Frage. Sie zu beantworten, war hier nicht mein Thema.

3. Der Begriff des Begriffs und die moderne Abstraktionstheorie

Den Worten am Schluß des vorigen Abschnitts mag man eine gewisse Skepsis anmerken, ob die von Frege geschaffene klassische Quantorenlogik auch alle sinnvollen Fragen der traditionellen Begriffslogik lösen könne (gleichgültig, ob *diese* dazu imstande war). Läßt sich mit ihrer Hilfe etwa ein „höherer" Begriff wie der Begriff der Zahl erhellen? Gelingt dies für den Begriff des Begriffes selbst? Eine gegenwärtig noch in der Diskussion befindliche Antwort auf diese Fragen will ich abschließend skizzieren.

Einen geeigneten Ausgangspunkt bildet Freges in den *Grundlagen der Arithmetik*[26] gegebene Antwort auf die erste Frage, wenn man unter „Zahl" den einfachsten Spezialfall, nämlich „Anzahl" versteht. Frege definierte, nach der Verwerfung eines gleich noch zu betrachtenden anderen Versuchs, die Anzahl, die einem Begriff $B(\xi)$ zukommt (d.h. die Anzahl der unter ihn fallenden Gegenstände) als *den Umfang des Begriffes „gleichzahlig dem Begriffe* $B(\xi)$". Gleichzahligkeit zweier Begriffe war dabei die Möglichkeit einer umkehrbar eindeutigen Zuordnung der unter den einen Begriff fallenden Gegenstände zu den unter den anderen Begriff fallenden Gegenständen. Offensichtlich ist bei dieser Anzahldefinition als bekannt vorausgesetzt, was (traditionell?) der Umfang eines Begriffes ist. Akzeptieren wir das einmal, so ist die Anzahl der unter einen Begriff $B(\xi)$ fallenden Gegenstände *gleich* der Anzahl der unter einen anderen Begriff $C(\xi)$ fallenden Gegenstände, wenn der Umfang des Begriffes „gleichzahlig dem Begriff $B(\xi)$" *gleich* ist dem Umfang des Begriffes „gleichzahlig dem Begriff $C(\xi)$".

Übernehmen wir aus der traditionellen Logik die Erklärung, daß zwei Begriffe den gleichen Umfang haben, wenn jeder unter den einen Begriff fallende Gegenstand auch unter den anderen fällt, so heißt die gegebene Erklärung der Gleichheit der Anzahl

[26] Frege (1884).

der unter $B(\xi)$ fallenden Gegenstände mit der Anzahl der unter den Begriff $C(\xi)$ fallenden Gegenstände: jeder zu dem Begriff $B(\xi)$ gleichzahlige Begriff $H(\xi)$ ist auch gleichzahlig zu dem Begriff $C(\xi)$. Dann ist insbesondere $B(\xi)$ selbst gleichzahlig zu $C(\xi)$, und die Anzahl der unter jeden dieser Begriffe fallenden Gegenstände ist der *Begriff zweiter Stufe* „gleichzahlig zu dem Begriff $B(\xi)$“ [oder: $C(\xi)$].

Praktisch funktioniert das ganz gut. Beispielsweise ist die Anzahl der unter den als $B(\xi)$ gewählten leeren Begriff $\neg\xi = \xi$ fallenden Gegenstände gleich Null, und ebensoviele, nämlich keine Gegenstände, fallen unter jeden zu diesem $B(\xi)$ gleichzahligen Begriff. Frege erklärt daher den Umfang des Begriffes „gleichzahlig zu dem Begriff $\neg\xi = \xi$“ zur Anzahl Null (0) (nicht, wie Georg Cantor Freges Vorschlag mißverstand, den Umfang des Begriffes $\neg\xi = \xi$ selbst!). Ebenso liefert der Begriff $\xi = 0$ die Anzahl Eins (1), und eine rekursive Erklärung die übrigen Anzahlen — dies ist Freges logizistische Anzahldefinition. Ihr wackliger Punkt ist natürlich, daß Frege die Wahl des ungewöhnlichen Begriffsumfangs nicht anders motivieren konnte als vom Ergebnis her — unter den möglichen Kandidaten hatte er eben die gewünschten Eigenschaften.

Der Fregeforscher Ignacio Angelelli hat dieses Verfahren mit Bezug auf eine fragwürdige methodologische Verallgemeinerung desselben in Carnap (1947) die „looking-around method“ genannt.[27] Wir gehen aus von einer Äquivalenzrelation $\xi \sim \zeta$ und setzen die Aussage ihres Bestehens zwischen zwei Gegenständen a und b, also $a \sim b$, gleich mit einer Identitätsaussage $\alpha a = \alpha b$ zwischen zwei vorerst nur hypostasierten „abstrakten“ Gegenständen αa und αb; dann sehen wir uns um, ob wir Entitäten „finden“, die dieser Gleichheit genügen. Finden wir welche, so erklären wir sie als durch diese Methode „eingeführt“. Auf diese Weise „findet“ Frege bestimmte Begriffsumfänge als seine Anzahlen — und muß Begriffsumfänge als solche dabei als existent voraussetzen. Und obwohl er auch die Gleichheit von Begriffsumfängen durch eine Äquivalenzrelation erklären konnte, verbietet sich für die Begriffsumfänge selbst eine Erklärung des gleichen Typus wie bei den Anzahlen wegen der offensichtlichen Zirkelhaftigkeit — die Erklärung müßte ja die Gestalt haben: „der *Umfang* des Begriffes $F(\xi) \leftrightharpoons$ der *Umfang* des Begriffes ...“. Allerdings zeigt ausgerechnet die Betrachtung von Freges Analyse der dann verworfenen alternativen Anzahldefinition und seiner (unbefriedigenden) Einführung der Begriffsumfänge in den *Grundgesetzen der Arithmetik* von 1893, daß es auch anders und besser geht.

Ernst Mach[28] und Giuseppe Peano[29] hatten ebenso wie in unserem Jahrhundert Hermann Weyl[30] Übergänge der von uns problematisierten Art, die bei Frege und Carnap ein „looking-around“ voraussetzen, im Sinne eines formalen Prozesses beschrieben, der gewisse Invarianzbedingungen erfüllt. Und hatte nicht Frege selbst für von der Richtungsgleichheit verschiedene Aussagen über Richtungen von Geraden (also über „Abstrakta“ ähnlich den Begriffsumfängen und Anzahlen) gefordert,[31] „dass die Ersetzbarkeit der Richtung einer Gerade durch die einer ihr parallelen gewahrt bleiben muss“? Offenbar ist damit gemeint, daß die Wahrheit einer wahren Aussage über die *Richtung* einer Geraden, für die in der Aussage ein Name genannt wird, erhalten bleiben soll, wenn wir den Namen der Geraden durch den einer zu ihr parallelen Geraden ersetzen (und in diesem Fall ist es ja auch so). Da Parallelität bei Geraden gerade die Äquivalenzrelation ist, von der wir ausgehen, könnten wir eine elementare[32]

[27]Angelelli (1979), vgl. auch, mit einer m.E. allzu einseitigen Zuordnung Freges zur „looking-around method“, Angelelli (1984).

[28]Mach (1883), insbesondere 203 ff. zum Begriff der Masse.

[29]Peano (1894), 45 ff. (= S. 167 ff. des Neudrucks in den *Opere Scelte*).

[30]Weyl (1918), 34, für den Mengenbegriff, sowie Weyl (1926), 9 ff., für das allgemeine Verfahren.

[31]Frege (1884), 77.

[32]Nicht elementare, d.h. logisch zusammengesetzte Aussagen über Richtungen müssen, damit das

Richtungsaussage $A(\vec{g})$ erklären als

$$A(\vec{g}) \Leftrightarrow \bigwedge_x (x \parallel g \rightarrow A(x)) \, .$$

Paul Lorenzen[33] hat, im Anschluß an Weyl, diesen Typus von Erklärung zur Einführung auch anderer „abstrakter Gegenstände" zu verwenden versucht, oder genauer, zur Einführung der *Rede über* abstrakte Gegenstände, denn die bei der „looking-around method" erforderliche Voraussetzung der *Existenz* der abstrakten Gegenstände, die wir in unsere Rede einführen wollen, entfällt ja jetzt gerade, da wir nur einen — durch die Doppelpfeilregel wiedergegebenen — Übergang zwischen zwei Arten von Aussagen erklären. Ebenso wie die *Annahme* der Existenz einschlägiger abstrakter Gegenstände entfällt freilich auch die *Leugnung* ihrer Existenz — nur liegt wie bei Existenzaussagen üblich die Beweislast bei dem, der sie vertritt.

Unter Hinweis auf die Literatur, in der dies alles schon dargestellt (und oft nur noch nicht richtig eingeordnet) ist,[34] läßt sich jetzt noch festhalten, wie sich ausgehend von einer durch (ein System von) Ableitbarkeitsregeln definierten Synonymitätsrelation $\xi \dashv\vdash \zeta$ zwischen einstelligen Aussageformen mit einem entsprechenden Schema

$$A(|B(\xi)|) \Leftrightarrow \bigwedge_X (X(\xi) \dashv\vdash B(\xi) \rightarrow A(X(\xi))$$

die Rede über *Begriffe* als Abstrakta aus einstelligen Aussageformen bezüglich der Äquivalenzrelation „$\dashv\vdash$" einführen läßt, durch

$$A(\in_x B(x)) \Leftrightarrow \bigwedge_X (X(\xi) \sim B(\xi) \rightarrow A(X(\xi))$$

die Rede über *Begriffsumfänge, Mengen oder Klassen* als Abstrakta aus ebensolchen einstelligen Aussageformen bezüglich der Äquivalenzrelation $A(\xi) \sim B(\xi) \leftrightharpoons \bigwedge_x (A(x) \leftrightarrow B(x))$, und auf ganz entsprechende Weise mit geeigneten Äquivalenzrelationen die Rede über Zahlen, über Wahrheitswerte, über Sachverhalte, über Strukturen usw. Offensichtlich exemplifizieren alle diese Beispiele das gleiche Schema

$$A(\alpha t) \Leftrightarrow \bigwedge_x (x \sim t \rightarrow A(t))$$

mit einer Äquivalenzrelation $\xi \sim \zeta$. Der Übergang von links nach rechts liefert eine *Erklärung* jeder elementaren Aussage über einen „abstrakten Gegenstand" αt: sie ist eine bezüglich der zugehörigen Äquivalenzrelation invariante Aussage über einen den abstrakten Gegenstand „darstellenden" Gegenstand t (der nicht notwendigerweise ein „Konkretum" ist, sondern einfach einer der Gegenstände, zwischen denen die Äquivalenzrelation definiert wurde). Umgekehrt eignet sich der Übergang von rechts nach links zur erstmaligen *Einführung* von Aussagen „über" bestimmte Abstrakta αt in unsere Terminologie. Welche Struktur dabei die Aussageformen haben, die aus den durch solche Abstraktionsschritte einführbaren elementaren Aussageformen zusammengesetzt werden können, ist durch das Abstraktionsschema nicht vorbestimmt, sondern muß jeweils mit dem Blick auf die beabsichtigten Anwendungen festgesetzt

---

Verfahren einwandfrei und insbesondere eindeutig durchführbar ist, in ihre elementaren Bestandteile zerlegt werden, für die dann der vorgeschlagene Übergang vorgenommen wird. Dies gilt entsprechend auch für die im folgenden genannten Erweiterungen auf andere abstrakte Gegenstände.

[33] Lorenzen (1962), Neudruck in Lorenzen (1974), 190–198.

[34] Vgl. Thiel (1972) sowie die Beiträge in Prätor (1989). Aus Vorträgen zu einer Tagung über „Methodische Abstraktionstheorie" in Weingarten 1991 stammen drei kontroverse Diskussionsbeiträge von G. Siegwart, C. Thiel und D. Hartmann in Band 47, Heft 2 (April–Juni 1993) der *Zeitschrift für philosophische Forschung*. Der Tagungsband selbst steht noch aus.

werden. Für die durchgeführten methodologiegeschichtlichen Ausführungen kam es mir darauf an, die bisher ganz vernachlässigte Möglichkeit der quantorenlogischen Zusammensetzung der beteiligten Aussageformen hervorzuheben.

Wie ich klarzumachen versucht habe, umfaßt diese Möglichkeit die traditionellen und gegenwärtig um relationenlogische und verbandstheoretische Ausdrucksmittel bereicherten Strukturierungsmöglichkeiten, ist also als eine Erweiterung derselben anzusehen. Frege hat mit seiner Charakterisierung des Begriffs als eines speziellen Typs von Funktion nicht den traditionellen Begriff des Begriffs abgeschafft, sondern lediglich in seiner Terminologie die heutige Auffassung formuliert, daß ein Begriff durch eine Aussageform dargestellt wird. Was „darstellen" dabei heißt, haben wir durch die Betrachtung des Abstraktionsschemas kennengelernt; welche Zusammensetzungs- oder allgemein Bildungsmöglichkeiten für die Aussageformen zu berücksichtigen sind (insbesondere wie weit man die Analyse der Begriffe — bei uns heißt das: der Aussageformen — treiben sollte), kann allein die intendierte Anwendung entscheiden.

Dennoch wäre es falsch, das Ausmaß der vorgenommenen Erweiterung herunterzuspielen: der Schritt von der Klötzchen-Auffassung der Merkmalslogik zur modernen, die Quantorenlogik einbeziehenden Begriffslehre hat trotz Fehlens scharfer Paradigmata den *Charakter* eines Paradigmawechsels, in dem eine neue (historisch übrigens *nicht* durch Freges Beitrag konstituierte) Auffassungsweise die alte in sich aufnimmt und neu interpretiert. Schon in Freges Darstellung erscheint die Rede von Merkmalen als Teilbegriffen in die Lehre von Begriffen als Funktionen (modern gesprochen: als Aussageformen) widerspruchsfrei integriert; die Pointe ist ihre Unterscheidung von Eigenschaften des Begriffes selbst. Mit dem langen Exkurs über die moderne Abstraktionstheorie wollte ich veranschaulichen, wie sich heute auch die Rede von *Begriffen*, die in unserem Jahrhundert lange Zeit in die logische Propädeutik abgedrängt schien, auf höherer Ebene wieder in die Logik selbst einholen läßt. Habe ich mit meiner Analyse auch nur in den Hauptpunkten recht, so hat die Krise der Merkmalslogik trotz des Sturzes der Klötzchen-Auffassung durchaus keinen katastrophalen Verlauf genommen. Sie markiert vielmehr den Übergang zu den subtileren, leistungsfähigeren und in den Wissenschaften überaus erfolgreichen quantorenlogischen Formen der Merkmalsverknüpfung. Was zunächst als Erneuerung und Kräftigung eines klassischen, aber doch sehr beschränkten logischen Themenkomplexes erschien, erweist sich angesichts der fast allgegenwärtigen Anwendungen als Entdeckung einer neuen „Denkform" im Sinne Hans Leisegangs (vgl. Leisegang (1928)), deren Bedeutung das bloße Interesse der Geschichtsschreibung der Logik weit übersteigt. Darauf aufmerksam zu machen, war die Absicht dieses Vortrags.

Literatur

ANGELELLI, I. (1979): Abstraction, Looking-Around and Semantics. In: A. Heinekamp/F. Schupp (eds.), *Die intensionale Logik bei Leibniz und in der Gegenwart. Symposion der Leibniz-Gesellschaft Hannover, 10. und 11. November 1978.* Franz Steiner, Wiesbaden, 108–123.

ANGELELLI, I. (1984): Frege and Abstraction. In: P. Weingartner/Ch. Pühringer (eds.), *Philosophy of Science/History of Science. A Selection of Contributed Papers of the 7th International Congress of Logic, Methodology and Philosophy of Science, Salzburg, 1983.* Anton Hain, Meisenheim/Glan (= *Philosophia Naturalis* 21, Heft 2–4), 453–471.

CARNAP, R. (1947): *Meaning and Necessity. A Study in Semantics and Modal Logic.* University of Chicago Press, Chicago etc.; [2]1956; deutsch: *Bedeutung und Notwendigkeit. Eine Studie zur Semantik und modalen Logik.* Springer, Wien/New York 1972.

DROBISCH, M.W. (1836): *Neue Darstellung der Logik nach ihren einfachsten Verhältnissen. Nebst einem logisch-mathematischen Anhange.* Leopold Voß, Leipzig; ab der stark veränderten 2. Auflage 1851 unter dem Titel *Neue Darstellung der Logik nach ihren einfachsten Verhältnissen, mit Rücksicht auf Mathematik und Naturwissenschaft.*

EISLER, R. (1929): *Wörterbuch der philosophischen Begriffe.* 4. Auflage. E.S. Mittler & Sohn, Berlin.

EXNER, F. (1843): *Über Leibnitzen's Universal-Wissenschaft.* Aus den Abhandlungen der k. böhm. Gesellschaft der Wissenschaften. In Commission bei Borrosch & André, Prag.

FINDLAY, J. (1963): *Meinong's Theory of Objects and Values.* 2nd edition. Clarendon Press, Oxford.

FOPPA, K. (1971): Begriffsbildung. *Historisches Wörterbuch der Philosophie.* Band 1. Schwabe & Co., Basel/Stuttgart, 787–788.

FREGE, G. (1879): *Begriffsschrift. Eine der arithmetischen nachgebildete Formelsprache des reinen Denkens.* Louis Nebert, Halle a.S.

FREGE, G. (1884): *Die Grundlagen der Arithmetik. Eine logisch mathematische Untersuchung über den Begriff der Zahl.* Wilhelm Koebner, Breslau. Unv. Neudruck: E. & H. Marcus, Breslau 1934; Centenarausgabe, mit ergänzenden Texten kritisch hg. v. Ch. Thiel, Felix Meiner, Hamburg 1986.

GABRIEL, G. (1989): Lotze und die Entstehung der modernen Logik bei Frege. In: Lotze (1989), XI–XXXV.

HERBART, J.F. (1836): Rezension von Drobisch (1836). *Göttingische gelehrte Anzeigen,* 128. Stück, 13. August 1836, 1267–1274.

HERBART, J.F. (1852): *Sämmtliche Werke,* ed. G. Hartenstein, Zwölfter Band: *Historisch-kritische Schriften.* Leopold Voß, Leipzig.

HEYDE, J.E. (1973): *Die Unlogik der sogenannten Begriffspyramide.* Horst Heiderhoff, Frankfurt am Main.

JOSEPH, H.W.B. (1906): *An Introduction to Logic.* ²1916 (revised). Clarendon Press, Oxford.

KLUGE, F. (1960): Begriff. In: ders., *Etymologisches Wörterbuch der deutschen Sprache.* 18. Aufl. Walter de Gruyter, Berlin, 60b.

LEIBNIZ, G.W. (1840): *Opera philosophica omnia quae exstant Latina Gallica Germanica omnia,* ed. J.E. Erdmann, Pars Prior. G. Eichler, Berlin (Reprint: Scientia, Aalen 1959).

LEISEGANG, H. (1928): *Denkformen.* Walter de Gruyter, Berlin/Leipzig; ²1951 Berlin.

LORENZEN, P. (1962): Gleichheit und Abstraktion. *Ratio* 4, 77–81.

LORENZEN, P. (1974): *Konstruktive Wissenschaftstheorie.* Suhrkamp, Frankfurt am Main (*stw* 93).

LOTZE, H. (1874): *Logik. Drei Bücher vom Denken, vom Untersuchen und vom Erkennen (= System der Philosophie. Erster Theil).* S. Hirzel, Leipzig. ²1880 ibid.

LOTZE, H. (1989): *Logik. Erstes Buch. Vom Denken (Reine Logik),* neu hg. v. G. Gabriel. Felix Meiner, Hamburg 1989.

MACH, E. (1883): *Die Mechanik in ihrer Entwickelung historisch-kritisch betrachtet.* F.A. Brockhaus, Leipzig.

MEIER, G.F. (1752a): *Vernunftlehre.* Gebauer, Halle.

MEIER, G.F. (1752b/1760): *Auszug aus der Vernunftlehre.* Gebauer, Halle.

PEANO, G. (1894): *Notations de logique mathématique. Introduction au Formulaire de Mathématique publié par la Rivista di Matematica* . (Keine Verlagsangabe; Druck: Charles Guadagnini), Torino; Neudruck 1958 in *Opere Scelte*, vol. II, ed. U. Cassina. Cremonese, Roma.

PRÄTOR, K. (1988): *Aspekte der Abstraktionstheorie. Ein interdisziplinäres Kolloquium*, Redaktion K. Prätor. Rader, Aachen.

PURCHOTIUS, E. (1730): *Institutiones Philosophicæ* [...] *Tomus Primus, Complectens Logicam, & Metaphysicam.* Apud Joannem Manfrè, Venetiis.

REVICZKY, A. (1757): *Elementa Philosophiæ Rationalis seu Logica* [...]. Typis Academicis Societatis Jesu, Tyrnaviæ.

THIEL, C. (1972): Gottlob Frege: Die Abstraktion. In: J. Speck (ed.), *Grundprobleme der großen Philosophen. Philosophie der Gegenwart I.* Vandenhoeck & Ruprecht, Göttingen, 9–44. [3]1985 (erw.), 9–46.

TRENDELENBURG, A. (1840): *Logische Untersuchungen.* I–II. Gustav Bethge, Berlin. 2., erg. Auflage 1862: S. Hirzel, Leipzig.

TWARDOWSKI, K. (1894): *Zur Lehre vom Inhalt und Gegenstand der Vorstellungen.* Alfred Hölder, Wien. Reprint 1982: Philosophia, München/Wien.

WARNER, A. (1966): Internationale Angleichung fachsprachlicher Wendungen der Elektrotechnik. Versuch einer Aufstellung phraseologischer Grundsätze für die Technik. *Beiträge der ETZ*, Heft 4.

WEYL, H. (1918): *Das Kontinuum. Kritische Untersuchungen über die Grundlagen der Analysis.* Veit & Comp., Leipzig; unv. Neudruck 1932: Walter de Gruyter, Berlin/Leipzig.

WEYL, H. (1926): *Philosophie der Mathematik und Naturwissenschaft. Erster Teil: Mathematik* (= *Handbuch der Philosophie*, 4. Lfg., Teil I). R. Oldenbourg: München/Berlin.

WUNDT, W. (1883): *Methodenlehre* (= *Logik. Eine Untersuchung der Principien der Erkenntniss und der Methoden wissenschaftlicher Forschung*, 2. Band). Ferdinand Enke, Stuttgart.

Ideas of Algebraic Concept Analysis

Frank Vogt and Rudolf Wille

Forschungsgruppe Begriffsanalyse, Fachbereich Mathematik, Technische Hochschule Darmstadt
Schloßgartenstraße 7, D–64289 Darmstadt, Germany

Summary: Algebraic Concept Analysis extends Formal Concept Analysis by algebraic methods. This extention is formalized by the introduction of algebraic and bialgebraic contexts. Formal contexts which may arise from data analysis can be embedded into algebraic or bialgebraic contexts; then their concepts can be described algebraically. This is demonstrated by four examples.

1. Algebraic Descriptions of Concepts

Algebraic Concept Analysis shall be understood as an extension of *Formal Concept Analysis* by algebraic methods of description and construction. Let us recall that *Formal Concept Analysis* is based on a set–theoretic model for concepts and conceptual hierarchies. The model mathematizes the philosophical understanding of a concept as a unit of thought consisting of two parts: its extension, covering all objects belonging to the concept, and its intension, comprising all attributes abstracting from the concept (cf. [3], [8], [10], [12]). This understanding of concepts reflects the idea of describing sets of objects by attributes and representing sets of attributes by objects.

In general, *Algebra* may be understood *as a language which provides "good" descriptions of contexts and facts in our scientific and technicologic world* (cf. [6]). Here, "good" can have several meanings, for instance: adequate, appropriate, constructive, economic, effective, practical etc. There are two basic types of algebraic descriptions: the *description by generation* and the *description by specification.* The basic idea of Algebraic Concept Analysis is to combine such algebraic descriptions with conceptual descriptions of Formal Concept Analysis; in particular, we are searching for "good" algebraic descriptions of concepts, i.e., of their extensions and intensions.

First, let us recall some basic notions of Formal Concept Analysis. A **formal context** is a triple (G, M, I) where G is a set (of objects), M is a set (of attributes), and $I \subseteq G \times M$ is a binary relation. For an object $g \in G$ and an attribute $m \in M$ we read gIm $(:\Longleftrightarrow (g,m) \in I)$ as *the object* g has *the attribute* m. By defining

$$C^I := \{m \in M | (g,m) \in I \text{ for all } g \in C\} \quad \text{and}$$
$$D^I := \{g \in G | (g,m) \in I \text{ for all } m \in D\}$$

for all subsets C of G and D of M, respectively, we establish a **Galois connection** between the power sets of G and M. The pairs (C, D) with $C \subseteq G$, $D \subseteq M$, $C^I = D$ and $D^I = C$ are called the **formal concepts** of the context (G, M, I). For a formal concept (C, D), the sets C and D are called the **extent** and the **intent** of the concept, respectively. The set $\mathfrak{B}(G, M, I)$ of all formal concepts of (G, M, I) together with the order relation defined by

$$(C_1, D_1) \leq (C_2, D_2) \quad :\Longleftrightarrow \quad C_1 \subseteq C_2 \quad (\Longleftrightarrow \quad D_2 \subseteq D_1)$$

forms a complete lattice which is called the **concept lattice** of (G, M, I) and denoted by $\underline{\mathfrak{B}}(G, M, I)$. More details on the basics of Formal Concept Analysis can be found in, e. g., [3] and [7].

The extents of (G, M, I) are exactly the subsets of G which are closed under the (conceptual) closure operator $X \mapsto X^{II}$ and hence form a closure system. Now, let us assume an algebraic structure on the set G of objects. Then the corresponding subalgebras form also a closure system on G and, if those two closure systems are equal, we can describe the extents algebraically and the subalgebras conceptually. The dual situation may be established by an algebraic structure on the set M of attributes. The sketched ideas give rise to the following definitions: A triple $(\boldsymbol{A}, B, J)$ is called an **algebraic context** if $\boldsymbol{A} := (A, F)$ is an algebra (of type $(\mathcal{F}, \sigma)$) with A as underlying set and F as family of operations on A and if (A, B, J) is a formal context whose extents are exactly the underlying sets of subalgebras of $\boldsymbol{A}$. Dually, the triple $(A, \boldsymbol{B}, J)$ is said to be a **dual algebraic context** if $(\boldsymbol{B}, A, J^{-1})$ is an algebraic context for the algebra $\boldsymbol{B} := (B, G)$ (of type $(\mathcal{G}, \tau)$). Finally, $(\boldsymbol{A}, \boldsymbol{B}, J)$ is called a **bialgebraic context** if $(\boldsymbol{A}, B, J)$ is an algebraic context and if $(A, \boldsymbol{B}, J)$ is a dual algebraic context.

As prototype of a bialgebraic context we consider the context $(\boldsymbol{V}, \boldsymbol{V}^*, \perp)$ where $\boldsymbol{V}$ is a finite dimensional vector space over a field $\boldsymbol{K}$, $\boldsymbol{V}^*$ is its dual space (i. e. the space of all linear maps from $\boldsymbol{V}$ into $\boldsymbol{K}$ with pointwise operations), and $\perp$ is defined by

$$v \perp \varphi \quad :\Longleftrightarrow \quad \varphi(v) = 0$$

for all $v \in V$ and $\varphi \in V^*$. The extents of this bialgebraic context are exactly the subspaces of $\boldsymbol{V}$ and the intents are exactly the subspaces of $\boldsymbol{V}^*$. The algebraic description of a subspace may be given by some of its bases (description by generation) whereas the conceptual description corresponds to the representation of a subspace as the solution set of a system of homogeneous linear equations.

The general definition of a bialgebraic context abstracts from this prototype example in several ways. In the example, both algebras are vector spaces but, in general, we do not require any correspondence between the types $(\mathcal{F}, \sigma)$ and $(\mathcal{G}, \tau)$ of $\boldsymbol{A}$ and $\boldsymbol{B}$, respectively. Another generalization is that we introduce algebraic and dual algebraic contexts for situations where we have an appropriate algebraic structure only on the set of objects or the set of attributes, respectively. Finally, we may extend the definitions to partial algebras. Then the subalgebras are understood as the closed subalgebras as they are defined in [1].

The ideas explained so far enrich the language of Formal Concept Analysis by algebraic terms. In a concrete situation of data analysis, this can help to obtain "better" descriptions of the concepts and to make them accessible for computations. However, we cannot hope to find always such useful algebraic structures like vector spaces on the sets G and M of a given formal context (G, M, I) so that this context becomes bialgebraic. Therefore, we weaken our ideas by requiring only the description of (G, M, I) within suitable algebraic structures. This leads to the following notions of embeddings: An **embedding** of a formal context (G, M, I) into a formal context (A, B, J) is a pair (α, β) where α is a map from G into A, β is a map from M into B, and the condition

$$gIm \iff \alpha(g)J\beta(m)$$

holds for all $g \in G$ and $m \in M$. An embedding (α, β) of (G, M, I) into an algebraic context $(\boldsymbol{A}, B, J)$ is called an **algebraic embedding** if

$$\alpha(C^{II}) = (\bar{\alpha}(C))^{JJ} \cap \alpha(G)\,, \text{ with } \bar{\alpha}(C) = \bigcup_{g \in C} \alpha(\{g\}^{II}) \tag{1}$$

holds for all $C \subseteq G$. This condition models that we wish to have the extent generated by C as the trace on G of the subalgebra of $\boldsymbol{A}$ which is generated by the elements of C. Since this is still too strong, we instead consider the subalgebra which is generated by the union of all extents which are generated by some single element of C. Let us remark that, in general, the union $\bar{\alpha}(C)$ may be a proper subset of $\alpha(C^{II})$. Dually, an embedding (α, β) of (G, M, I) into a dual algebraic context $(A, \boldsymbol{B}, J)$ is said to be a **dual algebraic embedding** if

$$\beta(D^{II}) = (\bar{\beta}(D))^{JJ} \cap \beta(M)\,, \text{ with } \bar{\beta}(D) = \bigcup_{m \in D} \beta(\{m\}^{II}) \tag{2}$$

holds for all $D \subseteq M$. An algebraic and dual algebraic embedding of (G, M, I) into a bialgebraic context $(\boldsymbol{A}, \boldsymbol{B}, J)$ is called a **bialgebraic embedding**.

2. Descriptions by Abelian Groups

In the following we explain the notions introduced in the preceding section by four examples. The first example is taken from mathematical music theory ([9]). A challenging problem of music theory is to find meaningful descriptions for chords. Most attempts to solve this problem for tonal music are based on descriptions of harmonic forms reflecting some structural relations. For a satisfactory solution of the description problem, an analysis of the inherent structure of the harmonic forms seems to be indispensable. We present such an analysis for the *harmonic forms of the diatonic scale*. Figure 1 shows a formal context whose objects and attributes are the harmonic forms of the diatonic scale (named as in [15] and represented by subsets of $\{1, \ldots, 7\}$ where $^{-}$ denotes complementation) and whose relation indicates

	1	12	13	14	123	124	125	126	135	$\overline{135}$	$\overline{126}$	$\overline{125}$	$\overline{124}$	$\overline{123}$	$\overline{14}$	$\overline{13}$	$\overline{12}$	$\overline{1}$
1 : unison	×	×	×	×	×	×	×	×	×	×	×	×	×	×	×	×	×	×
12 : second		×			×	×	×	×		×	×	×	×	×	×	×	×	×
13 : third			×		×	×		×	×	×	×	×	×	×	×	×	×	×
14 : fourth				×		×	×	×	×	×	×	×	×	×	×	×	×	×
123: fifth–seventh–free ninth chord					×						×		×	×	×	×	×	×
124: fifth–free seventh chord						×				×	×	×		×	×	×	×	×
125: third–seventh–free ninth chord							×				×	×	×		×	×	×	×
126: third–free seventh chord								×		×		×	×	×	×	×	×	×
135: triad									×	×	×		×		×	×	×	×
$\overline{135}$: compact seventh chord										×					×	×		×
$\overline{126}$: seventh–free ninth chord											×				×	×	×	×
$\overline{125}$: fifth–ninth–free eleventh chord												×			×		×	×
$\overline{124}$: third–free ninth chord													×		×	×	×	×
$\overline{123}$: fifth–free ninth chord														×		×	×	×
$\overline{14}$: ninth–free eleventh chord															×			×
$\overline{13}$: compact ninth chord																×		×
$\overline{12}$: seventh–free eleventh chord																	×	×
$\overline{1}$: compact eleventh chord																		×

Figure 1: The formal context of the harmonic forms of the diatonic scale

the inclusion of suitable representatives of the harmonic forms.

For this formal context we have a bialgebraic embedding into a bialgebraic context of the finite Abelian group $C_{125} \times C_4 \times C_9 \times C_{125}$ (where C_n denotes the cyclic group on $\{0, \ldots, n-1\}$ with addition modulo n). Before explaining a specific bialgebraic embedding, we discuss the bialgebraic context itself.

For a finite Abelian group G, the dual group G^* is defined as the group of all group homomorphisms from G into the unit circle $\mathbb{S}$ of complex numbers (i. e. the set of complex numbers with absolute value 1 together with the complex multiplication as group operation). The group operation on G^* is given by pointwise multiplication. If we define the binary relation $\perp$ between G and G^* by

$$g \perp \varphi \quad :\Longleftrightarrow \quad \varphi(g) = 1$$

for all $g \in G$ and $\varphi \in G^*$ then the context $(G, G^*, \perp)$ is bialgebraic.

Let us recall some more results about the dual group. For a finite Abelian group G, the dual group G^* is already isomorphic to G. For $G = C_n$, we can take the isomorphism $^* : C_n \longrightarrow C_n^*$ with $k \longmapsto k^*$ where k^* is the unique group homomorphism from C_n into $\mathbb{S}$ which maps $1 \in C_n$ to $\left(\exp\left(\frac{2\pi}{n} \cdot i\right)\right)^k$. Furthermore, the dual group of a direct product $C_m \times C_n$ is isomorphic to the direct product $C_m^* \times C_n^*$ of the dual groups. For more details, see [5].

x	$\alpha(x)$
1	(0,0,0,0)
12	(25,0,0,0)
13	(0,2,0,0)
14	(0,0,0,25)
123	(5,2,0,0)
124	(25,1,0,25)
125	(5,0,0,5)
126	(25,2,3,25)
135	(0,2,0,5)
$\overline{135}$	(25,1,3,1)
$\overline{126}$	(5,1,0,10)
$\overline{125}$	(5,1,1,5)
$\overline{124}$	(5,2,3,10)
$\overline{123}$	(1,1,3,25)
$\overline{14}$	(5,1,1,1)
$\overline{13}$	(1,1,3,1)
$\overline{12}$	(1,1,1,5)
$\overline{1}$	(1,1,1,1)

y	$\beta(y)$
1	$(1,1,1,1)^*$
12	$(5,1,1,1)^*$
13	$(1,2,1,1)^*$
14	$(1,1,1,5)^*$
123	$(25,2,1,1)^*$
124	$(10,0,1,5)^*$
125	$(5,1,1,20)^*$
126	$(10,2,3,5)^*$
135	$(1,2,1,25)^*$
$\overline{135}$	$(5,0,3,5)^*$
$\overline{126}$	$(25,0,1,25)^*$
$\overline{125}$	$(5,0,0,20)^*$
$\overline{124}$	$(25,2,3,25)^*$
$\overline{123}$	$(0,0,3,5)^*$
$\overline{14}$	$(25,0,0,0)^*$
$\overline{13}$	$(0,0,3,0)^*$
$\overline{12}$	$(0,0,0,25)^*$
$\overline{1}$	$(0,0,0,0)^*$

Figure 2: A bialgebraic embedding of the formal context in Figure 1

Coming back to our example, we can now state that the context $\mathbb{K}_G := (C_{125} \times C_4 \times C_9 \times C_{125}, C_{125}^* \times C_4^* \times C_9^* \times C_{125}^*, \perp)$ is bialgebraic. A bialgebraic embedding of the formal context of harmonic forms into $\mathbb{K}_G$ is given by the tables in Figure 2.

A line diagram of the concept lattice of the formal context in Figure 1 is shown in Figure 3. Additionally, the bialgebraic embedding is indicated within the line diagram. The concepts can be thought as subgroups of $C_{125} \times C_4 \times C_9 \times C_{125}$ or

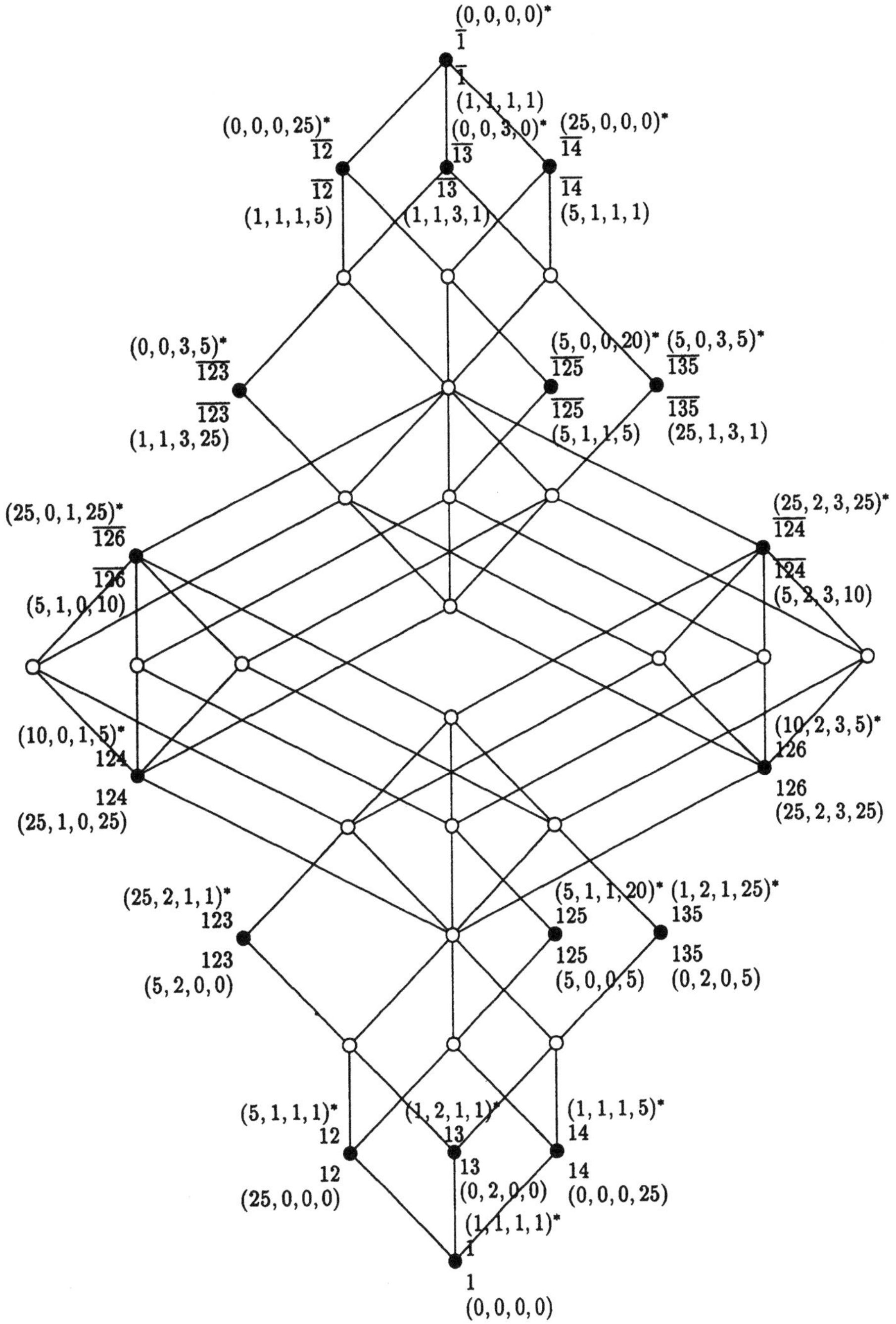

Figure 3: The concept lattice of the harmonic forms of the diatonic scale as a sublattice of the subgroup lattice of $C_{125} \times C_4 \times C_9 \times C_{125}$

of its dual group. These subgroups can be constructed by applying the conditions (1) and (2). For example, the subgroup $\boldsymbol{H}$ of $\mathbf{C}_{125} \times \mathbf{C}_4 \times \mathbf{C}_9 \times \mathbf{C}_{125}$ associated to the concept generated by the harmonic form "triad" (135) is generated by the group elements $(0,2,0,5)$ and $(0,2,0,0)$ (and $(0,0,0,25)$ which is not necessary for the generation process). Observe that the inclusion order of the subgroups exactly matches the hierarchy of concepts. In fact, the concept lattice is a sublattice of the subgroup lattice of $\mathbf{C}_{125} \times \mathbf{C}_4 \times \mathbf{C}_9 \times \mathbf{C}_{125}$.

Let us remark that there is also a bialgebraic embedding of the context of harmonic forms into the bialgebraic context of a vector space. This vector space must have at least dimension 10 whereas the finite Abelian group has only dimension 4. Since $\mathbf{C}_{125} \times \mathbf{C}_4 \times \mathbf{C}_9 \times \mathbf{C}_{125}$ is isomorphic to $\mathbf{C}_{500} \times \mathbf{C}_{1125}$, we have that two factors are enough if we use finite Abelian groups instead of vector spaces.

The context of harmonic forms is an example which allows to apply the notion of a bialgebraic embedding directly. Our next example shows a new aspect in the construction of bialgebraic embeddings: the introduction of further attributes. In [2], V. Duquenne describes a design for an experiment concerning the *access to lexical information*. The experiment tests how the recognition of words depends on the context in which the words appear. The subjects of the experiment are French who also speak English.

In the experiment, pairs (x, y) of an English word x and a French word y are shown to the subjects on a computer screen, first the word x for 100ms and then the word y. In between, pairs (x, z) are shown where x is again an English word but z is a meaningless string of letters which is derived from a French word. The task of the subjects is to press a button immediately after they have recognized that a French word (and not a meaningless string) has appeared on the screen. Then the reaction time is measured to give information about the connection between the reaction time and the following types of pairs (x, y):

c_1: x and y have a semantic connection (e. g. (DOG, CHAT)),

c_2: x and y have no semantic connection (e. g. (BOX, CHAT)),

c_3: x translates y (e. g. (MOON, LUNE)),

c_4: x does not translate y, and there is no semantic connection (e. g. (CORN, LUNE)).

Let $C_4 := \{c_1, c_2, c_3, c_4\}$ be the set of the four types.

For the experiment, the French words y are essential. The design described in [2] is based on a set $M_{60} := \{m_1, \ldots, m_{60}\}$ of 60 French words. A requirement is that each subject who sees a word m_i in a related pair (semantic connection (c_1) or translation (c_3)) must also see the same word in an unrelated pair (c_2 or c_4). Therefore, a set $B_2 := \{b_1, b_2\}$ of two blocks consisting of such pairs is introduced. On the other hand, for avoiding a learning process, each word m_i should never be shown more than twice to a subject. This motivates to partition the 60 words into classes $\{m_1, \ldots, m_{30}\}$ and $\{m_{31}, \ldots, m_{60}\}$. The design described so far can be represented by the following table.

	c_1	c_2	c_3	c_4
b_1	$m_1, \ldots, m_{15}$	$m_{16}, \ldots, m_{30}$	$m_{31}, \ldots, m_{45}$	$m_{46}, \ldots, m_{60}$
b_2	$m_{16}, \ldots, m_{30}$	$m_1, \ldots, m_{15}$	$m_{46}, \ldots, m_{60}$	$m_{31}, \ldots, m_{45}$

From the table we may read, for example, that the pairs (x, y) of type c_1 where y is one of the words $m_1, \ldots, m_{15}$, are in block b_1, whereas the corresponding pairs of type c_2 are in block b_2. The words $m_1, \ldots, m_{30}$ are used only in pairs of type c_1 and c_2 and, complementary, the words $m_{31}, \ldots, m_{60}$ are used only in pairs of type c_3 and c_4.

Furthermore, 16 subjects participate in the experiment which form the set $S_{16} := \{s_1, \ldots, s_{16}\}$. The experiment is performed in two sessions e_1 and e_2 so that we have the set $E_2 := \{e_1, e_2\}$. The next table shows which subjects are confronted with which blocks of pairs in the two sessions.

	e_1	e_2
b_1	$s_1, \ldots, s_8$	$s_9, \ldots, s_{16}$
b_2	$s_9, \ldots, s_{16}$	$s_1, \ldots, s_8$

Now, the entire experimental design can be described by the following table.

	b_1				b_2			
	$m_1 \ldots m_{15}$	$m_{16} \ldots m_{30}$	$m_{31} \ldots m_{45}$	$m_{46} \ldots m_{60}$	$m_1 \ldots m_{15}$	$m_{16} \ldots m_{30}$	$m_{31} \ldots m_{45}$	$m_{46} \ldots m_{60}$
$s_1 \vdots s_8$	c_1, e_1	c_2, e_1	c_3, e_1	c_4, e_1	c_2, e_2	c_1, e_2	c_4, e_2	c_3, e_2
$s_9 \vdots s_{16}$	c_1, e_2	c_2, e_2	c_3, e_2	c_4, e_2	c_2, e_1	c_1, e_1	c_4, e_1	c_3, e_1

The results of the experiment are evaluated by methods of *Analysis of Variance.* The factors for the analysis are given by a family of equivalence relations on the set of all triples (s_i, b_j, m_k). According to the table above, the choice of such a triple determines uniquely the type of the pair (x, m_k) and the session e_1 or e_2 during which the subject s_i has seen this pair. This yields two functions $c : S_{16} \times B_2 \times M_{60} \longrightarrow C_4$ and $e : S_{16} \times B_2 \times M_{60} \longrightarrow E_2$. For the evaluation of the experiment, the following five equivalence relations on $S_{16} \times B_2 \times M_{60}$ are basic:

S: $((s_i, b_j, m_k), (s_p, b_q, m_r)) \in S$ if $s_i = s_p$,

B: $((s_i, b_j, m_k), (s_p, b_q, m_r)) \in B$ if $b_j = b_q$,

M: $((s_i, b_j, m_k), (s_p, b_q, m_r)) \in M$ if $m_k = m_r$,

C: $((s_i, b_j, m_k), (s_p, b_q, m_r)) \in C$ if $c(s_i, b_j, m_k) = c(s_p, b_q, m_r)$,

E: $((s_i, b_j, m_k), (s_p, b_q, m_r)) \in E$ if $e(s_i, b_j, m_k) = e(s_p, b_q, m_r)$.

Additionally, the experimental design suggests three more equivalence relations:

O: $((s_i, b_j, m_k), (s_p, b_q, m_r)) \in O$ if s_i and s_p are in the same class of 8 subjects,

L: $((s_i, b_j, m_k), (s_p, b_q, m_r)) \in L$ if m_k and m_r are in the same class of 15 words,

T: $((s_i, b_j, m_k), (s_p, b_q, m_r)) \in T$ if m_k and m_r are in the same class of 30 words.

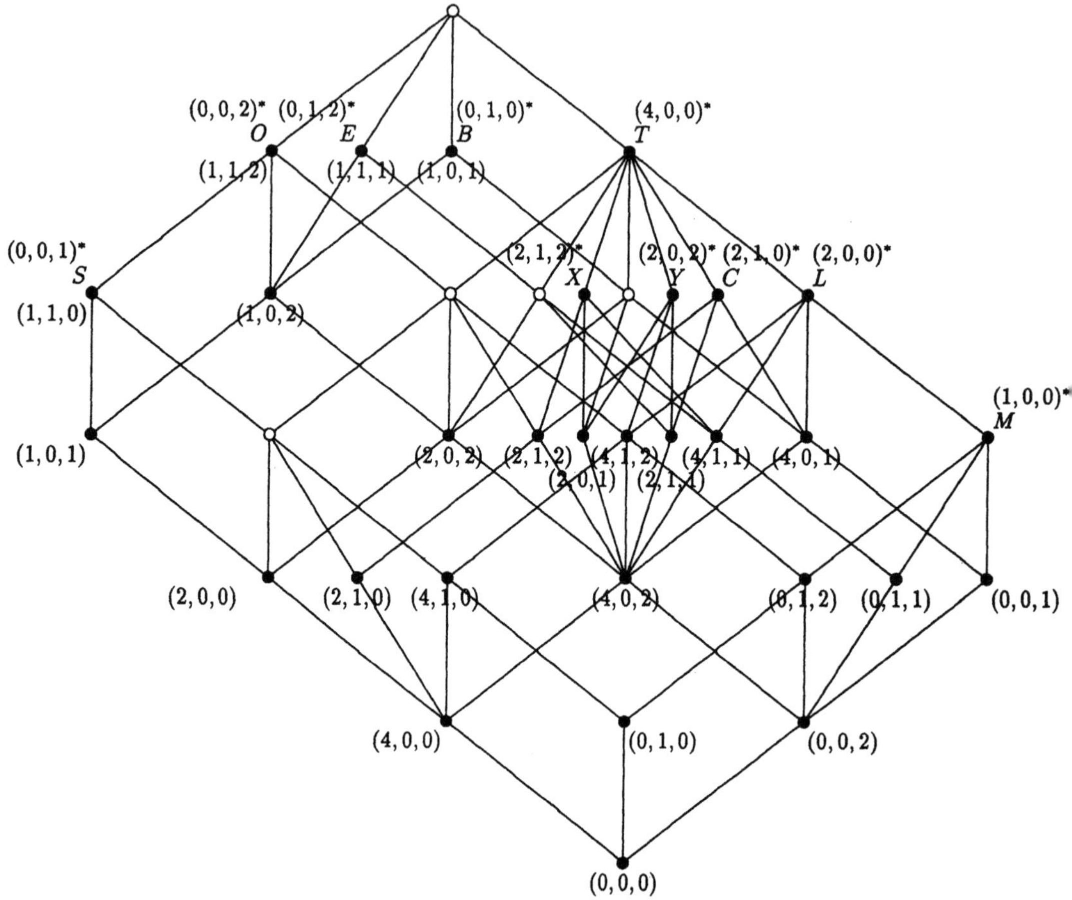

Figure 4: The lattice of equivalence relations of the described experimental design represented by subgroups of $\mathbf{C}_8 \times \mathbf{C}_2 \times \mathbf{C}_4$

The defined eight equivalence relations and their intersections yield the conceptual structure for the analysis of variances. They can be understood as the formal concepts of the formal context $(\mathfrak{P}_2(S_{16} \times B_2 \times M_{60}), \{O, S, B, T, L, M, C, E\}, \Delta)$ where $\mathfrak{P}_2(S_{16} \times B_2 \times M_{60})$ denotes the set of all two–element subsets of $S_{16} \times B_2 \times M_{60}$ and $\{(s_i, b_j, m_k), (s_p, b_q, m_r)\}\Delta Z$ holds if and only if $((s_i, b_j, m_k), (s_p, b_q, m_r)) \in Z$. The structure of the concept lattice suggests a bialgebraic embedding of this context into the bialgebraic context $(\mathbf{C}_8 \times \mathbf{C}_2 \times \mathbf{C}_4, \mathbf{C}_8^* \times \mathbf{C}_2^* \times \mathbf{C}_4^*, \perp)$, but this is not yet possible. We have to introduce two additional equivalence relations $X := (O \wedge C) \vee (E \wedge L)$ and $Y := (O \wedge L) \vee (E \wedge C)$. Then we can find a bialgebraic embedding of the formal context $(\mathfrak{P}_2(S_{16} \times B_2 \times M_{60}), \{O, S, B, T, L, M, C, E, X, Y\}, \Delta)$ into $(\mathbf{C}_8 \times \mathbf{C}_2 \times \mathbf{C}_4, \mathbf{C}_8^* \times \mathbf{C}_2^* \times \mathbf{C}_4^*, \perp)$ as it is indicated in Figure 4. Clearly, we did not list all the 1842240 elements of $\mathfrak{P}_2(S_{16} \times B_2 \times M_{60})$ in the diagram; we have only noted the elements of $C_8 \times C_2 \times C_4$ to which these two–element sets are mapped.

This example shows that it is sometimes necessary to introduce further attributes if one wants a bialgebraic embedding into a bialgebraic context of well-known algebraic structures. V. Duquenne in [2] introduced the additional equivalence relations X and Y because he considers the sublattice of the lattice of all equivalence relations on $S_{16} \times B_2 \times M_{60}$ which is generated by the original eight equivalence relations.

3. Non–linear Descriptions

The first two examples have shown bialgebraic embeddings into bialgebraic contexts of finite Abelian groups. Since such bialgebraic contexts have themselves bialgebraic embeddings into bialgebraic contexts of direct products of finite vector spaces, it is adequate to call algebraic descriptions by finite Abelian groups linear. In this section we want to demonstrate by two examples that non–linear descriptions might also be useful.

	v: vertices	e: edges	f: faces	$v-f$
t: tetrahedron	4	6	4	0
h: hexahedron	8	12	6	2
o: octahedron	6	12	8	−2
d: dodecahedron	20	30	12	8
i: icosahedron	12	30	20	−8

Figure 5: Geometric invariants of the regular polyhedra

Let us consider the (many–valued) context of the regular polyhedra in Figure 5 which shows the values of some important geometric invariants for the regular polyhedra (cf. [11]). By plain scaling with one–dimensional interordinal scales (cf. [3]), we derive the formal context $\mathbb{K}_P := (\{t,h,o,d,i\}, \{v,e,f,v-f\} \times \{-8,\ldots,30\} \times \{\leq,\geq\}, \diamond)$, where $g\diamond(m,j,\leq) :\Longleftrightarrow m(g) \leq j$ and $g\diamond(m,j,\geq) :\Longleftrightarrow m(g) \geq j$.

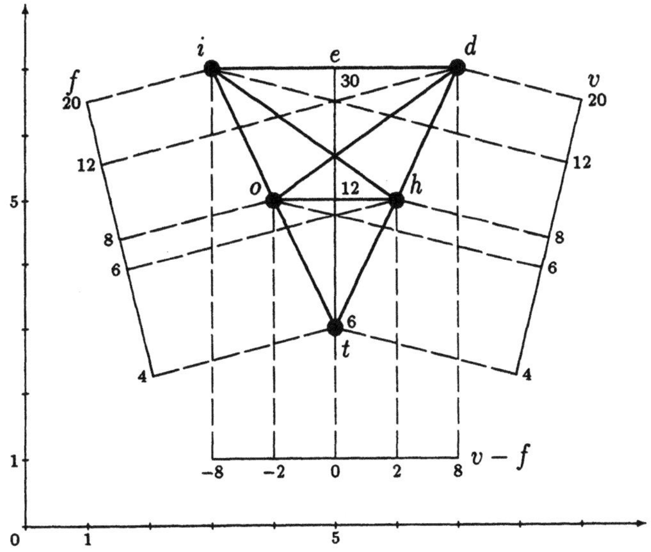

Figure 6: A convex representation of the context of regular polyhedra

The extents of $\mathbb{K}_P$ are exactly the convex subsets on the five points in the diagram shown in Figure 6; they are represented by the line segments in the diagram. Our aim is to find an algebraic embedding of $\mathbb{K}_P$ into an algebraic context. This suggests to create an algebraic context whose extents are exactly the convex subsets of $\mathbb{R}^2$.

Therefore we consider the algebra $\underline{\mathbb{R}^2} := (\mathbb{R}^2, (\beta_t)_{t\in[0,1]})$ where $\beta_t : (\mathbb{R}^2)^2 \longrightarrow \mathbb{R}^2$ denotes the barycentric operation $\beta_t(x,y) := tx+(1-t)y$. Obviously, the subalgebras of $\underline{\mathbb{R}^2}$ are exactly the convex subsets of $\mathbb{R}^2$. Let us denote the unit circle in $\mathbb{R}^2$ again by $\mathbb{S}$. Let $\bar{a} := (-a_2, a_1)$ for every vector $a := (a_1, a_2) \in \mathbb{S}$. We define a binary relation E between $\mathbb{R}^2$ and $\mathbb{S}\times\mathbb{R}\times\mathbb{R}\times\{<,>\}$ by

$$x\,E\,(a,r,s,<) \;:\Longleftrightarrow\; a\cdot x < r \text{ or } (a\cdot x = r \text{ and } \bar{a}\cdot x < s) \quad \text{and}$$
$$x\,E\,(a,r,s,>) \;:\Longleftrightarrow\; a\cdot x < r \text{ or } (a\cdot x = r \text{ and } \bar{a}\cdot x > s)\,,$$

where $\cdot$ denotes the inner product in $\mathbb{R}^2$. The condition $a\cdot x < r$ describes an open half plane in $\mathbb{R}^2$ which is obviously convex. Then a tuple $(a,r,s,.)$ is understood as a half plane in $\mathbb{R}^2$ which contains also an open half line at the boundary of the half plane. This open half line is described by the second condition of the disjunctions above. The described half planes are also convex and each convex subset of $\mathbb{R}^2$ can be represented as an intersection of such half planes. The context $\mathbb{K}_c := (\underline{\mathbb{R}^2}, \mathbb{S}\times\mathbb{R}\times\mathbb{R}\times\{<,>\}), E)$ is algebraic, and we get an algebraic embedding of $\mathbb{K}_P$ into $\mathbb{K}_c$ as it is indicated in Figure 6.

		subunit				superunit			
		♩, –	♩., –	𝅗𝅥, –	𝅗𝅥., –	–, ♩	–, ♩.	–, 𝅗𝅥	–, 𝅗𝅥.
1:	♪ ♪ ♪ ♪ ♪ ♪ ♪ ♪ ♪ ♪ ♪ ♪					×	×	×	×
2:	♩ ♩ ♩ ♩ ♩ ♩	×				×		×	×
3:	♩. ♩. ♩. ♩.		×				×		×
4:	𝅗𝅥 𝅗𝅥 𝅗𝅥	×		×				×	
6:	𝅗𝅥. 𝅗𝅥.	×	×		×				×
12:	𝅝.	×	×	×	×				

	subunit or superunit															
	♩, ♩	♩, ♩.	♩, 𝅗𝅥	♩, 𝅗𝅥.	♩., ♩	♩., ♩.	♩., 𝅗𝅥	♩., 𝅗𝅥.	𝅗𝅥, ♩	𝅗𝅥, ♩.	𝅗𝅥, 𝅗𝅥	𝅗𝅥, 𝅗𝅥.	𝅗𝅥., ♩	𝅗𝅥., ♩.	𝅗𝅥., 𝅗𝅥	𝅗𝅥., 𝅗𝅥.
1	×	×	×	×	×	×	×	×	×	×	×	×	×	×	×	×
2	×	×	×	×	×		×	×	×		×	×	×		×	×
3		×		×	×	×	×	×		×		×		×		×
4	×	×	×	×			×		×	×	×	×			×	
6	×	×	×	×	×	×	×	×				×	×	×	×	×
12	×	×	×	×	×	×	×	×	×	×	×	×	×	×	×	×

Figure 7: Equidistant rhythms in twelve–eighth time

With our last example we turn back to music. Figure 7 shows the context of subunits and superunits of equidistant rhythms in twelve–eighth time. Let r denote a rhythm and let (a,b) be a pair of notes. Then $rI(a,b)$ means that the rhythm r has the length of a as a subunit or the length of b as a superunit (non–exclusive). A suitable algebraic structure for an embedding is the distributive lattice $\boldsymbol{L}$ shown in Figure 8. For this lattice we obtain a bialgebraic context by defining $\mathcal{D}(\boldsymbol{L})$ to be the set of all intervals of $\boldsymbol{L}$ which have a join–irreducible element or 0 as least element and a meet–irreducible element or 1 as greatest element. With a suitable partial binary operation

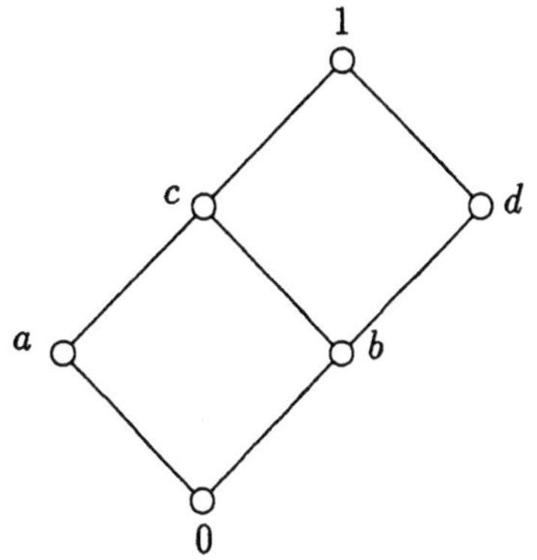

Figure 8: A finite distributive lattice

$*$ on $\mathcal{D}(\boldsymbol{L})$, the context $\mathbb{K}_{\mathbb{D}} := (\boldsymbol{L}, (\mathcal{D}(\boldsymbol{L}), *), \notin)$ becomes bialgebraic (cf. [4]). If we delete the reducible attributes (𝅗𝅥., −), (−, ♩), (♩, ♩.), (♩, 𝅗𝅥.), (♩., 𝅗𝅥), (𝅗𝅥, 𝅗𝅥.), and (𝅗𝅥., ♩) from the formal context of equidistant rhythms, then we get a bialgebraic embedding of this smaller context into $\mathbb{K}_{\mathbb{D}}$ (see Figure 9).

We can extend the bialgebraic embedding by inserting the seven deleted attributes as it is shown in Figure 9. Then we have no longer a bialgebraic embedding of the context, but still an embedding of the concept lattice. We may extend also the set $\mathcal{D}(\boldsymbol{L})$ by suitable attributes. This is possible without destroying the algebraic structure of $\mathcal{D}(\boldsymbol{L})$, hence the context remains bialgebraic. In a sense, the situation of this example is converse to the one in the second example: There we had to add some further attributes to the given context to find a bialgebraic embedding, here we have to extend the algebraic context itself.

4. Further Topics

In this paper, ideas of Algebraic Concept Analysis are discussed with regard to algebraic descriptions of formal contexts and concepts. But there are further topics which are of actual interest. Here we want to mention only the algebraic treatment of many–valued contexts which combine objects, attributes, and attribute values. The algebraic prototype of such contexts is again constructed by a finite dimensional vector space $\boldsymbol{V}$ over a field $\boldsymbol{K}$: the many-valued context $(\boldsymbol{V}, \boldsymbol{V}^*, \boldsymbol{K}, E)$ where

$$(v, \varphi, r) \in E \quad :\Longleftrightarrow \quad \varphi(v) = r$$

for all "objects" $v \in V$, all "attributes" $\varphi \in V^*$, and all "attribute values" $r \in K$. An extensive analysis of these "linear" contexts can be found in [13] and [14]. Of course, it is meaningful to replace the linear forms in those contexts by polynomials which opens a fruitful connection of Algebraic Concept Analysis to *Algebraic Geometry.*

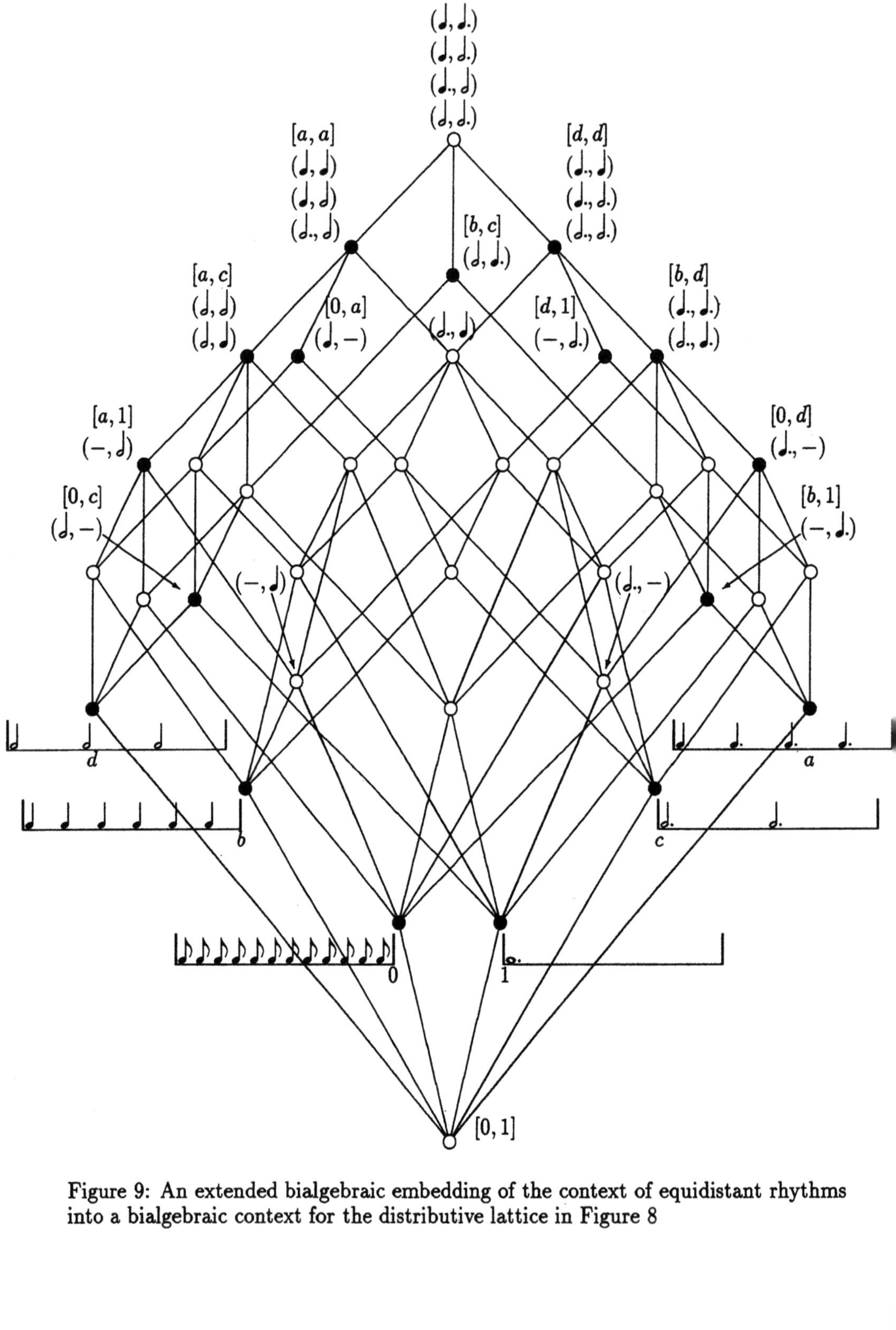

Figure 9: An extended bialgebraic embedding of the context of equidistant rhythms into a bialgebraic context for the distributive lattice in Figure 8

References:

[1] P. Burmeister. *A model theoretic oriented approach to partial algebras*, vol. 32 of *Mathematical Research*. Akademie–Verlag, Berlin, 1986.

[2] V. Duquenne. What can lattices do for experimental designs? *Mathematical Social Sciences* **11** (1986), 243-281.

[3] B. Ganter and R. Wille. Conceptual scaling. In: F. Roberts (ed.), *Applications of combinatorics and graph theory to the biological and social sciences.* Springer Verlag, New York, 1989, 139–167.

[4] F. Vogt. *Bialgebraic contexts for finite distributive lattices.* FB4–Preprint 1548, TH Darmstadt, 1993.

[5] F. Vogt. *Subgroup lattices of finite Abelian groups: structure and cardinality.* FB4–Preprint 1532, TH Darmstadt, 1993.

[6] R. Wille. Versuche der Restrukturierung von Mathematik am Beispiel der Grundvorlesung "Lineare Algebra". In: B. Artmann (ed.), *Beiträge zum Mathematikunterricht 1981.* Schroedel Verlag, Hannover, 1981, 102–112.

[7] R. Wille. Restructuring lattice theory: an approach based on hierarchies of concepts. In: I. Rival (ed.), *Ordered sets.* Reidel, Dordrecht-Boston, 1982, 445–470.

[8] R. Wille. Liniendiagramme hierarchischer Begriffssysteme. In: H.–H. Bock (ed.), *Anwendungen der Klassifikation: Datenanalyse und numerische Klassifikation.* Indeks–Verlag, Frankfurt, 1984, 32–51. Line diagrams of hierarchical concept systems (engl. Übersetzung). *Int. Classif.* **11** (1984), 77–86.

[9] R. Wille. Musiktheorie und Mathematik. In: H. Götze and R. Wille (eds.), *Musik und Mathematik.* Springer–Verlag, Berlin–Heidelberg–New York–Tokyo, 1985, 4–31.

[10] R. Wille. Bedeutungen von Begriffsverbänden. In: B. Ganter, R. Wille, and K. E. Wolff (eds.), *Beiträge zur Begriffsanalyse.* B. I.–Wissenschaftsverlag, Mannheim, 1987, 161–211.

[11] R. Wille. Geometric representation of concept lattices. In: O. Opitz (ed.), *Conceptual and numerical analysis of data.* Springer–Verlag, Berlin–Heidelberg, 1989, 239–255.

[12] R. Wille. Concept lattices and conceptual knowledge systems. *Computers & Mathematics with Applications* **23** (1992), 493–515.

[13] U. Wille. Eine Axiomatisierung bilinearer Kontexte. *Mitt. Math. Sem. Gie"sen* **200** (1991), 72–112.

[14] U. Wille. *An axiomatization for bilinear models.* Preprint 92/03, Geometrie und Diskrete Mathematik, Universität Gie"sen, 1992.

[15] F. A. Wolpert. *Neue Harmonik.* Heinrichshofen, Wilhelmshaven, 1972.

Conceptual Structures in Mathematical Logic and Their Formal Representation

Elisabeth Wette–Roch

Forschungsstelle Math. Logik, FB Informatik, Universität Kaiserslautern, Postfach 30 49, D–67653 Kaiserslautern, Germany

Summary: The paper presents an approach to reorganizing concepts in mathematical logic from the viewpoint of bibliographic documentation and information. This project aims for the development of a structured thesaurus of mathematical logic concepts. We show impacts of domain–specific preconditions on both the aspect of identifying conceptual structures and the aspect of formally describing them by suitable representation models. With respect to conceptual structuring, we argue for a methodical distinction of two levels of concept formation and discuss criteria for the selection of domain elements and classifying features. With respect to models of knowledge representation, we propose an integration of terminological concept descriptions, including definitions, into a broader frame approach.

1. Introduction

In the area of bibliographic documentation and information, an ordered collection of concepts and their interrelations is used in the form of a thesaurus to support the tasks of document indexing and retrieval [1]. Irrespective of the intended application, a thesaurus should fix a certain terminology including references to related terms. Here particular relationships like equivalence of concepts and concept generalizations or specializations play a central role; though a thesaurus may also give some unspecified association relations. Delimiting the domain of application to a scientific discipline, one may take advantage of its technical language and conceive an extended thesaurus by making such additional relations explicit.

The aims of the THESAURUS project described in this paper are directed to realize an approach like that for the field of mathematical logic. The project is related to the *Mathematical Logic* project of the Heidelberger Akademie der Wissenschaften which maintains a complete bibliography of mathematical logic [2]. Within the bibliography project, emphasis is given to structural viewpoints. They find their expression in the particular classification and are going to be accomplished by conceptual networks that may be derived from thesaurus relations. Further integration of the thesaurus into an electronically based information system will support utilizing scientific contents to improve data organization and retrieval.

With respect to these objectives, the conception of a structured thesaurus presupposes adequate ways to identify and represent structural aspects of mathematical logic: Conceptual structures, depending on the determination of organizing principles and structuring features, will have to be sufficiently complex to reflect the essentials of the scientific background. In addition, a formal representation of concept hierarchies will have to be chosen sufficiently simple to meet requirements for the realization of a corresponding knowledge base. The paper studies guidelines on conceptual structuring and on knowledge representation referring to both the specific domain and the specific application. Though quite opposite demands will have to be considered, we argue for a unifying approach.

---

[1] Deutsche Norm DIN 1463

[2] Müller et al. (1987)

2. Conceptual structures in mathematical logic

As an exact science mathematical logic deals with a well-structured domain of conceptual entities: Following modern mathematics, its methodology requires formal definitions to invent concepts and explicit proofs to justify theorems. But in general its subjects are even more abstract than those of mathematics since the origins of this discipline developed from reflections on the principles of mathematical reasoning.

According to the specific methodology, the objects of mathematical logic are characterized and interrelated by systems of propositions which in their turn determine the existing concept relationships. Such systems provide structure in multiple respects: First of all, each single proposition relates its conceptual components corresponding to its logical construction. At a higher level of abstraction, an order of propositions is given by their logical dependencies. Moreover, propositions may differ in their logical status: They may act as definitions or, in the case of axiomatic theories, as axioms for inventing concepts, they may be (proven) theorems of a special theory, or they may have the status of a hypothesis. Finally, even systems of propositions, when formalized, form their own structures.

2.1 Object-level versus description-level structures

The kind of conceptual structures that are of interest to our approach is to a great extent determined by the intended application. For the main matter of concern of a conceptual organization within mathematical logic for documentation purposes should not be to redefine already defined concepts or to (automatically) prove already proven theorems. Rather the task is to describe concepts in terms of common properties in a way that provides information about their characteristic interrelations. This means that the conceptual constructs of mathematical logic become the subjects of new investigations which will rely upon a basically different methodology directed to integrate scientific contents into a knowledge representation system.

Hence it seems convenient to distinguish two conceptual levels and their corresponding description languages, an object-level for concepts defined within mathematical logic, and a meta-level for describing structural features of those entities that are used as object-level concepts [3]. Common features of conceptual entities can be stated from quite different points of view: Distinct definitions may originate in the same idea, and they may invent concepts which are extensionally equivalent or have at least some closely related characterizations; concepts may be introduced for similar application purposes, and they may be of comparable importance in several theoretical settings. This includes that structuring may refer to specifications, like valuations, which are beyond scientific methods. But in general, statements about concept interdependencies based on meta-level descriptions should be well-founded in propositions that are valid at the object-level. In particular, they should allow classifications of concepts according to some "objective" criteria to support the representation of concept hierarchies.

According to our approach, a guideline for conceiving suitable meta-level descriptions will have to be found by considering the way humans refer to and treat knowledge of mathematical logic. For our intention is not to look for methods generating new results but to provide tools that support scientific activities. Analyzing the systematic organization of topics in relevant pieces of literature, we may observe at least two significant procedures: One of them consists of an identification of a variety of

---

[3]Frege in Patzig (ed.) (1969) gives logical reasons for the necessity to make a clear distinction between the use of concepts in a statement and statements about conceptual entities; see *Über Begriff und Gegenstand*, ibid. pp. 66ff.

central conceptual units by the use of proper names, like "Ackermann function", "μ–operator", "Peano arithmetic", or "Cut elimination"; likewise, important results or theoretical assumptions are named, e.g. as "Gödel's second incompleteness theorem", "Recursion theorem", or "Church's thesis", but also entire scientific enterprises like "Hilbert's program". The other procedure relies on abstractions used to classify conceptual entities according to manifold criteria and attaches names to classes of thereby related concepts like "Algorithm", "Recursive function", "First order theory", "Consistency proof", "Arithmetical relation", "Turing degree", or "Reducibility". [4]

Following such forms of reference to the subjects of mathematical logic, meta–level descriptions will be based on individual as well as on generic terms, some of them being well–defined within the scientific language [5], others being rather used intuitively to characterize certain (kinds of) theoretical objects or methods of investigation. But in any case, the above requirements concerning the correctness of concept–descriptions presuppose that definite relationships between the two different levels may be established. In general, description–level statements will be obtained by abstracting from the exact determination of concepts at the object–level. Therefore the resultant structures will only reflect certain parts or aspects of the underlying conceptual dependencies. However, as mathematical logic deals with notions at different stages of abstraction, a relation between both levels may be thought of as a partial mapping preserving structure in the sense that it provides translations of meta–level statements to valid propositions at the object–level.

2.2 Elements at the description level

As a first step, the systematic organization of mathematical logic concepts within a structured thesaurus requires the selection of significant conceptual units that may serve as the carriers of a representation of domain–specific knowledge. Our approach to this step is embedded in the framework of general tasks of the bibliographic enterprise *Mathematical Logic* which includes providing a comprehensive and representative vocabulary for the field and setting up a standardized nomenclature. It aims at compiling a list of "keywords" referring to concepts at the object–level that are suitable to form a base for the conception of description–level structures. The main intention is to classify such keywords in a way that shows the subsisting conceptual interdependencies correctly and – in certain respects – completely. So criteria that determine their choice may have an impact on the expressiveness or simplicity of the possible resulting representations of conceptual structures. We will give some basic considerations concerning adequacy.

First of all, the identification of keywords that should carry meaning presupposes a valuation of the concepts used in mathematical logic according to their methodical status. In view of the fact that the discipline has now split up into various partly connected, partly separated special fields, it does not seem convenient to make overall decisions. Rather a rough pre–classification of concepts reflecting their affiliation to the distinct topics should precede a sensible selection. For concepts that are central in some branches of mathematical logic may be basic or even irrelevant with respect to other ones, so relevance criteria only make sense relative to some special subject. To have an effect on standardizing the way to refer to the field, such criteria will finally have to be justified by analyzing the state of present knowlege, its historical background, and principle aspects of future developments. Apart from valuations of

---

[4] Explications of most of the concepts mentioned in this paragraph may be found in Barwise (1977).

[5] Conceptual classification e.g. in form of hierarchies is one of the typical methods within mathematical logic as well.

the scientific significance of concepts, some structural viewpoints are to be taken into account for an appropriate keyword selection. They mainly concern possibilities of a uniform and correct concept description, which may force the insertion or deletion of certain elements of the compiled lists.

For practical reasons, the process of keyword selection will have to be supported by some automatic procedure. Statistics derived from the bibliographic database of the *Mathematical Logic* project give reason for estimating the number of possible keywords to be in the order of magnitude of tens of thousands. To take further advantage of the information provided by the existing bibliography, common methods applied within the area of information retrieval are going to be evaluated with regard to their suitability for the determination of keywords [6]. Of course, a selection by statistical methods based on titles and abstracts of scientific publications is not tailored to meet the criteria for adequacy mentioned above. But the strength of such methods lies in their independence from individual, possibly inconsistent, decisions. In any case, it will be imperative to make results obtained in this way accessible to scientific revision.

2.3 Characteristics of conceptual orders

In a second step, the intended classification of keywords according to concept interdependencies requires an identification of characteristic structuring features suitable for constituting conceptual orders on the meta-level. Concept definitions of course provide such features but, in general, they show underlying relations between concepts rather implicitly. To obtain more expressive descriptions of concept classes allowing their hierarchical organization, some further explicit details derived from object-level theorems will have to be included, even if they cause redundancies. Apart from giving direct information about those relations that are central for the construction of concept hierarchies, it seems convenient to make use of any existing systematic approaches for concept characterization [7]. Moreover, a choice of structuring elements within the description language should be made with regard to a uniformization of representational devices.

Following the above criteria, we base concept classification on a collection of facts that constitute our knowledge of conceptual entities from mathematical logic within a certain theoretical environment. In particular, related notions will be taken into account to enrich the characterization of concept classes by structural means. To illustrate this procedure we will give an outline of a description of the notion of reducibility, which is central within the fields of recursion and complexity theory: Reducibilities determine binary relations between encodings of decision problems and provide the main tools for classifying such problems according to their difficulty. The intended meaning of the original notion is expressed by the (derived) formula $\mathrm{B} \in \mathrm{REC} \,\&\, \mathrm{A} \leq_{\mathcal{R}} \mathrm{B} \rightarrow \mathrm{A} \in \mathrm{REC}$, where REC denotes the class of decidable problems and $\leq_{\mathcal{R}}$ refers to some reducibility relation. But in a general setting there are several types of reducibility, and it is usual to specify them by quite separate formal definitions. Nevertheless, on the basis of some known theorems which allow the study of reducibilities induced by classes $\mathcal{R}$ of operators, a unifying approach to characterizing all these notions is possible in a way that provides tools for a comparison of their power [8]. Likewise, a standardized representation is obtained for the corresponding degree structures and for the closely related notions of hardness or completeness of a set within a class of sets.

[6] In our intended application, the selected keywords will be used especially for document indexing and retrieval purposes.

[7] E.g. the dependence of concept extensions on certain parameters may be stressed.

[8] For further details we refer to Wagner and Wechsung (1986), pp. 46–51, 84–96.

Organized by such means, concept classification may serve as a base for the representation of hierarchical structures according to certain distinctive features that provide an order relation. Apart from concept inclusion, a variety of relations of that kind is conceivable, and in general conceptual entities will be possible components of several distinct hierarchies. Moreover, if structural organizations of concepts exist at different abstraction levels, they may be related to each other and thereby form structures of some higher complexity. Here we will indicate a way to obtain such a composition of structures by deriving a possible classification scheme for reducibilities from the above example:

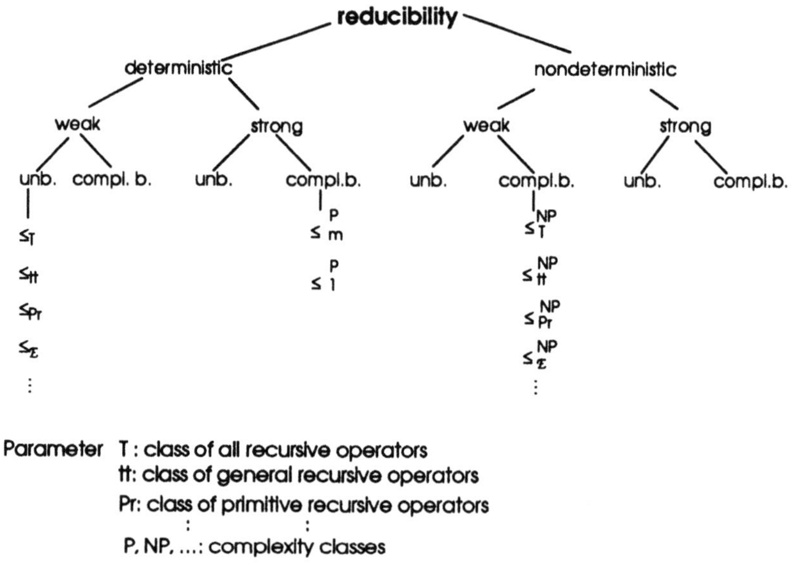

Fig. 1: Classification scheme for reducibilities

According to the chosen approach of description, an order of reducibility notions may on the one hand be obtained by classifying operator classes containing elements of certain general characteristics, e.g. functional or relational operators or (in addition) complexity bounded ones. On the other hand, the existing relations between single classes of operators determine interconnections of the induced special reducibility notions, which form a separate subsisting structure.

3. Knowledge based concept representations

A strong inherent conceptual structure is one of the domain-specific preconditions which one encounters in trying to form a systematic classification of mathematical logic concepts. Another precondition concerns the formal representation of such conceptual organizations: Mathematical logic indeed comprises a variety of formalisms to represent its subjects and reasoning methods. So an approach to impose additional structure on the contents of this science and to formally describe this structure in an adequate way is faced with the methods used within the discipline itself.

To be adequate for our purpose, a formal representation will have to support the consistent construction of meta-level conceptual structures and, in particular, the hierarchical organization of concepts by algorithmic procedures. Several formalisms to encode and reason about knowledge of special application fields have been evolved within the area of artificial intelligence. The most common ones are oriented around the frame approach developed by Marvin Minsky [9]. In this paper, we want to review two particular representation models and discuss their suitability with regard to our specific intended applications.

3.1 General frame–based systems

Frames provide structured representations of objects or object classes of some – quite arbitrarily eligible – application domain by means of attributes and their associated values. If only the composition of single frames is taken into account, frame languages reflect a rather traditional view of concept formation. In particular, frame systems support the description of classes by referring to some generic class ("genus proximum") and to certain special features ("differentia specifica") according to the aristotelian idea of the definition of a concept. But, in general, the organization of frames within a system allows their interconnection not only by the specialization or instantiation relation but also by some of the attributes that are part of the representation. Formally, attributes may be regarded as binary relations mapping certain (classes of) objects to single properties they possess within a specified range or to certain other (classes of) objects. So, after all, the frame approach makes available quite powerful tools for collecting in a standardized way descriptive information concerning objects, or classes, and their relations to each other, and for placing the described classes into taxonomies.

The particular power of frame representations becomes apparent when frames are taken as data structures within a knowledge–based system. For the structural organization of frames allows an application of several reasoning facilities that may directly support the development and maintenance of a knowledge base according to certain consistency criteria. Some of those facilities rely on the internal frame structure that allows a verification of specific (semantically founded) integrity constraints. However, the inference mechanism that is characteristic of frame systems, called "inheritance", is based on the taxonomic order of frames which has been set out and provides possibilities for sharing descriptive attribute specifications among them.[10] Via subclass links, it causes a cumulation of attributes and their associated values, and, via membership links, it enables a description of class members to be given by prototypes. In general, however, the correct determination of links in which frame taxonomies are grounded will not be supported by the system but will have to be given explicitly, at least if representations may be nested and interlinked up to a certain degree of complexity.

In view of the applicability of frame representations to a domain of conceptual entities, particularly following our approach, the question arises to what extent the desired meta–level descriptions can take advantage of the facilities that are available in a frame–based system. According to the specific representation paradigm, propositions of mathematical logic will have to be referred to by attributes and possible associated values which will form an essential part of the structuring features for concept description. Moreover, those attributes will have to be skilfully chosen to support the use of inheritance and integrity constraints. On these conditions, strategies for a systematized description of concepts by frames may consist of utilizing generic terms for prototype description and of utilizing concept specialization by indicating results

[9]The notion of a frame was first introduced in Minsky (1975).

[10]Further details concerning the components of typical frame–based representation facilities may be found e.g. in Fikes and Kehler (1985).

that refer to the described concepts at the highest possible level of generalization. To be concrete, we will review our example of the reducibility notion and suggest a frame representation according to the informal characterizations given above.

Reducibility

- representation: $\leq_{\mathcal{R}}$
- intended meaning: $B \in REC \;\&\; A \leq_{\mathcal{R}} B \rightarrow A \in REC$
- type: binary relation between sets
- parameter: $\mathcal{R}$; class of recursive operators
REC: class of recursive sets
- definition:
- characterization: $A \leq_{\mathcal{R}} B \leftrightarrow \exists \mathcal{O} \in \mathcal{R};\ C_A = \mathcal{O}(C_B)$
- derived notions: hard for a complexity class w.r.t. $\leq_{\mathcal{R}}$
complete in a compl. class w.r.t. $\leq_{\mathcal{R}}$
closure of a compl. class under $\leq_{\mathcal{R}}$
degree w.r.t. $\leq_{\mathcal{R}}$
- properties: reflexive
- theorems: ...

Deterministic Reducibility

- superconcept: reducibility
- properties: + transitive
- parameter: $\mathcal{R}$; class of recursive function operators
- theorems: A $\leq_{\mathcal{R}}$-complete in $\mathcal{A}$ & $\mathcal{B}$ closed
under $\leq_{\mathcal{R}} \Rightarrow A \in \mathcal{B} \leftrightarrow \mathcal{A} \subseteq \mathcal{B}$

Strong Deterministic Reducibility

- superconcept: deterministic reducibility
- parameter: $\mathcal{R}$; class of recursive functions

Fig. 2: Frame representation for reducibility notions

The example shows the possible use of generic terms by referring to some general characteristics of all reducibility relations at a top level of description. Accumulation of attributes by inheritance provides further simplified representations, and finally the several types of reducibility may be characterized by specifying a few parameters. The values attached to those parameters may be considered to be frames as well, though, in general, an exact determination of their interdependencies will be beyond the scope of the facilities of a frame language.

3.2 Terminological concept descriptions

Terminological knowledge representation, although partly founded in the frame approach, is focused on a different kind of structural analysis. Arising from foundational considerations concerning semantic networks, its intention is to formalize the notion of a structured concept. Corresponding representation systems like KL–ONE rely on a formal concept description language which is built on certain "concept structuring primitives"[11] referring to basic conceptual relationships. The common language constructs allow (partial) definitions of concepts in terms of others, based on "primitive" concepts, by the use of "roles" expressing binary relations between domain entities, by logical connectives, and by quantification. As a consequence, they may be semantically founded by a reduction e.g. to expressions of first order predicate logic. In fact, the KL–ONE language may be embedded within a fragment of first order logic which is restricted to at most binary predicate symbols and to the use of

[11] This term was suggested in Woods and Schmolze (1992); for further details see ibid. pp. 134ff.

precisely two different individual variables.[12] In particular, the semantic foundation of concept descriptions gives well-defined criteria for determining subsumption relationships, thus providing means for the conception of algorithms which perform an "automatic classification" of concepts by locating their position within a hierarchical structure.

Based on the KL-ONE language facilities, various special representation systems have been developed which mainly differ in the expressiveness of the language constructs they provide and in their specific sets of inferences. However, the main reasoning mechanisms provided by all of them perform subsumption determination and consistency checks for concept specifications.[13] So, investigations concerning the soundness or completeness of single inference algorithms with regard to the consisting set-theoretic semantics, as well as several results on the computational complexity of terminological reasoning, give rise to some well-founded comparative analysis of the expressive power of these systems.[14] From an overall point of view, there is a trade-off between the expressive power of representation languages and the decidability or even computational tractability of their corresponding concept subsumption problems. Whereas subsumption in the general KL-ONE approach is undecidable according to the result in Schmidt-Schauß (1989), some systems using a restricted terminology provide complete classification algorithms. Tractability of such algorithms, however, may only be obtained for some of the weakest concept languages which are hardly of any practical interest for comparatively complex application domains.

The limited expressive power of terminological representation systems allowing for tractable inferential algorithms seems to be an obstacle to possible applications of this knowledge representation approach to describe taxonomies within mathematical logic. For exact term definitions, even by means of the constructs available in more extensive KL-ONE-like languages, will only cover rather simple concept formations. Hence, in general, terminological descriptions of mathematical logic concepts will have to be performed at high levels of abstraction, by an extensive use of primitive concepts and of primitive roles which indicate conceptual interrelations. Resuming our example of reducibility notions, we may give such an abstract terminology concerning the related concepts of hardness and completeness.

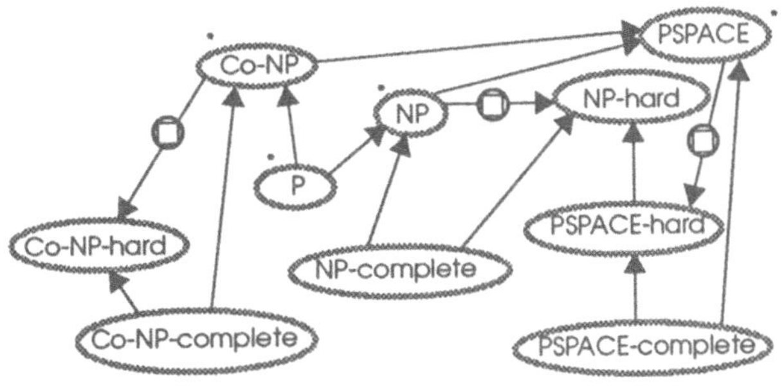

Fig. 3: Concept descriptions with definitions

[12] This result was given by Ward Henson (1991) in a manuscript at the University of Kaiserslautern.

[13] A systematic comparison of features of the most common terminological representation systems is given in Heinsohn et al. (1992).

[14] A survey of results in this field may be found in Woods and Schmolze (1992), pp. 159ff.

As usual for KL–ONE notations, the circled square denotes a binary primitive relation **R** having the intended meaning $\mathbf{R} \equiv \leq_m^{\mathrm{P}}$ of polynomial time m–reducibility. Now, sufficiently expressive terminological languages also permit role definitions by operations like intersection or disjunction, range restriction, and even substitution. But, similar to the case of concept description, these facilities for representing relations often are weak compared to the means that an exhaustive analysis of the subsisting definitions would require. More complex relations, e.g. the relation of m–reducibility, which asserts the existence of certain functions, will be inaccessible by terminological reasoning. Nevertheless, the distinction of some delimited definitional part within possibly broader descriptions of conceptual entities seems to be reasonable, as terminological knowledge is essential in the field of mathematical logic. Even partial definitions will indicate some important underlying structures, which, in addition, when formally represented with a semantical foundation may be taken advantage of for automatic concept classification.

3.3 Applicability and limitations

A well–founded assessment of the applicability of both approaches to knowledge representation in view of our intentions should not rely on scientific considerations only. Rather, specific aspects arising from the intended application by itself may provide additional criteria for judgements on what are suitable forms of concept description. Typical requirements that will have to be met by a bibliographic documentation and retrieval system must be identified and analyzed to determine the framework which is necessary for structuring and representing the domain knowledge involved. Here we give an outline of such a standard situation within bibliographic retrieval following again our sample example.

The most general precondition for a search of scientific literature may be assumed to consist of a – possibly vague – interest in some special field, in our case some topic in mathematical logic. Therefore, a general task of an advanced retrieval system for bibliographic information will be to provide tools to allow the formulation of queries by circumscriptions of the particular information needs in their contents rather than simply by (combinations of) single search terms. Certain types of reducibility for instance might be referred to as "Methods for comparing complexity classes". Moreover, as the system behaviour in general will require an iteration of the retrieval process, there will be the need for further tools to allow circumscriptions of query modifications aimed at broadening, narrowing or varying the set of retrieved documents. In our example, only strong methods for comparing complexity classes might be of special interest. To succeed in the goal of making appropriate selections or changes corresponding to the meaning of such queries, an adequate information system above all will have to comprehend the following: concept classification according to well–differentiated fields, and, within each field, concept specialization or generalization as well as specific semantic relations holding among concepts that are closely connected. References to conceptual relations of greater complexity, especially to those concerning concepts from different fields, will of course form an important constituent part of the system facilities to improve retrieval results, but they may be treated separately in view of their possible representation.

Such considerations indicate that the limited expressiveness of both presented models for knowledge representation, compared with the possibilities of complex conceptual constructions within mathematical logic, does not provide arguments against their applicability for representing mathematical logic concepts. For an application like bibliographic retrieval mainly refers to singled–out parts of the entire domain knowledge and so, in general, will not require the representation of complete concept definitions but rather a description of some conceptual "environments" that may be given on a certain abstraction level. This implies that, with regard to our application, knowledge representation aims at a modular characterization of conceptual entities by means that allow efficient mechanisms for the retrieval of related documents. A promising approach to this might be the attempt to integrate the preferences of both frame and terminological representation within a specific hybrid system. Thus, within each module, advantage could be taken of definitional knowledge, providing conceptual interrelations like those between reducibility, hardness, and completeness. On the other hand, broader descriptions might be covered by frame representations, which, in accordance with the structure of frame systems, could also provide links to interconnect the different modules. In this way, quite a powerful representation device tailored to the specific application is conceivable. Its limitations ultimately will arise from the complexity of concept descriptions strived for, especially in cases that include concept formations at different levels of abstraction.

4. Conclusions

To sum up, our task of developing a thesaurus for the field of mathematical logic involves the subtasks of conceiving a metalanguage for describing conceptual structures, a corresponding representation formalism, and retrieval tools that are based on this formalism. Adequate solutions will require these subtasks to be correlated in such a way that concept descriptions provide the semantic foundations for possible representations and that formal concept representations support bibliographic retrieval by possible queries about scientific contents. In this paper we argue for the feasibility of an appropriate integrative approach.

Conceptual structures and their formal representation will make up the core of a thesaurus realized in the form of a knowledge-based system. Therefore current activities of the THESAURUS project concentrate on structuring special parts of mathematical logic in accordance with our representation approach. The results are intended to be taken as paradigms for setting up a representation formalism which is suited for the tranfer to other fields. Future development of the bibliography project is directed to make conceptual structures available for retrieval and indexing purposes. For this, the thesaurus will be attached to the existing bibliographic system by links to the classification as well as to the representation of documents.

6. References:

BARWISE, J. (ed.) (1977): *Handbook of Mathematical Logic.* North-Holland Publ. Co., Amsterdam, New York.

BRACHMAN, R.J. (1979): On the Epistemological Status of Semantic Networks. In: N. V. Findler (ed.): *Associative Networks. Representation and Use of Knowledge by Computers.* Academic Press, New York, 3-50.

FIKES, R., and KEHLER, T. (1985): The Role of Frame-Based Representation in Reasoning. *Communications of the ACM, 28, 904-920.*

HEINSOHN, J. et al. (1992): *An Empirical Analysis of Terminological Representation Systems.* DFKI Research Report RR-92-16, Deutsches Forschungszentrum für Künstliche Intelligenz.

MINSKY, M. (1975): A Framework for Representing Knowledge. In: P. Winston (ed.): *The Psychology of Computer Vision*. McGraw–Hill, New York.

MÜLLER, G.H. et al. (eds.) (1987): *Ω–Bibliography of Mathematical Logic, Vols. I–VI.* Springer–Verlag, Heidelberg, New York.

PATZIG, G. (ed.) (1969): *Gottlob Frege. Funktion, Begriff, Bedeutung. Fünf logische Studien.* Vandenhoek & Ruprecht, Göttingen, 3rd edition.

SCHMIDT–SCHAUSS, M. (1989): Subsumption in KL–ONE is Undecidable. In: R. J. Brachman et al. (eds.): *Principles of Knowledge Representation and Reasoning. Proceedings of the 1st International Conference, Toronto, ON.* Morgan Kaufmann, Los Altos, 421–431.

WAGNER, K., and WECHSUNG, G. (1986): *Computational Complexity.* D. Reidel, Dordrecht, Boston.

WOODS, W.A., and SCHMOLZE, J.G. (1992): The KL–ONE Family. *Computers and Mathematics with Applications, 23/2-5, 133–177.*

Part II

Mathematical and Statistical Methods for Classification and Data Analysis

- Clustering and Discrimination
- Data Analysis Methods

Three-Mode Hierarchical Cluster Analysis of Three-Way Three-Mode Data

Thomas Eckes[1], Peter Orlik[2]

[1] Fachbereich Gesellschaftswissenschaften, Bergische Universität Wuppertal, Gaußstr. 20, D-42097 Wuppertal, Germany

[2] Fachrichtung Psychologie, Universität des Saarlandes, Im Stadtwald, D-66123 Saarbrücken, Germany

Summary: A method is proposed for the simultaneous hierarchical clustering of row, column, and block elements of a three-way three-mode data matrix. The procedure generalizes the two-mode error-variance approach (Eckes & Orlik, 1993) to the three-mode case. At each step of the agglomerative process, the algorithm merges those clusters whose fusion results in the smallest possible increase in an internal heterogeneity measure. Optionally, the procedure yields an overlapping cluster solution by assigning further row and/or column and/or block elements to a given number of clusters. An application to a data set drawn from object perception research illustrates the approach. Finally, several indications of three-mode clustering are discussed.

1. Introduction

This paper presents a discrete model and a corresponding algorithm, allowing the simultaneous representation of three modes (Tucker, 1964) or sets of entities (e.g., stimuli, responses, subjects, variables, experimental conditions). The type of input data considered are numerical values x_{ijk}, with $i = 1, \ldots n, j = 1 \ldots m$, and $k = 1 \ldots p$, indicating the relationships between three *different* sets of entities (i.e., the data are three-way three-mode). Such data typically result from ratings of n stimuli on m property scales by p subjects. Another instance of this type of data are measurements of n persons or objects on m variables or features under p experimental conditions or at p moments in time. Cattell's (1946, 1952) covariation chart (a graphically displayed design for data collection encompassing three modes, i.e., subjects, variables, and conditions) and its later extension to the data box (a comprehensive framework for data collection and analysis based on considering as many as ten modes; Cattell, 1966) alerted researchers to the empirical-analytical and conceptual advantages of three- and higher-mode data even before appropriate models for their analysis were available.

Currently, various spatial or continuous, in particular factor-analytic, models exist for analyzing such data (e.g., Harshman & Lundy, 1984; Kroonenberg & De Leeuw, 1980; Tucker, 1966; see, for reviews, Kiers, 1991; Krolak-Schwerdt, 1991; Snyder et al., 1984). However, until recently, there have been few *discrete* models for representing structure in three-way three-mode data (see, for a review, Arabie & Hubert, 1992). Several discrete models have been proposed that are only applicable to the special case where the modes corresponding to two ways are identical and where the data are symmetric in these two ways; that is, the data are three-way *two-mode*, as with "stimuli by stimuli by condition" similarity data (see Arabie et al., 1987). Prominent examples are the INDCLUS model (Carroll & Arabie, 1983) in which interobject similarities are represented as common features that are differentially weighted depending on the level of the third way, and the INDTREES model (Carroll et al., 1984) in which the levels of the mode corresponding to the first two ways are represented as terminal nodes of tree structures having the same topology but different height values or branch lengths per level of the third way. Note that these two models do *not* allow to represent both modes *jointly*.

More recently, De Soete and Carroll (1989) developed three different procedures for representing three-way three-mode data by one or more ultrametric trees. In the first, distances are computed from the complete data matrix weighting the levels of two modes (say, the last two) optimally; then, using these distances, the levels of the first mode are represented by the terminal nodes of an ultrametric tree. In the second, the data are optimally aggregated over one mode (say, the third); then the first two modes are jointly represented by the terminal nodes of an ultrametric tree. In the third, the first two modes are jointly represented for each level of the third mode by an ultrametric tree subject to the constraint that all p trees have the same topology (but possibly different height values).

The method proposed here yields a *three-mode* ultrametric tree representation and optionally allows constructing an overlapping cluster solution given a preselected number of disjoint three-mode clusters. Whereas De Soete and Carroll's third procedure represents three-mode three-way data by a series of two-mode ultrametric trees, the present approach jointly portrays the three modes in a *single* ultrametric tree. Our method is a straightforward generalization of the error-variance approach (Eckes & Orlik, 1993) for the simultaneous hierarchical and nonhierarchical clustering of n row and m column elements to the three-mode case. In the next section, some pertinent terminology is introduced and details of the algorithm are presented. Subsequently, the method is applied for purposes of illustration to a real data set from a study on the perception of simple geometric objects.

2. The error-variance approach to three-mode clustering

2.1 Definitions

Following Carroll and Arabie (1980) and Tucker (1964), a *mode* is defined as a particular set of entities. Modes will be denoted by capital letters A, B, and C. Entities are denoted by subscripts; for example, $A_i, i = 1, \dots, n$, could denote n stimuli (row elements), $B_j, j = 1, \dots, m$, could denote m rating scales (column elements), and $C_k, k = 1, \dots, p$ could denote p subjects (block elements). Below it is assumed that modes are not repeated; that is, the number of ways is identical to the number of modes.

A *three-mode array* is defined as the Cartesian product $A \times B \times C$ of three modes A, B, and C with triples or cells (A_i, B_j, C_k). A *three-mode data matrix* $\mathbf{X} = (x_{ijk})$ is an assignment of numerical values x_{ijk} (e.g., ratings, confusion frequencies, reaction times) to the elements (A_i, B_j, C_k) of a three-mode array. $\mathbf{X}$ contains nmp elements. The set of row elements of $\mathbf{X}$ is $\{A_i\}$, the corresponding sets of column and block elements are $\{B_j\}$ and $\{C_k\}$, respectively.

Let $A' = \{A_{i'}\}$ be a subset of A, $B' = \{B_{j'}\}$ be a subset of B, and $C' = \{C_{k'}\}$ be a subset of C. A *three-mode cluster* $\mathcal{L}_r$ is defined as the union of the three sets A', B', and C':

$$\mathcal{L}_r = A' \cup B' \cup C' = \{A_{i'}\} \cup \{B_{j'}\} \cup \{C_{k'}\}. \quad (1)$$

A *three-mode submatrix* is an assignment of numerical values $x_{i'j'k'}$ to elements of $A' \times B' \times C' = \{(A_{i'}, B_{j'}, C_{k'})\} \subset A \times B \times C$. The three-mode submatrix corresponding to $A' \times B' \times C'$ is $\mathbf{X}_r = (x_{i'j'k'})$ with $n_r m_r p_r$ elements, where n_r is the number of entities in A', m_r is the number of entities in B', and p_r is the number of entities in

C'.

The *union of two clusters* $\mathcal{L}_r = A' \cup B' \cup C'$, and $\mathcal{L}_s = A'' \cup B'' \cup C''$, where $A'' \subset A$, $B'' \subset B$, $C'' \subset C$, $A' \cap A'' = \oslash$, $B' \cap B'' = \oslash$, $C' \cap C'' = \oslash$, is a cluster $\mathcal{L}_t$ defined as follows:

$$\mathcal{L}_t = \mathcal{L}_r \cup \mathcal{L}_s = \{A' \cup A''\} \cup \{B' \cup B''\} \cup \{C' \cup C''\}. \tag{2}$$

Let there be two clusters $\mathcal{L}_r$ and $\mathcal{L}_s$ at a particular hierarchical level. The assignment of numerical values to the elements of a three-mode array $A^* \times B^* \times C^*$ with $A^* = A' \cup A''$, $B^* = B' \cup B''$, and $C^* = C' \cup C''$ yields a matrix $\mathbf{X}_t$ with $n_t m_t p_t$ elements. This matrix is decomposed into eight submatrices, two of which correspond to the clusters $\mathcal{L}_r = A' \cup B' \cup C'$ and $\mathcal{L}_s = A'' \cup B'' \cup C''$: $\mathbf{X}_r = (x_{i'j'k'})$ with $n_r m_r p_r$ elements and $\mathbf{X}_s = (x_{i''j''k''})$ with $n_s m_s p_s$ elements, respectively. The other six correspond to the sets $R_\alpha = A' \cup B' \cup C''$, $R_\beta = A' \cup B'' \cup C'$, $R_\gamma = A' \cup B'' \cup C''$, $R_\delta = A'' \cup B' \cup C'$, $R_\epsilon = A'' \cup B' \cup C''$, and $R_\eta = A'' \cup B'' \cup C'$: $\mathbf{X}_\alpha = (x_{i'j'k''})$ with $n_r m_r p_s$ elements, $\mathbf{X}_\beta = (x_{i'j''k'})$ with $n_r m_s p_r$ elements, $\mathbf{X}_\gamma = (x_{i'j''k''})$ with $n_r m_s p_s$ elements, $\mathbf{X}_\delta = (x_{i''j'k'})$ with $n_s m_r p_r$ elements, $\mathbf{X}_\epsilon = (x_{i''j'k''})$ with $n_s m_r p_s$ elements, $\mathbf{X}_\eta = (x_{i''j''k'})$ with $n_s m_s p_r$ elements, respectively.

Let $\mathcal{M}(A, B, C)$ be the set of all subsets of the union of sets A, B, and C. A *three-mode hierarchical clustering system* is defined as a subset Ψ of $\mathcal{M}(A, B, C)$, which satisfies the following conditions:

(1) $A \cup B \cup C \in \Psi, \oslash \notin \Psi$;
(2) $\{A_i\}, \{B_j\}, \{C_k\} \in \Psi$ for all $A_i \in A$, $B_j \in B$, $C_k \in C$;
(3) if $\mathcal{L}_r, \mathcal{L}_s \in \Psi$ with $\mathcal{L}_r \cap \mathcal{L}_s \neq \oslash$, then $\mathcal{L}_r \subseteq \mathcal{L}_s$ or $\mathcal{L}_s \subseteq \mathcal{L}_r$.

2.2 The algorithm

The present algorithm constructs a three-mode hierarchical clustering system with clusters having minimal internal heterogeneity. Specifically, a three-mode cluster is said to have low internal heterogeneity to the extent that its elements show strong inter-mode relations with as small a variance of corresponding numerical values as possible.

In the following, the input data are presumed to be carefully scored or normalized (see Milligan & Cooper, 1988) such that larger entries indicate a stronger relationship between the corresponding row, column, and block elements; that is, the data are interpreted as "similarities". In line with the above notion of cluster heterogeneity, the internal heterogeneity measure of a three-mode cluster $\mathcal{L}_r$ is expressed as

$$MSD_r = \frac{1}{n_r m_r p_r} \sum_{\substack{A_{i'} \in A' \\ B_{j'} \in B' \\ C_{k'} \in C'}} (x_{i'j'k'} - \mu)^2. \tag{3}$$

where μ is the maximum entry in the input matrix $\mathbf{X}$; that is, $\mu = max(x_{ijk})$ is the largest similarity. Thus, MSD_r is the mean squared deviation of entries $x_{i'j'k'}$ in the corresponding submatrix $\mathbf{X}_r$ from the maximum entry μ in $\mathbf{X}$.

In the definition of the MSD index, μ is chosen to be the maximum entry since it has the advantage of providing an easy-to-interpret measure of the internal heterogeneity of a newly formed three-mode cluster. Specifically, the present choice ensures that the cluster elements will show relatively strong, homogeneous inter-mode relations. In addition, it is in accordance with the traditional error-variance rationale in one-mode hierarchical clustering: the fusion value of the cluster formed at the lowest level, defined as the increase in the error variance within the cluster, is zero.

Thus, our approach parallels Ward's (1963) one-mode agglomerative algorithm which starts with merging two entities whose fusion results in the minimum increase in the within-cluster error-sum-of-squares measure.

The fusion rule may now be specified as follows. At each step of the agglomerative algorithm, a particular subset of mode A is merged with particular subsets of modes B and C such that the increase in the internal heterogeneity measure of the resulting three-mode cluster is as small as possible.

The most general case is given when two existing three-mode clusters are to be merged. According to the definition of the union of two clusters, the squared deviations are to be computed only for those entities of modes A, B, and C which belong to the sets R_α, R_β, R_γ, R_δ, R_ϵ, and R_η, respectively. This case may be formally expressed as follows.

A three-mode cluster $\mathcal{L}_r = A' \cup B' \cup C'$ is to be merged with a three-mode cluster $\mathcal{L}_s = A'' \cup B'' \cup C''$ into a new cluster $\mathcal{L}_t$. The resulting increase in the internal heterogeneity measure is

$$\begin{aligned} \Delta_t \quad = \quad & \frac{1}{n_t m_t p_t - n_r m_r p_r - n_s m_s p_s} \times \\ & \left(\sum_{\substack{A_{i'} \in A' \\ B_{j'} \in B' \\ C_{k''} \in C''}} (x_{i'j'k''} - \mu)^2 + \sum_{\substack{A_{i'} \in A' \\ B_{j''} \in B'' \\ C_{k'} \in C'}} (x_{i'j''k'} - \mu)^2 + \right. \\ & \sum_{\substack{A_{i'} \in A' \\ B_{j''} \in B'' \\ C_{k''} \in C''}} (x_{i'j''k''} - \mu)^2 + \sum_{\substack{A_{i''} \in A'' \\ B_{j'} \in B' \\ C_{k'} \in C'}} (x_{i''j'k'} - \mu)^2 + \\ & \left. \sum_{\substack{A_{i''} \in A'' \\ B_{j'} \in B' \\ C_{k''} \in C''}} (x_{i''j'k''} - \mu)^2 + \sum_{\substack{A_{i''} \in A'' \\ B_{j''} \in B'' \\ C_{k'} \in C'}} (x_{i''j''k'} - \mu)^2 \right). \end{aligned} \tag{4}$$

To see more clearly that Δ_t is the *increment* to the internal heterogeneity index produced by merging three-mode clusters $\mathcal{L}_r$ and $\mathcal{L}_s$, equation (4) is rewritten in terms of the MSD measures of the clusters involved:

$$\Delta_t = \frac{1}{n_t m_t p_t - n_r m_r p_r - n_s m_s p_s} (n_t m_t MSD_t - n_r m_r MSD_r - n_s m_s MSD_s). \tag{5}$$

When single-element subsets $\{A_{i'}\}$, $\{B_{j'}\}$, and $\{C_{k'}\}$ are to be merged with each other or these subsets are to be merged with an existing three-mode cluster $\mathcal{L}_r$, the corresponding increments are computed as special cases of (4).

As a conventional criterion for determining the number of three-mode clusters present in a given data matrix, a marked increase in fusion values can be considered indicative of the formation of a relatively heterogeneous cluster. Thus, a decision as to the number of clusters can be reached analogously to the fusion criterion in the one-mode error-sum-of-squares clustering method (Ward's [1963] method).

Since the number-of-clusters problem has no satisfactory solution based on a single criterion (see, e.g., Everitt, 1979; Milligan & Cooper, 1985), one may find it helpful to use another criterion related to the MSD index above. For a given cluster $\mathcal{L}_r$, this criterion, which we call the *centroid effect ratio* (CER), is:

$$CER_r = \frac{\bar{x}_r^2}{s_r^2 + \bar{x}_r^2}. \tag{6}$$

The CER index measures the contribution of the mean "cluster effect size", expressed as the squared mean $\bar{x}_r^2$, to the total "cluster effect size", expressed as the sum of the within-cluster variance s_r^2 and $\bar{x}_r^2$. CER is directly related to the well-known "coefficient of variation" (V): $CER = 1/(1+V)$. Thus, a cluster having a relatively low value of the CER index can be interpreted as having a small average inter-mode relation between entities relative to a large within-cluster variability. Preliminary evidence suggests that a value of the CER index smaller than 80% is indicative of a relatively low cohesiveness among the elements of the respective cluster. This cluster may then be excluded from further consideration.

As a final option, the algorithm allows constructing overlapping clusters to circumvent the problem characterizing hierarchical clustering methods in general: Once an element of the set or sets to be classified is a member of a cluster at some hierarchical level it can no longer be merged with any other cluster at the same level of the agglomerative process. The procedure is as follows: Having reached a decision as to the number of (disjoint) clusters, each row, column, and block element of the input matrix not already belonging to a given cluster is considered in turn a possible candidate for joining it. An element may be added if the corresponding Δ_t index does not show a marked increase and/or the centroid effect ratio of the resulting cluster remains sufficiently high. Once again, values of the CER index above 80% may be considered a lower limit for including the respective element into the cluster. In any case, the decision as to the inclusion of a given element into some cluster is left to the user.

3. An illustrative application

In this section we present an application of our approach to a reasonably small real data set drawn from research concerning the semantics of object perception. This application is only suggestive of a range of substantive applications where a three-mode clustering analysis seems to be the method of choice. For some further, more complex but also more space-consuming, applications the interested reader is referred to Eckes (1993) and Eckes and Hassebrauck (1993).

Stimulus materials were three squares differing in color (purple, vermilion) and size (1 cm side length = "small", 3 cm side length = "large"): a purple small square (ps), a purple large square (pl), and a vermilion large square (vl). These squares were fixed in the middle of DIN A6 size index cards. Stimuli were rated by subjects on a set of bipolar scales (with steps −2, −1, 0, +1, +2) in one of two experimental conditions. In the first, a sample of eight subjects rated the original colored stimuli ("color condition"); in the second, an independent sample of eight subjects rated photocopied (i.e., colorless) versions of the original stimuli ("no color condition"). The poles of the rating scales were: happy—sad, reserved—gaudy, lively—quiet, calming—arousing, massive—dainty, mean—generous, large—small, and narrow—wide. In a previous study (Orlik, 1965), the first four of these proved to be particularly sensitive to the

Tab. 1: Example Data.

	Color			No color		
Scales	ps	vl	pl	ps	vl	pl
happy—sad	0.625	−1.625	1.125	−0.375	0.375	0.125
reserved—gaudy	−1.500	1.625	−0.875	−1.125	0.375	0.250
lively—quiet	0.875	−2.000	1.000	−0.500	1.000	0.625
calming—arousing	−0.625	1.500	−0.375	0.125	−0.250	−0.375
massive—dainty	0.875	−1.250	−1.250	1.500	−1.250	−1.375
mean—generous	−0.750	1.250	0.875	−0.125	0.875	0.750
large—small	1.625	−1.125	−0.875	1.750	−1.375	−1.125
narrow—wide	−1.625	1.125	0.875	0.750	1.125	1.000

Note. ps = purple small square, vl = vermilion large square, pl = purple large square.

color dimension ("color scales") and the last four proved to be particularly sensitive to the size dimension ("size scales"). Scales were presented to subjects in a random order, with poles of each rating scale systematically varied.

Hence, in this study, three modes were involved: the set of three geometric stimuli, the set of eight bipolar rating scales, and the set of two experimental conditions. In the following, it is shown how the three-mode clustering approach proposed here represents the structure experimentally built into the data (see Table 1).

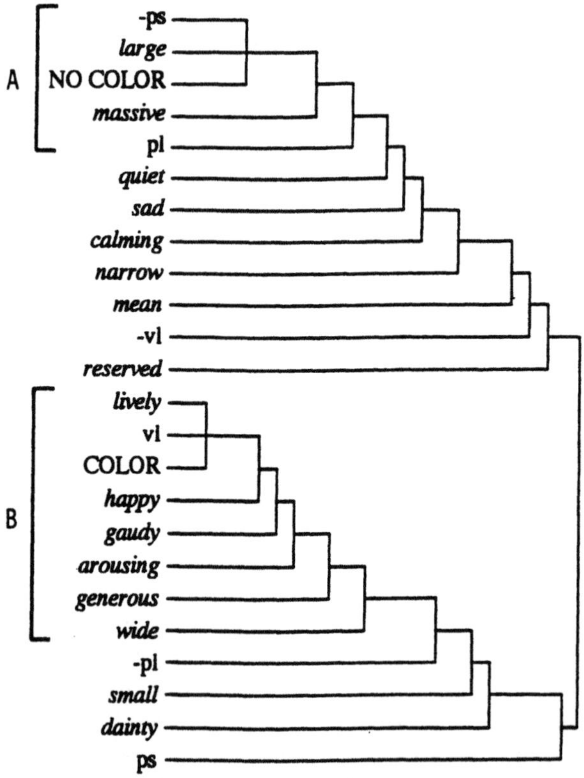

Fig. 1: Three-mode hierarchical clustering representation for the example data.

Tab. 2: Two Cluster Nonoverlapping and Overlapping Solution for the Example Data.

Original elements			Added elements		
Features	Stimuli	Condition	Features	Stimuli	Condition
		Cluster A			
large *massive*	−ps pl	NO COLOR	*generous* *wide*	vl	COLOR
		Cluster B			
lively *happy* *gaudy* *arousing* *generous* *wide*	vl	COLOR	*large* *massive*	−ps	—

Note. Dash indicates that no elements have been added.

In a preparatory step of the analysis, a row- and columnwise reflection of the data within each block was performed; that is, entries were first duplicated for each row and each column and then the duplicated entries were rescored by multiplying with −1. This reflection procedure ensured (a) that each pole of a given bipolar rating scale constituted a feature of its own which could be clustered separately from the respective opposite pole, and (b) that each stimulus judged had a contrast stimulus which could be clustered separately from it (i.e., it is assumed that the stimuli are implicitly perceived in a bipolar way based on the size dimension). Figure 1 presents the hierarchical clustering solution for the augmented $16 \times 6 \times 2$ input matrix.

The succession of fusion values suggested the selection of two three-mode clusters. Cluster A comprises features and stimuli closely associated with each other in the no color condition: features *large* and *massive* are linked to −ps and pl. Cluster B contains features and stimuli closely associated with each other in the color condition: features *lively*, *happy*, *gaudy*, *arousing*, *generous*, and *wide* describe stimulus vl.

The two cluster solution gives some indication as to the differential usage of rating scales in the presence of colored and colorless stimuli, respectively. Both feature elements of Cluster A, representing the no color condition, belong to the a priori category of size scales; similarly, Cluster B, representing the color condition, contains a subset of the size scales *and* a subset of the color scales. However, due to the nonoverlapping nature of the present solution, a given element that is classified as a member of one cluster cannot also be a member of the other. Thus, it could be that some of the color features would be joined together with Cluster B (the no color condition cluster) if clusters were allowed to overlap. To test this possibility, an overlapping two cluster solution was constructed using the optional procedure outlined above. Each row, column, and block element of the input matrix not belonging to the other cluster was considered in turn a candidate for joining it. An element was added to a given cluster when the corresponding value of the CER index did not fall below 80%. The overlapping solution is shown together with the nonoverlapping solution in Table 2.

The overlapping cluster solution perfectly recovers the built-in structure of the three-mode data set: subjects in the no color condition differentiated exclusively along the four size scales, contrasting pl with ps; subjects in the color condition differentiated both along the four size scales and along the four color scales, contrasting vl with ps. In sum, when clusters are allowed to overlap, the structure inherent in the data is brought out most clearly.

4. Concluding remarks

A three-mode clustering analysis may be indicated for at least three reasons. First, the data are intrinsically three-mode; that is, the respective sets of entities are inextricably linked to each other. A typical example can be found in personality research when data are gathered that interrelate persons, behaviors, and situations. If a researcher is interested in certain kinds of persons behaving in certain kinds of ways in certain kinds of situations (Bem, 1983), a three-mode analysis may be the method of choice. Second, the modes are attributed equal structural status within the analysis. For instance, in research using the semantic differential technique (such as in the study reported here), the structure of the object set may be of much the same interest as the structure of the person set and the structure of the scale set (Osgood et al., 1957). Third, the interpretation of the structure within a given set is alleviated by the simultaneous consideration of the structure within the other sets. In cognitive psychology, for example, a three-mode research strategy of data collection and data analysis may allow to follow the subjects' own usage of the stimulus or feature elements involved and to represent the structure of the subjects' cognition, perception or memory organization in a more faithful way than would be possible if the researcher had to rely on her or his own intuitions or on some auxiliary analytical methods when interpreting some solution. The present three-mode clustering approach allows the analysis of data when any of these indications, or some combination of these, is given.

References:

ARABIE, P., CARROLL, J.D., and DESARBO, W.S. (1987): *Three-Way Scaling and Clustering*. Sage, Newbury Park.

ARABIE, P., and HUBERT, L.J. (1992): Combinatorial Data Analysis. *Annual Review of Psychology, 43, 169–203.*

BEM, D.J. (1983): Constructing a Theory of the Triple Typology: Some (Second) Thoughts on Nomothetic and Idiographic Approaches to Personality. *Journal of Personality, 51, 566–577.*

CARROLL, J.D., and ARABIE, P. (1980): Multidimensional Scaling. *Annual Review of Psychology, 31, 607–649.*

CARROLL, J.D., and ARABIE, P. (1983): INDCLUS: An Individual Differences Generalization of the ADCLUS Model and the MAPCLUS Algorithm. *Psychometrika, 48, 157–169.*

CARROLL, J.D., CLARK, L.A., and DESARBO, W.S. (1984): The Representation of Three-Way Proximity Data By Single and Multiple Tree Structure Models. *Journal of Classification, 1, 25–74.*

CATTELL, R.B. (1946): *The Description and Measurement of Personality.* World Book, New York.

CATTELL, R.B. (1952): The Three Basic Factor-Analytic Research Designs—Their Interrelations and Derivatives. *Psychological Bulletin, 49, 499–520.*

CATTELL, R.B. (1966): The Data Box: Its Ordering of Total Resources in Terms of Possible Relational Systems. In: R.B. Cattell (ed.): *Handbook of Multivariate Experimental Psychology.* Rand McNally, Chicago, 67—128.

DE SOETE, G., and CARROLL, J.D. (1989): Ultrametric Tree Representations of Three-Way Three-Mode Data. In: R. Coppi and S. Bolasco (eds.): *Multiway Data Analysis.* North-Holland, New York, 415–426.

ECKES, T. (1993): Multimodale Clusteranalyse: Konzepte, Modelle, Anwendungen. In: L. Montada (ed.): *Bericht über den 38. Kongreßder.Deutschen Gesellschaft für Psychologie in Trier 1992* (Vol. 2). Hogrefe, Göttingen, 166–176.

ECKES, T., and HASSEBRAUCK, M. (1993): Multimodale Analysen in der physischen Attraktivitäts-Forschung. In: M. Hassebrauck and R. Niketta (eds.): *Physische Attraktivität.* Hogrefe, Göttingen, 95–121.

ECKES, T., and ORLIK, P. (1993): An Error Variance Approach to Two-Mode Hierarchical Clustering. *Journal of Classification, 10, 51–74..*

EVERITT, B.S. (1979): Unresolved Problems in Cluster Analysis. *Biometrics, 35, 169–181.*

HARSHMAN, R.A., and LUNDY, M.E. (1984): The PARAFAC Model for Three-Way Factor Analysis and Multidimensional Scaling. In: H.G. Law, C.W. Snyder, J.A. Hattie and R.P. McDonald (eds.): *Research Methods for Multimode Data Analysis.* Praeger, New York, 122–215.

KIERS, H.A.L. (1991): Hierarchical Relations Among Three-Way Methods. *Psychometrika, 56, 449–470.*

KROLAK-SCHWERDT, S. (1991): *Modelle der dreimodalen Faktorenanalyse.* Lang, Frankfurt/Main.

KROONENBERG, P.M., and DE LEEUW, J. (1980): Principal Component Analysis of Three-Mode Data By Means of Alternating Least Squares Algorithms. *Psychometrika, 45, 69–97.*

MILLIGAN, G.W., and COOPER, M.C. (1985): An Examination of Procedures for Determining the Number of Clusters in a Data Set. *Psychometrika, 50, 159–179.*

MILLIGAN, G.W., and COOPER, M.C. (1988): A Study of Standardization of Variables in Cluster Analysis. *Journal of Classification, 5, 181–204.*

ORLIK, P. (1965): Eine Modellstudie zur Psychophysik des Polaritätsprofils. *Zeitschrift für Experimentelle und Angewandte Psychologie, 12 , 615–647.*

OSGOOD, C.E., SUCI, G.J., and TANNENBAUM, P.H. (1957): *The Measurement of Meaning.* University of Illinois Press, Urbana.

SNYDER, C.W., LAW, H.G., and HATTIE, J.A. (1984): Overview of Multimode Analytic Methods. In: H.G. Law, C.W. Snyder, J.A. Hattie and R.P. McDonald (eds.): *Research Methods for Multimode Data Analysis.* Praeger, New York, 2–35.

TUCKER, L.R. (1964): The Extension of Factor Analysis to Three-Dimensional Matrices. In: N. Frederiksen and H. Gulliksen (eds.): *Contributions to Mathematical Psychology.* Holt, Rinehart and Winston, New York, 109–127.

TUCKER, L.R. (1966): Some Mathematical Notes on Three-Mode Factor Analysis. *Psychometrika, 31, 279–311.*

WARD, J.H. (1963): Hierarchical Grouping to Optimize an Objective Function. *Journal of the American Statistical Association, 58, 236–244.*

The Testing of Data Structures with Graph-Theoretical Models

Erhard Godehardt, Armin Horsch

Arbeitsgruppe Biometrie der Klinik für Thorax- und Kardiovaskularchirurgie,
Heinrich Heine-Universität,
Moorenstr. 5 (Gebäude 12.41), D-4000 Düsseldorf, Germany

Summary: In sociometric research, group structures like cliques in school classes have to be analyzed. Here, we can use combinatorial models like directed graphs (or digraphs) to describe, for example, the result of a questionnaire where persons choose other persons in a group. We present test procedures to test the randomness of group structures found by such a procedure of choices. These procedures use the frame of directed graphs or undirected multigraphs and allow to analyse and test group structures when choices are not necessarily independent. We also present a more general probability model for random multigraphs which can be used to test the homogeneity of a sample of multidimensional metric data with a given dependence structure between the different variables.

1. Introduction

Graph-theoretical models are useful in many applications where the structures of data sets have to be explored. In cluster analysis, they are used to define types of clusters theoretically, to uncover clusters of the desired type, and to test the "randomness" of the clusters detected (that means more exactly to test the hypothesis of homogeneity of a data set against the alternative of heterogeneity). If the similarity between the pairs of objects of the data set is based on a distance measure, then we use undirected graphs (or just "graphs"), where the n objects are the vertices and the edges between some vertices are to be drawn if and only if the corresponding objects are similar enough (Bock (1980), Godehardt (1990, 1991a, 1993), Ling (1973)).

In sociometric research, where the group structures like cliques in a school class have to be analyzed, we also can use graphs and combinatorial models for them to test the randomness of groups (and in some sense their stability, too). Here, we usually have no symmetric similarity measure defining the clusters since in a group, a person i may choose a person j as companion, but not vice versa. Therefore directed graphs (abbreviated as "digraphs") are used for the description of within-group and between-group relations. Again, the persons (that means the objects of research) are represented by vertices of a—now directed—graph, and the fact that a person i elects a person j is indicated by an arc pointing from i to j.

As for random graphs and their use in cluster analysis, a well founded probability theory for random digraphs exists which can be used to test the randomness of sociometric networks. In order to describe and evaluate structures in social groups, different kinds of random graph models and random digraph models have been discussed in the past (see for example Capobianco (1983), Palka (1986), Jaworski (1993)).

The information on the social structure is typically gained by asking questions like "Who are your five best friends in this group?" or "Who are your friends in this group?". These two kinds of questions lead to different types of models, the fixed choice and the variable choice model (see Capobianco (1983)).

Analysis of a group structure can be based on criteria which become obvious when

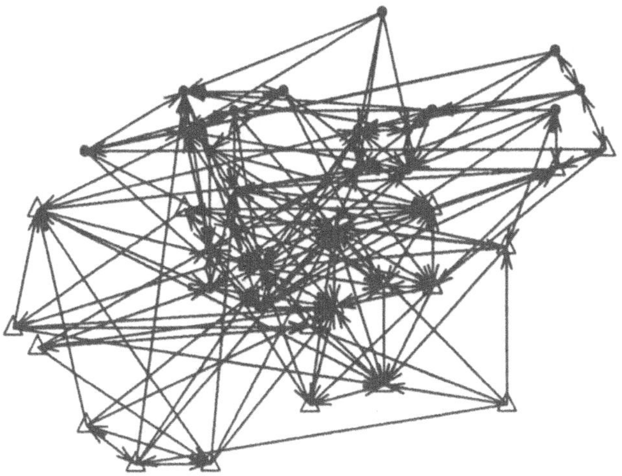

Figure 1: *Digraph of the holiday example;*
△ denotes a boy, •: denotes a girl; no initials of names are given.

looking at the corresponding digraph, for example we can answer the following questions. How many 2-cycles (pairs of persons with mutual choices) are there? How many persons with maximum number of choices indegree (that is the number of incoming arcs) are in the group? Are there any persons with indegree 0? Another important (maybe the most important) question is, whether the "obvious" structure is real, which means that a group has a structure which is very unlikely to appear at random. Thus, analyzing properties of social groups can be done by using random graph theory.

In the following two chapters, we first give an example from our own sociometric research work and then discuss probability models for random digraphs and give some exact and asymptotic results on the distributions of 2-cycles and indegrees for random digraphs under these probability models. At the end of the paper, we describe models for random multigraphs and their relations to the random digraph models.

2. An example from sociometric research

The following example is from a holiday camp with $n = 40$ boys and girls who were between 10 and 14 years old. One week after the start of the camp, every child had to name $f = 5$ other children (not including himself or herself) with whom the child mostly liked to be together. No accumulation of more than one vote to the same chosen person was allowed (Godehardt (1974, 1991b)). The result can be represented as a digraph. The children are represented by labelled vertices and the votes by arcs; an arc showing from child i to child j shows that i had chosen j. Figure 1 shows the result for our example.

In this example, we have $n = 40$ children with $f = 5$ votes. This gives a total of $N = fn = 200$ arcs in Figure 1. We found $k = 2$ children who had not been elected as partners from any other child in that group, one boy and one girl (such persons are called "hedgehogs" by sociologists). We also found three popular persons ("popular" in the context of Capobianco (1983)); one was elected by 14 other children, another one by 15 other children, and one person had 18 votes (the distribution of choices—or indegrees—is given in Godehardt (1991b)). From Figure 1, we can find 36 pairwise choices, the underlying graph of these matched pairs or matchings is shown in Figure 2. The question is whether such constellations like two hedgehogs or 36 matched pairs or three persons with at least 14 votes may be expected under the assumption of random choices or not.

Assuming that the persons are a "random sample" with no special sympathies or antipathies between all possible pairs of persons which would lead to a subgroup pattern, we can accept that every person makes his or her f choices "at random". Thus, the probability that a person i chooses a given other person is $1/(n-1)$, and the probability for a given f-tuple to be chosen from person i is $1/\binom{n-1}{f}$. This assumption means that the digraph of choices can be considered as a realization of a graph drawn at random from all possible digraphs with outdegree $f = 5$ for all vertices. The number X of hedgehogs, the number W of rulers, who have been chosen by everyone else (Palka (1986)), and the number V of pairwise choices then are random variables which can be used as test statistics because under the hypothesis of random choices, we can derive the exact and asymptotic distributions of these random variables. Using the sieve formula (formula of inclusion and exclusion) we get

$$P(X = k) = \binom{n}{k} \sum_{\mu=k}^{n-1-f} (-1)^{\mu-k} \binom{n-k}{\mu-k} \binom{n-\mu}{f}^{\mu} \binom{n-1-\mu}{f}^{n-\mu} \binom{n-1}{f}^{-n}$$

for the distribution of the number of hedgehogs. The expectation of this number is

$$\mathrm{E}\,X = n \binom{n-1}{f} \binom{n-2}{f}^{n-1} \binom{n-1}{f}^{-n} = n \left\{ \frac{n-1-f}{n-1} \right\}^{n-1}.$$

The vertices representing the rulers in the digraph of choices must have the indegree $n-1$. The distribution of their number can be found by looking at the complement of the original digraph. Every person has *not* elected $n-1-f$ other persons; instead of the choices these "non choices" between the persons are represented by arcs. Thus, the outdegree of every vertex in the complement digraph is $n-1-f$, and the vertices of the rulers must have indegree 0 now. From this we get

$$\begin{aligned} P(W = k) &= \binom{n}{k} \sum_{\mu=0}^{f-k} (-1)^{\mu} \binom{n-k}{\mu} \binom{n-k-\mu}{f-k-\mu+1}^{k+\mu} \\ &\times \binom{n-k-\mu-1}{f-k-\mu}^{n-k-\mu} \binom{n-1}{f}^{-n}. \end{aligned}$$

The expected number of rulers is

$$\mathrm{E}\,W = n \binom{n-1}{f} \binom{n-2}{f-1}^{n-1} \binom{n-1}{f}^{-n} = n \left\{ \frac{f}{n-1} \right\}^{n-1}.$$

In our holiday camp example, we have $X = 2$ hedgehogs and get $P(X = 0) = 0.82473, P(X \leq 1) = 0.98614, P(X \leq 2) = 0.99931, P(X \leq 3) = 0.99998$ as the probability to get no vertex, at most one, two or three vertices with indegree 0 in

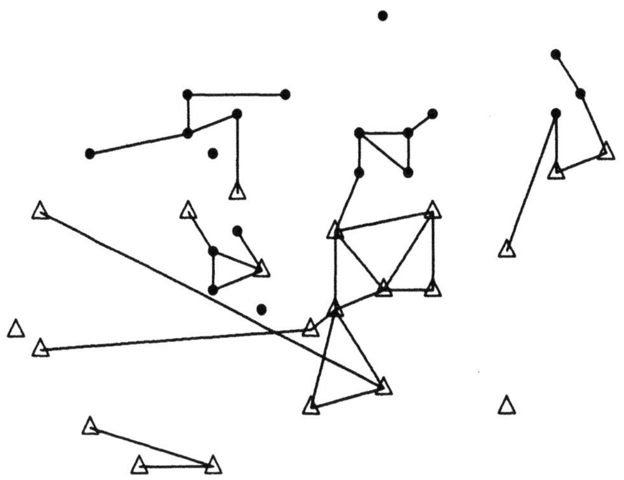

Figure 2: *Matched pairs graph from the digraph of the holiday example; △ denotes a boy, •: denotes a girl; no initials of names are given.*

a random digraph where the outdegree of every vertex is 5. Using this random variable X as test statistic for the hypothesis of random choices, for a probability $\alpha = 0.05$ for the error of the first kind, we can reject this hypothesis within this group of 40 persons and can assume that during the first days of the stay in the camp subgroups have been formed.

The exact, finite distribution of the random variable V, the number of mutual or pairwise choices in random digraphs of order n, is known but rather complicated and not useful for numerical computations on random digraphs if the number of vertices is not very small. Therefore, we only give its expectation and standard deviation:

$$\mathrm{E}\,V = \frac{n}{n-1}\frac{f^2}{2}, \qquad \mathrm{Var}\,V \sim \mathrm{E}\,V.$$

If the order n of the random digraph grows, then the distribution of V tends to a Poisson distribution with parameter $f^2/2$. Here and throughout the paper, if we speak of "convergence of random variables", then we always deal with convergence in distribution. For example a sequence of random variables X_n is said to be asymptotically Poisson distributed with mean λ if $P(X_n = k) \to e^{-\lambda}\lambda^k/k!$ as $n \to \infty$. We denote this fact by writing $X_n \xrightarrow{d} \mathrm{P}(\lambda)$ as $n \to \infty$. If a random variable X is exactly Poisson distributed, then we write $X_n \sim \mathrm{P}(\lambda)$.

In the holiday camp example, we may expect about $5^2/2 = 12.5$ pairwise choices (matchings or cycles of length 2). We have got, however, 36 pairs as Figure 2 shows. The probability for 36 or more pairs is close to 0 if we assume random choices. Therefore, we can reject this hypothesis of independent mutual choices if we use the number of 2-cycles as test statistic, too.

3. Random digraphs with independent choices of images

The model discussed in the previous chapter has two weak points.

1. Every person has a fixed number f of choices.

2. The weights per choice are the same (that means that we assume that it is the same whether person number i elects person j at the first place, at the second place, ..., at the f-th place).

One can consider two other probability models to overcome the first point mentioned. For example, we may assume that n persons make N votes altogether. Thus, the number d_i^+ of votes for a person i, that is the outdegree of the corresponding vertex i, is a random variable with the expectation $\mathrm{E}\, d_i^+ = N/n$. This assumption leads to a probability model, where a digraph is chosen at random from all possible digraphs with N arcs, and is equivalent to the case $t = 2$ of a probability model for random multigraphs which is discussed for example in Godehardt (1990, 1993) and also later in this paper. In the second probability model for random digraphs, we may assume the independence of choices in the following way (see Jaworski (1987)).

Definition 1. Let $\mathcal{P} = \{p_d : d \in \{0, 1, \dots, n-1\}\}$ be a probability distribution, and let $D(n, \mathcal{P})$ denote a random digraph with vertex set $\mathcal{V} = \{1, 2, \dots, n\}$ such that for every vertex $i \in \mathcal{V}$, independently of other choices, $P(d_i^+ = d) = p_d, d = 0, 1, \dots, n-1$ holds and for every $S \subseteq \mathcal{V} - \{i\}$ with $|S| = d$

$$P\,(S \text{ is the set of images of } i) = p_d \binom{n-1}{d}^{-1}.$$

In the special case, when the distribution is degenerated, that is when every vertex has outdegree f, we simply write $D(n, f)$. •

For the remainder of this paper, d_i^+ denotes the outdegree and d_i^- the indegree of a vertex i. In addition we denote for every vertex $i \in \mathcal{V}$ its outdegree expectation by $\mathrm{E}\, d_i^+ = \sum_d d p_d = f$ and its variance by $\mathrm{Var}\, d_i^+ = \sigma^2$. Let $X^+(d)$ denote the number of vertices of outdegree d in a random digraph $D(n, \mathcal{P})$, and let $X = (X^+(0), \dots, X^+(n-1))$. Definition 1 implies that X follows a multinomial distribution, $X \sim \mathrm{M}(n, p_0, \dots, p_{n-1})$, and $X^+(d)$ is binomially distributed with expectation $\mathrm{E}\, X^+(d) = np_d$ and variance $\mathrm{Var}\, X^+(d) = n\, p_d\, (1 - p_d)$ for every $d \in \{0, 1, \dots, n-1\}$.

Now let us look at the distribution of the indegree d_j^- of a vertex $j \in \mathcal{V}$. Let us fix a vertex j and let $i \to j$ denote the fact that there is an arc from vertex i to vertex j. Denote by $X_{i,j}$ the respective random variables $(1 \le i, j \le n)$, that is

$$X_{i,j} = \begin{cases} 1, & i \longrightarrow j \\ 0, & \text{otherwise} \end{cases} .$$

Then

$$\begin{aligned} P(X_{i,j} = 1) &= \sum_{d \ge 1} P\,(X_{i,j} = 1 \mid d_i^+ = d)\; P(d_i^+ = d) \\ &= \sum_{d \ge 1} \binom{n-2}{d-1} \binom{n-1}{d}^{-1} p_d = \frac{1}{n-1} \sum_{d \ge 1} d\, p_d \\ &= \frac{1}{n-1} \mathrm{E}\, d_i^+ = \frac{f}{n-1} \end{aligned}$$

Since each vertex $i \in \mathcal{V} - \{j\}$ can choose j as its image independently and with the same probability $f/(n-1)$, we get

$$P\,(d_j^- = d) = \binom{n-1}{d} \left(\frac{f}{n-1}\right)^d \left(1 - \frac{f}{n-1}\right)^{n-d-1},$$

that is, for fixed f and n, the indegree of each vertex is binomially distributed, a fact which already has been noticed in Jaworski (1987). Obviously, $\mathrm{E}\, d_j^- = f$ and $\mathrm{Var}\, d_j^- = f - f^2/(n-1)$, and Poisson and normal approximations for the binomial distribution can be applied. It should be mentioned that generally the indegrees are not independent. Actually, the only case where they are independent is when the outdegrees are binomially distributed. However, then we deal with the binomial type of random graphs (Jaworski (1987)), which we are going to study in the next chapter.

Focussing on the number of 2-cycles in random digraphs, we show how important the assumption of the independence of choices is. In the following, we derive the formulas for the expectation and variance of this random variable (see the corresponding results for the special case $D(n, f)$ in Smit (1979, 1985), and also Jaworski (1993)).

Let V denote the number of cycles of length 2 in a random digraph $D(n, \mathcal{P})$ of order n with an expected outdegree f for each vertex and let $Y_{i,j}$ be the indicators of a 2-cycle between i and j. Obviously, $Y_{i,j} = X_{i,j}\, X_{j,i}$, and

$$\mathrm{E}\, Y_{i,j} = P(X_{i,j} = 1)\, P(X_{j,i} = 1) = \frac{f^2}{(n-1)^2}.$$

Now, the expectation of V is

$$\mathrm{E}\, V = \sum_{(i,j)} \mathrm{E}\, Y_{i,j} = \binom{n}{2} \frac{f^2}{(n-1)^2} = \frac{f^2}{2} \cdot \frac{n}{n-1}.$$

To prove a formula for the variance of V, let us notice that

$$\begin{aligned} \mathrm{E}\, V^2 &= \mathrm{E} \left(\sum_{(i,j)} Y_{i,j} \right)^2 = \mathrm{E} \left(\sum_{(i,j)} Y_{i,j}^2 + \sum_{(i,j) \neq (l,k)} Y_{i,j} Y_{l,k} \right) \\ &= \binom{n}{2} \frac{f^2}{(n-1)^2} + \binom{n}{2} \binom{n-2}{2} \mathrm{E}\, (Y_{1,2} Y_{3,4}) + n(n-1)(n-2)\, \mathrm{E}\, (Y_{1,2} Y_{1,3}). \end{aligned}$$

The random variables $Y_{1,2}$ and $Y_{3,4}$ are independent, thus $\mathrm{E}\,(Y_{1,2}Y_{3,4}) = f^4/(n-1)^4$. Due to the independence we get

$$\mathrm{E}\, (Y_{1,2} Y_{1,3}) = \mathrm{E}\, X_{2,1}\, \mathrm{E}\, X_{3,1}\, E(X_{1,2} X_{1,3}) = \frac{f^2}{(n-1)^2} \mathrm{E}\, (X_{1,2} X_{1,3}).$$

Furtheron,

$$\begin{aligned} \mathrm{E}\, (X_{1,2} X_{1,3}) &= \sum_{d \geq 2} P\left(X_{1,2} = 1, X_{1,3} = 1 \mid d_1^+ = d \right)\, P(d_1^+ = d) \\ &= \sum_{d \geq 2} \binom{n-3}{d-2} \binom{n-1}{d}^{-1} p_d = \frac{1}{(n-1)(n-2)} \sum_{d \geq 2} d(d-1) p_d \\ &= \frac{1}{(n-1)(n-2)} (f^2 - f + \sigma^2). \end{aligned}$$

Putting it all together, we obtain

$$\mathrm{Var}\, V = f^4 \frac{n}{2(n-1)^3} - f^3 \frac{n}{(n-1)^2} + f^2 \frac{n}{2(n-1)} + f^2 \sigma^2 \frac{n}{(n-1)^2}$$

for the variance of V.

From these formulas we see that $\mathrm{E}\,V \sim f^2/2$. The same holds for $\mathrm{Var}\,V$ if $f = o(n)$ and $\sigma^2 = o(n)$. Actually, for f fixed and $\sigma^2 = o(n)$, all 2-cycles are disjoint almost surely (with probability tending to 1 as $n \to \infty$). To see this, it is enough to show that for the random variable Y, the number of vertices of degree at least 2 in the matched-pairs graph of $D(n, \mathcal{P})$, which is obtained by replacing the 2-cycles by edges and removing all other arcs, $\mathrm{E}\,Y \to 0$ holds. This is easily verified since

$$\mathrm{E}\,Y = \binom{n}{3}\,\mathrm{E}\,Y_{1,2}Y_{1,3} = f^4\frac{n}{6(n-1)^2} - f^3\frac{n}{6(n-1)^2} + f^2\sigma^2\frac{n}{6(n-1)^2} = o(1).$$

Hence, the following holds.

Theorem 1. For finite f, $\sigma^2 = o(n)$, and $\lambda = f^2/2$, we get $V \xrightarrow{d} \mathrm{P}(\lambda)$ as $n \to \infty$. •

Proof: With $\mathrm{E}_k V$ denoting the k-th factorial moment of V ($k = 1, 2, \ldots$), we only have to prove that $\mathrm{E}_k V \to \lambda^k = \left(f^2/2\right)^k$ for $n \to \infty$ according to a result from Bollobás (1985, p. 23, Theorem 20). We have $\mathrm{E}_k V = \sum^* \mathrm{E}\left(Y_{i_1,j_1} Y_{i_2,j_2} \cdots Y_{i_k,j_k}\right)$, where the summation $\sum^*$ is extended over all k-tuples $((i_1, j_1), (i_2, j_2), \ldots, (i_k, j_k))$ of unordered pairs of vertices. Now, since the 2-cycles are disjoint almost surely under our assumptions,

$$\mathrm{E}_k V = \binom{n}{2}\binom{n-2}{2}\cdot\ldots\cdot\binom{n-2(k-1)}{2}\left(\frac{f^2}{(n-1)^2}\right)^k + o(1) = \left(\frac{f^2}{2}\right)^k + o(1)$$

is easily verified. ■

Using the same argument as in the last proof together with another theorem from Bollobás (1985, p. 24, Theorem 22), one can show the normal approximation for V if $f = f(n) \to \infty$ but is growing not too fast.

Generally, we can say that the results given here for the probability model for random digraphs as defined in Definition 1 are more appropriate for planning and analyzing sociometric experiments than the model we used in Chapter 2 since there is no need for identical outdegrees for each vertex. This allows questions like "With which children in this camp do you want to play?" instead of "Name exactly five children with whom you want to play."

4. Graph and multigraph models for classification

The matched-pairs graph of Figure 2 and the last remarks show that for some sociometric problems, we can use random graphs instead of random digraphs. In Jaworski (1993), it is shown how a random graph can be used to study the underlying matched-pairs graph of a random digraph (the probability p of drawing an edge in this graph is the probability of a 2-cycle in the original digraph). Since under a *cluster*, we often understand a maximal collection of suitably similar objects drawn from a larger collection or sample, usually classification procedures are based on *similarities*, or *dissimilarities* (*distances*), respectively, which must be defined or calculated between every pair of objects. Here, graph theoretical concepts are helpful, too.

The n objects of a data set to be clustered are interpreted as points (or *vertices*) $1, \ldots, n$ of a graph. Two vertices are connected by a line (or an *edge*) if and only if the related objects are similar enough, i.e., if their mutual distance d_{ij} is not greater than a user-defined threshold d. The *components* of such a graph $G = G(d)$ (the maximal subsets of vertices where any two vertices are interlinked by a sequence of edges) are known as *single-linkage clusters*, and the *cliques* (the maximal subsets of vertices where any two vertices are directly linked together by an edge) become the *complete-linkage clusters.* Advantages and disadvantages of graph-theoretic concepts for cluster analysis have been discussed for example in Bock (1980), Godehardt (1990), Ling (1973).

Using the concept of random graphs (similar to that of random digraphs), we can derive test statistics to test the "randomness" of the clusters found. That means that we look whether properties found in the similarity graph, are probable under the assumption of randomness in the process of drawing the edges or not. The following two probability models of random graphs are usually used. The first model was introduced by P. Erdős and A. Rényi in 1959 (see Erdős (1960)).

Uniform model. Consider the family of all graphs on n labelled vertices and with N edges. A random graph $G_{n,N}$ is an element drawn at random—uniformly—from this family. Equivalently, one may obtain $G_{n,N}$ by choosing N edges at random among the $\binom{n}{2}$ possible edges, so that all $\binom{\binom{n}{2}}{N}$ possible choices are supposed to be equiprobable. •

The second probability model, a random graph $G_{n,p}$, is due to Gilbert (1959).

Binomial model. For each pair of n labelled vertices make a random choice, independent of each other, whether or not to join the vertices of the pair by an edge with the common probability p. Equivalently, one may erase edges, with common probability $q = 1 - p$, from the complete graph. •

In the binomial model, $G_{n,p}$, the number N of edges is a random variable with expectation $\mathrm{E}\,N = \binom{n}{2}\,p$, and the decisions of drawing the different edges are independent. In the uniform model, $G_{n,N}$, the number N of edges is fixed. In this model, the probability of drawing a specific edge depends on the number of edges which have already been drawn, that is, the choices are not independent of each other. Thus, for fixed n, the probability for a random graph property depends on the probability model we choose. If we are interested in asymptotic results, N in $G_{n,N}$ and p in $G_{n,p}$ are usually functions of n. If we put $\binom{n}{2}\,p(n) = N(n)$, then sequences of random graphs $(G_{n,N})_{n\to\infty}$ and of random graphs $(G_{n,p})_{n\to\infty}$ show the same asymptotic behaviour. This property has been proved in Bollobás (1985) and Luczak (1990).

There is a well elaborated theory with many results for the "typical structure" of random graphs based on these two probability models. The use of these results in cluster analysis together with the benefits and constraints and problems they imply has been discussed thoroughly, too (see for example Godehardt (1990, 1991a, 1993), Ling (1973)). Often, especially for medical data, the scale levels vary considerably between the different items, i.e. between the dimensions of the data vectors. It then is questionable if not impossible to compute overall (or global) similarities s_{ij} or distances d_{ij}, respectively, between the elements of a data set. The structure of a data set consisting of n multidimensional vectors can be described better by a multigraph than by a graph. We therefore introduced a certain type of multigraphs which can be used to describe this structure and to test the homogeneity within a data set (that means the hypothesis of "randomly generated" clusters).

Definition 2. Let t graphs with identical vertex set $\mathcal{V}$ be superposed one upon the other. This gives an *undirected, completely labelled multigraph*, where t edges, labelled by their two end points and by the number of their respective layers, may join the same pair of vertices. Let $G_t = (\mathcal{V}, \mathcal{E})$ be an undirected, completely labelled multigraph with vertices $i \in \mathcal{V}$, and edges $(i,j)_l \in \mathcal{E}$. Every subset $\mathcal{E}_{ij} = \{(i,j)_1, \ldots, (i,j)_t\} \cap \mathcal{E} \neq \emptyset$ is called a *connection* between i and j. Let s and u be two natural numbers with $1 \leq s \leq u \leq t$. If $s \leq |\mathcal{E}_{ij}| \leq u$ holds (that means if at least s and at most u edges link the vertices i and j together) then $\mathcal{E}_{ij}$ is called an *(s,u)-connection.* •

Examples of such multigraphs are given in Figure 3; the odd lines show the layers of multigraphs G_5. With Definition 2, we define for natural numbers $1 \leq s \leq u \leq t$, the following mapping from multigraphs to graphs: The *(s,u)-projection* $\tilde{G}$ of a multigraph G_t is the graph with the same vertex set and exactly those edges (i,j) for which $s \leq |\mathcal{E}_{ij}| \leq u$ holds in the multigraph. This mapping $G_t \rightarrow \tilde{G}$ lets us generalize the definitions commonly used in graph theory to the multigraphs as defined above. Since these generalizations are intuitively clear, we give them in table form:

$$\begin{aligned}
\text{Multigraph } G_t &\rightarrow (s,u)-\text{projection } \tilde{G} \\
(s,u)\text{-connection } \mathcal{E}_{ij} &\rightarrow \text{edge } (i,j) \\
(s,u)\text{-degree of a vertex} &\rightarrow \text{degree of a vertex} \\
(s,u)\text{-isolated vertex} &\rightarrow \text{isolated vertex} \\
(s,u)\text{-path} &\rightarrow \text{path} \\
(s,u)\text{-component} &\rightarrow \text{component} \\
(s,u)\text{-tree} &\rightarrow \text{tree} \\
(s,u)\text{-cycle} &\rightarrow \text{cycle.}
\end{aligned}$$

For $u = s$, and $u = t$, respectively, we get special cases; we call the (s,s)-projection *proper s-projection*; the (s,t)-projection is simply an *s-projection.* In the second case, we also speak of *s-connections, s-isolated vertices,*..., and so on. Figure 3 shows s-projections of multigraphs G_5 in the even lines.

There are several ways to generalize the uniform model and the binomial model of random graphs to probability models of random multigraphs. In Godehardt (1990, 1993), we assumed that N edges out of the $t\binom{n}{2}$ possible ones are drawn randomly without replacement. This gives $\binom{t\binom{n}{2}}{N}^{-1}$ as the probability for any multigraph G_t with N edges, a generalization of the uniform model, which we called Model (A1). In the same manner, performing $t\binom{n}{2}$ independent Bernoulli experiments with probability p for drawing an edge, Model (A2) was introduced as a generalization of the binomial model. Another approach is to adopt the classical probability models of random graphs for each layer of the multigraph separately. This gives us the following models, (B1) and (B2), respectively, for random multigraphs.

Model (B1). Let t random graphs $G_{n,N_1}, \ldots, G_{n,N_t}$ be chosen independently of each other. Superposition of all these random graphs gives a random multigraph $G_{t,n,(N_1,\ldots,N_t)}$ with $N_1 + \cdots + N_t = N$ edges. The probability for any multigraph G_t with N_l edges in the l-th layer, $l = 1, \ldots, t$, is $\prod_{l=1}^{t} \binom{\binom{n}{2}}{N_l}^{-1}$. •

Model (B2). Let t random graphs $G_{n,p_1}, \ldots, G_{n,p_t}$ be chosen independently of each other. Superposition of these random graphs gives a random multigraph $G_{t,n,(p_1,\ldots,p_t)}$. The probability for any multigraph G_t with N_l edges in the l-th layer, $l = 1, \ldots, t$, is $\prod_{l=1}^{t} p_l^{N_l} (1-p_l)^{\binom{n}{2}-N_l}$. •

For $p_1 = \cdots = p_t = p$, the probability models (A2) and (B2) are identical and for this case, we will denote the random multigraph by $G_{t,n,p}$. For $N_1 \sim \cdots \sim N_t \sim N/t$, it is not difficult to prove that (A1) and (B1) are asymptotically equivalent. Since a multigraph according to Definition 2 consists of t graphs, it follows from Bollobás (1985) and Luczak (1990) that the probability models (B1) and (B2) are asymptotically equivalent for $N_l \sim \binom{n}{2} p_l$, $l = 1, \ldots, t$. Therefore, we may restrict our consideration to Model (B2). The corresponding results can be immediately rewritten for random multigraphs of the model (B1) by replacing p_l by $N_l / \binom{n}{2}$.

The following random variables are defined on the probability spaces of random multigraphs $G_{t,n,(p_1,\ldots,p_t)}$:

$$T_{i,j,l}(G_{t,n,(p_1,\ldots,p_t)}) = \begin{cases} 1, & (i,j)_l \in \mathcal{E} \\ 0, & \text{otherwise} \end{cases},$$

which indicates whether there is an edge joining i and j in the l-th layer, and

$$U_{(s,u),i,j}(G_{t,n,(p_1,\ldots,p_t)}) = \begin{cases} 1, & s \le T_{i,j,.}(G_{t,n,(p_1,\ldots,p_t)}) \le u \\ 0, & \text{otherwise} \end{cases},$$

which shows whether two vertices i and j are (s,u)-connected (the "."-subscript denotes summation over the respective index). The random variable $U_{(s,u),i,.}$ is the (s,u)-degree of the vertex i, and $V_{(s,u)} = \frac{1}{2} U_{(s,u),.,.}$ is the number of (s,u)-connections in a random multigraph.

With $\mathcal{T} = \{1, 2, \ldots, t\}$, we get from (B2)

$$p^*_{(s,u)} = \sum_{\substack{\mathcal{S} \subset \mathcal{T} \\ s \le |\mathcal{S}| \le u}} \prod_{l \in \mathcal{S}} p_l \prod_{m \in \mathcal{T} - \mathcal{S}} (1 - p_m)$$

as the probability that an edge is in the (s,u)-projection of $G_{t,n,(p_1,\ldots,p_t)}$. Therefore, the (s,u)-projections of random multigraphs are random graphs $G_{n,p^*_{(s,u)}}$ with expected number of edges equal to $\binom{n}{2} p^*_{(s,u)}$. The probability $p^*_{(s,u)}$ and therefore the number of edges in the projection are both independent of the upper value u if n is large enough and $\max p_l \to 0$ holds. For $s = u = t$, $p^*_{(t,t)} = \prod_{l=1}^t p_l$ follows. If $s = 1$, then $p^*_{(1,t)} = 1 - \prod_{l=1}^t (1 - p_l)$, and for $\max p_l \to 0$, we have $p^*_{(1,t)} \sim \sum_{l=1}^t p_l$.

Let us consider the case $p_1 = \cdots = p_t = p$. Then, $p^*_{(s,u)} = \sum_{k=s}^u \binom{t}{k} p^k (1-p)^{t-k}$. If $u = t$, then $p^*_{(s,u)}$ is given by the tail of a binomial distribution with parameters t and p. Therefore, we can state the following theorem.

Theorem 2. For sequences $(G_{t,n,p})_{n \to \infty}$ of random multigraphs with t layers, n vertices, and an edge probability $p = p(n)$, given by

$$p(n) = \frac{1}{\binom{t}{s}^{1/s}} \left(\frac{f(n)}{n} \right)^{1/s}$$

with $f(n)/n \to 0$, for large n, the (s,u)-projection is a random graph $G_{n,p^*_{(s,u)}}$ with edge probability $p^*_{(s,u)} \sim f(n)/n$. This result holds for all u with $s \le u \le t$ independently of u. •

In random graph theory, a crucial role is played by threshold functions. The definition of a threshold function means that for large n almost every graph has a property $\mathcal{A}$ if the number of edges grows faster than defined by this function, and the probability of this property $\mathcal{A}$ tends to 0 if the number of edges grows slower. The notation "threshold function" has been introduced in Erdős (1960) for edge functions $N(n)$ of random graphs $G_{n,N}$. Here, we modify the definition slightly so that it is for the part $f(n)$ of $p(n)$ only; this makes the definition useful for completely labelled multigraphs, too. For analogous properties in graphs and multigraphs, respectively, the threshold function $f(n)$ remains the same. Moreover, this definition can be used for all models, the uniform models for $G_{t,n,N}$ and $G_{t,n,(N_1,\dots,N_t)}$, and the binomial models for $G_{t,n,p}$ and $G_{t,n,(p_1,\dots,p_t)}$, since the same function $f(n)$ occurs in the edge functions and in the probabilities.

The following two theorems give examples for threshold functions. In Theorem 3, $f(n) = c/n$ is the threshold function for the property that an (s,u)-projection contains an edge. In Theorem 4, $f(n) = \log n + c$ is the threshold for the property that the (s,u)-projections are connected. If the edge probability grows faster than given by $f(n) = \log n + c$, then the (s,u)-projection is almost surely connected for any integer u with $s \le u \le t$; and if this probability grows slower than the edge probability given by $f(n) = \log n + c$, then the (s,u)-projection almost surely consists of several components. For the s-projection, both theorems have been proved in earlier papers (see Godehardt (1991a, 1993)); obviously, they follow directly from the more general Theorem 2 and the corresponding results for $G_{n,p}$.

Theorem 3. For sequences $(G_{t,n,p})_{n\to\infty}$ of random multigraphs with

$$p = \frac{1}{\binom{t}{s}^{1/s} n^{2/s}} \left(c + o(1)\right)^{1/s} \qquad (c > 0),$$

the expected number of the (s,u)-connections tends to a positive limit for $n \to \infty$: $\mathrm{E}\, V_{(s,u)} \to c/2$. The distribution of $V_{(s,u)}$ tends to a Poisson distribution with expectation $c/2$ for $n \to \infty$ and any u, $s \le u \le t$. We expect only (s,u)-components of size 2 and (s,u)-isolated vertices for large n in such multigraphs. •

Theorem 4. For sequences $(G_{t,n,p})_{n\to\infty}$ of random multigraphs with

$$p = \frac{1}{\binom{t}{s}^{1/s} n^{1/s}} \left(\log n + c + o(1)\right)^{1/s},$$

the expected numbers of (s,u)-isolated vertices remain finite. The distribution tends to a Poisson distribution with expectation $\mu = e^{-c}$ for $n \to \infty$ and any u, $s \le u \le t$. A random multigraph $G_{t,n,p}$ almost surely consists of a "giant" (s,u)-component and some (s,u)-isolated vertices. The probability that $G_{t,n,p}$ is completely (s,u)-connected, tends to $e^{-e^{-c}}$. •

The application of this multigraph model to cluster analysis is as follows. We combine the dimensions of the data vectors to be clustered to *blocks* (for example, we can combine all binary components of the data vectors to a block and all continuous items form another block etc.). That gives t blocks. For each such block, we calculate *local* similarities or distances (using, for example, the matching coefficient or Tanimoto's distance for the block of binary data, and the Euclidean distance for the block of continuous data). Thus, we get t local distances for every pair of objects and can define a multigraph G_t with the n objects as vertices as follows: For every block l, we define a threshold d_l; this gives a vector $\vec{d}^T = (d_1,\dots,d_t)$ of t local distance thresholds. We superpose the n vertices in t layers according to the t blocks so that every block of dimensions of the data vectors is represented by a layer in the multigraph. Now, two vertices i and j are linked together by an edge $(i,j)_l$ in the l-th layer if the corresponding objects are similar enough in the l-th block of variables, that means if $d_{ijl} \leq d_l$. This allows t edges connecting two vertices directly, and we get an undirected, completely labelled multigraph $G_{t,n,(N_1,\dots,N_t)}$ with $N_l = N_l(d_l)$ edges per layer ($l = 1,\dots,t$), or a $G_{t,n,(p_1,\dots,p_t)}$, where the probabilities p_l for drawing edges can be estimated by $p_l = N_l / \binom{n}{2}$ for $1 \leq s \leq t$.

Single-linkage clusters are now defined as the components of the s-projection of such a multigraph $G_t(\vec{d}^T)$ generated from the data and a threshold vector. It has been discussed in Godehardt (1991a, 1993) how to use the probability models to derive test statistics for testing the hypothesis of homogeneity within a data sample. The main problems when applying these models are:

- For each layer l of the multigraph, we need a random graph G_{n,N_l} or G_{n,p_l}. That means that in each layer, the triangle inequality for data sets based on similarity or distance measures is violated.
- The different layers may not be independent (depending on the choice of the random variables in the data set). This holds especially if we consider each measurement or dimension of the data set as a layer in the multigraph.

In the following chapter, we consider a probability model where the choice of an edge in a given layer may depend on the fact whether some edges join the respective vertices in other layers or not. This is a first attempt to eliminate the second disadvantage given above.

5. A multigraph model with dependence structure

Modelling dependencies between the layers of a random multigraph can be achieved by using an urn model given in Johnson (1969), based on a model introduced by Pólya and Eggenberger in 1923: Let vertices i and j be fixed. Suppose an urn containing a white balls and b black balls. One ball is drawn at random and then replaced, together with r balls of the same color. This is done sequentially t times, defining random indicator variables $T_{i,j,l}$ taking the value 1 (an edge between i and j is drawn in the l-th layer) if a white ball is drawn in the l-th step, and 0 otherwise. Then the total number of successes $T_{i,j,.}$ follows a Pólya-Eggenberger distribution with parameters $\alpha = a/r$, $\beta = b/r$, and t. Putting $p = \alpha/(\alpha+\beta)$ and $\varphi = 1/(\alpha+\beta)$ gives

$$P\left(T_{i,j,.} = k\right) = \binom{t}{k} \frac{p^{[k,\varphi]}\,(1-p)^{[t-k,\varphi]}}{1^{[t,\varphi]}}.$$

In the formula above we use the notation of ascending factorial powers, that means $a^{[m,x]} = a\,(a+x)\cdots(a+(m-1)x)$.

This implies the dependence structure

$$P\left(T_{i,j,k+l+1} = 1 | T_{i,j,1} = \ldots = T_{i,j,k} = 1, T_{i,j,k+1} = \ldots = T_{i,j,k+l} = 0\right) = \frac{p+k\varphi}{1+(k+l)\varphi}$$

for $k+l < t$ and $\varphi \geq 0$. Conversely, let $T_{i,j,1},\dots,T_{i,j,t}$ be exchangeable indicators with this dependence structure, then the distribution of $T_{i,j,.}$ is again a Pólya-Eggenberger distribution. We will call this distribution a *beta-binomial* distribution, due to the fact that the distribution of $T_{i,j,.}$ can also be obtained as a binomial distribution $\mathrm{B}(t,p)$, where p has previously been drawn according to a beta distribution with parameters α and β. The corresponding multigraph model will be called the beta-binomial model, a random multigraph drawn according to this model will be denoted by $G_{t,n,p,\varphi}$. Obviously, for $\varphi = 0$ we get the binomial model.

From the beta-binomial model, we get

$$p^*_{(s,u)} = \sum_{k=s}^{u} \binom{t}{k} \frac{p^{[k,\varphi]}\,(1-p)^{[t-k,\varphi]}}{1^{[t,\varphi]}}$$

as the probability for an edge in the (s,u)-projection. In Model (B2), the edge probability $p^*_{(s,u)}$ in the (s,u)-projection, does not depend on u for $p \to 0$. That means that we can expect only a neglegible amount of vertices which are stronger than s-connected. In the beta-binomial model, this probability however grows in u as the following theorem shows.

Theorem 5. For sequences $(G_{t,n,p,\varphi})_{n\to\infty}$ of random multigraphs with t layers, n vertices, $\varphi > 0$, and edge probability

$$p = p(n) = b_{t,(s,u)}(\varphi)\, f(n)/n$$

with $b_{t,(s,u)}(\varphi)$ defined by

$$1/b_{t,(s,u)}(\varphi) = \sum_{k=s}^{u} \binom{t}{k} \frac{(k-1)!\,\varphi^{k-1}}{\prod_{l=t-k}^{t-1}(1+l\varphi)},$$

the (s,u)-projection is a $G_{n,p^*_{(s,u)}}$ with edge probability $p^*_{(s,u)} \sim f(n)/n$ as $n \to \infty$. •

Comparing Theorems 2 and 5, we see that for $s \geq 2$, we need much more edges (or higher edge probability p) to get the same asymptotic properties for the (s,u)-projection of a random multigraph with independent layers for any u. For $s = 1$, however, the probability to get an edge in the $(1,u)$-projection is $p^*_{(1,u)} \sim tp$ for all u in the independence case. For the beta-binomial model ($\varphi > 0$), we get $p^*_{(1,1)} \sim tp/(1+(t-1)\varphi) < tp$ for the edge probability in the proper 1-projection; and for the 1-projection, we get $p^*_1 \approx tp$ for small values of φ and $p^*_1 \approx p$ for big values of φ. Since $p^*_{(1,u)}$ is increasing in u in the beta-binomial case, we need more edges for the same properties in the $(1,u)$-projection than in the binomial case.

Like Theorems 3 and 4, Theorems 6 and 7 both are generalizations of well known results from Erdős (1960). They are direct consequences of Theorem 5.

Theorem 6. For sequences $(G_{t,n,p,\varphi})_{n\to\infty}$ of random multigraphs with t layers, n vertices, $\varphi > 0$, and edge probability $p = p(n)$ given by

$$p(n) = b_{t,(s,u)}(\varphi)\,\frac{1}{n^2}\,(c + o(1)),$$

the expected number of (s,u)-connections tends to a positive limit for $n \to \infty$: $\mathrm{E}\,V_{(s,u)} \to c/2$. The distribution of $V_{(s,u)}$ tends to a Poisson distribution with expectation $c/2$ for $n \to \infty$. We expect only (s,u)-components of size 2 and (s,u)-isolated vertices for large n in such multigraphs. •

Theorem 7. For sequences $(G_{t,n,p,\varphi})_{n\to\infty}$ of random multigraphs with t layers, n vertices, $\varphi > 0$, and edge probability $p = p(n)$ given by

$$p(n) = b_{t,(s,u)}(\varphi)\,\frac{1}{n}\,(\log n + c + o(1)),$$

the expected numbers of (s,u)-isolated vertices remain finite. The distribution of these numbers tends to a Poisson distribution with expectation $\mu = e^{-c}$ for $n \to \infty$. With probability tending to 1, such a random multigraph consists of a "giant" (s,u)-component and some (s,u)-isolated vertices. The probability that such a random multigraph is completely (s,u)-connected, tends to $e^{-e^{-c}}$. •

Figure 3 shows the effect of φ in our beta-binomial model. For this figure, we chose $t = 5$, $n = 15$ and a probability $p = 0.2$ for drawing edges in each layer. The parameter φ varies: $\varphi = 0, 0.2, 0.5, 1, 10$. We generated the random multigraphs $G_{5,15,0.2,\varphi}$ of Figure 3 using *Mathematica 2.1*. The effect of growing φ, that means of growing dependence between the layers when drawing edges is clearly shown. This figure therefore illustrates the fact that for $s \geq 2$ we need fewer edges per layer in the beta-binomial model with $\varphi > 0$ than in the binomial model to get the same expected number of edges in the s-projection. For $s = 1$, the reverse result holds.

6. Final remarks

Especially in medical research, where it is often tolerated that objects can differ in some variables if they are similar enough in the other variable, the concept of the multigraph model is appropriate. Choosing s, $1 \leq s \leq t$, here gives an s-projection which defines the clusters in the sample (for example, as the components of the s-projection). The problem of dependent layers, which will prevent us from using Model (B1) or (B2) for testing the hypothesis of homogeneity (that means of random clusters) is—at least partly—solved by the beta-binomial model. Because of the need of exchangeable indicators for this model, however, we must propose $p_1 = \cdots = p_t = p$ for the edge probability in each layer. Moreover, exchangeable indicators imply the same pairwise dependence structure between the layers.

For each layer l of the multigraph, we need a random graph $G_{n,p}$. That means that in each layer, the triangle inequality is violated by the model if we want to test the homogeneity of a data set based on similarity or distance measures (the same problem as with Models (B1) and (B2)). For metric data, we therefore will consider the uniform distribution of the data points as the hypothesis of homogeneity. Instead of the random drawing of edges giving a $G_{n,p}$ in each layer, the edge set in each layer is defined by a threshold vector $\vec{d}^T$, which is exactly the way how we generate the multigraph from a data set. Therefore this will give us a probability model which is more appropriate for metric data. However: What are the properties of random graphs and multigraphs generated by this model? How can we model dependencies between the layers in this model?

For random digraphs generated by choices between the members of a group, the beta-binomial model allows us to test the independence of choices. To every digraph we can attach a multigraph with two layers as follows. If an arc shows from vertex i to vertex j with $i < j$, then an edge is drawn in the first layer of that multigraph, and if this holds with $i > j$, then an edge is drawn in the second layer. From the assumption of independent choices it follows that we expect the same number of edges in each layer (or the same probability for an edge in each layer), and the expected number of edges in the 2-projection is given by the formula of the expected number of pairs under the hypothesis of independent choices. From a real data set like the holiday camp example, we then can estimate the edge probabilities p_1 and p_2 for the layers (with $p_1 = p_2$) and test the hypothesis $\varphi = 0$ of independent mutual choices.

Acknowledgements: The results presented in Chapter 5 (without Theorems 2–4) are part of Armin Horsch's doctorial dissertation whose research is supported by DFG grant

$\varphi = 0$

$\varphi = 0.2$

$\varphi = 0.5$

$\varphi = 1$

$\varphi = 10$

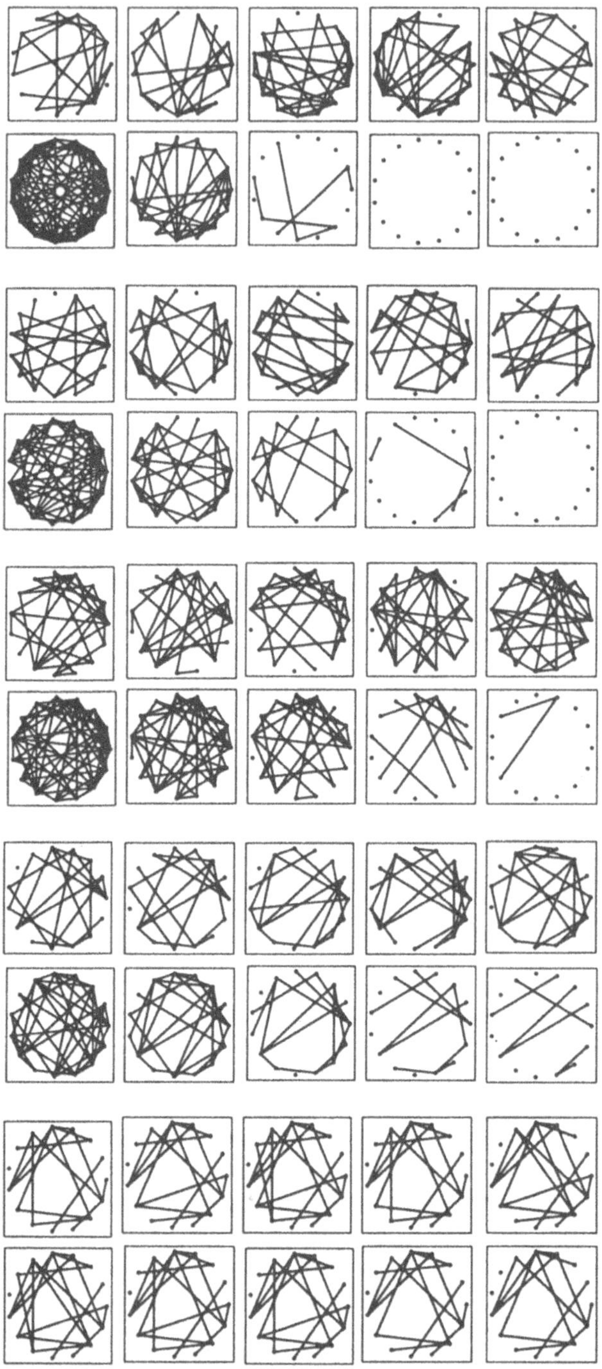

Figure 3: *The effect of φ in the beta-binomial model: The columns of the odd lines show the five layers of random multigraphs $G_{5,15,0.2,\varphi}$, the columns in the even lines show the respective 1-,..., 5-projections of that multigraph for $\varphi = 0, 0.2, 0.5, 1, 10$.*

no. Go 490/4-1. The authors wish to thank Jerzy Jaworski from the Adam-Mickiewicz-University in Poznań, Poland, and Jerzy Szymański from the Technical University in Poznań, for many helpful discussions during their stays in Düsseldorf which were supported by DFG grant no. Go 490/4-1, too. The chapters on random digraphs are based on discussions between the authors, Jerzy Jaworski, Jerzy Szymański and Torsten Reineke from the Technical University in Aachen.

References

BOCK, H.H. (1980): Clusteranalyse – Überblick und neuere Entwicklungen. *OR Spektrum, 1, 211–232.*

BOLLOBÁS, B. (1985): *Random Graphs.* Academic Press, London – New York – Tokyo.

ERDŐS, P., RÉNYI, A. (1960): On the Evolution of Random Graphs. *Publications of the Mathematical Institute of the Hungarian Academy of Sciences, 5, 17–61.*

GILBERT, E.N. (1959): Random Graphs. *Annals of Mathematical Statistics, 30, 1141–1144.*

CAPOBIANCO, M., PALKA, Z. (1983): The Distribution of Popular Persons in a Group. *Social Networks, 5, 383–393.*

GODEHARDT, E. (1974): Graphen in der Soziometrie. *(Unpublished data and presentation).*

GODEHARDT, E. (1990): *Graphs as Structural Models: The Application of Graphs and Multigraphs in Cluster Analysis (Advances in Systems Analysis, Vol. 4).* Friedr. Vieweg & Sohn, Braunschweig – Wiesbaden.

GODEHARDT, E. (1991a): Multigraphs for the Uncovering and Testing of Structures. In: H.H. Bock, P. Ihm, (eds.): *Classification, Data Analysis, and Knowledge Representation.* Springer, Berlin – Heidelberg – New York, 43–52.

GODEHARDT, E. (1991b): Zufallsgraphen in der Soziometrie. In: D.P.F. Möller, O. Richter, (eds.): *Analyse dynamischer Systeme in Medizin, Biologie und Ökologie: 4. Ebernburger Gespräch, Bad Münster am Stein, April 1990.* Springer, Berlin – Heidelberg – New York, 254–258.

GODEHARDT, E. (1993): Probability Models for Random Multigraphs with Applications in Cluster Analysis. *Annals of Discrete Mathematics, 55, 93–108.*

JAWORSKI, J., PALKA, Z. (1993): Cliques in a Social Group. *(to appear).*

JAWORSKI, J., SMIT, I.H. (1987): On a Random Digraph. *Annals of Discrete Mathematics, 33, 111–127.*

JOHNSON, N.L., KOTZ, S. (1969): *Discrete Distributions.* John Wiley & Sons, New York – Chichester – Brisbane.

LING, R.F. (1973): A Probability Theory of Cluster Analysis. *Journal of the American Statistical Association, 68, 159–164.*

ŁUCZAK, T. (1990): On the Equivalence of Two Basic Models of Random Graphs. In: M. Karoński, J. Jaworski, A. Ruciński, (eds.): *Random Graphs '87.* John Wiley & Sons, New York – Chichester – Brisbane, 151–157.

PALKA, Z. (1986): Rulers and Slaves in a Random Social Group. *Graphs and Combinatorics, 2, 165–172.*

SMIT, I.H. (1979): *The Distribution of the Number of 2-Cycles in Certain Kinds of Random Digraphs.* Rapport No. 116, Wiskundig Seminarium, Vrije Universiteit, Amsterdam.

SMIT, I.H. (1985): Matchmaking between two Collections. In: M. Karoński, A. Ruciński, (eds.): *Random Graphs '83.* North-Holland, Amsterdam – New York – Oxford, 231–242.

Geometric Approach to Evaluating Probabilities of Correct Classification into two Gaussian or Spherical Categories

Dorit Krause, Wolf-Dieter Richter

Fachbereich Mathematik, Universität Rostock,
Universitätsplatz 1, 18051 Rostock, Germany

Summary: The problem of classification is considered within a linear model approach. Two competing hypotheses concerning one-dimensional expectations are reflected by so-called "hypothesis", suitable subspaces of the model space. The distances between either the sample point or its projection onto the extended model space and the hypotheses spaces are equivalently used to decide between the competing hypotheses. The event of making a correct statistical decision is then described by suitable subsets of the sample space or of the extended model space, respectively. The probability of the sets is calculated by using geometric representation formulas for Gaussian or spherical distributions in the respective spaces. In this way we get the explicit numerical values for the probabilities of a correct classification.

1. Introduction

Statistical decisions such as estimating parameters and testing hypotheses are exhaustively studied for the general linear model. Geometric methods are, among the most often used analytical methods, frequently applied tools for evaluating or approximating probability distributions of various types of statistics. On the other hand, geometric methods are often exploited only for illustrating analytical solutions of statistical problems and relatively seldom really used as the main tools of investigation. Although already famous statisticians like R.A.Fisher, M.S.Bartlett, H.Scheffé and others derived many classical results of mathematical statistics by the help of geometric argumentations, up to now there does not seem to be a textbook combining a clear geometric approach for deriving statistical decisions with the strong measure theoretical foundation for describing the probabilistic properties of these decisions. However, there are several books written in a somewhat closely related spirit such as, e.g., those of Anderson (1984), Eaton (1983) and Saville and Wood (1991). A quite close connection between the geometric orthogonal projection methods for estimating parameters and the analytical maximum likelihood method is given by the Gauss-Markov theorem for the linear model with Gaussian error variables. A consistent geometric approach to this area of classical statistics allows the understanding of a probabilistic evaluation of statistical decisions by measuring certain subsets of the sample space such like, e.g., confidence regions or critical test regions. These regions are statistically well motivated and each of them can be derived by the help of a suitably chosen optimality criterion. The Gaussian or the spherical distribution measures of many regions occuring in statistics can be relatively easy calculated by using a geometric measure representation formula given in Richter(1991). In the just mentioned paper some examples from classical statistics for Gaussian samples are discussed as well as their extensions to spherical samples. In Richter (1992) and in Richter and Steinebach (1993) certain consequences for the vector algebraic or geometric approach to the ANOVA model are discussed. Clearly, the geometric approach does not only apply to decisions like estimating parameters and testing hypotheses, but also to ordering, classifying, selecting and other decisions. The aim of the present paper is to show this strategy for the problem of classifying an observation vector into two Gaussian or spherical distributions and to give as an result an explicit calculation of the probability of a correct classification. As a first step into this direction, we shall consider here only the case of one dimensional classification parameters and known variances but with repeated measurements in the observation variable.

2. The model and preliminary results

The statistical structure S_n which will be considered in this paper, can be interpreted as follows. An investigator makes n_1, n_2 and n_3 one-dimensional measurements of individuals from certain categories or populations Π_1, Π_2 and Π_3, respectively. Categories Π_1 and Π_2 are assumed to be distinguishable in the sense that their expectations μ_1 and μ_2 satisfy $\mu_1 \neq \mu_2$ and with respect to Π_3 it is assumed that its (expectation, variance)-vector coincides with either that of Π_1 or that of Π_2. One could imagine Π_3 to be a copy of either Π_1 or Π_2.
To be more specifically, the vector $Y_{(n)}$ in our model is assumed to be distributed according to a distribution $\Phi_{\mu,\Sigma;g}$ with a density

$$\begin{aligned} f\left(y_{(n)}\right) &= C(g,n)\,(\det\Sigma)^{-1/2}\, g\left((y_{(n)}-\mu)^T\Sigma^{-1}(y_{(n)}-\mu)\right) \\ &= \tilde{g}\left(\sum_{i=1}^{3}\sigma_i^{-2}\sum_{j=1}^{n_i}\left(y_j^{(i)}-\mu_i\right)^2\right). \end{aligned}$$

It is well known from, e.g., Tong (1990), Fang (1990) and Kai-Tai and Yao-Ting (1990), that all marginal distributions of $\Phi_{\mu,\Sigma;g}$ are elliptically contoured, but not necessarily all with the same g. Any k-dimensional marginal distribution having $\sigma^2 I_k$ as covariance matrix is called a spherical distribution, $\sigma^2 > 0$. Here, $\sigma^2 = -2\phi'(0)$, where $\phi(\|t\|^2)$ denotes the characteristic function of the respective spherical distribution.
Notice that within the class of spherical distributions with densities the independence of a distribution's marginals is equivalent to their uncorrelatedness if the density generating function g of the whole vector coincides with that of the respective Gaussian distribution, i.e., if $g(c) = \exp\{-c/(2\sigma^2)\}$, $c > 0$, $\sigma^2 > 0$.
Let us now consider the statistical structure

$$\begin{aligned} S_n &= \left(\mathbb{R}^n,\ \mathcal{B}^n,\ \left\{\Phi_{\mu,\Sigma;g} : \mu = \begin{pmatrix} 1_{n_1} & 0_{n_1} & 0_{n_1} \\ 0_{n_2} & 1_{n_2} & 0_{n_2} \\ 0_{n_3} & 0_{n_3} & 1_{n_3} \end{pmatrix}\begin{pmatrix} \mu_1 \\ \mu_2 \\ \mu_3 \end{pmatrix},\ (\mu_1,\mu_2)\in\mathbb{R}^2,\ \mu_1\neq\mu_2,\right.\right. \\ &\qquad \mu_3\in\{\mu_1,\mu_2\},\ \Sigma = \begin{pmatrix} \sigma_1^2 I_{n_1} & 0 & 0 \\ 0 & \sigma_2^2 I_{n_2} & 0 \\ 0 & 0 & \sigma_3^2 I_{n_3} \end{pmatrix},\ n = n_1+n_2+n_3, \\ &\qquad \left.\left. \sigma_3^2 = I\{\mu_3=\mu_1\}\sigma_1^2 + I\{\mu_3=\mu_2\}\sigma_2^2;\ 0 < \int_0^\infty r^{n-1} g(r^2)\,dr < \infty\right\}\right), \end{aligned}$$

where $\Phi_{\mu,\Sigma;g}$ denotes an elliptically contoured probability distribution with expectation μ, covariance matrix Σ and density generating function g, $1_{n_i} = (1,\ldots,1)^T \in \mathbb{R}^{n_i}$, $0_{n_i} = (0,\ldots,0)^T \in \mathbb{R}^{n_i}$, I_{n_i} denotes the $n_i \times n_i$ - unit matrix and $I\{\mu_3 = \mu_i\}$ stands for

$$I_{\left\{\binom{x_1}{x_2}\in\mathbb{R}^2:\, x_1=x_2\right\}}\left(\begin{pmatrix}\mu_3\\ \mu_i\end{pmatrix}\right) = \begin{cases} 1 & \text{if } \mu_3 = \mu_i \\ 0 & \text{if } \mu_3 \neq \mu_i \end{cases}.$$

Let

$$Y_{(n)} = \left(Y^{1}_{(n_1)}{}^{T}, Y^{2}_{(n_2)}{}^{T}, Y^{3}_{(n_3)}{}^{T}\right)^{T}, \; n_1 + n_2 + n_3 = n$$

denote a random sample vector which induces the statistical structure S_n and satisfies the following *sample-model-equation*:

$$Y_{(n)} = \begin{pmatrix} 1_{n_1} & 0_{n_1} & 0_{n_1} \\ 0_{n_2} & 1_{n_2} & 0_{n_2} \\ 0_{n_3} & 0_{n_3} & 1_{n_3} \end{pmatrix} \begin{pmatrix} \mu_1 \\ \mu_2 \\ \mu_3 \end{pmatrix} + \mathcal{E}_{(n)} =: \mathcal{X}\beta + \mathcal{E}_{(n)}, \tag{1}$$

where $\mathcal{E}_{(n)}$ is a random vector in $\mathbb{R}^n$ with an elliptically contoured distribution and having expectation zero and covariance matrix

$$\Sigma = \begin{pmatrix} \sigma_1^2 I_{n_1} & 0 & 0 \\ 0 & \sigma_2^2 I_{n_2} & 0 \\ 0 & 0 & \sigma_3^2 I_{n_3} \end{pmatrix}.$$

As usual the set $\mathcal{M}$ of all possible values of the expectation of Y_n will be called the *model space*, i.e.

$$\mathcal{M} = \left\{ \begin{pmatrix} 1_{n_1} \\ 0_{n_2} \\ 0_{n_3} \end{pmatrix} \mu_1 + \begin{pmatrix} 0_{n_1} \\ 1_{n_2} \\ 0_{n_3} \end{pmatrix} \mu_2 + \begin{pmatrix} 0_{n_1} \\ 0_{n_2} \\ 1_{n_3} \end{pmatrix} \mu_3 : \begin{pmatrix} \mu_1 \\ \mu_2 \end{pmatrix} \in \mathbb{R}^2, \mu_1 \neq \mu_2, \mu_3 \in \{\mu_1, \mu_2\} \right\}.$$

The smallest linear subspace $\widetilde{\mathcal{M}}$ of the *sample space* $\mathbb{R}^n$ which includes the model space $\mathcal{M}$, i.e.

$$\widetilde{\mathcal{M}} = \left\{ \begin{pmatrix} \mu_1 1_{n_1} \\ \mu_2 1_{n_2} \\ \mu_3 1_{n_3} \end{pmatrix} : \begin{pmatrix} \mu_1 \\ \mu_2 \\ \mu_3 \end{pmatrix} \in \mathbb{R}^3 \right\}$$

will be called *extended model space*. Consider the following sets which are almost subspaces of $\mathcal{M}$:

$$\mathcal{M}_{1/3} = \left\{ \begin{pmatrix} \mu_1 1_{n_1} \\ \mu_2 1_{n_2} \\ \mu_1 1_{n_3} \end{pmatrix} : \mu_1 \neq \mu_2 \right\} \quad \text{and} \quad \mathcal{M}_{2/3} = \left\{ \begin{pmatrix} \mu_1 1_{n_1} \\ \mu_2 1_{n_2} \\ \mu_2 1_{n_3} \end{pmatrix} : \mu_1 \neq \mu_2 \right\}.$$

If a concrete sample, i.e. a realization $y_{(n)} = \left(y^{1}_{(n_1)}{}^{T}, y^{2}_{(n_2)}{}^{T}, y^{3}_{(n_3)}{}^{T}\right)^{T}$ of $Y_{(n)}$ belongs to either $\mathcal{M}_{1/3}$ or $\mathcal{M}_{2/3}$ then it will be natural for the investigator to interprete this as that μ_3 coincides with μ_1 or μ_2, respectively, although, obviously, this conclusion can be false. Note that $\mathcal{M} = \mathcal{M}_{1/3} \cup \mathcal{M}_{2/3} \subset \widetilde{\mathcal{M}}$, $\mathcal{M}_{1/3} \cap \mathcal{M}_{2/3} = \emptyset$, since $\mu_1 \neq \mu_2$. The problem which the investigator wants to decide is whether

$$H_{1/3} : \mu_3 = \mu_1 \quad \text{or} \quad H_{2/3} : \mu_3 = \mu_2 \tag{2}$$

holds true. Whatever n_1, n_2, n_3 are, so that $n = n_1 + n_2 + n_3$, these hypotheses can be reflected in the sample space $\mathbb{R}^n$ as

$$H_{1/3}^{(n)} : EY_{(n)} \in \mathcal{M}_{1/3} \quad \text{or} \quad H_{2/3}^{(n)} : EY_{(n)} \in \mathcal{M}_{2/3}, \tag{3}$$

respectively. In this two-decision problem it seems to be natural that one will preferably assume $H_{1/3}^{(n)}$ (or $H_{1/3}$) to be true instead of $H_{2/3}^{(n)}$ (or $H_{2/3}$) if $y_{(n)}$ lies closer to $\mathcal{M}_{1/3}$ than to $\mathcal{M}_{2/3}$ in some sense. Allowing for different definitions what is thought to be the dissimilarity from $y_{(n)}$ to $\mathcal{M}_{i/3}$ (i=1,2), we shall make use of the *decision functions* $d_i \mid \mathbb{R}^n \longrightarrow \{1,2\}$, $i \in \{1,2,3,4\}$, for deciding between $H^{(n)}_{1/3}$ and $H^{(n)}_{2/3}$:

$$\begin{aligned}
d_1(y_{(n)}) &= 2 - I\{dist(y_{(n)}, \mathcal{M}_{1/3}) < c\, dist(y_{(n)}, \mathcal{M}_{2/3})\}, &(4)\\
d_2(y_{(n)}) &= 2 - I\{dist(\Pi_{\widetilde{\mathcal{M}}} y_{(n)}, \mathcal{M}_{1/3}) < c\, dist(\Pi_{\widetilde{\mathcal{M}}} y_{(n)}, \mathcal{M}_{2/3})\}, &(5)\\
d_3(y_{(n)}) &= 2 - I\{\|\Pi_{1^{-0+}}\, y_{(n)}\| < c\,\|\Pi_{1^{0-+}}\, y_{(n)}\|\}, &(6)\\
d_4(y_{(n)}) &= 2 - I\{dist(\Pi_{\widetilde{\mathcal{M}}} y_{(n)}, \Pi_{\mathcal{M}_{1/3}} y_{(n)}) < c\, dist(\Pi_{\widetilde{\mathcal{M}}} y_{(n)}, \Pi_{\mathcal{M}_{2/3}} y_{(n)})\}, &(7)
\end{aligned}$$

$c > 0$ a constant. Here, *dist* means euclidean distance, Π_x and $\Pi_{\mathcal{N}}$ stand for orthogonal projection onto the direction of x or onto the subspace $\mathcal{N} \subset \mathbb{R}^n$, respectively, and $1^{-0+} = \left(-1^T_{n_1}/n_1,\, 0^T_{n_2},\, 1^T_{n_3}/n_3\right)^T$, $1^{0-+} = \left(0^T_{n_1},\, -1^T_{n_2}/n_2,\, 1^T_{n_3}/n_3\right)^T$. Note that decision rule d_3 is motivated by the circumstance that the least squares estimators of $1^{-0+}(\mu_3 - \mu_1)$ and $1^{0-+}(\mu_3 - \mu_2)$ are $(1/n_1 + 1/n_3)\,\Pi_{1^{-0+}}\, y_{(n)}$ and $(1/n_2 + 1/n_3)\,\Pi_{1^{0-+}}\, y_{(n)}$, respectively.

More specifically, write $y_{(n)} = (y^{(1)}_1, \ldots, y^{(1)}_{n_1}, y^{(2)}_1, \ldots, y^{(2)}_{n_2}, y^{(3)}_1, \ldots, y^{(3)}_{n_3})^T$, and put

$$\bar{y}^{(i)}_{\bullet} = \left(\sum_{k=1}^{n_i} y^{(i)}_k\right)/n_i,\ i = 1,2,3, \qquad \bar{y}^{(i/j)}_{\bullet\bullet} = \left(\sum_{k=1}^{n_i} y^{(i)}_k + \sum_{k=1}^{n_j} y^{(j)}_k\right)/\,(n_i + n_j)\,,\ i \neq j.$$

Then we can rewrite d_1, d_2, d_3 into the following explicit forms:

$$\begin{aligned}
d_1(y_{(n)}) &= 2 - I\Big\{ \sum_{i=1,3}\sum_{k=1}^{n_i} \left(y^{(i)}_k - \bar{y}^{(1/3)}_{\bullet\bullet}\right)^2 + \sum_{k=1}^{n_2}\left(y^{(2)}_k - \bar{y}^{(2)}_{\bullet}\right)^2 \\
&\qquad < c^2\Big[\sum_{i=2,3}\sum_{k=1}^{n_i}\left(y^{(i)}_k - \bar{y}^{(2/3)}_{\bullet\bullet}\right)^2 + \sum_{k=1}^{n_1}\left(y^{(1)}_k - \bar{y}^{(1)}_{\bullet}\right)^2\Big]\Big\}, &(8)\\
d_2(y_{(n)}) &= 2 - I\Big\{n_1\left(\bar{y}^{(1)}_{\bullet} - \bar{y}^{(1/3)}_{\bullet\bullet}\right)^2 + n_3\left(\bar{y}^{(3)}_{\bullet} - \bar{y}^{(1/3)}_{\bullet\bullet}\right)^2 \\
&\qquad < c^2\Big[n_2\left(\bar{y}^{(2)}_{\bullet} - \bar{y}^{(2/3)}_{\bullet\bullet}\right)^2 + n_3\left(\bar{y}^{(3)}_{\bullet} - \bar{y}^{(2/3)}_{\bullet\bullet}\right)^2\Big]\Big\}, &(9)\\
d_3(y_{(n)}) &= 2 - I\Big\{\frac{n_1}{n_1+n_3}\left(\bar{y}^{(3)}_{\bullet} - \bar{y}^{(1)}_{\bullet}\right)^2 < c^2\,\frac{n_2}{n_2+n_3}\left(\bar{y}^{(3)}_{\bullet} - \bar{y}^{(2)}_{\bullet}\right)^2\Big\}. &(10)
\end{aligned}$$

Using the theorem of Pythagoras, it is not very hard now to prove the following theorem concerning equivalences between the four *decision rules* d_1 up to d_4.

Theorem 1: $d_2(y_{(n)}) \equiv d_3(y_{(n)}) \equiv d_4(y_{(n)})$ for all $y_{(n)} \in \mathbb{R}^n$ and
$d_1(y_{(n)}) \equiv d_2(y_{(n)})$ for all $y_{(n)} \in \mathbb{R}^n$ if $c = 1$.

A general motivation why to derive various representation formulas for the same decision rule consists in the fact that several data reduced forms are suited to fast decisions in special concrete situations, whilst the more formal representations are more appropriate for studying the probabilistic behaviour of the decision rule. It is well known, see e.g. in Anderson (1984), that evaluating probabilities of correct classification is rather complicated in the case of finite sample sizes n_1, n_2 and n_3. This is true as well both for the Bayesian classification rule and for the maximum likelihood classification rule. Note that if we put $c = 1$ and $n_3 = 1$ then our decision rule in the form d_3 coincides with formula (32) of § 6.5 in Anderson (1984) for the maximum likelihood rule if we put there $p = 1$ for the dimension p of the observation variable and cancel S^{-1}. Recall that the class of Bayesian decision rules under a suitable additional condition coincides with the class of admissible decision rules if no prior probabilities for the two categories are known. Further it is known, see Anderson (1984, Chapter 6), that the probabilities of misclassification behave asymptotically in the same way when using either the maximum likelihood rule or the plug-in version of a decision rule which is optimal for known parameters.

3. Main results

From Theorem 1 it follows that we can make our decision at least in two equivalent ways, either based upon the whole sample $Y_{(n)}$ or based upon the reduced statistic

$$Z = \begin{pmatrix} Z_1 \\ Z_2 \end{pmatrix} = \begin{pmatrix} \bar{Y}_{\bullet}^{(3)} - \bar{Y}_{\bullet}^{(1)} \\ \bar{Y}_{\bullet}^{(3)} - \bar{Y}_{\bullet}^{(2)} \end{pmatrix}. \tag{11}$$

If $(\mu_3, \sigma_3^2) = (\mu_i, \sigma_i^2)$, i.e. if $H_{i/3}$ holds true, then

$$CC_i = Y_{(n)}^{-1}\left(\left\{y_{(n)} \in \mathbb{R}^n : d_j(y_{(n)}) = i\right\}\right) \tag{12}$$

describes the event of correct classification into the population (hypothesis) $i \in \{1,2\}$, when using d_j, $j \in \{1,2,3,4\}$. From Theorem 1, we obtain

$$CC_i = Z^{-1}(CC_i^*), \tag{13}$$

$$CC_1^* = \left\{\begin{pmatrix} z_1 \\ z_2 \end{pmatrix} \in R^2 : \frac{n_1}{n_1+n_3} z_1^2 < c^2 \frac{n_2}{n_2+n_3} z_2^2 \right\}, \tag{14}$$

$$CC_2^* = \left\{\begin{pmatrix} z_1 \\ z_2 \end{pmatrix} \in R^2 : \frac{n_1}{n_1+n_3} z_1^2 > c^2 \frac{n_2}{n_2+n_3} z_2^2 \right\}. \tag{15}$$

Consequently, under $H_{i/3}$, the probability of correct classification is

$$P(CC_i) = P^Z_{(\mu_i,\sigma_i^2)}(CC_i^*),\ i \in \{1,2\}, \tag{16}$$

i.e., as expected, we have transformed the original n-dimensional problem into a two-dimensional problem. The geometric method yields in all dimensions both the respective distribution measure and the event set to be measured with it. Note that

$$P^Z_{(\mu_i,\sigma_i^2)} = \Phi_{\nu_i,\Gamma_i;g} \tag{17}$$

with

$$\nu_1 = \begin{pmatrix} 0 \\ \mu_1 - \mu_2 \end{pmatrix},\ \nu_2 = \begin{pmatrix} \mu_2 - \mu_1 \\ 0 \end{pmatrix},\ \Gamma_i = \begin{pmatrix} \sigma_1^2/n_1 & 0 \\ 0 & \sigma_2^2/n_2 \end{pmatrix} + \frac{\sigma_i^2}{n_3}\begin{pmatrix} 1 & 1 \\ 1 & 1 \end{pmatrix}. \tag{18}$$

It is immediately seen from (18) that the covariance matrix Γ_i becomes asymptotically degenerated if, e.g., n_1 and n_2 are approaching infinity while n_3 remains bounded. In this asymptotic case, the probability measure of Z would therefore be concentrated on a certain one-dimensional subspace of R^2. This means that if we are given relative large sample sizes n_1, n_2 in the training samples within the populations Π_1 and Π_2 and if the sample size n_3 in the population Π_3, which is to be assigned to either Π_1 or Π_2, is relatively small then the problem of evaluating probabilities of correct classification can essentially be reduced to a one-dimensional case. Theorems 2 and 3 below, however, are devoted to the general two-dimensional case, i.e. they deal with arbitrary sample sizes n_1, n_2 and n_3. Moreover, Theorem 2 will be devoted to the special case of a Gaussian random experiment S_n while Theorem 3 deals with the general spherical case. This means, in Theorem 2, we put $g(c) = \exp\{-c/2\}, c > 0$ while Theorem 3 applies without this restriction.

Theorem 2: If in the Gaussian case $g(c) = \exp\{-c/2\}, c > 0$, $H_{i/3}$ is true, then the probability of correct classification allows the following integral representation:

$$P(CC_i) = \int_0^\infty r \exp\{-r^2/2\}\, \mathcal{F}(CC_i^{**}, r)\, dr, \tag{19}$$

where $\mathcal{F}(CC_i^{**}, r)$ is a certain function essentially depending on the parameters $|\mu_1 - \mu_2|$, n_1, n_2, n_3, σ_1^2, σ_2^2 and c, given in all details by formula (32) below.

Note that the function $\mathcal{F}$ reflects a geometric property of the class of spherical distribution measures. This geometric property has been discussed to a certain extend, e.g., in Richter(1991) and further papers cited therein. It consists essentially in that all spherical measures are invariant in a certain sense with respect to orthogonal transformations. This circumstance will be several times exploited in the proof of Theorem 2 below (see section 4). Therefore it is not surprising that the probabilities of correct classification allow a quite similar representation in the general spherical case, too.

Theorem 3: If in the case of a general density function g $H_{i/3}$ is true and the marginal distributions of both subsample vectors $Y^1_{(n_1)}$ and $Y^2_{(n_2)}$ from $Y_{(n)}$ are spherical distributions with density generating functions g_1 and g_2, respectively, then the probabilities of correct classification into the two subpopulations allow the following integral representations:

$$P(CC_i) = \left(\int_0^\infty r g_i(r^2)\, dr\right)^{-1} \int_0^\infty r\, g_i(r^2)\, \mathcal{F}(CC_i^{**}, r)\, dr,\ i = 1, 2, \tag{20}$$

where $\mathcal{F}$ is exactly the same function as in Theorem 2.

The formulas (19) and (20) will be the basis for a numerical study in order to derive tables of probabilities of correct classifications. Such tables will be published elsewhere.

4. Proofs of Theorems 2 and 3

After having reduced the n-dimensional problem of calculating probabilities of correct classification to a two-dimensional problem, see (16), (17), (18), the main idea to prove Theorems 2 and 3 is now to apply the geometric representation for spherical measures given in Richter (1991) and in an earlier paper mentioned therein. To this end we have to transform both the covariance matrix Γ_i to the unit matrix and the expectation vector ν_i to the zero element. Note that along with a suitable transformation T concerning the random vector Z we have to transform the region CC_i^* in the same way. The transformation T will be split into three partial transformations T_1, T_2, T_3, i.e. $T = T_1 \circ T_2 \circ T_3$.

T_1 : Norming the variances

The first step to simplify the covariance matrix Γ_i consists in norming the variances of Z, i.e. in applying the transformation $T_1 \mid I\!R^2 \longrightarrow I\!R^2$ with

$$T_1\left(\begin{pmatrix} z_1 \\ z_2 \end{pmatrix}\right) = \begin{pmatrix} z_1/\sqrt{V(Z_1)} \\ z_2/\sqrt{V(Z_2)} \end{pmatrix}, \begin{pmatrix} z_1 \\ z_2 \end{pmatrix} \in I\!R^2.$$

Note that T_1 depends on i because the variances $V(Z_1)$ and $V(Z_2)$ depend on whether $H_{1/3}$ or $H_{2/3}$ is true. Proceeding in this way, we get the random vectors $U_i := T_1 Z$:

$$U_i = \begin{pmatrix} Z_1/\sqrt{\sigma_1^2/n_1 + \sigma_i^2/n_3} \\ Z_2/\sqrt{\sigma_2^2/n_2 + \sigma_i^2/n_3} \end{pmatrix}.$$

Recall that i stands for the two ways of classification, i.e. $i \in \{1,2\}$.
Define

$$a := \left(\frac{1}{n_2} + \frac{\sigma_1^2}{\sigma_2^2 n_3}\right), \; b := \left(\frac{1}{n_1} + \frac{\sigma_2^2}{\sigma_1^2 n_3}\right), \; d := \frac{\mu_1 - \mu_2}{\sqrt{\frac{\sigma_2^2}{n_2} + \frac{\sigma_1^2}{n_3}}} \text{ and } e := \frac{\mu_2 - \mu_1}{\sqrt{\frac{\sigma_1^2}{n_1} + \frac{\sigma_2^2}{n_3}}}$$

and put $c^2 = \sigma_1^2/\sigma_2^2$.
The respectively transformed areas $T_1(CC_i^*)$ to which the $U_i's$ belong in the cases of correct classifications to i are given by

$$\begin{aligned} CC_1^{T_1} &= T_1(CC_1^*) \\ &= \left\{\begin{pmatrix} u_1 \\ u_2 \end{pmatrix} \in I\!R^2 : \frac{n_2+n_3}{n_2 n_3} {u_1}^2 < a\,{u_2}^2\right\}, \\ CC_2^{T_1} &= T_1(CC_2^*) \\ &= \left\{\begin{pmatrix} u_1 \\ u_2 \end{pmatrix} \in I\!R^2 : b\,{u_1}^2 > \frac{n_1+n_3}{n_1 n_3} {u_2}^2\right\}, \end{aligned}$$

i.e., starting from (16) we arrived at

$$P(CC_i) = P(T_1(Z) \in T_1(CC_i^*)) = P(U_i \in CC_i^{T_1}), i \in \{1,2\}. \tag{21}$$

Note that the U_i's have the following expectation vectors and covariance matrices:

$$U_1 : \begin{pmatrix} 0 \\ d \end{pmatrix}, \begin{pmatrix} 1 & \varrho_1 \\ \varrho_1 & 1 \end{pmatrix} \text{ with } \varrho_1 = \frac{\sigma_1\sqrt{n_1 n_2}}{\sqrt{(n_1+n_3)(n_3\sigma_2^2 + n_2\sigma_1^2)}}, \tag{22}$$

$$U_2 : \begin{pmatrix} e \\ 0 \end{pmatrix}, \begin{pmatrix} 1 & \varrho_2 \\ \varrho_2 & 1 \end{pmatrix} \text{ with } \varrho_2 = \frac{\sigma_2\sqrt{n_1 n_2}}{\sqrt{(n_2+n_3)(n_3\sigma_1^2 + n_1\sigma_2^2)}}. \tag{23}$$

T_2: Centering the expectations

Applying the transformation $T_2 \mid I\!R^2 \longrightarrow I\!R^2$ with

$$T_2\left(\begin{pmatrix} z_1 \\ z_2 \end{pmatrix}\right) = \begin{cases} (z_1, z_2 - d)^T & \text{if } i = 1 \\ (z_1 - e, z_2)^T & \text{if } i = 2 \end{cases}, \begin{pmatrix} z_1 \\ z_2 \end{pmatrix} \in I\!R^2,$$

i.e. introducing the random vectors

$$V_1 = T_2(U_1) = \begin{pmatrix} U_{11} \\ U_{12} - d \end{pmatrix}, \; V_2 = T_2(U_2) = \begin{pmatrix} U_{21} - e \\ U_{22} \end{pmatrix}$$

we get the zero element for the expectation vectors of V_1 and V_2. From (21) it follows now:

$$P(CC_i) = P(T_2(U_i) \in T_2(CC_i^{T_1})) = P(V_i \in CC_i^{T_2}) \tag{24}$$

where

$$\begin{aligned} CC_1^{T_2} &= T_2(CC_1^{T_1}) \\ &= \left\{ \begin{pmatrix} v_1 \\ v_2 \end{pmatrix} \in \mathbb{R}^2 : \frac{n_2+n_3}{n_2 n_3} v_1^2 < a\,(v_2+d)^2 \right\} \\ CC_2^{T_2} &= T_2(CC_2^{T_1}) \\ &= \left\{ \begin{pmatrix} v_1 \\ v_2 \end{pmatrix} \in \mathbb{R}^2 : b\,(v_1+e)^2 > \frac{n_1+n_3}{n_1 n_3} v_2^2 \right\}. \end{aligned}$$

T_3: Principal axis transformation to uncorrelated components

Finally we put

$$W_i = T_3(V_i) =: K_i \cdot V_i \tag{25}$$

with

$$K_i = \frac{1}{\sqrt{2}} \begin{pmatrix} \frac{1}{\sqrt{1+\varrho_i}} & \frac{1}{\sqrt{1+\varrho_i}} \\ \frac{1}{\sqrt{1-\varrho_i}} & -\frac{1}{\sqrt{1-\varrho_i}} \end{pmatrix}, \; i \in \{1,2\}. \tag{26}$$

Note that the matrices K_i are chosen in such way that both W_1 and W_2 have the unit matrix $\left(\begin{smallmatrix} 1 & 0 \\ 0 & 1 \end{smallmatrix}\right)$ as covariance matrices. From (24) it follows now:

$$P(CC_i) = P(T_3(V_i) \in T_3(CC_i^{T_2})) = P(W_i \in CC_i^{**}), \tag{27}$$

where

$$\begin{aligned} CC_1^{**} &= T_3(CC_1^{T_2}) \\ &= \left\{ \begin{pmatrix} w_1 \\ w_2 \end{pmatrix} \in \mathbb{R}^2 : \frac{n_2+n_3}{n_2 n_3} \left(\frac{1}{2}\sqrt{2(1+\varrho_1)}\, w_1 + \frac{1}{2}\sqrt{2(1-\varrho_1)}\, w_2 \right)^2 \right. \\ &\qquad \left. < a \left(\frac{1}{2}\sqrt{2(1+\varrho_1)}\, w_1 - \frac{1}{2}\sqrt{2(1-\varrho_1)}\, w_2 + d \right)^2 \right\}, \end{aligned} \tag{28}$$

$$\begin{aligned} CC_2^{**} &= T_3(CC_2^{T_2}) \\ &= \left\{ \begin{pmatrix} w_1 \\ w_2 \end{pmatrix} \in \mathbb{R}^2 : b \left(\frac{1}{2}\sqrt{2(1+\varrho_2)}\, w_1 + \frac{1}{2}\sqrt{2(1-\varrho_2)}\, w_2 + e \right)^2 \right. \\ &\qquad \left. > \frac{n_1+n_3}{n_1 n_3} \left(\frac{1}{2}\sqrt{2(1+\varrho_2)}\, w_1 - \frac{1}{2}\sqrt{2(1-\varrho_2)}\, w_2 \right)^2 \right\}. \end{aligned} \tag{29}$$

After having made the three transformations T_1, T_2 and T_3, the probability of correct classification if $H_{i/3}$ is true can be written as

$$P(CC_i) = \Phi_{\left(\begin{smallmatrix} 0 \\ 0 \end{smallmatrix}\right), \left(\begin{smallmatrix} 1 & 0 \\ 0 & 1 \end{smallmatrix}\right); g}(CC_i^{**}), \; i \in \{1,2\} \tag{30}$$

for the given density generating function g. Now we are in a position to apply the geometric measure representation formula in Richter (1991). This yields immediatly formula (20) with

$$\mathcal{F}(CC_i^{**}, r) = \omega(r^{-1} CC_i^{**} \cap S_2(1)), \; r > 0, \tag{31}$$

where ω denotes the uniform probability distribution on the unit sphere in $\mathbb{R}^2$, i.e. on the circle $S_2(1) = \left\{\binom{x_1}{x_2} \in \mathbb{R}^2 : x_1^2 + x_2^2 = 1\right\}$. In other words, formula (31) means that we have to divide the arclength of $r^{-1}CC_i^{**} \cap S_2(1)$ by the arclength of the circle $S_2(1)$, i.e. by 2π, for all $r > 0$.

Note that CC_i^{**}, for $i \in \{1,2\}$, is a well defined double cone in $\mathbb{R}^2$ with the zero point in its inner part, and its apex therefore being different from the origin; i.e. CC_i^{**} is a set of points which are situated between two straight lines for which their increase is defined by the vectors $l_{1,i}$ and $l_{2,i}$ and the local vector to the apex of the double cone, which is different from the zero vector, is defined by q_i. Denote by $s_{min,i}$ and $s_{max,i}$ the minimal and maximal distances from the origin to the lines $\binom{w_1}{w_2} = t_j l_{j,i} + q_i$, $t_j \in \mathbb{R}$, $j \in \{1,2\}$, $i \in \{1,2\}$. From some geometric considerations we get now:

$$\mathcal{F}(CC_i^{**}, r) = \begin{cases} 1 & : r \leq s_{min,i} \\ 1 - \frac{1}{\pi}\arccos\frac{s_{min,i}}{r} & : s_{min,i} < r < s_{max,i} \\ 1 - \frac{1}{\pi}\arccos\frac{s_{min,i}}{r} - \frac{1}{\pi}\arccos\frac{s_{max,i}}{r} & : (s_{max,i} \leq r \leq \|q_i\|) \wedge \\ & \quad ((l_{1,i}, q_i)(l_{2,i}, q_i) > 0) \\ 1 - \frac{1}{\pi}\arccos\frac{s_{min,i}}{r} + \frac{1}{\pi}\arccos\frac{s_{max,i}}{r} & : (s_{max,i} \leq r \leq \|q_i\|) \wedge \\ & \quad ((l_{1,i}, q_i)(l_{2,i}, q_i) \leq 0) \\ \frac{1}{\pi}\arccos\frac{(l_{1,i}, l_{2,i})}{\|l_{1,i}\|\,\|l_{2,i}\|} & : r > \|q_i\|. \end{cases} \tag{32}$$

From this expression, the probability $P(CC_i)$, equations (19) and (20), can be calculated by numerical integration.

Acknowledgement The authors are grateful to the referee for giving several useful hints which led to a better presentation of the material.

References:

ANDERSON, T.W. (1984): *An Introduction to Multivariate Statistical Analysis.* Wiley & Sons, New York.

EATON, M.L. (1983): *Multivariate Statistics. A Vector Space Approach.* Wiley & Sons, New York.

FANG, K.-T., KOTZ, S., and NG, K.-W. (1990): *Symmetric Multivariate and Related Distributions.* Chapman and Hall, London.

KAI-TAI, F., and YAO-TING, Z. (1990): *Generalized Multivariate Analysis.* Science Press: Beijing and Springer-Verlag, Berlin.

RICHTER, W.-D. (1991): Eine geometrische Methode in der Statistik. *Rostocker Mathematisches Kolloquium, 44, 63–72.*

RICHTER, W.-D. (1992): *Vector algebraic approach to ANOVA for spherically distributed pseudo balanced samples.* Preprint, Universität Rostock.

RICHTER, W.-D., and STEINEBACH, J. (1993): *A geometric approach to finite sample and large deviation properties in two-way ANOVA with spherically distributed error vectors.* Preprint, Universität Marburg.

SAVILLE, D.J., and WOOD, G.R. (1991): *Statistical Methods: The Geometric Approach.* Springer-Verlag, New York.

TONG, Y.L. (1990): *The Multivariate Distribution.* Springer-Verlag, New York.

Jackknife Estimates of Similarity Coefficients Obtained from Quadrat Sampling of Species

J. Krauth

Department of Psychology, University of Düsseldorf
Universitätsstrasse 1, D-40225 Düsseldorf, F.R.G.

Abstract: Heltshe (1988) derived exact expressions for the jackknife estimates of two coefficients of similarity when quadrat sampling procedures are used. Furthermore, estimates of the variances were derived that allow the construction of approximate confidence intervals. We derive similar results for three coefficients proposed by Sokal and Sneath (1963), Ochiai (1957) and Dice (1945) besides the simple matching and the Jaccard coefficient considered by Heltshe.

1. Measures of Similarity Which are Useful in Ecological Biology

Janson and Vegelius (1981) list twenty coefficients of similarity which are used in ecological biology. In an effort to find a rationale for selecting an appropriate measure, the authors formulated six criteria which a measure should meet in order to be selected. In order to describe those of the twenty coefficients which meet all six criteria we use the following notation. We consider the case of m sites $L_1, \ldots, L_m$ and the total number (n) of species $S_1, \ldots, S_n$ which are found at the sites $L_1, \ldots, L_m$. For comparing two of the m sites, say L and L', we define by A, B, C, and D the numbers of species found or absent at site L or L', respectively. The four frequencies A, B, C, and D are shown in a 2×2 contingency table (cf. Table 1). A similarity measure for two sites L and L' is denoted by $r(L, L')$.

Tab. 1: Contingency table for the frequencies of species observed for two sites

		Site L	
		Present	Absent
Site L'	Present	A	B
	Absent	C	D

We consider five well-known similarity coefficients (which fulfill the axioms of Janson and Vegelius (1981), except r_M):

(1) the coefficient $r_O(L, L') = \dfrac{A}{\sqrt{A+B}\,\sqrt{A+C}}$ of Ochiai (1957),

(2) the coefficient $r_D(L, L') = \dfrac{2A}{2A+B+C}$ of Dice (1945),

(3) the coefficient $r_J(L, L') = \dfrac{A}{A+B+C}$ of Jaccard (1908),

(4) the coefficient $r_S(L, L') = \dfrac{A}{A+2B+2C}$ of Sokal and Sneath (1963, p. 129),

(5) the simple matching coefficient $r_M(L, L') = \dfrac{A+B}{A+B+C+D}$ of Sokal and Michener (1958).

2. Estimation of Similarity Coefficients

Heltshe (1988) pointed out that the similarity of sites for biological communities which can be fully censused can be determined exactly without sampling error. For large communities similarity is estimated from a sample, and this estimate is subject to sampling error. The assumption of a multinomial sampling model on which the calculations of Janson and Vegelius are based is justified only if the population is infinite and the identification of each individual is independent of other individuals sampled (Smith and Grassle (1977, p. 285)). If organisms occupy a continuum of space, e.g. in the marine environment, it is very difficult if not impossible to sample them individually. Here, quadrat sampling methods using random samples of space are usually applied. Because biological communities usually are not homogeneous in space but exhibit patchiness and clumping of species and individuals, the multinomial assumption and the asymptotic standard errors derived from it are inappropiate. For this reason Heltshe (1988,1989) derived explicit expressions for jackknife estimates of the Jaccard coefficient r_J and the simple matching coefficient r_M. In addition Heltshe derived estimates of the variances of the jackknife estimators.

Heltshe reported the results of extensive simulation studies of the behavior of the jackknife estimates in the presence of spatially clumped populations mimicking the patchiness and clumping of biological species. It seems that on the whole the jackknife procedure results in less biased estimates as compared to standard estimators. Further, it is possible to derive approximate confidence intervals. The estimation bias for the similarity coefficients seems to be independent of quadrat size if the sampling area is fixed.

3. Jackknife Estimates of Similarity Coefficients

For comparing two of the m sites, say L_1 and L_2, we assume that a random sample of t_1 quadrats is taken from L_1 and a random sample of t_2 quadrats from L_2. For each quadrat we record the presence or absence of the $H = A + B + C + D$ species. We denote by r^o the similarity coefficient calculated on the basis of all t_1 sampled quadrats from site L_1 and all t_2 sampled quadrats from site L_2. By r_j^i $(j = 1, 2, \;\; i = 1, \ldots, t_j)$ we denote the similarity coefficient which is based on the pool of all $(t_j - 1)$ quadrats from site L_j with the exception of quadrat i and the pool of all quadrats from the other site. The usual jackknife procedure applied to the two–sample problem will not eliminate the $O(1/t_1)$ and $O(1/t_2)$ bias terms, but these bias terms are eliminated in the following modification of the usual jackknife (Cox and Hinkley (1974, p.264)):

$$r^J = (t_1 + t_2 - 1)r^o - (t_1 - 1)\bar{r}_1 - (t_2 - 1)\bar{r}_2$$

with

$$\bar{r}_j = \sum_{i=1}^{t_j} r_j^i / t_j \quad (j = 1, 2).$$

This estimate was also used by Heltshe (1988).

Defining the pseudovalues

$$r_j^{i*} = (t_j - 0.5)r^o - (t_j - 1)r_j^i \quad (j = 1, 2, \ i = 1, \ldots, t_j)$$

we get

$$r^J = \frac{1}{t_1}\sum_{i=1}^{t_1} r_1^{i*} + \frac{1}{t_2}\sum_{i=1}^{t_2} r_2^{i*}.$$

An estimate of the variance of r^J is given by (Farewell, 1978; Smith, 1985)

$$(s^J)^2 = \frac{1}{t_1}s_1^2 + \frac{1}{t_2}s_2^2$$

where

$$s_j^2 = \frac{1}{t_j - 1}\sum_{i=1}^{t_j}(r_j^{i*} - \bar{r}_j^*)^2 \quad (j = 1, 2)$$

with

$$\bar{r}_j^* = \frac{1}{t_j}\sum_{i=1}^{t_j} r_j^{i*} = (t_j - 0.5)r^o - (t_j - 1)\bar{r}_j \quad (j = 1, 2).$$

Actually, we find

$$s_j^2 = \frac{1}{t_j - 1}\sum_{i=1}^{t_j}(r_j^{i*} - \bar{r}_j^*)^2 = (t_j - 1)\sum_{i=1}^{t_j}(r_j^i - \bar{r}_j)^2 \quad (j = 1, 2)$$

which differs by a factor of $(t_j - 1)^2$ from the variance estimates in Heltshe (1988) who used the r_j^i instead of the pseudovalues r_j^{i*}. A correction was given in Heltshe (1989).

In many cases, the pseudovalues r_j^{i*} defined above can be treated as approximately independent identically distributed normal variables even if the original r_j^i values are not (Gray and Schucany (1972, pp. 137-162); Zahl (1977)). Instead of checking whether conditions guaranteeing this approximate normality hold, a more pragmatic solution would be the use of the goodness-of-fit criteria for pseudovalues proposed by Zahl (1977).

Assuming approximate normality for the pseudovalues we derive approximate $(1-\alpha)$ confidence intervals by means of

$$r^J \pm t(1 - \tfrac{\alpha}{2}; t_1 + t_2 - 2)s^J$$

where $t(1 - \frac{\alpha}{2}; t_1 + t_2 - 2)$ denotes the $(1 - \frac{\alpha}{2})$ percentile of a t distribution with $t_1 + t_2 - 2$ degrees of freedom.

4. Jackknife Estimates of r_O, r_D, r_J, r_S and r_M

Just as in Heltshe (1988) we define a unique species as a species that occurs in one and only one quadrat at either site. Further we define the frequencies f_{juv} of r_j^i where u is the number of unique species in the removed quadrat i at one site but present at the other site, and v is the number of unique species in the removed quadrat i at one site and absent at the other site. If $H = A + B + C + D$ denotes the total number of species found at both sites we have

$$\sum_{u=0}^{H}\sum_{v=0}^{H} f_{juv} = t_j \quad (j = 1, 2).$$

For calculating r_j^i we observe that the removal of a quadrat implies that in the original formulas for r_j^i A is replaced by $A - u$, B by $B + u$, C by $C - v$, and D by $D + v$. This yields for the similarity measures (1)-(5):

$$\begin{aligned} r_j^i &= \frac{A - u}{\sqrt{A + B}\sqrt{A + C - u - v}} && \text{for } r_O, \\ r_j^i &= \frac{2A - 2u}{2A + B + C - u - v} && \text{for } r_D, \\ r_j^i &= \frac{A - u}{A + B + C - v} && \text{for } r_J, \\ r_j^i &= \frac{A - u}{A + 2B + 2C + u - 2v} && \text{for } r_S, \\ r_j^i &= \frac{A + D - u - v}{A + B + C + D} && \text{for } r_M. \end{aligned}$$

With the frequencies f_{juv} defined above and writing $r_j^i = g_j(u, v)$ we get

$$\bar{r}_j = \frac{1}{t_j}\sum_{u=0}^{H}\sum_{v=0}^{H} g_j(u, v) f_{juv} \quad (j = 1, 2),$$

$$\begin{aligned} s_j^2 &= (t_j - 1)\sum_{u=0}^{H}\sum_{v=0}^{H}(g_j(u, v) - \bar{r}_j)^2 f_{juv} \\ &= (t_j - 1)\left\{\sum_{u=0}^{H}\sum_{v=0}^{H}(g_j(u, v))^2 f_{juv} - \frac{1}{t_j}\Big(\sum_{u=0}^{H}\sum_{v=0}^{H} g_j(u, v) f_{juv}\Big)^2\right\} \quad (j = 1, 2). \end{aligned}$$

Substituting $\bar{r}_j$ into r^J we get the jackknife estimate of r and substituting s_j^2 into $(s^J)^2$ we get the variance estimate of r^J. Obviously, we have $g_1(0, 0) = g_2(0, 0) = r^o$.

5. Example

For illustration we examined the same data as Heltshe (1988). Two sites were considered and from each site five quadrats were randomly sampled. The data concern the presence (1) or absence (0) of $n = H = 25$ algal species as given in Tab. 2. We have chosen the same sequence of species as in Tab. 12 of Heltshe (1988), i.e. species 1 corresponds to Chaetomorpha linum, species 2 corresponds to C. melagonium, etc. Because only two sites are considered we have $A = 17, B = 3, C = 5, D = 0$. Furthermore, we observe $f_{101} = 2, f_{102} = 1, f_{100} = 2, f_{201} = 1, f_{200} = 4$. Our jackknife estimates (r^J) agree for the Jaccard coefficient and the simple matching coefficient with those calculated by Heltshe (1988) while the estimates of the standard deviations (s^J) are larger by a factor $t_1 - 1 = t_2 - 1 = 4$ than the estimates of Heltshe

Tab. 2: Algal species data from Heltshe (1988)

Species	Site 1 Quadrats					Site 2 Quadrats					Unique Species
	1	2	3	4	5	1	2	3	4	5	
1	1	1	1	1	1	1	1	1	1	1	
2	1	1	1	1	1	1	1	1	1	1	
3	0	1	0	0	0	0	0	0	0	0	x
4	0	0	0	1	0	0	0	0	0	0	x
5	0	0	1	0	0	0	0	0	0	0	x
6	0	1	1	0	1	1	0	1	1	1	
7	1	1	1	1	1	0	1	0	1	1	
8	0	1	0	0	1	0	1	1	1	0	
9	1	1	1	0	1	1	1	0	1	1	
10	1	1	1	1	1	1	1	1	1	1	
11	1	1	1	1	1	1	1	1	1	1	
12	1	1	1	1	1	1	1	1	1	1	
13	1	1	1	1	1	1	1	1	1	1	
14	0	0	0	0	0	0	1	0	0	0	x
15	0	1	0	0	1	0	0	0	0	0	
16	0	0	0	0	0	1	1	0	0	0	
17	1	1	1	1	1	1	1	1	1	1	
18	1	1	1	1	1	1	1	1	1	1	
19	1	1	1	1	1	0	1	0	1	1	
20	0	0	0	0	0	0	0	0	1	1	
21	1	1	1	1	1	1	1	1	1	1	
22	1	0	1	0	0	0	0	1	1	0	
23	1	1	1	1	1	1	1	1	1	1	
24	0	0	1	0	0	0	0	0	0	0	x
25	1	1	1	1	1	1	1	1	1	1	

(1988) but are identical to Heltshe's (1989) estimates (cf. Tab. 3). At the end of Tab. 3 the lower and the upper bounds of the approximate 95% confidence intervals for the population similarity coefficients are given using $t = t(0.975; 8) = 2.306$. These intervals are rather large due to the small sample sizes of $t_1 = t_2 = 5$.

References:

COX, D.R. and HINKLEY, D.V. (1974): *Theoretical Statistics.* Chapman and Hall, London.

DICE, L.R. (1945): Measures of the amount of ecological association between species. *Ecology, 26, 297-302.*

FAREWELL, V.T. (1978): Jackknife estimation with structured data. *Biometrika, 65, 444-447.*

GRAY, H.L. and SCHUCANY, W.R. (1972): *The Generalized Jackknife Statistic.* Marcel Dekker, New York.

HELTSHE, J.F. (1988). Jackknife estimate of the matching coefficient of similarity. *Bio-*

Tab. 3: Jackknife estimates of similarity coefficients for the algal data

	f	r_O	r_D	r_J	r_S	r_M
$g_1(0,1)$	2	0.829515	0.829268	0.708333	0.548387	0.720000
$g_1(0,2)$	1	0.850000	0.850000	0.739130	0.586207	0.760000
$g_1(0,0)$	2	0.810443	0.809524	0.680000	0.515152	0.680000
$g_2(0,1)$	1	0.829515	0.829268	0.708333	0.548387	0.720000
$g_2(0,0)$	4	0.810443	0.809524	0.680000	0.515152	0.680000
$\bar{r}_1$		0.825983	0.825517	0.703159	0.542657	0.712000
$\bar{r}_2$		0.814258	0.813473	0.685667	0.521799	0.688000
r^J		0.733025	0.729756	0.564696	0.378542	0.520000
s_1^2		0.004339	0.004557	0.009681	0.013901	0.017920
s_2^2		0.001164	0.001248	0.002569	0.003535	0.005120
$(s^J)^2$		0.001101	0.001161	0.002450	0.003487	0.004608
s^J		0.033175	0.034071	0.049497	0.059053	0.067882
$r^J - ts^J$		0.656524	0.651189	0.450556	0.242366	0.363464
$r^J + ts^J$		0.809527	0.808323	0.678835	0.514718	0.676536

metrics, 44, 447-460.

HELTSHE, J.F. (1989). Correction to "Jackknife estimate of the matching coefficient of similarity" by J.F. Heltsche; 44, 447–460, June 1988. *Biometrics, 45, 1347.*

JACCARD, P. (1908): Nouvelles recherches sur la distribution florale. *Bulletin de la Société Vaudoise des Sciences Naturelles, 44, 223-270.*

JANSON, S. and VEGELIUS, J. (1981): Measures of ecological association. *Oecologica, 49, 371–376.*

OCHIAI, A. (1957): Zoografic studies on the soleoid fishes found in Japan and its neighbouring regions. *Bulletin of the Japanese Society of Scientific Fisheries, 22, 526-530.*

SMITH, E.P. (1985): Estimating the reliability of diet overlap measures. *Environmental Biology of Fishes, 13, 125-138.*

SMITH, W. and GRASSLE, J.F. (1977): Sampling properties of a family of diversity measures. *Biometrics, 33, 282-292.*

SOKAL, R.R. and MICHENER, C.D. (1958): A statistical method for evaluating systematic relationships. *University of Kansas Scientific Bulletin, 38, 1409-1438.*

SOKAL, R.R. and SNEATH, P.H.A. (1963): *Principles of Numerical Taxonomy.* Freeman, San Francisco.

ZAHL, S. (1977). Jackknifing an index of diversity. *Ecology, 58, 907-913.*

Clustering Techniques in the Computing Environment XploRe

Hans-Joachim Mucha

Institute of Applied Analysis and Stochastics (IAAS)
Mohrenstr. 39, D-10117 Berlin, Germany

Summary: Cluster analysis attempts to detect structures in the data. The K-means method and hierarchical clustering algorithms are well-known. These methods are available now in *XploRe:* A computing environment for nonparametric regression and data analysis (Härdle 1990). Moreover, *XploRe* offers adaptive clustering methods which seem to be a little bit intelligent. They are based on adaptive distances which are very important too for multivariate plots (Mucha 1992b). The cluster analysis should be accompanied by multivariate graphics in order to support the interpretation of clustering results. Both the adaptive multivariate graphic based on principal components analysis (PCA) and the dynamic graphic can be carried out by *XploRe*.

1. Introduction

Clustering algorithms are used frequently in ecology, marketing, geology, social sciences, economics, archaeology, ornithology, etc. They attempt to detect structures in the data. At least cluster analysis provides an useful reduction or description of the data. There are several synonyms like segmentation in the field of market research, or taxonomy in the biological sciences. Generally, clustering techniques divide a set of points into subsets (clusters) in such a manner that similar points belong to the same cluster, whereas dissimilar ones are allocated into different clusters.
The (unknown) structure of row points (and column points, respectively) of a (I^*J)-data table $\mathbf{X}=(x_{ij})$, $i = 1,2,...,I, j = 1,2,...,J$, can be investigated. $\mathbf{X}$ may contain various types of values: 0-1-values, measurements, frequencies, percentages, and a mixture of them. Without loss of generality we consider the clustering of row points (observations) here only. Analyzing row points of $\mathbf{X}'=(x_{ji})$ instead of $\mathbf{X}$, the cluster analysis of variables is practicable often in the same way as clustering the observations (without any further considerations in the case of a contingency table $\mathbf{X}$, or a 0-1-matrix $\mathbf{X}$). With *XploRe*, the transposition of a matrix (or other data management) is quite easy to do. As a result of cluster analysis we get either a partition $P(I,K)$ of I row points into K clusters $(I \ll K)$ or a sequence of partitions, i.e. a so-called hierarchy. A partition is a categorical variable $\mathbf{p}$ which allocates an integer (state) $p_i \in \{1,2,...,K\}$ to every row point $\mathbf{x}_i$. Distances between points are the base of cluster analysis. Let's focus on metric-scaled variables chiefly. In that circumstance the squared weighted Euclidean distance

$$d_Q^2(\mathbf{x}_i, \mathbf{x}_{i'}) = (\mathbf{x}_i - \mathbf{x}_{i'})'\mathbf{Q}(\mathbf{x}_i - \mathbf{x}_{i'}) = \|\mathbf{x}_i - \mathbf{x}_{i'}\|_Q^2$$

between two observations $\mathbf{x}_i$ and $\mathbf{x}_{i'}$ is a well-known dissimilarity measure used in cluster analysis and PCA, respectively. Here $\mathbf{Q}$ is diagonal. Usually, the weights $q_{jj} = 1/s_j^2$ (called standard weights; hereby very different scales of the variables become comparable one with another), or $q_{jj} = 1$ (called trivial weights; the scales of the variables should be already comparable one with each other) are used by the statistical software packages *SAS, SPSS, S-plus*, etc., where s_j^2 is the total variance of the variable j:

$$s_j^2 = \frac{1}{M} \sum_{i=1}^{I} m_i (x_{ij} - \overline{x}_j)^2.$$

Here $\overline{x}_j$ is the total mean value of the variable j, i.e.

$$\overline{x}_j = \frac{1}{M} \sum_{i=1}^{I} m_i x_{ij},$$

and m_i is the mass of the observation $\mathbf{x}_i$ (we say, the i-th observation is weighted by its respective mass m_i). Accordingly M is the total mass of all I observations. Further, we focus on the K-means method (MacQueen 1967) in a special version, the so-called exchange method (Hartigan and Wang 1979, Späth 1985, Mucha 1992b). This method minimizes the sum of the within-cluster variances

$$V_K = \sum_{k=1}^{K} \sum_{i=1}^{I} \delta_{ik} m_i d_Q^2(\mathbf{x}_i, \overline{\mathbf{x}}_k)$$

by moving points from one cluster to another. The indicator function δ_{ik} is 1 if $\mathbf{x}_i$ (often we write shortly: i) comes from cluster k, or 0 otherwise. Furthermore, the element $\overline{x}_{kj}$ of the vector $\overline{\mathbf{x}}_k$ is the mean value of the variable j in the cluster k:

$$\overline{x}_{kj} = \frac{1}{n_k} \sum_{i=1}^{I} \delta_{ik} m_i x_{ij}.$$

Herein with n_k we denote the mass of the cluster k, which is equal to the sum of the masses of all observations belonging to the cluster k. As above introduced the indicator function δ_{ik} is 1 if the observation i comes from cluster k, or 0 otherwise. With the aim of interpreting the clustering results a graphical representation in a low-dimensional space is a helpful tool. In the case of the standard weights $q_{jj} = 1/s_j^2$ the PCA of the correlation matrix (instead of the covariance matrix) is computed. As a result each variable becomes an equal importance in cluster analysis as well as in PCA. Moreover, the use of the inverse total variances ($q_{jj} = 1/s_j^2$) as weights in a cluster analysis contradicts to the fundamental assumption that there are several populations (clusters) with different parameters each we look for. As a final consequence of using these standardizations the detection of an underlying cluster structure in $\mathbf{X}$ becomes difficult or even impossible.

2. The K-means method in XploRe

What is *XploRe*? *XploRe* is an interactive computing environment for e*Xplo*ratory *Re*gression and data analysis (Härdle 1990). Parameter free methods in high dimensions are feasible through additive models and massive use of automatic smoothing methods. *XploRe* has highly interactive graphics. Windows of different types (text, static two-dimensional, dynamic three-dimensional, Flury faces, box plots,...) can be linked to highlight and mask certain points. One can get a battery of windows. Moreover, *XploRe* is an open system, which offers a programming language as well as a macro language. New commands and macros concerning cluster analysis are available now in *XploRe* (release 3.1: March 1993). Generally, the input parameters of the *XploRe*-commands and *XploRe*-macros are located on the right hand side of the equal sign, whereas the output parameters are located at the left hand side of the equal sign. The K-means method is called by the command

$$(\mathbf{p}\ \mathbf{C}\ \mathbf{V}\ \mathbf{e}) = kmeans(\mathbf{X}\ \mathbf{b}\ \{\mathbf{w}\}\ \ \{\mathbf{m}\}),$$

where the vectors $\mathbf{w}$ (column weights $w_j = q_{jj}$) and $\mathbf{m}$ (row weights, or masses) are optional input parameters. As usual the input data matrix is denoted by $\mathbf{X}$. It consists of I rows (row points) and J columns (column points). An initial partition $\mathbf{b}$ of the I points into K clusters can be, for instance, a categorical variable containing random generated integers; for example a vector computed by the *XploRe*-command

$$\mathbf{b} = ceil(uniform(rows(\mathbf{X})). * K),$$

where K denotes the number of clusters which have to be chosen in advance. Otherwise, for example, $\mathbf{b}$ can contain $I - K$ zeros (that means the corresponding points do not effect the initial clusters) and the K values $1, 2, ..., K$, which define the clusters (so-called seed points). The output of the *kmeans*-command consists of the final partition $\mathbf{p}$ of the I points into K clusters (which minimizes the sum of within-cluster variances), the matrix $\mathbf{C}$ of the cluster mean values (the so-called centroids; K rows and J columns), the matrix $\mathbf{V}$ of the within cluster variances divided by the corresponding weight (mass) of the cluster (K rows and J columns), and the vector $\mathbf{e}$ containing the weights (masses) of the K clusters. In the case of row weights $m_i = 1, (i = 1, 2, ..., I)$, the vector $\mathbf{e}$ contains the number of observations per cluster. The partition vector $\mathbf{p}$ is very useful for the (dynamic) multivariate graphic which is available in *XploRe*. Strictly speaking one can either transform $\mathbf{p}$ into a so-called mask vector $\mathbf{f}$ (which contains the codes of K different colors) using the command

$$\mathbf{f} = vtocc(80. + \mathbf{p}),$$

or into a mask vector $\mathbf{g}$ (which contains the codes of K different symbols; this is a makeshift only) by typing the command

$$\mathbf{g} = vtocc(31 + 16. * \mathbf{p}).$$

In order to create a (colored) graphical representation these mask vectors are used by the *XploRe*-command *show*.

Example: The Iris-data of Fisher (1936) consist of 3 classes which are known beforehand. As a result of this knowledge the performance of the K-means clustering can be evaluated in an easy way. For further details about this data set see Mucha (1992a,b). Figure 1 shows the final result of the K-means clustering of $I = 150$ observations into $K = 3$ clusters in the plane of the first two principal components (here the correlation matrix was used in the PCA). First the data are read in from an *ASCII*-file named *irisdata.dat*. The following *XploRe*-commands have to be executed:

```
X=read(irisdata)
w=1./var(X)
b=ceil(uniform(rows(X)).*3)
(p C V e)=kmeans(X b w)
U=X./trn(sqrt(var(X)))
S=pca(U)
S=S[,1:2] ~vtocc(31+16.*p)
show(S s2d)
```

The error rate is a quite terrible one: 16.7% (i.e., 25 errors are counted). The reason for this failure is the use of unsuitable variable weights: The standard weights $1/s_j^2$ in the squared weighted Euclidean distance.

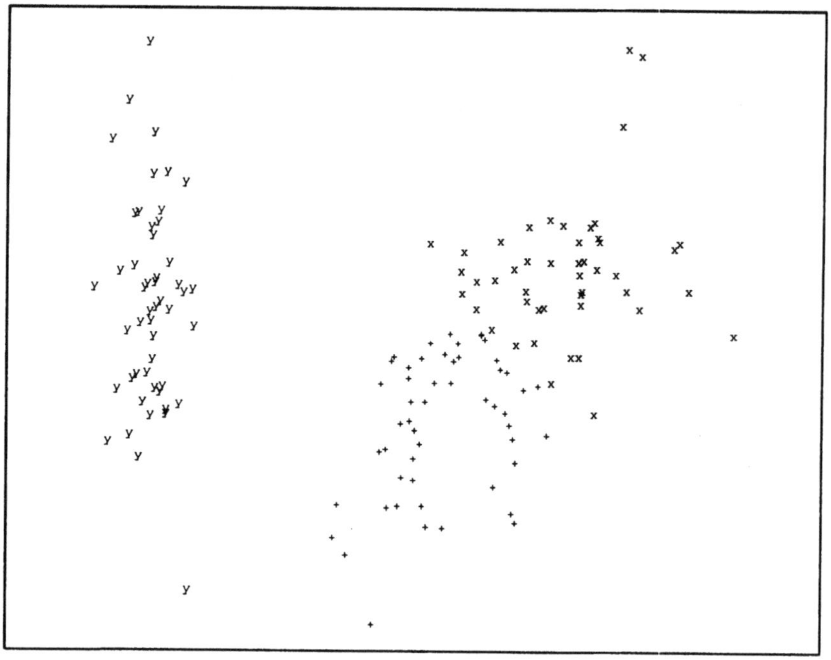

Figure 1: PCA-plot of the Iris-data after an "usual" K-means clustering. Generally, the class membership is marked by colors instead of symbols.

Note: The number of row points (observations) I, the number of variables J as well as the number of clusters K (but: $K < I$) are not limited by the command *kmeans*.

3. The adaptive K-means method

As mentioned above, the usual weights used in the Euclidean distance lead often to a bad result in cluster analysis and PCA, respectively. A better one can be obtained in almost every case by using specific or adaptive weights of variables (Mucha 1992b). But, which is the best among these so-called weighted Euclidean distances?

Considering a contingency table $\mathbf{X} = (x_{ij})$ the chi-square distance (χ^2-distance)

$$d_W^2(\mathbf{y}_i, \mathbf{y}_{i'}) = (\mathbf{y}_i - \mathbf{y}_{i'})'\mathbf{W}(\mathbf{y}_i - \mathbf{y}_{i'}) = \|\mathbf{y}_i - \mathbf{y}_{i'}\|_W^2$$

is the appropriate dissimilarity measure between two row profiles $\mathbf{y}_i = \mathbf{x}_i z_{i+}$ and $\mathbf{y}_{i'} = \mathbf{x}_{i'} z_{i'+}$. Here $z_{i+} = 1/x_{i+}$ and $z_{i'+} = 1/x_{i'+}$ are the inverse of the row total x_{i+} of the row point i and the inverse of the row total $x_{i'+}$ of the point i' , respectively. The special weights are given by $w_{jj} = x_{++}/x_{+j}$, with the grand total x_{++} and the column total $x_{+j}, j = 1, 2, ..., J$. The decomposition of the total inertia

$$T = \sum_{i=1}^{I} d_W^2(\mathbf{y}_i, \mathbf{a}) \frac{x_{i+}}{x_{++}}$$

has to be performed in order to minimize the sum of within-clusters inertia. Here, **a** is the average row profile with the elements $a_j = x_{+j}/x_{++}, j = 1, 2, ...J$. For example, the K-means method minimizes the sum of the within-clusters inertia

$$V_K = \sum_{k=1}^{K} \sum_{i=1}^{I} \delta_{ik} d_W^2(\mathbf{y}_i, \mathbf{b}_k) \frac{x_{i+}}{x_{++}}$$

for K clusters. The indicator function δ_{ik} is 1 if the row profile $\mathbf{y}_i$ comes from cluster k, or 0 otherwise. Obviously the vector $\mathbf{b}_k$ is the average row profile of all row points within the cluster k. The i-th row point is weighted by its respective mass $x_{i+}/x_{++}, i = 1, 2, ..., I$. With the help of the matrix language of *XploRe*, one can use the command *kmeans* to carry out a partitioning cluster analysis of the rows (as well as of the columns) of a contingency table **X** (for example into 5 clusters):

```
w=sum(sum(X))./sum(X)
m=sum(trn(X)/sum(sum(X)
Y=X./sum(trn(X))
b=ceil(uniform(rows(Y)).*5)
(p C V e)=kmeans(Y b w m),
```

where the vectors **w** (with the elements $w_j = x_{++}/x_{+j}, j = 1, 2, ..., J$) and **m** (with $m_i = x_{i+}/x_{++}, i = 1, 2, ..., I$) have to be computed before calling *kmeans*. How to see above, the matrix **Y** of the row profiles can be computed in an easy way by *XploRe*. Moreover Greenacre (1988) describes the hierarchical clustering of row points (as well as column points) of a contingency table using Ward's method and the χ^2-distance. The clustering method by Ward (1963) provides an optimum decomposition of the total inertia given above within the meaning, that the sum of within-clusters inertia becomes minimal. Usually, the K-means method provides a better partition of I points into K clusters in the sense of minimum V_K than Ward's method. The latter is available in the computing environment *XploRe* too (see below).

In conformity with the χ^2-distance the correspondence analysis (named dual scaling by Nishisato, 1980) provides an appropriate graphical representation of the row and column points of a contingency table **X**. Moreover, the clusters (represented by their average row profiles $\mathbf{b}_k$) should be placed into the plot. Either the *XploRe*-macro of correspondence analysis or the matrix language itself (in the frame of a weighted PCA see Mucha 1992b) can be used in order to obtain appropriate multivariate graphics.

Let us go back to metric-scaled data values. In order to increase the stability of cluster analysis (for a description of this difficult topic see Mucha 1992b) specific weights (for instance, in the case of nonnegative data values: $q_{jj} = 1/\overline{x}_j^2$, or the χ^2-distance), or adaptive weights in the distance formula should be applied instead of the usual weights $q_{jj} = 1/s_j^2$ or $q_{jj} = 1$. For example, the simple adaptive weights $q_{jj} = 1/\overline{s}_j^2$ can be used in the squared weighted Euclidean distance, where $\overline{s}_j$ is the pooled within-clusters standard deviation of the variable j

$$\overline{s}_j^2 = \frac{1}{M} \sum_{k=1}^{K} \sum_{i=1}^{I} \delta_{ik} m_i (x_{ij} - \overline{x}_{kj})^2.$$

The indicator function δ_{ik} is defined in the usual way. Because of simplicity one can use M equals the sum of all weights m_i, $i = 1, 2, ..., I$, in the denominator instead of $M - K$, i.e. the denominator is independent from the number of clusters K.
The pooled standard deviations can't be computed in cluster analysis in advance because the cluster structure is unknown usually. Considering a random partition the pooled standard deviations $\overline{s}_j$ are nearly equals the total ones s_j. Therefore, starting with the weights $q_{jj} = 1/s_j^2$ and a random initial partition $P^0(I, K)$, the K-means method computes a (local) optimum partition $P^1(I, K)$ of I observations into K clusters. In a repeated K-means clustering (with a new random initial partition or another arbitrary partition) the above given weights $q_{jj}^{(1)} = 1/\overline{s}_j^2$ are used, where $\overline{s}_j$ is the pooled standard deviation over K clusters regarding to the new partition $P^1(I, K)$. After carrying out the second K-means run we get a new partition $P^2(I, K)$. The new weights $q^{(2)}$, which correspond to the partition $P^2(I, K)$, are used in a next K-means clustering, and so on. We repeat this procedure as long as no changes in the partition or concerning the adaptive weights are pointed out.
Figure 2 shows the final result of the adaptive K-means clustering in the plane of the first two principal components (here the covariance matrix of the weighted variables was used). It should be emphasized that both figure 1 and 2 are scaled graphics.
The adaptive K-means method can be performed by the *XploRe*-macro ***adaptive*** which looks like the command ***kmeans***. Additionally to the output parameters of the command ***kmeans*** the final adaptive weights **q** of the variables are prepared for further use. The number of clusters K replaces the ***kmeans***-parameter **b**:

$$(\mathbf{p}\ \mathbf{C}\ \mathbf{V}\ \mathbf{e}\ \mathbf{q}) = \mathit{adaptive}(\mathbf{X}\ K\ \mathbf{w}\ \mathbf{m}),$$

Furthermore, the macro ***adaptive*** shows the adaptive (dynamic) graphic using the original variables as well as using the scores from the PCA. It is quite easy to write your own macros using other adaptive weights than $q_{jj} = 1/\overline{s}_j^2$.
Other statistical software offers at least the "usual" K-means method which makes an adaptive K-means algorithm by means of the macro language possible. For instance, the ***kmeans***-function of *S-plus*, written by Hartigan and Wong (1979), can be used for adaptive clustering by writing a new *S-plus*-function of the parameters data matrix **X**, number of clusters K, and maximum number N of calls of the ***kmeans***-function:

$$\mathit{adapt.kmean}\ <-\ \mathit{function}(\mathbf{X}, K, N)\ \{...\}\ .$$

However, the ***kmeans***-function of ***S-plus*** requires an already transformed data matrix (no weights of the variables are allowed) and initial cluster centers as input parameters. Therefore, another self-made function is called by ***adapt.kmean***: ***clus.stat*** computes the cluster centroids and pooled within-clusters standard deviations $\bar{\mathbf{s}} = (\bar{s}_1, \bar{s}_2, ..., \bar{s}_J)'$ by using the *S-plus*-function *apply* (Mucha forthcoming). Afterwards the raw data $\mathbf{X}$ are transformed by the *S-plus*-command

$$\mathbf{Z} < -\mathbf{X}\% * \%\ diag\ (1/\bar{\mathbf{s}})$$

Späth (1985) recommended a scale-invariant adaptive cluster analysis which is based on the more general distances

$$d^2_{Q_k}(\mathbf{x}_i, \mathbf{x}_{i'}) = (\mathbf{x}_i - \mathbf{x}_{i'})'\mathbf{Q}_k(\mathbf{x}_i - \mathbf{x}_{i'})$$

between the two observations $\mathbf{x}_i$ and $\mathbf{x}_{i'}$ belonging both to the cluster k. Here $\mathbf{Q}_k$ $(k = 1, 2, ..., K)$ are (different) positive definite matrices. However, there are some serious disadvantages like the lack of a geometrical presentation as well as a huge amount of parameters which have to be estimated. Keep in mind that the number of parameters is K times a quadratic increase with the number of variables J.

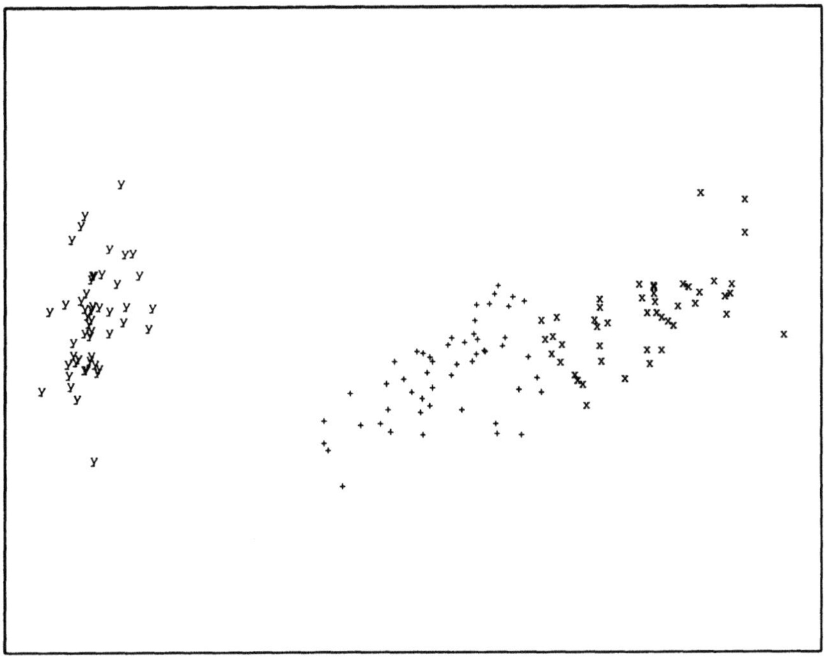

Figure 2: Scaled PCA-plot of the Iris-data based on adaptive weights $q_{jj} = 1/\bar{s}_j^2$. Here, the number of misclassifications decreases from 25 (figure 1) to 6.

4. The hierarchical cluster analysis in XploRe

Before computing a hierarchy a distance matrix $\mathbf{D} = (d_{ii'})$ have to be prepared. This is much more convenient than computing a distance value from the data matrix if it is just needed. The distance matrix contains the dissimilarity values between each pair of row points. The computation of a distance matrix $\mathbf{D}$ can be done by typing

$$\mathbf{D} = distance(\mathbf{X}\ \ \text{metric}\ \{\mathbf{w}\}),$$

where **w** (column weights) is an optional input parameter. As usual the data matrix is denoted by **X**. The following values of metric are supported in *XploRe*: Euclid (Euclidean distance), L1 (absolute metric, Manhattan metric), Maximum (supremum norm), Cosine, Chisquare (χ^2-distance), Tanimoto, Jaccard, and Matching (simple matching coefficient). The corresponding formulae are described in detail by Mucha (1992b). The last three coefficients are based on binary data (0–1–data); because there is an internal temporary dichotomization the input matrix **X** may contain arbitrary data values (if $x_{ij} < 1$ then $x_{ij} = 0$; if $x_{ij} \geq 1$ then $x_{ij} = 1$).
Otherwise, one can read in an external distance matrix or transform similarities (for example a correlation matrix) into distances. Furthermore, distance matrices can be computed by using the macro language of *XploRe* (Mucha forthcoming). However, this costs a lot of computer time if the number of points exceeds 20.
With the help of the matrix language of *XploRe*, further distance measures can be computed (see the example below). Otherwise one can handle mixed data in an easy but nevertheless flexible way by adding two (or more than two) distance matrices to a new one. For example $\mathbf{D}_B$ contains the *simple matching* distances (which are computed on the basis of the binary variables only) and $\mathbf{D}_L$ contains the *L1* distances based on metric variables (Mucha 1992b).
About eight hierarchical ascending clustering methods are available under the computing environment *XploRe*. The methods are called by typing the command

$$(\mathbf{p}\ \mathbf{t}) = agglom(\mathbf{D}\ \text{method}\ f\ \ \{\mathbf{m}\}),$$

where **m** (row weights, masses) is an optional input parameter. Usually, the distance matrix (denoted by **D**) is the result of the command *distance* and the *XploRe*-language. The clustering methods, which can be chosen by the user, are the following: Single (single linkage, or nearest neighbor), Complete (complete linkage, or furthest neighbor), Mean (simple average linkage), Median (median method), Average (average linkage), Centroid, Ward (Ward's minimum variance method), and Lance (flexible method of Lance and Williams). The latter uses a parameter f (Mucha 1992b). Bock (1974) and Gordon (1981) give a detailed description of both the distance measures and the clustering algorithms. A dendrogram can be presented by using the command

$$show(\mathbf{t}\ s2d),$$

where the input vector **t** is the result of the command *agglom* above. Figure 3 shows the dendrogram of the Ward's clustering of the Iris-data. The following commands are used (where **X** already contains the data matrix):

```
w=1./var(X)
D=distance(X euclid w)
D=D.*D
(p t)=agglom(D ward)
h=matrix(rows(t)).+110)
show(t ~ h s2d)
```

The vector **h** is a mask vector which contains the symbol "*" here. *XploRe* shows the right hand part of the large dendrogram only. Every terminal node of the dendrogram is a cluster, named 1,2,... from the top to the bottom. The number of clusters is chosen automatically to K =minimum$(20, I)$. A serious question that might turn on is, how to get a single partition into R clusters from the hierarchy ($R < K$)? Because the vector **p** contains the corresponding partition of I points into K clusters the *XploRe*-function *recode* performs any chosen partition on the basis of the dendrogram. For instance, if one is interested in a partition **u** of the I =150 observations of the Iris-data into 3 clusters then the following command is recommended:

$$\mathbf{u} = recode\ (\#(5\ \ 15\ \ 20)\ \mathbf{p})$$

Although the dendrogram looks very nice the error rate of **u** is very high: 17.3% (26 misclassifications). If the final weights of the adaptive K-means clustering (see sub chapter 3) are used instead of the inverse variances then the Ward method works very well: 5 misclassifications are counted only.

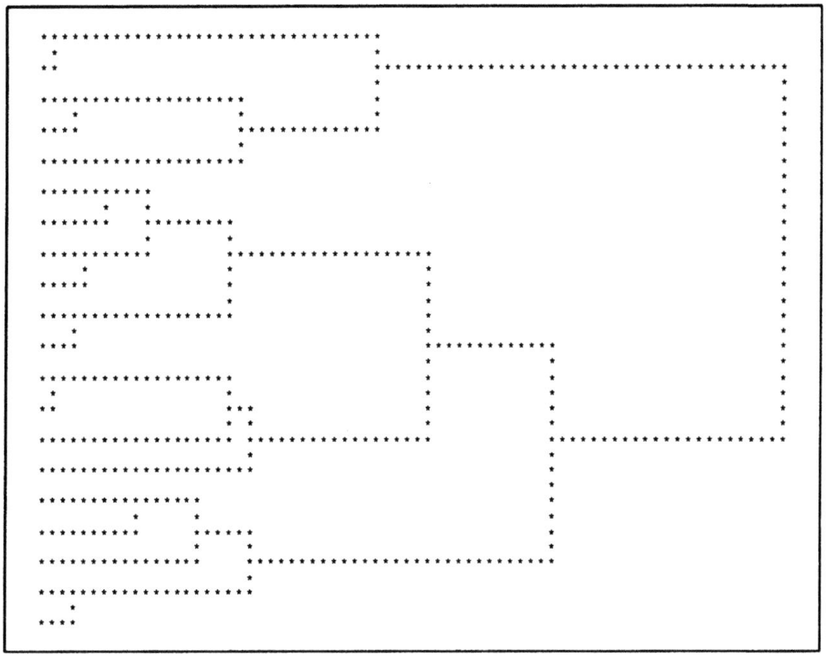

Figure 3: Ward's cluster-dendrogram based on the weights of inverse variances.

5. The investigation of the stability of adaptive weights

As mentioned above the true cluster structure is unknown in practice. For this reason we determine the stability (i.e. the quality) of clustering results in simulation studies. The stability of results of cluster analysis can be measured by the Rand's value or other measures of correspondence between partitions (Rand 1971, Hubert and Arabie 1985) . For a detailed description of this topic in the context of simulation studies see Mucha (1992b). This book contains also some simulation results. For example the adaptive K-means clustering of the well-known Iris-data was carried out. A stable partition $P(I, 3)$ is reached after four cycles. Finally, six misclassifications are

counted only (see figure 2). It was shown by the author that this solution is very stable against random selection of observations (called internal stability) as well as against disturbance of data values by random errors (called external stability) as well as against random weighting the observations.
But, in the case of unknown true cluster structure the question arises, how can the stability be investigated in cluster analysis? Considering the internal stability for instance, each simulation l starts with a random selection of $I/2$ observations. Afterwards the adaptive K-means clustering method computes a partition $R^{(l)}(I/2,3)$ as well as the adaptive weights of variables in the above described manner. The partition $P(I,3)$ of the whole data set which is computed by the adaptive K-means method too, is compared with the partition $R^{(l)}(I/2,3)$ on the basis of the measures of correspondence between partitions (Hubert and Arabie 1985). The simulation results are collected for further graphical and numerical presentations.

References:

BOCK, H.H. (1974): Automatische Klassifikation. *Vandenhoeck and Ruprecht*, Göttingen

FISHER, R.A. (1936): The use of multiple measurements in taxonomic problems. *Ann. Eugenics 7*, 179–188

GORDON, A.D. (1981): Classification. *Chapman and Hall*, London

GREENACRE, M.J. (1988): Clustering the rows and columns of a contingency table. *J. Classif. 5*, 39–52

HÄRDLE, W. (1990): Applied Nonparametric Regression. *University Press*, Cambridge

HARTIGAN, J.A., and WONG, M.A. (1979): A K-Means clustering algorithm. *Applied Statistics 28*, 100-108

HUBERT, L.J., and ARABIE, P. (1985): Comparing partitions. *J. Classif. 2*, 193–218

MACQUEEN, J.B. (1967): Some methods for classification and analysis of multivariate observations. *Proceedings 5th. Berkeley Symp. on Math. Statist. and Prob., (Eds.: LeCam, L.M. and Neyman, J.), Vol.1, 281–297, University of California Press*, Berkeley

MUCHA,H.-J.(1992a): Specific metrics for cluster analysis and principal components analysis. In: Faulbaum, F. (Ed.): SoftStat'91 Advances in Statistical Software 3. *Gustav Fischer*, Stuttgart, 249–258

MUCHA, H.-J. (1992b): Clusteranalyse mit Mikrocomputern. *Akademie Verlag*, Berlin

MUCHA,H.-J. (forthcoming): Clustering Techniques. *Springer-Verlag*, Berlin

NISHISATO, S. (1980): Analysis of Categorical Data: Dual Scaling and its Applications. *University of Toronto Press*, Toronto

RAND, W.M. (1971): Objective criteria for the evaluation of clustering methods. *Journal American Statistic Ass. 66*, 846–850

SPÄTH, H. (1985): Cluster Dissection and Analysis. Theory, FORTRAN Programs, Examples. *Ellis Horwood Limited*, Chichester

WARD, J.H. (1963): Hierarchical grouping methods to optimize an objective function. *Journal of the American Statistical Association 58*, 236-244

On the Application of Discriminant Analysis in Medical Diagnostics

Klaus-D. Wernecke

Medical Faculty (Charité) , The Humboldt-University of Berlin
Schumannstraße 20/21 , D - 10098 Berlin

Summary: In the construction of statistically based approaches for medical decision-making we have to take into consideration that the data resulting in the process of medical diagnostics may incorporate different levels of information which have to be treated as a whole. The paper deals with problems arising in context with the development of such decision aids and gives recommendations, which discrimination models to use in the respective situations, how the quality of discrimination and especially of feature selection are to be evaluated, which approaches in the treatment of mixed data can be used and which computer software for the corresponding calculations is available.

1. Introduction and examples

With three typical examples from medical diagnostics we present a review on discrimination methods, propose a procedure of cross-validated feature selection, consider methods for the estimation of the classification error and give an exploratory strategy for solving problems of discrimination in practice.

1.1 Diagnostic of glaucoma

A difficult problem in the ophthalmological diagnostics is the differentiation between glaucoma and ocular hypertension. For the purpose of differential diagnosis a lot of clinical examinations are performed which result among others, in continuous data and categorical observations. In order to analyze the usefulness of such examinations we investigated the data of two groups of patients (cf. Wernecke (1991a)):
- 58 patients suffering from glaucoma
- 41 patients with ocular hypertension.

Among others the following continuous features were included in the investigations:
- daily intraocular pressure
- brachialis blood pressure (bp), ophthalmica blood pressure (op)
- perfusion pressure (difference between bp and op).

Further categorical observations completed the diagnostics:
- visual fields according to Goldmann, Tuebinger and Octopus
- 6 different anamnestic findings.

1.2 Early diagnostics of cerebellopontine angle tumors

In the surgical therapy of specific inner ear tumors it is of great importance to recognize the eligible tumor in time because otherwise a successfull operation may be doubtful. A considerable contribution to early diagnostics of these tumors is the recording of auditory brain stem potentials. In order to differentiate between cochlear and retrocochlear lesions we examined three groups of patients (cf. Wernecke (1991b)):
- 70 patients with cochlear hearing loss
- 35 patients with cerebellopontine tumours
- 50 patients with other brain-stem lesions.

1.3 Statistically aided diagnostics in case of neuroborreliosis in childhood

The correct diagnosis of neuroborreliosis (NB) in children is known to be difficult. Clinical data of three patient groups were investigated in order to develop a score supporting diagnostic/therapeutic decisions (cf. Millner et al. (1992)):
- 39 patients with liquor-manifest NB (positive B. burg. CSF-titre, definitive NB)
- 62 patients without NB (control group)
- 28 patients with serum-manifest NB (positive B. burg. serum-titre, probable NB).
In the usual practice the clinical parameters used for diagnosis always get a separate valuation, i.e. also very suspected interactions between the parametres are not taken into consideration and additionally, till now is was not known which of the usually considered parameters are really relevant.
Eight parameters were included into the examination:
-Cephalea, Stiff Neck, Paresis, CSF-Cells, CSF-IGG, CSF-Alb, CSF-Prot, SB-IGG.

2. Choice of the statistical model

In all the mentioned applications we have supposed that each patient belongs to one of K groups $c_1, \dots, c_K$, each characterized by a distribution density $p_i(\vec{x})$. The aim was to construct a discrimination rule which allocates every observation vector (patient) $\vec{x} = (x_1, x_2, ..., x_p)^T$ into exactly one group. Given a multivariate training sample $\{\mathbf{x}\} = (\vec{x}_1, \vec{x}_2, ..., \vec{x}_n)$ of observation vectors in the structure $\vec{x}_{ij} = (x_{ij1}, x_{ij2}, ..., x_{ijp})^T$; $i = 1, \dots, K$; $j = 1, \dots, n_i$; $n = \sum_{i=1}^{K} n_i$. According to Bayes formula we allocate an observation $\vec{x} = (x_1, \dots, x_p)^T$ to the class c_k with the greatest posterior probability of membership:

$$P(c_k|\vec{x}) = P(c_k)p(\vec{x}|c_k)/[\sum_{i=1}^{K} P(c_i)p(\vec{x}|c_i)] = \pi_k p_k(\vec{x})/[\sum_{i=1}^{K} \pi_i p_i(\vec{x})] \qquad (1)$$

($P(c_k) = \pi_k$ priors, $p_k(\vec{x})$ probability functions), this implies the decision rule (identification):

$$\mathcal{E}_{\mathcal{I}}(\vec{x}) = m \text{ if } P(c_m|\vec{x}) = \max_k P(c_k|\vec{x}). \qquad (2)$$

If we consider, additionally losses l_{ik} ($0 \leq l_{ik} \leq 1$) for a misclassification from c_i into c_k we get for all possible misallocations into class c_k a posterior loss of

$$L_k(\vec{x}) = \sum_{i=1}^{K} l_{ik} P(c_i|\vec{x}) = \sum_{i=1}^{K} l_{ik} \pi_i p_i(\vec{x}) \qquad (3)$$

and therefore a new decision rule:

$$\mathcal{E}_{\mathcal{L}}(\vec{x}) = m \text{ if } L_m(\vec{x}) = \min_k L_k(\vec{x}). \qquad (4)$$

Allocation according to (4) implies a minimal total loss TL , i.e. in the case of continuous observations $\vec{x}$:

$$TL = \sum_{i,m=1}^{K} \pi_i l_{im} \int_{R_m} p_i(\vec{x}) d\vec{x} \qquad (5)$$

($p_i(\vec{x}) = f_i(\vec{x})$ probability densities)
is minimized by the acceptance regions for the class c_m given by

$$R_m = \{\vec{x} \mid L_m(\vec{x}) = \min_k L_k(\vec{x}) = \min_k \sum_{i=1}^{K} \pi_i l_{ik} f_i(\vec{x})\}. \qquad (6)$$

Before you can apply an adequate method for discrimination some questions have to be answered:
- Which discrimination method should be used?
- What measures have to be taken in order to guarantee a suffficiently high quality of the data, for a special method in consideration?
- Which criterion can evaluate the quality of the decision rule?
- How is it possible to extract the essential features from the given data set?
- Which software is available and what requirements on the data structure have to be fulfilled?

All our applications have different scales of measurement. While in the first problem there are both continuous and categorial observations, we have time dependent examinations in the diagnosis of cerebellopontine tumours, in the diagnosis of neuroborreliosis especially categorial data are of interest.

At present, our problem consists in the adaptation of various models for discrimination taking into account the different scales of measurement or data structure.

For continuous data (like the pressure values in the example from the ophthalmology) the common linear discriminant analysis (LDA) can be utilized. If the data are categorically scaled, we need the frequencies $h_i(x_1, \ldots, x_p)$ which the feature combinations appear in every class (where for $\vec{x} = (x_1, \ldots, x_p)^T$ all combinations of category levels have to be inserted) for the construction of the decision rule. Such frequencies are usually arranged in contingency tables as shown in Table 1 for the criteria 'cephalea' and 'paresis' in the diagnosis of neuroborreliosis for a specified class. After estimating

Paresis	Cephalea		
	no (1)	little (2)	heavy (3)
no (1)	$h_1(1,1)$	$h_1(1,2)$	$h_1(1,3)$
present (2)	$h_1(2,1)$	$h_1(2,2)$	$h_1(2,3)$

Table 1: Contingency table of two features

the corresponding cell probabilities $p_i(x_1, \ldots, x_p)$ with the help of specified statistical methods (most simply by the relative frequencies $h_i(x_1, \ldots, x_p)/n_i$), an observation vector $\vec{x}$ is ordered into that class c_i for which the (estimated) probability $p_i(\vec{x})$ is maximal. In general we have to cope with p-dimensional contingency tables considering p features. With a total number of cells of $g = \prod_{v=1}^{p} b_v$ the estimation of the unknown probabilities $p_i(x_1, \ldots, x_p)$ is not easy, especially if the number b_v of levels per feature is great (cf. Wernecke et al. (1989)).

A different discrimination method have to be used if time series underly the data vectors. The well-known simple algorithms (like the LDA) fail here because a lot of features per patient (e.g. 100 sample points) have to be treated with just a comparatively small number of observations (e.g. 35 or 50 patients). Adequate methods of discrimination apply here essentially the theory of spectral analysis using the (discrete) Fourier–transformation (cf. Wernecke (1991b)).

2.1 Discrimination of mixed data

Besides the previously discussed models for the discrimination of continuous (normally or arbitrarily distributed) variables of different structure (independent observations or time series) or of categorical features there are a lot of practical applications requiring allocation rules coping with both continuous and categorical variables.

Especially in the biological field such applications arise rather often, since we incorporate besides objective measurements almost always also certain subjective evaluations of living objects. Thus a physician will always use a complex of characteristics of dif-

ferent graduation in making the diagnosis.
We made various proposals (Wernecke et al. (1986), Wernecke (1991a) and Wernecke (1992)) for a discrimination scheme by using the adequate discriminator for every data type (supposing we have a number of various discriminators at our disposal) and by coupling logically the different discrimination to a new (common) discriminator. This procedure covers the dependencies between the different types of data merely in an indirect way (through the various discriminators) but is applicable to a greater number of both continuous and categorical features.
We look at the allocations $(a_{1l}, ..., a_{nl})^T$ of n objects achieved by each of the seperate discriminators D_l (l = 1, ..., L) suited to the respective scale of measurement (see Table 2). This corresponds to replacing the initial vector $\vec{x}_j = (x_{j1}, ..., x_{jp})^T$ with p

Objekts	D_1	D_2	D_3	...	D_L
1	a_{11}	a_{12}	a_{13}	...	a_{1L}
2	a_{21}	a_{22}	a_{23}	...	a_{2L}
...	...	...	...	...	...
n	a_{n1}	a_{n2}	a_{n3}	...	a_{nL}

Table 2: Allocations with L separate discriminators ($a_{jl} \in [1, 2, ..., K]$)

variables by the vectors $\vec{a}_j = (a_{j1}, ..., a_{jL})^T$ with L variables (j = 1, ..., n) and so we may apply an appropriate categorical discrimination model (e.g. the actuary or the Birch model). In a next step we do not only consider the (categorical) classification in the particular classes, but the posterior probabilities p_{jlk} of every observation as well, which can be easily calculated from the model in consideration. For every observation we get a scheme according to Table 3. Regarding these values p_{jlk} as realizations of

Discriminator	classes				
D_1	p_{j11}	p_{j12}	p_{j13}	...	p_{j1K}
D_2	p_{j21}	p_{j22}	p_{j23}	...	p_{j2K}
...	...	...	...	...	...
D_L	p_{jL1}	p_{jL2}	p_{jL3}	...	p_{jLK}

Table 3: Scheme of posterior probabilities p_{jlk} (j = 1, 2, ..., n)

new $L \times K$ variables with vectors

$$\vec{w}_j = (w_{j1}, w_{j2}, ..., w_{jr})^T = (p_{j11}, ..., p_{j1K}, p_{j21}, ..., p_{j2K}, ..., p_{jL1}, ..., p_{jLK})^T \quad (7)$$

as new 'observation vectors' (with $r = L \times K$), we may now apply suitable discrimination methods to these continuous variables (e.g. the linear discriminant analysis).

2.2 Discrimination methods

In the past there were a lot of publications about a multitude of various discrimination methods (cf. Titterington et al. (1981), Krishnaih and Kanal (1982), Lachenbruch (1975), Knoke (1982) and others). Although Fisher's discriminant analysis (Fisher (1936)) is sufficient for a lot of practical applications (and there are also theoretical reasons for its universality - cf. Läuter (1992)) there are applications where it is advisable to apply other discrimination methods. We have dealt with various discrimination methods (Wernecke (1983), Wernecke (1989), Wernecke et al. (1989)) and their mutual relationship in several papers and have discussed their implementation in the program-package ISP (Interactive Scientific Processor, ©Data Vision, Klosters, Switzerland) (cf. Wernecke and Nagel (1992)). Here we can give only a brief list of the most common methods.

- Linear Discriminant Analysis (for multivariate normal distributions with equal covariance matrices) – Fisher (1936)
- Quadratic Discriminant Analysis (multivariate normal distributions with arbitrary covariance matrices) – Lachenbruch (1975)
- Logistic Discrimination (via logistic regression) (multivariate normal distributions with equal covariance matrices, multivariate independent dichotomous distributions, multivariate dichotomous distributions according to log-linear models, combinations from these) – Anderson (1972), Anderson (1973)
- Discriminant Analysis with Kernel Estimators (arbitrary continuous distributions, also applicable to categorical features) – Meisel (1972), Habbema et al. (1974)
- Discrimination with Categorical Data (multiplicative Birch-model with different interactions, additive Lancaster-model, independence and actuary model) – Birch (1963), Bishop et al. (1975), Streitberg (1990), Wernecke et al. (1989)
- Ridge-Discrimination (via ridge-regression) (continuous distributions with strong collinearities in the variables) – Hoerl and Kennard (1970)
- Discrimination with Time Series (continuous distributions) – Shumway (1982), Wernecke (1991b)
- Discrimination of Mixed Data (data consisting of both continuous and categorical features) – Wernecke et al. (1986), Wernecke (1991a), Wernecke (1992)
- Regularized Discriminant Analysis (treatment of unequal covariance matrices and collinearities) – Friedman (1989), Greene et al. (1989)
- Quadratic Logistic Discrimination (different dependence structure in the classes) – Anderson (1975)
- Decomposition in two Normal Distributions – Krishnaiah and Kanal (1982)
- Nearest–Neighbor–Discrimination – Fukunaga (1972)
- Distribution Free Discriminant Analysis – Bunke and Fischer (1985)
- Robust Estimations of the Model Parameters – Hoaglin et al. (1983).
- Classification and Regression Trees – Breimann (1984)

In principle in all mentioned methods a feature selection is possible. We realize it preferably under the criterion of the minimal discrimination error (estimated with the help of leaving-one-out, cf. section 3) or the optimal test value (smallest critical α) in building-up or reduction steps.

3. Assessment of the quality of the discrimination rule

The aim of every discrimination is to find a classification rule for which the error probability (for a false allocation of patients) turns out as small as possible.
In the case of decision rule (4) the error probability is exactly the minimal total loss, i.e. Bayes' rule means allocation with minimal error probability. Given a training sample $\{\mathbf{x}\} = (\vec{x}_1, \ldots, \vec{x}_n)$ we ask for the error rate with which a certain vector $\vec{x}_j = (x_{j1}, \ldots, x_{jp})^T$ (patient) of unknown class-membership will be allocated on the

basis of the trained classifier. That corresponds to the definition of Lachenbruch's actual or true error rate $E_T(\vec{x}) = E_T(x_1, ..., x_p)$ (Lachenbruch (1975)) and follows immediately from equation (5) replacing R_m by $\hat{R}_m$, i.e.

$$E_T(x_1, x_2, ..., x_p) = \sum_{i,m=1}^{K} \pi_i l_{im} \int_{\hat{R}_m} f_i(\vec{x}) d\vec{x} \tag{8}$$

where

$$\hat{R}_m = \{\vec{x} \mid \hat{L}_m(\vec{x}) = \min_k \hat{L}_k(\vec{x}) = \min_k \sum_{i=1}^{K} \pi_i l_{ik} \hat{f}_i(\vec{x})\}. \tag{9}$$

$\hat{R}_m$ is the set of all observation vectors $\vec{x}$ which were classified into class c_m with the trained classifier (e.g. via a plug-in-rule with an estimated denstity function $\hat{f}_i(\vec{x})$).

3.1 Parametric estimators of the discrimination error

Under specified assumptions on the underlying distributions the discrimination error may be determined 'parametrically' (among others McLachlan (1974), McLachlan (1980), Okamoto (1963), Schaafsma and Vark (1977)).

3.2 Distribution-free error estimation methods

Various methods are known for the estimation of error rates in the case of unknown distributions. In almost all commercial program packages the so-called resubstitution is implemented (cf. Smith (1947)) where the discriminator is formed by the elements of the training sample and tested with the same elements later on:

$$E(R) = \sum_{l=1}^{n} e_l \quad \text{where} \quad e_l = \begin{cases} 0 \ , & \text{if } \vec{x}_l \text{ correctly allocated} \\ 1 \ , & \text{if } \vec{x}_l \text{ wrongly allocated.} \end{cases} \tag{10}$$

In general the resubstitution estimate E(R) of E_T is too optimistic and should therefore not be used. An adequate (unbiased, but not efficient) estimation of the true error rate is guaranteed by the so-called leaving-one-out method introduced by Lachenbruch where in the estimation step one data point is eliminated in turn, and classified with the discrimination rule obtainded from the other n - 1 observations (Lachenbruch (1967), Lachenbruch and Mickey (1968)).
Toussaint (Touissant and Sharpe (1975)) proposed a compromise in which, instead of one element of the training sample a group of s (e.g. s=10) observation vectors is removed at the same time. This procedure is designated as π – method. In both methods, the error is estimated from objects which do not belong to the training sample (true error of discrimination, see above). Usually the error declaration is complemented by a classification matrix which includes important information about the manner of discrimination. Occasionally, conditional error rates are estimated for the classification in specified classes, what is of particular interest in medicine (number of wrongly positive or wrongly negative classifications).
Thus for our glaucoma example (using LDA and 9 continuous features) we obtained the matrices in Table 4 and Table 5 with an error estimation (leaving-one-out) of 15.15%.

4. Feature selection

In the early diagnostics of cerebellopontine tumors we wanted to find out whether , in principle, the brain stem potential can be actually used as a diagnostical aid. In the other two applications it was particularly important to select those features which

From class	Into class		Total
1	46	12	58
2	3	38	41

Table 4: Allocation matrix for LDA (Glaucoma, p = 9)

From class	Into class		Total
1	0.79	0.21	1.00
2	0.07	0.93	1.00

Table 5: Conditional allocation matrix for LDA (Glaucoma, p = 9)

were essential for the diagnosis in question, i.e. to clear up which examinations or measurements can save the patient.
This is typical for many practical applications of discriminant analysis where the question arises to select a set of features $x_{i_1}, x_{i_2}, ..., x_{i_q}$ $(q \leq p)$ out of the given p characteristics $x_1, x_2, ..., x_p$ with the property that this feature set includes the primary information about the differentiation between the classes. In many applications this problem is even the primary one and the question for the 'best' discriminator (with the minimal error rate) moves in the background. In this connection we have to decide what criterion should be used for feature selection and how it is possible to evaluate the stability of the obtained set of characteristics.
From the literature a lot of selection methods are known (Siadlecka et al. (1988)) using different selection criteria. In the context of discriminant analysis it seems to be meaningful to use as selection criterion the respective estimation of the true error rate E_T equation (8) (compare also McLachlan (1976)). That means that we look for that feature set for which the corresponding estimation $\hat{E}_T$ is the minimal one. This is adequate for the statistical method used and easy to realize on the computer (see also Habbema and Hermans (1977)).
Customarily, we proceed step by step by a building-up or reduction procedure eliminating or adding one feature in turn.
Reduction procedure:
Eliminate that feature x_r from the p given ones as redundant for which it holds that

$$\hat{E}_T(x_1, x_2, ..., x_{r-1}, x_{r+1}, ..., x_p) = min! \tag{11}$$

This is the characteristic which reduces the estimated true error rate at most. If the error estimation remains constant (or increases only slightly) the eliminated feature is most dispensible, i.e. that of the same or the least increasing error rate.
Building-up procedure:
In the same way the relevance of features to be added can be assessed:
Add that feature x_r to the p previous ones, for which

$$\hat{E}_T(x_1, x_2, ..., x_p, x_r) = min! \tag{12}$$

holds.
Thus, we add that characteristic which again reduces the error estimation at most.

4.1 Validation of the feature selection

In a comparison of different criteria of feature selection using Bayes discriminators we showed (Wernecke (1983)) that if the ratio of sample size n_i per class to the number of features p, remains under some value, z, say:

$$\frac{n_i}{p} < z \quad ; \quad 1 < z < 3 \quad ; \quad i = 1, 2, ..., K \tag{13}$$

the feature selection leads to discriminators with remarkably (downwards) biased error estimations. On the other hand, the error rates for all p given features and the inspected sample sizes yielded substantially larger values. Consequently, any feature selection has the serious disadvantage that it results quite independently from the structure of the given sample and thus has not to be representative for the underlying population. For this reason it is absolutely necessary to reduce the inherent random fluctuations e.g. by imbedding the selection in a cross-validation procedure. Thereby the given sample is reduced repeatedly (e.g. 10 times) by a certain number s of observations (e.g. s = n/10) and with the remaining rest (of 9n/10 each), the feature selection process is carried out (cf. π - method and Wernecke and Läuter (1986)). In the end we decide for this number of features (and that features) which appeared most frequently in the validation steps (cf. Fig. 1).
In our glaucoma example with p = 9 continuous features we obtained after 10 validation steps:

Number of selected features:	2	3	1
Corresponding frequencies:	6	3	1

Name of selected features:	1	2	8	3
Corresponding frequencies:	10	9	4	1

Thus, most frequently 2 features have been selected (6 times) and therefore we decide for the variables 1 (always choosen) and 2 (9 times choosen) because of their frequencies. With this **recommended** feature set the estimated (recommended) error rate amounts to 15.15% (using leaving-one-out). A further error rate results from counting the total number of misallocations of that observations removed in the validation steps. This **cross-validated** error estimation is an upper bound for the true error rate. The relationship between both estimators is important for assessing the stability of the feature selection:
Large differences in both error rates point to unstable feature sets, small differences mean stabile selection of characteristics. If a certain number of features (definite features) show up again and again in the validation steps they can be considered as particularly stable. In the case of the mentioned glaucoma example the cross-validated error estimation amounts to 18.18%.

5. Computer programs

Programs for the implementation of a linear discriminant analysis are included in all known program packages. Because the LDA yields acceptable results in a lot of practical situations (cf. Krzanowski (1977) – for a theoretical explanation of its universality see Läuter (1992)), this method should be sufficient for the most routine applications. Modern products (like BMDP or SAS) offer also adequate estimations of the discrimination error obtained from Jackknife - or cross-validation methods. Unfortunately, a possibility for validating the feature selection is not included in these

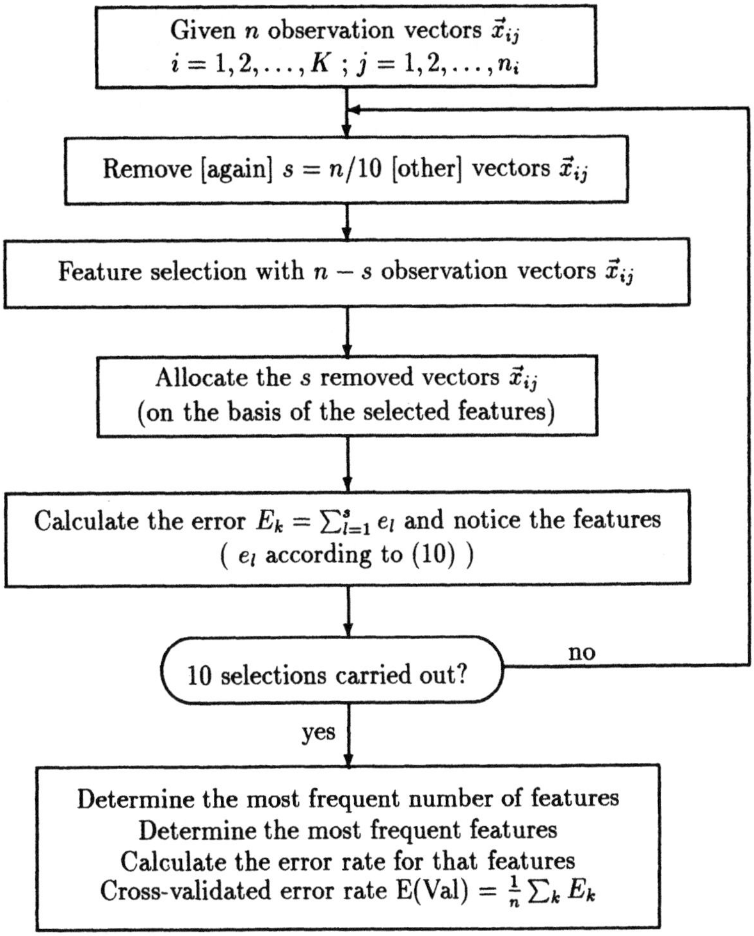

Figure 1: Validation of feature selection

products. In several packages (SAS, BMDP or SPSS) you find also the quadratic discriminant analysis, as a rule in connection with the multivariate test on the equality of the covariance matrices. For the implementation of a logistic discrimination (via the logistic regression) the respective module for the logistic regression can be used in the mentioned packages (here GLIM has to be mentioned additionally). But it is more favourable to proceed by the formulas of Cox (1970) and Anderson (1973), respectively, which can be included into the existing program packages. For the discrimination with kernel functions the sophisticated system ALLOC80 (Hermans et al. (1982)) is available. Programs for the linear discriminant analysis with terms of higher order (Truett et al. (1967)) and for the location model by Kranzkowski (1975) are as yet not included in the common packages and are only available by the authors.

6. An exploratory strategy for solving problems of discrimination

The beginning of every multivariate analysis consists in a comprehensive exploratory univariate structure analysis of the data. In this first step the questions relating to the scale of measurement of the data, the cell-frequencies of multi-dimensional contingency tables, the structure of the continuous data (time series – independent measurements), the distribution of the continuous observations, the univariate outliers and the relevance of certain variables (pre-selection of features) are answered. The results of this phase are essential for revealing the multivariate structure of the data and for the evaluation of the importance of the variables (pre-selection of features).

From the first phase we may have obtained a general view of the multivariate structure of the data material with the help of a graphical system. In the next step we check the material for multivariate outliers as well as influential observations. For these purposes various plots (plots of residuals, normalized and studentized residual plots, Jackknife residuals, leverages, Cook's distance plots, differences in the estimated regression coefficients, hat matrix) are helpful. In the case of categorical features additional information about the data such as a conjecture on the interaction structure is important. If continuous variables are expected to be univariately normally distributed you can test for a multivariate normal distribution (Anderson (1984)). Subsequently you could test for equality of the covariance matrices (cf. among others Brown (1939), Scheffé (1959), Nagao (1973) or Fahrmeir and Hamerle (1984)). An inspection of possible collinearities in the continuous variables (cf. Hoerl and Kennard (1970)) builds the conclusion of the preliminary phase for discrimination. In dependence on the results of the described steps you have to choose between different discrimination methods.

But the proposed procedure should not be understood as a dogma. Especially the mentioned test statistics may lead to problems. Thus the test for equality of the covariance matrices proves to be very sensitive in a lot of practical situations. More suitable may possibly react robust test procedures (O'Brien (1992)).

Another problem is to define an effective stopping criterion for the feature selection. The application of F-test values seems to be problematical, because the F-criterion tends rather to well-separated populations than that of bad-separated ones (Habbema and Hermans (1977)). For this reason we prefer the estimated true error rate as a stopping - and selection - criterion. But here also we recommend to observe the whole selection process for obtaining a comprehensive view of the data structure.

7. Evaluation of the discrimination results

The evaluation of the obtained discrimination results can be established on the basis of different criteria:

- Estimated true error rate (leaving-one-out-method)
- Costs of misclassifications (including losses)
- Error estimation using resubstitution
- Combined error estimations (cf. Wernecke and Kalb (1983))
- Cross-validation
- Judgement based on a test-sample.

8. Implementations in ISP

The commercial program packages offer a rather modest spectrum of algorithms in the field of discrimination. Thus the question arises how, e.g., some of the above mentioned discrimination models can be realized with the computer. It does not seem to be meaningful (especially for a statistician working in a particular field of application) to write your own new programs in a chosen programing language. Therefore, there is a demand for a higher, modular language which allows an uncomplicated realization

of tasks of multivariate statistics. That is quite the situation for an effective application of the Interactive Scientific Processor (ISP). The processor ISP offers within a higher command language a great variety of modules which are particulary appropriate for multivariate statistics. The user can use these modules (macros) separately or - connected by certain language elements of ISP- in every desired combination. In this way, you may easily develop your own macros.
Additionally, it is of special advantage that you can develop, supported by the system, comfortable menus for users of your new macros which facilitate the selection of the methods - from the univariate analysis over the multivariate exploratory data analysis to the specification of the finally chosen discrimination model.
A lot of such procedures has been developed by the author, but cannot be described here for space restrictions. Further information can be obtained from the author.

References:

ANDERSON, J.A. (1972): Seperate sample logistic discrimination. *Biometrika, 59, 19–35*

ANDERSON, J.A. (1973): *Logistic discrimination with medical applications.* In: T. Cacoullos (ed.): *Discriminant analysis and applications.* Academic Press, New York

ANDERSON, J.A. (1975): Quadratic logistic discrimination. *Biometrika, 62, 149–154*

ANDERSON, T.W. (1984): *An introduction to multivariate analysis.* John Wiley & Sons, New York

BIRCH, M.W. (1963): Maximum likelihood in three-way contingency tables. *J. Royal Statist. Soc., Ser. B, 25, 220–233*

BISHOP, Y. M. M., FIENBERG, S., and HOLLAND, P.W. (1975): *Discrete Multivariate Analysis: Theory and Practice.* MIT Press, Cambridge

BREIMANN, D. (1984): *Classification and Regression Trees.* Wadsworth, Monterey

BROWN, G. W. (1939): On the power of the L1–test for equality of several variances. *Ann. Math. Statist., 10, 119–168*

BUNKE, O., and FISCHER, K. (1985): Some fundamentals and procedures of parametric distribution-free and discrete discriminant analysis. *Statistics, 16, 185–201*

COX, D.R. (1970): *The analysis of binary data.* Methuen, London

FAHRMEIR, L., and HAMERLE, A. (1984): *Multivariate statistische Methoden.* Walter de Gruyter, Berlin, New York

FISHER, R. A. (1936): The use of multiple measurements in taxonomic problems. *Ann. Eugenics, 7, 179–188*

FRIEMAN, J.H. (1989): Regularized discriminant analysis. *JASA, 84, 165–175*

FUKUNAGA, K. (1972): *Introduction to Statistical Pattern Recognition.* Academic Press, New York London

GREENE, T., and RAYENS, W.S. (1989): Partially pooled covariance matrix estimation in discriminant analysis. *Commun. Statist. Theory Meth., 18 (10), 3679*

HABBEMA, J. D. F., HERMANS, J., and VAN DEN BROEK (1974): A stepwise discriminant analysis program using density estimation. In: *Compstat 1974 — Proc. in Computational Statistics.* Physica, Wien

HABBEMA, J. D. F., and HERMANS, J. (1977): Selection of variables in discriminant analysis by F–statistics and error rate. *Technometrics, 19*

HERMANS, J., HABBEMA, J. D. F., KASANMOENTALIB, T. K. D., and RAATGEVER, J. W. (1982): *Manual for the 'ALLOC 80' discriminant analysis program.* Dept. of Med. Stat., University of Leiden.

HOAGLIN, D.C., MOSTELLER, F. and TUCKEY, J.W. (1983): *Understanding robust and exploratory data analysis.* Wiley, New York

HOERL, E.H., and KENNARD, R.W. (1970): Ridge regression: Applications to nonorthogonal problems. *Technometrics, 12 (1)*

KNOKE, J. D. (1982): Discriminant analysis with discrete and continuous variables. *Biometrics, 38, 191–200*

KRISHNAIAH, P. R., and KANAL, L. N. (1982): *Handbook of Statistics* North Holland, Amsterdam, New York, Oxford, Tokyo

KRZANOWSKI, W. J. (1975): Discrimination and classification using both binary and continuous variables. *J. Americ. Stat. Ass., 70, 782–790*

KRZANOWSKI, W. J. (1977): The performance of fisher's linear discriminant function under non-optimal conditions. *Technometrics, 19, 191–200*

LACHENBRUCH, P. A. (1967): An almost unbiased method of obtaining confidence intervals for the probability of misclassification in discriminant analysis. *Biometrics, 23, (4), 235–241*

LACHENBRUCH, P. A., and MICKEY, M. R. (1968): Estimation of error rates in discriminant analysis. *Technometrics, 10, 1–11*

LACHENBRUCH, P. A. (1975): *Discriminant Analysis.* Hafner Press, New York,

LÄUTER, J. (1992): *Stabile multivariate Verfahren (Diskriminanzanalyse Regressionsanalyse Faktoranalyse).* Akademie Verlag, Berlin

McLACHLAN, G. J. (1974): An asymptotic unbiased technique for estimating the error rate in discriminant analysis. *Biometrics, 30, 2*

McLACHLAN, G. J. (1976): A criterion for selecting variables for the linear discriminant function. *Biometrics, 32, 529–534*

MCLACHLAN, G. J. (1980): Estimation of the error of misclassification on the criterion of asymptotic mean square error. *Technometrics, 22, 621–627*

MEISEL, W. S. (1972): *Computer–oriented approaches to pattern recognition.* Academic Press, New York, London

MILLNER, M., MUELLEGGER, R.R., and WERNECKE, K.-D. (1992): How to classify children with suspected neuroborreliosis and a positive serum b.burgdorferi titer? *Proc. 6th Congress Intern. Child. Neurol. Assoc.*

NAGAO, H. (1973): On some the criteria for covariance matrix. *Ann. Statist., 1, 700–709*

O'BRIEN, P. C. (1992): Robust procedures for testing equality of covariance matrices. *Biometrics, 48, 819–829*

OKAMOTO, M. (1963): An asymptotic expansion for the distribution of the linear discriminant function. *Ann. Math. Statist. 34, 1286–1301, [Korrektur: Ann.Math.Statist. 39, 1968, 1358-1359]*

SCHAAFSMA, W., and VARK, G. N. (1977): Classification and discrimination problems with applications. *Statistica Neerlandica, 31, 1*

SCHEFFÉ, H. (1959): *The analysis of variance.* John Wiley & Sons, Inc. New York

SHUMWAY, R. H. (1982): Discriminant analysis for time series. In: P. R. Krishnaiah and L. N. Kanal, editors, *Handbook of Statistics, volume 2, pages 1–42.* North Holland Publ. Coop.

SIADLECKA, K., SIADLECKI, W., and SKLANSKY, J. (1988): An overview of mapping techniques for exploratory pattern analysis. *Pattern Recognition, 21, 411–429*

SMITH, C.A.B. (1947): Some examples of discrimination. *Ann. Eugen., 13*

STREITBERG, B. (1990): Lancaster interactions revisited. *The Annals of Statistics, 18, 1878–1885*

TITTERINGTON, D.M., MURRAY, G.D., MURRAY, L.S., SPIEGELHALTER, D.J., SKENE, A.M., HABBEMA, J.D., and GELPKE, G.J. (1981): Comparison of discrimination applied to a complex data set of head injured patients. *J.R.Stat.Soc. A, 1144, 145–175*

TOUSSAINT, G. T., and SHARPE, P. M. (1975): An efficient method for estimating the probability of misclassification applied to a problem in medical diagnosis. *Comput. Biol. Med., 4, 269–278*

TRUETT, D., CORNFIELD, and KANNEL, W.B. (1967): A multivariate anlysis of the risk of coronary heart disease in framingham. *Journal of Chronic Diseases, 20, 511–524*

WERNECKE, K.-D. (1983): *Fehlerschätzung und Merkmalsselektion bei Bayes'schen Klassifikatoren.* Habil.-Schrift , Martin-Luther-Universität Halle-Wittenberg

WERNECKE, K.-D., and KALB, G. (1983): Further results in estimating the classification error in discriminance analysis. *Biom. J., 25,247–258*

WERNECKE, K.-D., and LÄUTER, J. (1986): A remark to the feature selection in discriminant analysis. *Paper presented at the 7th International Summer School on Problems of Model Choice and Parameter Estimation in Regression Analysis. November 26th–December 2nd. Holzhau*

WERNECKE, K.-D., UNGER, S., and KALB, G. (1986): The use of combined classifiers in medical functional diagnostics. *Biom. J., 28, 81–88*

WERNECKE, K.-D. (1989): Discrimination methods for different data strucures - a survey. *Proc. 3rd Intern. Sem. – Stat. Methods of Identification. 12.6.– 16.6.1989 Sobotka. Poland*

WERNECKE, K.-D., HAERTING, J., KALB, G., and STÜRZEBECHER; E. (1989): On model choice in discrimination with categorical variables. *Biom. J., 31 (3), 289–294*

WERNECKE, K.-D. (1991a): A new approach for the discrimination of mixed data. *Biom. J., 33 (7), 893–896*

WERNECKE, K.-D. (1991b): On discrimination of (stationary) stochastic processes. *Biom. J., 33, 17–39*

WERNECKE, K.-D. (1992): A coupling procedure for the discrimination of mixed data. *Biometrics, 48, 479–506*

WERNECKE, K.-D., and NAGEL, M. (1992): Diskriminanzanalyseverfahren unter ISP. In: *Methoden und Werkzeuge für die exploratorische Datenanalyse in den Biowissenschaften. Ausgewählte Arbeiten von einer Arbeitstagung - veranstaltet von den deutschsprachigen Regionen der IBS.* Fischer, Stuttgart und Jena

Similarity Searching in Databases of Three-Dimensional Chemical Structures

Peter Willett

Department of Information Studies, University of Sheffield,
Sheffield S10 2TN, United Kingdom

Summary: The measurement of similarity between pairs of chemical substances plays an increasingly important part in chemical research. This paper starts by providing a brief overview of the methods that have been developed over the last few years for similarity searching and clustering in databases of two-dimensional chemical structures. It then goes on to provide a more detailed account of current research into methods for similarity searching in databases of three-dimensional chemical structures, with particular reference to the work being carried out on the atom-mapping method for similarity searching.

1. Introduction

The fine chemicals industry makes extensive use of systems for the storage and retrieval of information pertaining to chemical structures. Until recently, chemical information systems have focused on the storage and retrieval of two-dimensional (2-D) representations of molecular structure, but these systems are being rapidly enhanced by the development of techniques for the representation and searching of three-dimensional (3-D) representations of molecular structure (Willett (1991)).

One of the most important facilities in a chemical information system is *similarity searching*, where a database is scanned to identify those molecules that are structurally most-similar to a user-defined query molecule. A similarity search involves the user inputting a *target* molecule, e.g., a lead compound that had been shown to exhibit some beneficial chemical or biological property. The similarity is calculated between the target and each compound in the database using some quantitative measure of inter-molecular structural similarity. The database is sorted into order of decreasing similarity with the target, and interesting compounds from near the top of this ranking can then be used as the basis for subsequent searches (Willett (1987)). The primary rationale for such a searching mechanism is the *similar property principle* of Johnson and Maggiora (1990), which states that structurally-similar molecules are expected to exhibit similar properties; thus, if a biologically-active molecule is used as the target in a similarity search, then the top-ranked molecules are expected to have a greater probability of exhibiting the same activity than would be the case if they were selected at random. Accordingly, given a single compound that has been shown to be active in a biological test system, similarity searching provides an excellent way of exploring a database to identify additional active compounds.

In this paper, we discuss the development of techniques for 3-D similarity searching. Section 2 provides a brief introduction to chemical information systems and to the methods that are available for 2-D similarity searching. Section 3 then discusses current work in Sheffield on the *atom-mapping* method for 3-D similarity searching, and Section 4 reviews other current research in this area. The paper concludes by considering the implementation of 3-D similarity searching in large chemical databases.

2. Similarity searching in 2-D

2.1 Chemical information systems

Chemical database systems have traditionally been used for the storage and retrieval of 2-D chemical structure diagrams (Ash *et al.* (1991)). The most important retrieval facility offered by a chemical information system is that of *substructure searching*, which identifies all of those molecules in a database that contain some user-defined partial structure, e.g., some substructural moiety that is believed to be related to a useful chemical property. Molecules are normally represented by labelled graphs, in which the nodes and edges of a graph are used to represent the atoms and bonds, respectively, of a 2-D chemical structure diagram. The use of a graph-theoretic description for a chemical molecule means that a substructure search can be effected using a *subgraph isomorphism* algorithm, which allows one to determine whether one graph is contained within another graph. Subgraph isomorphism is known to belong to the class of NP-complete problems, and is thus normally effected by a two-level search procedure, in which the time-consuming subgraph-isomorphism stage is preceded by a rapid *screening search* that eliminates the great bulk of the database.

Screens are small, atom-, bond- or ring-centred fragment substructures, the presences of which in a molecule are recorded in a bit string. The screen search involves checking the bit string representing each of the database structures for the presence of those screens that are present in the query substructure, i.e., that are denoted by non-zero bits in the query bit string. The atom-by-atom search is then carried out for those few molecules that match the query at the screen level.

2.2 Similarity searching

The last few years have seen substantial interest in the development of similarity-searching systems, in which the compounds in a database are ranked in order of decreasing similarity with an input query structure (rather than identifying those compounds that contain the query as in substructure searching).

The implementation of similarity searching requires an appropriate measure of intermolecular structural similarity, and many such measures are available (Johnson (1989)). The most widely-used approach expresses the degree of structural similarity between a pair of compounds in terms of a simple association coefficient (the Tanimoto coefficient is normally used) based on the substructural fragments that are common, and that are not common, to the two molecules that are being compared. This approach makes use of the fragment-occurrence data that is encoded in the bit-strings that are used for the screening stage of a substructure search. Extensive studies over several years have demonstrated that the use of such data results in similarity and clustering methods that are effective in operation and that can be used to make reasonably accurate predictions of molecular properties using structurally-based similarities (Johnson (1990), Willett (1987)). In addition to being effective in operation, such similarities can be calculated very efficiently using an inverted-file, nearest-neighbour algorithm that permits interactive searching of databases containing hundreds of thousands of structures (Willett (1990)). Fragment-based similarity searching is thus now a standard retrieval facility in chemical-structure information systems (Ash *et al.* (1991)).

2.3 Clustering

Fragment-based similarities can also be used as the input to a clustering method to produce groups, or clusters, of structurally-related molecules. An extended study has been carried out of the effectiveness of over thirty hierarchic and non-hierarchic methods for the clustering of chemical structures, with the various classifications being evaluated by the extent to which they were able to predict correctly a range of physical, chemical and biological properties. This comparison (Willett (1987)) suggested that non-hierarchic methods gave results that were comparable with, or superior to, those given by the computationally more-demanding hierarchic methods, and identified the non-hierarchic method due to Jarvis and Patrick (1973) as the most generally useful for this application area.

The main use of clustering is to assist in the selection of compounds for random biological screening, which is an important method for the identification of potential lead compounds in drug-discovery programmes (Ash *et al.* (1991)). Given a clustering of the structures in a database, a screening programme can be carried out by systematically selecting one compound from each cluster in turn. If a molecule that is tested proves to be active, then the other members of its class are tested, since they are likely to exhibit the same activity; alternatively, attention can be shifted to the next cluster. This approach to the selection of compounds focuses on those compounds that can be clustered together; alternatively, a screening programme can be based on the testing of those compounds that do *not* cluster readily when a clustering method is applied to a dataset (since this implies that they are markedly dissimilar to the rest of the database and are thus likely to provide more useful information on testing than a compound that is known to resemble closely other members of the database) (Johnson *et al.* (1989)).

Cluster analysis is now widely used in the selection of compounds for inclusion in biological screening programmes: examples include the systems that have been developed at the National Cancer Institute (Hodes (1989)), at Pfizer Central Research (U.K.) (Willett *et al.* (1986)) and at The Upjohn Company (Johnson (1990)). The use of cluster analysis in 2-D chemical information systems is reviewed in a recent paper by Barnard and Downs (1992).

3. Similarity searching in 3-D: the atom-mapping method

3.1 Introduction

The fragment-based approach to similarity and clustering has three useful characteristics that have led to it being widely used in chemical information systems: the approach is highly effective, despite the simplicity of the structural representations that are used; efficient algorithms are available that allow the methods to be implemented on very large files, containing tens or hundreds of thousands of structures; and the requisite fragment occurrence data is already widely used in chemical information systems. However, the approach is limited to 2-D representations of molecular structure, and the increasing availability of 3-D substructure searching systems (Willett (1991)) has led to interest in the provision of complementary facilities for 3-D similarity searching. In this section, we give an overview of work in Sheffield over the last three years on one particular 3-D similarity-searching method; other such methods are discussed in Section 4.

3.2 The method

Pepperrell and Willett (1991) have reported a detailed comparison of four methods for 3-D similarity searching. The comparison suggested that the *atom mapping* method was the most effective of the methods that were tested, and was sufficiently rapid in execution to permit the searching of large databases of 3-D structures. The workings of the atom-mapping method are detailed in the remainder of this section.

Assume that a molecule, A, contains $N(A)$ non-hydrogen atoms. The distance matrix for A, DA, is then an $N(A) \times N(A)$ matrix such that the IJ-th element, $DA(I, J)$, contains the distance between the I-th and J-th atoms in A, and similarly for the matrix DB representing the $N(B)$-atom molecule B; in the following, we shall assume that $N(A) \leq N(B)$. The first stage of the procedure compares each atom from A with each atom from B to identify the extent to which each pair of atoms lies at the centre of comparable volumes of 3-D space, as defined by the geometric arrangement of the atoms that are contained within these volumes.

Consider the I-th atom in A, $A(I)$: then the I-th row of the distance matrix DA contains the distances from I to all of the other atoms in A. This set of distances is compared with each of the rows, J, from the distance matrix DB to identify those distances that are the same in the two molecules to within some user-defined tolerance, e.g., $\pm$ 0.5 Å, and that involve the same elemental type(s) in A and B. Let there be $C(I, J)$ such distance matches when $A(I)$ is compared with $B(J)$; then the similarity, $S(I, J)$, between $A(I)$ and $B(J)$ is given by the Tanimoto coefficient $C(I, J)/(N(A) + N(B) - C(I, J))$. $S(I, J)$ is the IJ-th element of an $N(A) \times N(B)$ matrix, S, that contains the similarities between all pairs of atoms, $A(I)$ and $B(J)$, from A and B. This matrix is referred to subsequently as the *atom match matrix*.

Once the atom match matrix has been created, the inter-atomic similarities contained within it are used to establish a set of equivalences of the form $A(I) \leftrightarrow B(J)$. A greedy algorithm is used to establish the set of equivalences, as follows:

1. Sort the elements of the atom match matrix into order of decreasing similarity.
2. Scan the atom match matrix to find the remaining pair of atoms, one from A and one from B, that have the largest calculated value for $S(I, J)$.
3. Store the resulting equivalences as a tuple of the form $\{A(I) \leftrightarrow B(J); S(I, J)\}$.
4. Remove $A(I)$ and $B(J)$ from further consideration.
5. Return to Step 2 if it is possible to map further atoms in A to atoms in B.

The overall degree of similarity between A and B is then calculated as the mean of the similarities over all of the atoms in A, using the information in the tuples that is stored in Step 3 of the algorithm above.

Most similarity measures are *global* in character in that they return a number that describes the overall degree of similarity between a pair of objects. This is nearly always the case with non-chemical applications of similarity and also applies to the fragment-based measures that are used in 2-D similarity-searching systems, where the calculated similarities do not provide any information as to which particular parts of the molecules that are being compared are responsible for the observed degree of similarity. Such measures are less appropriate in the context of 3-D similarity searching, since biological activity at a receptor site is determined by the presence of a particular set of atoms in a particular geometrical arrangement, i.e., the pharmacophore (Dean (1987)). The atom mapping method is global in nature since it provides a single, real-valued number that measures the overall degree of similarity between the target and each of the database structures. However, it can also be regarded as a *local* similarity measure since the atom-match matrix provides information about the structural equivalences that apply to individual pairs of atoms, and about the contribution of each of these pairs to the overall, global similarity.

3.3 Validation of the method

The significance of the atom-mapping ranking that results when a biologically-active 3-D target structure is searched against a database has been investigated by comparing the activities or inactivities of the top-ranking molecules with those of randomly-selected molecules. A detailed simulation study (Pepperrell *et al.* (1991)) showed that the rankings resulting from the atom-mapping method were, indeed, significantly superior to those resulting from a random ranking of a database, thus validating the use of the method to identify those molecules in a database that are structurally most similar to an input target structure. Pepperrell *et al.* (1991) have also reported an investigation of the extent to which variations in the various components of the procedure affect the rankings that are produced when a similarity search is carried out. They found that changes in the similarity coefficient or in the way that the atoms were characterised had very little effect on the performance of the method, and a similar conclusion was reached when valence-angle, rather than inter-atomic distance, information was used to calculate the similarities in the atom match matrix. Pepperrell *et al.* (1991) concluded that the atom-mapping method is not crucially dependent on the precise way in which it is implemented and, accordingly, that it provides a robust and generally-applicable approach to the calculation of molecular similarity in databases of 3-D chemical structures.

3.4 Implementation of the method

A complexity analysis of the atom mapping method shows that the time requirement for the comparison of two molecules containing $N(A)$ and $N(B)$ atoms is of order $O(N(A)^3)$ if $N(A) \approx N(B)$ (with the computation being dominated by the calculation of the atom-match matrix) (Pepperrell *et al.* (1990)). It is thus clear that extremely efficient algorithms are required if the atom mapping method is to be used to search large databases of 3-D structures. Accordingly, we have investigated several upperbound strategies that can reduce the numbers of molecules that need to be compared with the target structure during a similarity search, whilst still ensuring the retrieval of the true *nearest neighbours*, i.e., those molecules that are structurally most similar to the target compound. The overall reduction in computation depends on the size of the file, on the number of nearest neighbours required, and on the number of atoms in the target structure: our experiments would suggest that it is possible to reduce the run-time by 60-80% of the time required for a sequential scan when none of these upperbound strategies are used (Pepperrell *et al.* (1990)). Improvements in the speed of searching can also be obtained by the use of parallel computer hardware, and we have recently described three different ways in which atom mapping can be implemented on a massively-parallel processor (Artymiuk *et al.* (1992)).

A prototype search system has been implemented at ICI Agrochemicals, using the upperbound procedures on a file of 4,500 structures (Pepperrell *et al.* (1990)). A typical search of this file for the 50 nearest neighbours of an input target structure requires about 60 seconds, using a Fortran 77 program running on an Evans and Sutherland ESV 3 workstation. Thus, similarity searches on a corporate database containing a quarter of a million structures would be expected to take about one hour using comparable equipment, the precise time depending on the target structure and the search parameters that are used. An example of the search output from this system is shown in the accompanying diagram.

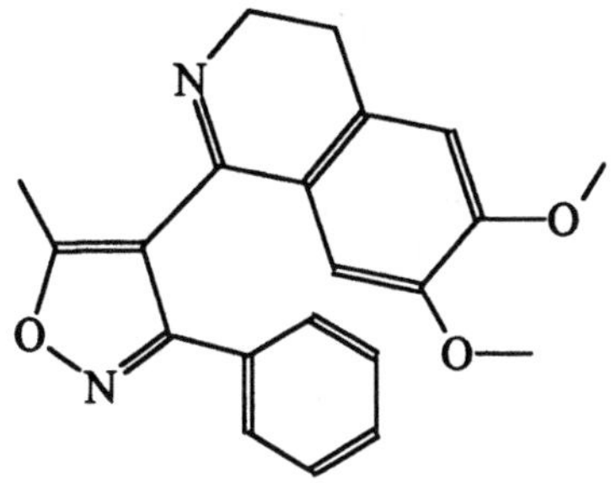

Target

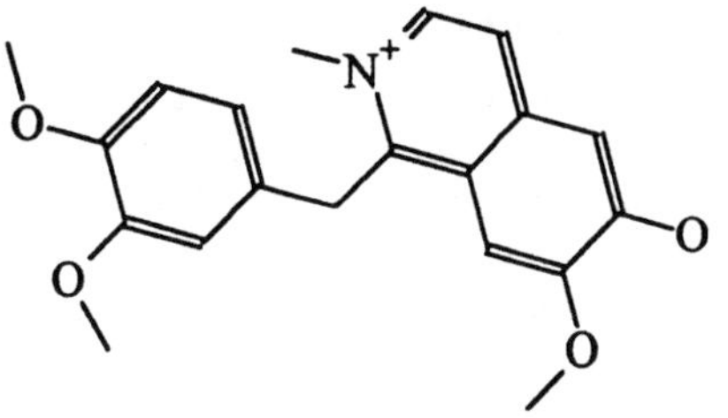

(1)

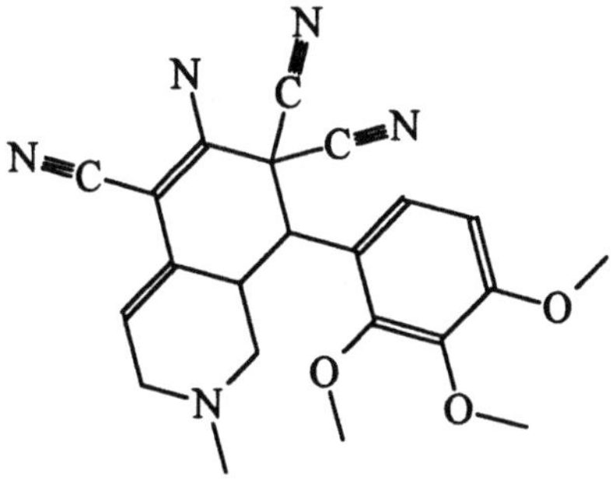

(2)

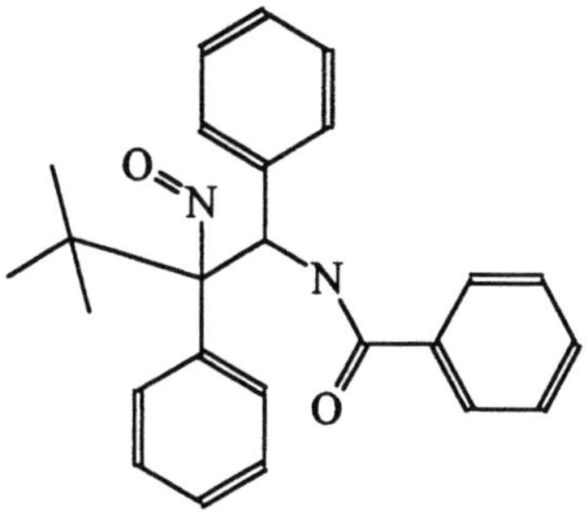

(3)

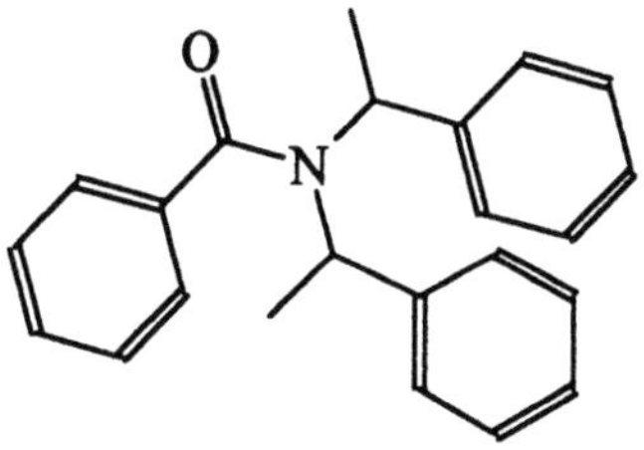

(4)

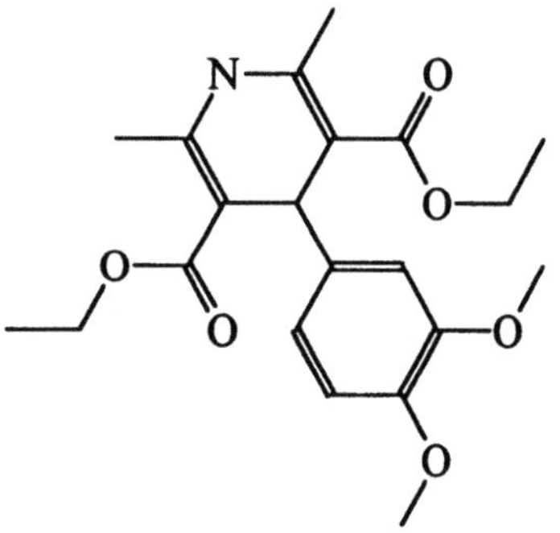

(5)

Top 5 nearest neighbours resulting from an atom-mapping search for the target molecule in the top left-hand corner of the figure

4. Similarity searching in 3-D: other methods

4.1 Introduction

We have noted in Section 2 that 2-D similarity searching generally involves fragment-based similarity measures: there is much less consensus as to what should form the basis for similarity searching in databases of 3-D structures, although most of the systems that have been described to date have considered either the steric or the electrostatic factors that are the most important determinants of biological activity (Dean (1987)).

There is generally a trade-off between the *effectiveness* of a similarity-searching technique, i.e., the extent to which the technique is able to identify structures that the chemist also perceives as resembling an input target molecule, and its *efficiency*, i.e., the associated computational requirements. This trade-off means that many of the quantum-mechanical and molecular-modelling procedures that have been used for measuring molecular similarity are far too time-consuming to be considered for use in a database-searching environment, where a target molecule must be matched against many thousands of database structures.

4.2 Methods based on overlapping

Steric similarities can be estimated using procedures based on the matching of atoms or of superimposed volumes. For example, Dean and co-workers have discussed the matching of pairs of atoms, one from a target structure and one from a database structure, so as to minimise the sum of the squared distance errors (Danziger and Dean (1985), Papadopoulos and Dean (1991)). Meyer and Richards (1991) measure the similarity of a pair of molecules by the extent to which their volumes overlap, and describe an efficient algorithm to obtain good, but not necessarily optimal, overlaps for pairs of structurally-related molecules.

The molecular volume provides one direct, and readily-calculable description of molecular shape. More sophisticated approaches to the quantification of molecular shape, and hence to the measurement of resemblance between pairs of structures, have been reported in a series of papers by Mezey and his co-workers, who have developed a range of topological similarity measures; an overview of this work is presented by Mezey (1992). Such approaches provide detailed insights into the similarity relationships that exist between pairs of structures but their computational requirements may be too great for searching large numbers of structures unless some alternative procedure is used to identify a subset of the database. A much simpler approach involves characterising molecular shape by a single number or ratio; examples of this have been described by Cano and Martinez-Ripoll (1992) and by Petitjean (1992), but there have not been any evaluations, to date, of their search effectiveness.

Carbo *et al.* (1980) suggested that the similarity between two molecules could be estimated by the overlap of the molecules' electron charge clouds. This idea has been taken up by several workers, using both electron densities and molecular electrostatic potentials (Burt *et al.* (1990), Hermann and Herron (1991), Hodgkin and Richards (1987), Manaut *et al.* (1991), Richard (1991)). A molecule is positioned at the centre of a 3-D grid and the electrostatic potential is calculated by taking the product of the potentials at each point in the grid. The similarity between a pair of molecules is then estimated by comparing the potentials at each grid point and summing over the entire grid, with a suitable normalising factor to bring the similarities into the range -1.0 to +1.0. Good *et al.* (1992) have recently reported an alternative approach in which the potential distribution is approximated by a series of Gaussian functions that can be processed analytically, rather than numerically as in the conventional approach, with a substantial increase in the speed of the similarity calculation and with only a minimal effect on its accuracy. This elegant idea removes one of the main limitations of field-based approaches to 3-D similarity searching, but still requires the identification of an appropriate alignment of the two molecules that are being compared.

Field-based similarity searching in large databases will be feasible only with the identification of an appropriate alignment procedure. Exhaustive searching for alignments is completely infeasible unless extremely coarse grids are used, in which case the calculated similarities are unlikely to reflect accurately the true degree of resemblance of the molecules that are being compared. Even stochastic searching procedures can run for extended periods if one wishes to conduct a detailed examination of the alignment space. For example, Kearsley and Smith (1990) have described a method, called SEAL (Steric and Electrostatic ALignment), which uses a Monte-Carlo procedure to optimize the alignment of two rigid 3-D molecules and which bases the matching on atomic partial charges and steric volumes. The method generates high-quality alignments but the comparison of a pair of molecules requires tens of minutes of CPU time on an IBM 3090 with a vector-processing unit.

The procedures discussed above are all far too slow for database-searching applications, unless coupled with an initial screening procedure that can eliminate the bulk of the database from the time-consuming exact calculation; an example of such a two-stage approach is provided by the SPERM program of van Geerestein *et al.* (1990), which is discussed in detail below. Future developments in computer hardware may enable similarity procedures such as these to be used for database searching; meanwhile, more efficient similarity algorithms are starting to be developed that use inter-atomic distance information. This provides a simple and obvious way of exploring the steric similarities between pairs of molecules, and several distance-based, similarity-searching techniques have been reported that seem to be sufficiently fast in execution to be used on databases of non-trivial size and to provide effective measures of structural resemblance. These approaches are discussed further below.

4.3 Methods based on distance information

The first investigation of 3-D similarity searching using inter-atomic distance information was that reported by Pepperrell and Willett (1991), which resulted in the development of the atom-mapping method, as detailed in Section 3 above. An alternative use of inter-atomic distance information has been reported by Bemis and Kuntz (1992). Here, a structure is decomposed into all possible three-atom substructures (so that a molecule containing N heavy atoms will produce $N(N-1)(N-2)/6$ such substructures). The procedure represents a molecule by a frequency distribution, the 64 elements of which are initially set to zero. A three-atom substructure is selected, and the sum of the squared inter-atomic distances calculated over all of the pairs of atoms in the chosen substructure; this sum is then used to increment one of the elements of the frequency distribution. The procedure is repeated for all of the possible three-atom substructures, and the resulting frequency distribution is then used to provide a simple characterisation of the shape of the chosen molecule. The similarity between a pair of molecules is obtained either by comparing hash values that are derived from the elements of the corresponding frequency distributions, or by comparing the distributions directly. As described, the Bemis-Kuntz procedure is limited to the comparison of pairs of molecules that differ in size by one heavy atom, at most. However, the methodology would seem to be extendible to more disparate sets of structures and to be sufficiently fast in operation for searching large files (even if use is made of the full frequency distributions, rather than the hash codes that were used in Bemis and Kuntz's experiments). That this is, indeed, the case is demonstrated by the work of Nilakantan *et al.* (1993), who have used a similar atom-triplet representation for shape-based similarity searching in a file of a quarter of a million compounds at Lederle Laboratories, with search times of 2-3 CPU hours on a VAX 8650.

The distance-based measures discussed thus far are all global in nature (although, as noted previously, atom-mapping does provide some local similarity information). The three remaining procedures, called DOCK, CLIX and SPERM, are all local similarity measures that consider the precise alignment of the target structure in the calculation of similarity, and are thus much slower in operation.

The DOCK program (Kuntz *et al.*, 1982) is designed to retrieve those molecules from a database that are geometrically most complementary in shape to a protein binding site, and that might thus be putative ligands for this site. The approach assumes that the geometry of the binding site is known, typically from crystallographic or NMR analysis, so that it can be described by a set of spheres that are complementary to the grooves and ridges in the receptor's surface and that fill the available binding site. The atoms comprising a putative ligand are represented by a similar set of spheres, and the shape similarity of the ligand to the site is then determined by the extent to which it is possible to overlap, or to *dock*, the two sets of spheres. The output from the program is a list of the best-matching equivalences, these being assumed to correspond with the preferred orientations of the ligand in the site; these orientations can then be checked by using crystallographic studies of the bound ligand. The original DOCK program (Kuntz *et al.*, 1982) considered only the docking of a single ligand into a site, but the procedure was soon generalised to permit the searching of a whole database of 3-D structures, with the structures being ranked in decreasing order of their likelihood of fitting the receptor site on steric grounds (DesJarlais *et al.* (1988), DesJarlais *et al.* (1990)).

Lawrence and Davis (1992) have recently reported a program, called CLIX, that performs a similar function to DOCK but that uses information from the GRID program of Goodford (1985), which identifies interaction sites, i.e., regions of high affinity for chemical probes on the molecular surface of a bonding site. CLIX takes a 3-D structure from the Cambridge Structural Database (the main database of experimentally-determined 3-D structures) and then exhaustively tests whether it is possible to superimpose a pair of the candidate's substituent chemical groups with a pair of corresponding favourable interaction sites proposed by GRID (only non-hydrogen atoms are considered). All possible combinations of ligand pairs and GRID binding-site pairs are tested; if a match is obtained, then the candidate ligand is rotated about the two pairs of groups and checked for steric hindrance and coincidence of other candidate atomic groups with appropriate GRID sites.

The SPERM (Superpositioning by PERMutations) program has been developed by van Geerestein *et al.* (1990) to provide a computationally-efficient means of quantifying the degree of shape similarity between pairs of 3-D molecules. Following previous work by Dean and co-workers (Chau and Dean (1987), Dean and Chau (1987), Dean *et al.* (1988)), the shape of a molecule in SPERM is described by mapping a specified property onto each of 32 points on a sphere surrounding the molecule, these points being the 12 vertices of an icosahedron and the 20 vertices of a dodecahedron oriented such that its vertices lie on the vectors from the centre of the sphere through the mid-points of the icosahedral faces. In fact, not one but two properties are used to characterise a molecule, these being the distances from each hedral point to the nearest surface of the molecule and to the surface of the molecule along the vector linking that point to the centre of the sphere. A database molecule is matched with an input target molecule by orienting the sphere representing the former molecule so as to give the greatest degree of agreement between the two sets of property values, this orientation being obtained using an algorithm due to Bladon (1989). The molecules in a database are ranked in decreasing order of the resulting similarities, and some number of the top-ranked molecules are then input to the electrostatic similarity measure described by Hodgkin and Richards (1987). This is based on the superposition of a pair of electron-density grids and is about three orders of magnitude slower than the sphere-matching procedure; it is thus typically applied to just the top 250 structures in the ranking. The orientation with the best Hodgkin-Richards score for a particular molecule is then taken to be the shape similarity for that molecule with the target structure.

4.4 Methods based on property information

The descriptions of the similarity methods thus far have focused on the use of distance information; however, several of them can also be used to allow property-based similarity searching.

The atom-mapping procedure that has been presented in Section 2 assumes that some atom, $A(I)$, in the target structure is considered for mapping to some atom, $B(J)$, in a database structure if, and only if, both $A(I)$ and $B(J)$ are of the same elemental type. However, there is no *a priori* reason why this should be so, and it is thus possible, if the user so wishes, for the program to match molecules on the basis of the arrangement of some specific atomic property (or properties) in 3-D space (Pepperrell *et al.* (1990)). Specifically, methods have been developed that allow an atom to be characterised by its hydrogen-bonding characteristics, its partial charge, its atomic radius, or a combination of these three types of property (rather than the elemental type as in the basic form of the method). It is also possible to specify weights when the elemental, hydrogen-bonding or partial-charge classes are used, which allows the user to designate certain atoms as being of greater importance than others. If very high weights are assigned to some atoms, it is possible to obtain an output in which the top-ranked structures will contain all (or most) of the highly-weighted atoms, so that one can execute a form of ranked 3-D substructure search.

Atomic physical properties can also be encompassed within the SPERM program, where the matching algorithm can be used with any property for which values can be calculated at the 32 vertices, e.g., electrostatic or hydrophobic factors; indeed, searching using electrostatic potential is a standard option of the program that can be used instead of the distance criteria described in the previous section. The most recent version of DOCK augments the steric matching-scores with electrostatic and molecular-mechanics interaction energies for the ligand-receptor complex. The inclusion of the additional information provides more accurate dockings but this is at the expense of considerably-enlarged run-times, which might be too great to allow rapid searching of a large database.

An alternative use of property-based searching has been reported recently by Fisanick *et al.* (1992), who describe experiments using a set of 6,000 3-D structures for which molecular-property data are available. Each structure is represented by a total of 29 computed properties (these including molar refractivity, van der Waals' surface area, dipole moment, and the heat of formation, *inter alia*). The standard deviation across the whole database is calculated for each such property. The overall similarity between the target structure and a database structure is incremented by 1, 2, 3 or 4 if the property value for the target structure is within 1.0, 0.75, 0.5 or 0.25 standard deviations, respectively, of the corresponding value for the database structure; this simple matching procedure is repeated for some or all of the properties. Initial experiments suggest that the most-similar structures resulting from a database search do, indeed, closely resemble the target molecule in structural terms.

5. Conclusions

Structural chemistry provides an extremely fruitful area of application for similarity and clustering methods (Willett (1987)). Similarity searching is already carried out on a routine basis on databases that contain hundreds of thousands of 2-D structures, and there is now considerable interest in the development of comparable methods for 3-D similarity searching. This paper has provided an overview of this work, with special reference to the atom-mapping method that has been developed in our laboratory in Sheffield.

Several of the methods that have been discussed in this paper are already in use in pharmaceutical and agrochemical organisations, but are currently used much less than 2-D similarity searching owing to the substantial computational costs: it is difficult to carry out highly-interactive, browsing-type searches if even a small database requires a CPU-day or more to identify the nearest neighbours for a user's target structure. It is hence to be expected that the next few years will see the development of new software and/or hardware techniques that can increase the speed of 3-D similarity searching (either by improvements to currently-available similarity methods or by the development of new methods, such as the genetic-algorithm approach to molecular recognition that has been suggested recently by Payne and Glen (1993)). Such improvements are particularly needed if full use is to be made of the highly effective, but very inefficient, overlap-based similarity methods that have been discussed in Section 4.2. Only if it proves possible to identify methods that are sufficiently fast in operation for interactive 3-D similarity searching, will it then be appropriate to consider the clustering of databases of 3-D structures (using either the Jarvis-Patrick method discussed in Section 2.3 or some alternative method that may be more appropriate for this sort of data).

Acknowledgements. I acknowledge the contributions of David Bawden, Andrew Brint, Geoffrey Downs, Catherine Pepperrell, Robin Taylor, Peter Walsh, David Wild and Vivienne Winterman to the work reported here, and Yvonne Martin for helpful discussions of much of the work that is reviewed in Section 4. Funding for the research on chemical similarity-searching in Sheffield has been provided by the European Communities Joint Research Centre, the Health and Safety Executive, ICI Agrochemicals, Pfizer Central Research, the Science and Engineering Research Council, and Tripos Associates.

References:

ARTYMIUK, P.J., BATH, P.A., GRINDLEY, H.M., PEPPERRELL, C.A., POIRRETTE, A.R., RICE, D.W., THORNER, D.A., WILD, D.J. and WILLETT, P. (1992): Similarity Searching in Databases of Three-Dimensional Molecules and Macromolecules. *Journal of Chemical Information and Computer Sciences, 32*, 617–630.

ASH, J.E., WARR, W.A. and WILLETT, P. (eds.) (1991): *Chemical Structure Systems.* Ellis Horwood, Chichester.

BARNARD, J. and DOWNS, G.M. (1992): Clustering of Chemical Structures on the Basis of Two-Dimensional Similarity Measures. *Journal of Chemical Information and Computer Sciences, 32*, 644–649.

BEMIS, G.W. and KUNTZ, I.D. (1992): A Fast and Efficient Method for 2D and 3D

Molecular Shape Description. *Journal of Computer-Aided Molecular Design, 6*, 607–628.

BLADON, P. (1989): A Rapid Method for Comparing and Matching the Spherical Parameter Surfaces of Molecules and Other Irregular Objects. *Journal of Molecular Graphics, 7*, 130–137.

BURT, C., RICHARDS, W. G.; HUXLEY, P. (1990): The Application of Molecular Similarity Calculations. *Journal of Computational Chemistry, 11*, 1139–1146.

CANO, F.H. and MARTINEZ-RIPOLL, M. (1992): On Shape. *Journal of Molecular Structure (Theochem), 258*, 139–158.

CARBO, R., LEYDA, L. and ARNAU, M. (1980): How Similar is a Molecule to Another? An Electron Density Measure of Similarity Between Two Molecular Structures. *International Journal of Quantum Chemistry, 17*, 1185–1189.

CHAU, P.-L. and DEAN, P.M. (1987): Molecular Recognition: 3D Surface Structure Comparison by Gnomic Projection. *Journal of Molecular Graphics, 5*, 97–100.

DANZIGER, D.J. and DEAN, P.M. (1985): The Search for Functional Correspondences in Molecular Structure Between Two Dissimilar Molecules. *Journal of Theoretical Biology, 116*, 215–224.

DEAN, P.M. (1987): *Molecular Foundations of Drug-Receptor Interaction*; Cambridge University Press: Cambridge, 1987.

DEAN, P.M., CALLOW, P. and CHAU, P.-L. (1988): Molecular Recognition: Blind Searching for Regions of Strong Structural Match on the Surfaces of Two Dissimilar Molecules. *Journal of Molecular Graphics, 6*, 28–34.

DEAN, P.M. and CHAU, P.-L. (1987): Molecular Recognition: Optimized Searching Through Rotational 3-Space for Pattern Matches on Molecular Surfaces. *Journal of Molecular Graphics, 5*, 152–158.

DESJARLAIS, R.L., SEIBEL, G.L., KUNTZ, I.D., FURTH, P.S., ALVAREZ, J.C., ORTIZ DE MONTELLANO, P.R., DECAMP, D.L., BABE, L.M. AND CRAIK, C.S. (1990): Structure-Based Design of Nonpeptide Inhibitors Specific for the Human Immunodeficiency Virus-1 Protease. *Proceedings of the National Academy of Sciences of the USA, 87*, 6644–6648.

DESJARLAIS, R.L., SHERIDAN, R.P., SEIBEL, G.L., DIXON, J.S, KUNTZ, I.D. and VENKATARAGHAVAN, R. (1988): Using Shape Complementarity as an Initial Screen in Designing Ligands for a Receptor Binding Site of Known Three-Dimensional Structure. *Journal of Medicinal Chemistry, 31*, 722–729.

FISANICK, W., CROSS, K.P. and RUSINKO, A. (1992): Similarity Searching on CAS Registry Substances. 1. Global Molecular Property and Generic Atom Triangle Geometric Searching. *Journal of Chemical Information and Computer Sciences, 32*, 664–674.

GOOD, A.C., HODGKIN, E.E. and RICHARDS, W.G. (1992): The Utilization of Gaussian Functions for the Rapid Evaluation of Molecular Similarity. *Journal of Chemical Information and Computer Sciences, 32*, 188–191.

GOODFORD, P.J. (1985): A Computational Procedure for Determining Energetically Favorable Binding Sites on Biologically Important Macromolecules. *Journal of Medicinal Chemistry, 28*, 849–857.

HERMANN, R.B. and HERRONN, D.K. (1991): OVID and SUPER: Two Overlap Programs for Drug Design. *Journal of Computer-Aided Molecular Design, 5*, 511–524.

HODES, L. (1989): Clustering a Large Number of Compounds. 1. Establishing the Method on an Initial Sample. *Journal of Chemical Information and Computer Sciences, 29*, 66–71.

HODGKIN, E.E. and RICHARDS, W. G. (1987): Molecular Similarity Based on Electrostatic Potential and Electric Fields. *International Journal of Quantum Chemistry: Quantitative Biological Symposium, 14*, 105–110.

JARVIS, R.A. and PATRICK, E.A. (1973): Clustering Using a Similarity Measure Based on Shared Nearest Neighbours. *IEEE Transactions on Computers, C-22*, 1025–1034.

JOHNSON, M.A. (1989): A Review and Examination of the Mathematical Spaces Underlying Molecular Similarity Analysis. *Journal of Mathematical Chemistry, 3*, 117–45.

JOHNSON, M.A. (1990): Similarity-Based Methods for Predicting Chemical and Biological Properties: a Brief Overview from a Statistical Perspective. In: D. Bawden and E.M. Mitchell (eds.): *Chemical Information Systems. Beyond the Structure Diagram.* Ellis Horwood, Chichester, 149–159.

JOHNSON, M.A., LAJINESS, M. and MAGGIORA, G. (1989): Molecular Similarity: a Basis for Designing Drug Screening Programs. *Progress in Clinical Biological Research Series, 291*, 167–171.

JOHNSON, M.A. and MAGGIORA, G.M. (eds.) (1990): *Concepts and Applications of Molecular Similarity.* Wiley, New York.

KEARSLEY, S.K. and SMITH, G.M. (1990): An Alternative Method for the Alignment of Molecular Structures: Maximising Electrostatic and Steric Overlap. *Tetrahedron Computer Methodology, 3*, 615–633.

KUNTZ, I.D., BLANEY, J.M., OATLEY, S.J., LANGRIDGE, R. and FERRIN, T.E. (1982): A Geometric Approach to Macromolecule-Ligand Interactions. *Journal of Molecular Biology, 161*, 269–288.

LAWRENCE M.C. and DAVIS, P.C. (1992): CLIX: a Search Algorithm for Finding Novel Ligands Capable of Binding Proteins of Known Three-Dimensional Structure. *Proteins: Structure, Function and Genetics, 12*, 31–41.

MANAUT, F., SANZ, F., JOSE, J. and MILESI, M. (1991): Automatic Search for Maximum Similarity between Molecular Electrostatic Potential Distributions. *Journal of Computer-Aided Molecular Design, 5*, 371–380.

MEYER, A.M. and RICHARDS, W.G. (1991): Similarity of Molecular Shape. *Journal of Computer-Aided Molecular Design, 5*, 427–439.

MEZEY, P.G. (1992): Shape-Similarity Measures for Molecular Bodies: a Three-Dimensional Topological Approach to Quantitative Shape-Activity Relations. *Journal of Chemical Information and Computer Sciences, 32*, 650–656.

NILAKANTAN, R., BAUMAN, N. and VENKATARAGHAVAN, R. (1993): New Method for Rapid Characterization of Molecular Shapes: Applications in Drug Design. *Journal of Chemical Information and Computer Sciences, 33*, 79–85.

PAPADOPOULOS, M.C. and DEAN, P.M. (1991): Molecular Structure Matching by Simulated Annealing. IV. Classification of Atom Correspondences in Sets of Dissimilar Molecules. *Journal of Computer-Aided Molecular Design, 5*, 119–133.

PAYNE, A.W.R. and GLEN, R.C. (1993): Molecular Recognition Using a Binary Genetic Search Algorithm. *Journal of Molecular Graphics, 11*, 74–91.

PEPPERRELL, C.A., POIRRETTE, A.R., WILLETT, P. and TAYLOR, R. (1991): Development of an Atom Mapping Procedure for Similarity Searching in Databases of Three-Dimensional Chemical Structures. *Pesticide Science, 33*, 97–111.

PEPPERRELL, C.A., TAYLOR, R. and WILLETT, P. (1990): Implementation and Use of an Atom Mapping Procedure for Similarity Searching in Databases of 3-D Chemical

Structures. *Tetrahedron Computer Methodology, 3*, 575–593.

PEPPERRELL, C.A. and WILLETT, P. (1991): Techniques for the Calculation of Three-Dimensional Structural Similarity Using Inter-Atomic Distances. *Journal of Computer-Aided Molecular Design, 5*, 455–474.

PETTITJEAN, M. (1992): Applications of the Radius-Diameter Diagram to the Classification of Topological and Geometrical Shapes of Chemical Compounds. *Journal of Chemical Information and Computer Sciences, 32*, 331–337.

RICHARD, A.M. (1991): Quantitative Comparison of Molecular Electrostatic Potentials for Structure-Activity Studies. *Journal of Computational Chemistry, 12*, 959–969.

VAN GEERESTEIN, V., PERRY, N.C., GROOTENHUIS, P.D.J. and HAASNOOT, C.A.G. (1990): 3D Database Searching on the Basis of Ligand Shape Using the SPERM Prototype Method. *Tetrahedron Computer Methodology, 3*, 595–613.

WILLETT, P. (1987): *Similarity and Clustering in Chemical Information Systems.* Research Studies Press, Letchworth.

WILLETT, P. (1990): Algorithms for the Calculation of Similarity in Chemical Structure Databases. In: Johnson, M.A. and Maggiora, G.M. (eds.) *Concepts and Applications of Molecular Similarity.* Wiley, New York, 43–63.

WILLETT, P. (1991): *Three-Dimensional Chemical Structure Handling.* Research Studies Press, Taunton.

WILLETT, P., WINTERMAN, V. and BAWDEN, D. (1986): Implementation of Non-Hierarchic Cluster Analysis Methods in Chemical Information Systems: Selection of Compounds for Biological Testing and Clustering of Substructure Search Output. *Journal of Chemical Information and Computer Sciences, 26*, 109–118.

Feature Generation and Classification of Time Series [1]

Christel Wisotzki, Fritz Wysotzki

Fraunhofer-Institut für Informations-und Datenverarbeitung
Einrichtung für Prozeßoptimierung
Kurstraße 33, D - 10117 Berlin, Germany

Summary: In the paper the realisations of time series are approximated by spline functions in order to reduce the number of features. The approximation procedure defines the number and the location of the joints. A special prototype classifier for features having different dimensions is developed.

1. Introduction

The present paper deals with problems connected with supervised classification for time series. The given values of the time series can be measurements with noise or the result of a filter procedure. The number of values to be treated is assumed to be so large that the application of a classification method to the time series is impossible without former treatment.

Developing classification methods for time series, the feature generation and the feature reduction have a large importance. In the present paper, time series are approximated by spline functions where the algorithm defines the number and the distribution of the joints. The approximation algorithm is described in chapter 2.

In chapter 3 some possibilities of feature generation are mentioned. The approximating spline function can be described by relatively few parameters which build the feature vector. In general, such feature vectors have different dimensions for different time series. If the dimensions are identical, the feature vectors can be treated by any classification method (for instance decision tree method or decision surface method).

In chapter 4 a prototype classifier is introduced. Using this classifier there are no problems with different feature vector dimensions.

2. Approximation

A finite number of realisations $[x_i]_{i=0}^{m-1}$ of a time series on equidistant measurement points is supposed to be given. Without loss of generality, $x_i = x(i)$ for $i = 0, 1, \ldots, m-1$ can be assumed. The given values are to be approximated by a spline function where the number and the positions of the joints depend on the given data.

Let $C = (i_0, i_1, \ldots, i_{n-1}, i_n)$ be a partition of the time interval $[0, m]$ in n subintervals $[i_j, i_{j+1}]$ $(j = 0, 1, \ldots, n-1)$ with

$$0 = i_0 < i_1 < \ldots < i_{n-1} < i_n = m. \tag{1}$$

Let $S_{q,r}(C)$ denote the set of all functions z for which the following conditions hold:

[1]This work was supposed by the German Ministry of Research and Technology within the project WISCON

- z is a polynomial of the qth degree in each of the subintervals $[i_{j-1}, i_j)$ for $j = 1, 2, \ldots, n-1$ and $[i_{n-1}, m-1]$;
- The derivatives $z^{(k)}$ exist and are continuous for $k = 0, 1, \ldots, r$, if $r > 0$ or z is piecewise continuous if $r = 0$.

The approximation problem can be formulated in the following way:

A partition $C = (i_0, i_1, \ldots, i_{n-1}, i_n)$ with (1) and a function $z_C \in S_{q,r}(C)$ are to be determined where z_C has to be optimal in the following sense:

$$\sum_{i=0}^{m-1} [z_C(i) - x_i]^2 = \min_{z \in S_{q,r}(C)} \{\sum_{i=0}^{m-1} [z(i) - x_i]^2\}. \tag{2}$$

The mentioned partition corresponds to a clustering of the measurement points $x_0, x_1, \ldots, x_{m-1}$, where points with consecutive indices only belong to the same cluster. The clustering has to be suitable corresponding to the error functional

$$\Phi(C) = \sum_{i=0}^{m-1} [z_C(i) - x_i]^2. \tag{3}$$

The problem of clustering with a fixed number of clusters has been considered in [1]. There necessary conditions for the optimality of a clustering for several error functionals (L_∞-distance, Sobolev distance) are given.

Obviously, the error functional vanishes for a correspondingly large number of clusters n, i.e. a general optimization does not use. For that reason, a method consisting of two stages is used. At the first stage an initial partition and the number of clusters are determined. At the second stage this initial partition is optimized corresponding to the error functional (3) where the number of clusters is now fixed.

Stage 1

A hierarchical agglomerative method (see [2]) is used for the first stage. Beginning with the finest clustering C_m where every point forms a cluster, in every step two clusters are merged. A sequence of clusterings $C_m, C_{m-1}, \ldots, C_1$ arises. The error $\Phi(C_k)$ of the clustering C_k is obtained by solving the optimization problem (2) and using (3). For the error sequence the relation

$$0 = \Phi(C_m) \leq \Phi(C_{m-1}) \leq \ldots \leq \Phi(C_1)$$

holds. The merging of two clusters is to carry out in such a way that it causes in each step a minimal growth in the error functional. This process can be broken with the help of some of the following criteria:

$$n \leq n_0 \tag{4}$$

$$\Phi(C_{n-1}) \geq \delta \tag{5}$$

$$\rho \cdot \Phi(C_{n-1}) \leq \Phi(C_n) \tag{6}$$

The maximal number of clusters n_0, the error bound δ and the bound ρ for the error rate are numbers to be given. The criterion (6) means that merging two clusters, the clustering error must not increase more than by the factor $1/\rho$.

Stage 2

In the second stage an exchange method (see [2]) is used for optimization. Step by step it tests wether the change of a boundary point of one cluster to the neighbouring cluster causes a reduction of the error functional. If it is true, the point is moved to the corresponding neighbouring cluster. After a finite number of steps there are no more possible reductions of the error functional, i. e. a local minimum of the error functional is reached.

3. Feature Generation

If a curve is approximated by piecewise polynomials of the qth degree, the approximating curve can be completely described by $(q+1) \cdot n$ parameters (n - number of polynomial pieces). There are several possibilities to define corresponding features. For simplicity, the case of piecewise linear, piecewise continuous approximations ($r = 0, q = 1$) is considered only.

Let $C = (i_0, i_1, \ldots, i_{n-1}, i_n)$ be the obtained clustering and z_C the approximating function fulfilling the condition (2). The set of triples

$$[z_C(i_j), z_C(i_{j+1} - 1), i_{j+1} - i_j]_{j=0}^{n-1}$$

can be chosen as a feature vector. It is possible to replace the values $z_C(i_{j+1} - 1)$ by the slopes of the corresponding straight line pieces. However, the straight line parameters defining the crossings with the axes are not recommended to be chosen as features. They increase along the time axis, and for this reason their use would cause instabilities.

Furthermore, one has to note that a classifier usually needs objects having a fixed number of features. This demand can only be realized by using (4) as break criterion at the first stage of the approximation procedure. Therefore, a constant number of clusters n_0 is obtained.

4. Prototype Classifier

The problem of different dimensions of the feature vectors described at the end of the previous chapter vanishes when the prototype classifier is used.

Let a training set be given by N objects

$$\Theta = (X^1, X^2, \ldots, X^N),$$

L classes

$$\Omega = (K_1, K_2, \ldots, K_L)$$

and a mapping defining to which class each of the given objects belongs

$$X^l \rightarrow K^l \in \Omega, \quad l = 1, 2, \ldots, N.$$

For every class a prototype is formed in the following way: Let I_j denote the set of the indices of all objects of the training set belonging to the class K_j, i. e.

$$I_j = \{l | K^l = K_j\}$$

and z_l denote the function approximating X^l. The prototype P_j of the class K_j is defined by

$$P_j(t) = |I_j|^{-1} \cdot \sum_{l \in I_j} z_l(t),$$

i. e. by the average of the approximating functions of all objects of the training set belonging to the corresponding class. $|I_j|$ is the number of the elements of the set I_j.

The prototypes are piecewise polynomial too, but, in general, the number of subintervalls increases if the number of objects taking part in the prototype formation increases, because, in general, the numbers and locations of the subintervals of different objects differ. If it is necessary, the number of subintervals can be reduced by the use of the approximation algorithm of chapter 2 to the prototypes.

Let $P_1, P_2, \ldots, P_L$ denote the prototypes of all classes and $X = [x_i]_{i=0}^{m-1}$ be an object to be classified. X is assigned to that class to which prototype its approximating function has the smallest distance. As a distance the R^m-distance

$$r^2(z, P) = \sum_{i=0}^{m-1} [z(i) - P(i)]^2 \tag{7}$$

or the $L_2[0, m-1]$-distance

$$R^2(z, P) = \int_0^{m-1} [z(t) - P(t)]^2 dt \tag{8}$$

can be used.

Using (7) the function values in all m measurement points have to be calculated. For (8) the values at the joints only are necessary. Therefore, for large m the distance (8) is prefered.

In the C++ program TSC for DOS-computers the approximation algorithm of the second chapter is realized for the case $r = 0, q = 1$ and connected with the prototyp classifier. The break criteria (4), (5) and (6) can be used. Several examples were calculated in order to test the algorithm where good results were obtained.

References:

PAVLIDIS, T. (1977): *Structural Pattern Recognation.* Springer-Verlag Berlin Heidelberg New York.

STEINHAUSEN, D. and LANGER, K. (1977): *Clusteranalyse.* Walter de Gruyter Berlin New York

TRIPAT:
a Model for Analyzing Three-Mode Binary Data

S. Krolak-Schwerdt[1], P. Orlik[1], B. Ganter[2]

[1] Fachrichtung Psychologie, Universität des Saarlandes, 66123 Saarbrücken, Germany

[2] Fachbereich Mathematik, Technische Universität Dresden, 01069 Dresden, Germany

Summary: A discrete, categorical model is presented for three-mode (conditions by objects by attributes) data arrays with binary entries $x_{ijk} \in \{0,1\}$. Basically, the model attempts a simultaneous classification of the entities or elements of the three modes in a number of common clusters. Clusters are defined by three-mode submatrices of maximum size with entries $x_{ijk} = 1$. In performing a discrete representation of the data structure, the model may be classified as a non-hierarchical clustering procedure. It involves a reorganization of the data array such that the final clustering solution is interpreted directly on the data, and it allows for overlapping as well as nonoverlapping clusters. The method is similar to three-mode component models such as CANDECOMP and SUMMAX in the model function to predict the data. An application concerning recall data in a study of social perception is provided.

1. Introduction

In this paper, a structural model is presented for three-mode data arrays $X = (x_{ijk})$ with binary entries $x_{ijk} \in \{0,1\}$. Generally, we will refer to such an array as a conditions by objects by attributes matrix. In applications, natural objects, concepts or variables will be designated as objects, the features they possess or lack, as attributes, and different experimental conditions, instructions or subjects will be referred to as conditions.
As an example, data of this type are acquired in cognitive psychology, when subjects list characteristics of specified objects or the recall of attributes is registered. Illustrative data from this domain also encompass sociometric data collected in various situational contexts or self concept data (Kelly, 1955). The main structure of the data consists of complex cross-relations between the objects, attributes and conditions.

If the structure present in the data is to be portrayed by a discrete representation, one approach would be to compute a distance measure for pairs of objects from the data matrix and subsequently to cluster analyze the object set. However, for three-way arrays there are several serious difficulties with the distance method. In deriving a distance measure one mode (i.e. the object set) is emphasized as the primary organizing principle though the attributes and/or conditions may reveal the main data structure. As Eckes and Orlik (1991) have noted, in many empirical studies there is no a priori basis for preferring the analysis of objects to the analysis of attributes or another set of entities. Unless distances of binary patterns for subjects represent the primary structure, where clusters are interpreted in terms of subjects' responses on the objects, the distance approach induces a loss of information in the current data design by cross-cutting two sets of entities in the derived proximity matrix.

On the other hand, some current models are designed for two-way two-mode data arrays such as GENNCLUS (DeSarbo, 1982) or the INDCLUS model (Carroll & Arabie, 1983) which assumes a three-way two-mode (subjects by stimuli by stimuli)

proximities matrix, and thus provide generalizations so as to present in a clustering solution differences among conditions or other sources of data. However, given the present three-mode data design these models imply that one mode serves to derive a proximity measure condensing the data into a two-mode format or require the data to be at least on the interval level.

When the data are binary matrices, an approach to be preferred would be a model which can be used with no assumptions about similarity or distance and which induces a three-mode classification to be interpreted directly on the data. The TRIPAT model to be presented in the following is designed to adopt to these requirements. The model produces a clustering of the objects which simultaneously indicates the attributes and conditions associated with the object grouping and which operates directly on the 0-1-pattern of the data array.

2. Definitions

Since a substantial portion of the exposition to follow will refer to the term of a 'mode' and of a 'concept', several definitions will be given at the outset.
As defined by Carroll and Arabie (1980), a *mode* is a set of entities (i.e., a set of objects, a set of variables, a set of conditions ...). Modes will be denoted by capital letters A, B, C and the entities corresponding to a mode are denoted by a_i, $i = 1, \ldots, l$, b_j, $j = 1, \ldots, m$ and c_k, $k = 1, \ldots, n$.

A *three-mode array* is the cartesian product $A \times B \times C$, where each mode comprises a different set of entities. A *three-mode data array* is the assignment of numerical values to the lmn elements of the array. In the following, it will be assumed, that mode A refers to a set of conditions, mode B to that of objects and C to that of attributes. The binary entries $x_{ijk} \in \{0,1\}$ of the three-mode data array will be interpreted as the presence of attribute c_k for object b_j under condition a_i, if $x_{ijk} = 1$, and as the absence of attribute c_k otherwise. The data structure will be presented by (A, B, C, V), where $V \subseteq A \times B \times C$ is a ternary relation, and the presence of attribute c_k for object b_j under condition a_i will be denoted by $(a_i, b_j, c_k) \in V$. Furthermore, it will be assumed that for each condition a_i, each object b_j and each attribute c_k there is at least one entry $x_{ijk} = 1$.

A *three-mode cluster* D_r, $r = 1, \ldots, t$, is the union of three subsets $A_r \subseteq A$, $B_r \subseteq B$, $C_r \subseteq C$,

$$D_r = A_r \cup B_r \cup C_r \ ,$$

where $\mid A_r \mid = l_r$, $\mid B_r \mid = m_r$, $\mid C_r \mid = n_r$.

Moreover, the model to be presented below refers to the notion of a 'concept' as defined by Formal Concept Analysis (Wille, 1982). The basic data type of Formal Concept Analysis is a *context* (A, B, I) which consists of two modes A and B, and a binary relation $I \subseteq A \times B$ between A and B. If mode A represents a set of objects and mode B a set of attributes, then $(a_i, b_j) \in I$ indicates the presence of attribute b_j for object a_i which in turn is equivalent to an entry $x_{ij} = 1$ in the corresponding two-mode data array. Given two subsets $A_0 \subseteq A$, $B_0 \subseteq B$, the set

$$A_0^I := \{b_j \in B \mid (a_i, b_j) \in I \text{ for all } a_i \in A_0\}$$

comprises all attributes of B the objects in A_0 possess. In an analogous way

$$B_0^I := \{a_i \in A \mid (a_i, b_j) \in I \text{ for all } b_j \in B_0\}$$

contains all objects of A which have the attributes of B_0 in common.

A *(formal) concept* is a pair (A_0, B_0) with $A_0 \subseteq A$, $B_0 \subseteq B$, satisfying the following condition:

$$A_0^I = B_0 \text{ and } B_0^I = A_0.$$

In the two-way data array, concepts exhibit the form of maximal rectangles $A_0 \times B_0 \subseteq I$. A_0 and B_0 are called the *extent* and the *intent* of the concept (A_0, B_0), respectively.

3. The TRIPAT model

The three-mode classification procedure proposed in the following is a discrete, categorical representation of objects, attributes and conditions. Basically, the model attempts a simultaneous classification of the entities of the three modes in a number of common clusters D_r. In performing a discrete representation of the corresponding data structure, the model may be classified as a non-hierarchical clustering procedure. It allows for overlapping as well as nonoverlapping clusters.
Beyond its relation to clustering procedures, TRIPAT is similar to component models for analyzing three-way data, especially to three-mode canonical decomposition (Carroll & Chang, 1970). In both procedures, equivalent model functions to predict the data are postulated with the exception that in the TRIPAT model parameters are restricted to binary values $\{0, 1\}$. Disregarding the very different data entries and measurement characteristics both models assume, TRIPAT in some sense may be termed a discrete counterpart to trilinear component models.

Similarly to a subgroup of clustering procedures like the 'bond energy algorithm' (McCormick, Schweitzer & White, 1972) and the 'modal block method' (Hartigan, 1976), which have been called 'direct methods' by Eckes and Orlik (1991), TRIPAT involves a reorganization of the three-mode data array such that the final clustering solution is interpreted directly on the data array and expresses the interaction between the three modes.

In TRIPAT, the rationale to construct a three-mode cluster may be characterized in the following way:
The procedure successively determines three-mode submatrices of maximum size from the data array whose entries entirely have values $x_{ijk} = 1$. In permuting rows, columns and planes of the data array these submatrices represent blocks or bricks with $x_{ijk} = 1$ if the entry belongs to the corresponding brick. The adjacent cluster is formed by the union of the entities associated with the rows, columns and planes of the brick. By interpretation, all objects of a cluster have each attribute in every condition of the same cluster in common, and no other entities besides those represented in a cluster can be added exhibiting the same property.

Formally, a *brick* (in V) is a subset $W_r \subseteq V$, which is the product of subsets, i.e.

$$W_r = A_r \times B_r \times C_r \qquad \text{where } A_r \subseteq A, B_r \subseteq B, C_r \subseteq C.$$

A brick W_r is *maximal* if there is no brick $W_{r'}$ with $W_r \subseteq W_{r'}, W_r \neq W_{r'}$.

The qualification of a three-mode submatrix to form such a brick derives from the notion of a concept in the following way:
For a given relation $V \subseteq A \times B \times C$ we define a formal context

$$\mathbb{K}_V := (A, B \times C, V')$$

by

$$(a_i, (b_j, c_k)) \in V' :\Longleftrightarrow (a_i, b_j, c_k) \in V.$$

The intents of this context are subsets of $B \times C$, thus relations themselves. For an arbitrary set $A_r \subseteq A$ we obtain the context

$$\mathbb{K}_{A_r} := (B, C, A_r^{V'}).$$

Proposition 1 *$W_r = A_r \times B_r \times C_r \subseteq V$ is a maximal brick if and only if the following two conditions hold*

1. $A_r = (B_r \times C_r)^{V'}$.
2. *$B_r \times C_r$ is a concept of $(B, C, A_r^{V'})$*

Proof '$\Rightarrow$': The first condition implies that $A_r \times (B_r \times C_r) \subseteq V'$, from which it follows that $W_r \subseteq V$, i.e. that W_r is a brick. The maximality of A_r also follows from the first condition, the maximality of B_r and C_r from the second one.
'$\Leftarrow$': If W_r is a brick, then $(B_r \times C_r)^{V'} \subseteq A_r$. If W_r is maximal, then equality holds. Similarly one concludes that $B_r \times C_r$ is a maximal rectangle in $A_r^{V'}$. □

The algorithm associated with the above proposition produces, for an arbitrary ternary relation, all maximal bricks. It is a modified version of a procedure that solves the respective problem in the case of binary relations, and which is frequently used in Formal Concept Analysis (see Ganter, 1987). Adapted to the present problem of a three-mode design, the algorithm proceeds as follows: we successively generate all extents of the context $\mathbb{K}_V$, and for each such extent A_r we check all concepts (B_r, C_r) of $\mathbb{K}_{A_r}$ whether they satisfy condition 2 of the proposition. Thus the current procedure is in fact a two-stage algorithm where each stage corresponds to one of the properties of the above proposition.

Unless the data array X represents trivial cases (as an example, if all objects possess all attributes in every condition), there is usually more than one brick to be found in a data set. The basic characteristic of the TRIPAT method to select one brick as defining a cluster D_r from the set of bricks is specified by the criterion

$$| W_r | = l_r m_r n_r = max .$$

That is, among all clusters whose associated elements by combination satisfy the condition of a brick the one of largest volume is chosen. Thus, by the TRIPAT criterion subgroups are required to account for the largest possible number of cross-relations within the three-mode data design.

Once a cluster D_r has been found whose elements constitute a brick of largest volume, the method proceeds iteratively by eliminating the brick selected in the previous phase from the data array. That is, in defining a relation $\tilde{V} \subseteq A \times B \times C$, where $(a_i, b_j, c_k) \in \tilde{V} :\Longleftrightarrow (a_i, b_j, c_k) \in V$ and $(a_i, b_j, c_k) \notin W_r$, $(A, B, C, \tilde{V})$ is analyzed in the next step for bricks, and the elements associated with the brick of largest volume in $\tilde{V}$ make up the subsequent cluster D_{r+1}.

In terms of the data array X, the successive elimination of bricks already chosen may be represented by calculating the remaining residuals $\tilde{x}_{ijk}$,

$$\tilde{x}_{ijk} = \begin{cases} 0 & \text{if } a_i, b_j, c_k \in D_r \\ x_{ijk} & \text{otherwise} \end{cases}$$

and continuing the analysis with $\tilde{X} = (\tilde{x}_{ijk})$. The updating of the data array in terms of a residual matrix insures that each specified combination (a_i, b_j, c_k) is part of only one selected brick. As a consequence, though two different clusters D_r and $D_{r'}$ may overlap in general, they never encompass the same combination of an object, attribute and condition.

The various steps of the TRIPAT algorithm are thus described as follows:

(o) $r = 1$, $\tilde{V} = V$.

(i) In $(A, B, C, \tilde{V})$ all bricks $W_r = A_r \times B_r \times C_r$ are formed.

(ii) The brick W_r of largest volume, $\mid W_r \mid = l_r m_r n_r = max$ defines the entities of cluster D_r.

(iii) $(A, B, C, \tilde{V})$ is updated by eliminating the brick previously chosen from the current data array.

(iv) With $(A, B, C, \tilde{V})$ steps (i)-(iii) are repeated, where $r = r + 1$.

(v) After t iteration cycles (completing steps (i)-(iv)):

$$\tilde{x}_{ijk} = 0 \qquad \text{for } i = 1, \ldots, l; j = 1, \ldots, m; k = 1, \ldots, n$$

and the algorithm terminates.

As step (v) indicates, the determination of t clusters yields a residual matrix of zero entries or, equivalently, a representation of the data structure accounting for all intermodal cross-relations present in the data. Furthermore, as long as the preliminary assumptions concerning the data array are met (i.e., there is at least one entry $x_{ijk} = 1$ associated with each object, attribute and condition), the resulting clustering solution is exhaustive in that each entity in every mode is at least in one cluster.

By the concept of a residual matrix in each iteration cycle the proposed method refers to one of the computing techniques of classical factor analysis (see, e.g., Horst, 1965). Thus, the principle of successively removing variation of the data accounted for by the current factor has its counterpart in TRIPAT by eliminating cross-relations accounted for by the current brick. Beyond mere computational details, it is the model function to predict the data TRIPAT has in common with a particular subgroup of component methods to decompose a three-way data matrix, which have been termed trilinear methods by Kruskal (1984). In contrast to principal component analysis, which refers to an eigenvalue- eigenvector decomposition of cross-products, trilinear component methods directly decompose the three-mode array and they involve one set of components that are common to all three modes. To show the link between these models and TRIPAT more specifically, let $P = (p_{ir})$ denote an $l \times t$ binary matrix characterizing the grouping of the conditions in TRIPAT

$$p_{ir} = \begin{cases} 1 & \text{if } a_i \in D_r \\ 0 & \text{otherwise} \end{cases}$$

and let $Q = (q_{jr})$, $S = (s_{kr})$ denote $m \times t$ and $n \times t$ binary matrices indicating the associated objects and attributes in an analogous way, that is

$$q_{jr} = \begin{cases} 1 & \text{if } b_j \in D_r \\ 0 & \text{otherwise} \end{cases}$$

$$s_{kr} = \begin{cases} 1 & \text{if } c_k \in D_r \\ 0 & \text{otherwise.} \end{cases}$$

Then the TRIPAT model can be written as

$$\hat{x}_{ijk} = \sum_{r=1}^{t} p_{ir} q_{jr} s_{kr} \ ,$$

where the $\hat{x}_{ijk}$ specify the data predicted by the model. In matrix notation

$$\hat{X} = PI(Q' \otimes S') \ ,$$

where $\otimes$ is the Kronecker product and I denotes the three-mode identity matrix. If it were not for the constraint that $p_{ir}, q_{jr}, s_{kr} \in \{0,1\}$, this model would be formally identical to trilinear component models such as CANDECOMP (Carroll & Chang, 1970) and SUMMAX (Orlik, 1980), which represent the relative weight of the entities on the respective component r' by coefficients $\pi_{ir'}, \varrho_{jr'}, \sigma_{kr'}, \hat{x}_{ijk} = \sum_{r'=1}^{t'} \pi_{ir'} \varrho_{jr'} \sigma_{kr'}$. However, by TRIPAT the possession of attributes and the contribution of various conditions is required to be all or none instead of continous shifts in the compositions of the subgroups.

The TRIPAT model outlined above covers various model specifications which permit either overlapping or nonoverlapping representations in particular modes. The general case consists of possibly overlapping configurations in each mode, where

$$\sum_{r=1}^{t} p_{ir} \geq 1 \qquad \text{for } i = 1, \ldots, l,$$

$$\sum_{r=1}^{t} q_{jr} \geq 1 \qquad \text{for } j = 1, \ldots, m,$$

$$\sum_{r=1}^{t} s_{kr} \geq 1 \qquad \text{for } k = 1, \ldots, n.$$

In this case, TRIPAT involves a 'complete covering' of the data array; that is, there is an exact reproduction of the data, $\sum_{r=1}^{t} p_{ir} q_{jr} s_{kr} = x_{ijk}$.

Alternatively, clusters might be restricted to nonoverlapping representations for specified modes. To accommodate theoretical considerations or research domains implying mutually exclusive subgroups, the following options might be chosen:

$$\sum_{r=1}^{t} p_{ir} = 1 \qquad \text{and/or} \qquad \sum_{r=1}^{t} q_{jr} = 1 \qquad \text{and/or} \qquad \sum_{r=1}^{t} s_{kr} = 1 \ .$$

In these various model specifications the decision to select a nonoverlapping classification within one mode does not predetermine any choice as to the remaining modes. Thus, any combination of parameter specifications across different modes might be chosen. With respect to the algorithm which has been outlined above for the general case of overlapping configurations in each mode, mutually exclusive subgroups are obtained by an adjustment of the residuals in each iteration cycle. As an example, $\tilde{V}$ is defined as $(a_i, b_j, c_k) \in \tilde{V} :\Longleftrightarrow (a_i, b_j, c_k) \in V$ and $a_i \notin D_r$, if conditions have to be portrayed by a nonoverlapping representation. In an analogous way, $\tilde{V}$ is adjusted to the corresponding choice of nonoverlapping object or attribute configurations.

4. An Application

To provide an illustration of the TRIPAT model, data were analyzed from a study investigating the effects of different goals in social cognition. Materials were 56 attributes describing person characterizations in ordinary language. The attributes were collected from a study by Eckes (1991), and they consist of verbal descriptions of the physical appearance, political beliefs of persons, attitudes, behaviors, traits and situations which are generally expected to be preferred by different types of persons. The attributes in part refer to common sex-role stereotypes (i.e., 'typical man', 'typical woman'), partly they are associated with subcategories of persons ('career woman', 'soft man') whose characteristics are discrepant from the superordinate gender specific stereotypes. Attributes which are unspecified may appear in every person category (for example, the attribute 'wears a beard' within the male categories). Subjects participating in the study received the attributes clustered by person categories in the presentation list with the goal of (a) recalling the attributes, (b) forming impressions about the target persons, (c) predicting the behavior of each target person, or (d) classifying the person categories along an evaluative dimension. Actually, subjects under each instructional set had to recall the presented attributes subsequently. By the 'continuum model' of social perception (Fiske & Neuberg, 1990) it was expected that subjects under judgment instructions should mainly recall the attributes defining the subtypes, whereas behavior prediction and learning goals should not direct subjects' attention to one specified person category. Thus, within the second subgroup of goals a uniform amount of recalled items under different person categories was expected. With respect to the attribute types, specifically the expectancy of subsequent behavior predictions should increase the recall of situational contexts.

To analyze the free-recall data, the proportion of items recalled correctly was registered as a function of goals, attribute type and person categories. These factors also define the three modes of the subsequent analysis. Items with recall rates above the median of the relative frequency distribution were scored as 1 and items below the median were scored as 0, respectively. The dichotomous recall data which are displayed in Table 1 served as input data for the TRIPAT analysis.

Table 1: The dichotomous recall data

	Recall			Impression Formation			Behavior Prediction			Evaluation		
	A	B	C	A	B	C	A	B	C	A	B	C
Physical appearance	1	1	1	1	0	1	1	1	1	1	1	1
Political beliefs	1	0	1	1	0	1	1	0	1	1	0	1
Attitudes	0	0	1	0	0	1	0	0	0	0	0	0
Behavior	0	0	1	0	0	1	0	0	0	1	0	0
Traits	0	0	1	0	0	1	0	1	0	0	1	0
Situations	0	0	1	0	1	1	1	0	1	0	0	1

A denotes characteristics of superordinate stereotypes,
B unspecified with respect to person categories,
C characteristics of subtypes

In order to explore common effects as well as possibly distinctive effects of the experimentally induced goal types, the TRIPAT model specification allowing for overlapping representations in each mode was selected. Table 2 shows the resulting classification which consists of eight clusters.

Table 2: The TRIPAT solution

	Cluster							
	1	2	3	4	5	6	7	8
Recall	1	1	0	0	0	0	0	1
Impression Formation	1	1	0	0	0	0	1	0
Behavior Prediction	1	0	1	1	0	0	0	0
Evaluation	1	0	1	0	1	1	0	0
Superordinate stereotypes	1	0	0	1	0	1	0	0
Unspecified	0	0	1	0	0	0	1	1
Subtypes	1	1	0	1	1	0	0	0
Physical appearance	1	0	1	0	0	0	0	1
Political beliefs	1	0	0	0	0	0	0	0
Attitudes	0	1	0	0	0	0	0	0
Behavior	0	1	0	0	0	1	0	0
Traits	0	1	1	0	0	0	0	0
Situations	0	1	0	1	1	0	1	0
Cluster size	16	8	4	2	1	1	1	1

The first cluster contains the entire set of experimental conditions in mode A, the general stereotypes and the subcategories of target persons in mode B, and attributes refering to physical appearance and political beliefs in mode C. By interpretation, the composition of the cluster indicates that subjects' attention was directed to the characteristics of the outer appearance and political orientation irrespective of any specific goal instruction.

With respect to the hypotheses underlying the study, the main results are summarized in clusters 2 and 3. These clusters group the experimental conditions into a subset of goals which primarily induce an increased recall of situations, traits, behaviors and attitudes characterizing the subtypes, and a distinct subset of goals initiating the superior recall of unspecified physical attributes and unspecified traits. Surprisingly, the theoretically expected effects of impression formation coincide with those of the recall instruction implying that judgment goals have much in common with learning goals, whereas the effects of behavior prediction are more similar to those of general evaluations.
Finally, the fourth cluster indicates that behavior prediction has the unique effect of attending to preferred situations ascribed to both stereotypical targets and the subtypes.

The last row of Table 2 gives the size of the brick in the data array associated with each cluster. By the TRIPAT algorithm, cluster sizes successively decrease in each classification cycle. Unless theoretical considerations contradict a certain amount of information loss, one approach would be to define a minimum cluster size in advance and to accept clusters exceeding the minimum limit as the final solution. As to the underlying research hypotheses, a cluster size of 2 was sufficient in the present example, and thus the remaining clusters of Table 2 were omitted from further interpretations.

5. Concluding remarks

In this paper, a structural model is presented yielding a discrete, categorical representation of objects, attributes and conditions simultaneously. To summarize briefly, the basic idea of the model consists of decomposing a three-mode binary data array into a number of submatrices of largest size with entries $x_{ijk} = 1$. The resulting clustering solution encompasses subgroups of objects having each attribute in every condition of the corresponding cluster in common. Different clusters may overlap or may comprise mutually exclusive entities in particular modes. The development of the model is primarily motivated by applications in implicit belief systems such as self concept data (Kelly, 1955). In this domain, a differentiation of an individuum's views of specified role- and construct-patterns into a variety of contextual subsystems is part of the theoretical discourse. However, this is only suggestive of a range of possible applications.

With respect to the main intention of the model, TRIPAT is similar to the HICLAS method (De Boeck & Rosenberg, 1988). In HICLAS, subsets of objects are defined by the attributes they share and thus, both models refer to equivalent specifications of a cluster. However, HICLAS uses a two-mode binary data array as input and aims at the recovering the underlying structure of object-attribute-relations. Furthermore, in defining an order relation between different subsets of objects, HICLAS is more in line with the hierarchical subconcept-superconcept ordering employed in Formal Concept Analysis.

Another structural model, TRIPAT in some sense is related to, is blockmodeling (Arabie, Boorman & Levitt, 1978). In constructing blockmodels, rectangular submatrices, termed 'blocks', of binary $l \times m$ or $m \times m \times l$ data arrays are derived, whose entries mostly consist of 1's and 0's, respectively. Thus, the underlying data types may be characterized as either two-mode or three-way two-mode restricting the planes of the array to be quadratic. In applications, various density levels for the number of 1's constituting a block may be chosen. In comparison, TRIPAT does not allow for continous shifts in the number of entries $x_{ijk} = 1$ within a block or brick, but rather specifies the possession of attributes and conditions within a block to be thorough. Moreover, TRIPAT is more concerned with internally homogeneous subgroups with respect to the corresponding attributes and conditions, and different subgroups are generally not required to be isolated or unrelated in all attributes and conditions.

Finally, the TRIPAT model outlined in this paper includes as a special case the analysis of two-mode objects by attributes data arrays. The corresponding two-mode approach consists of successively finding rectangles in the data of largest size with entries $x_{ijk} = 1$, and defines clusters as object- attribute-combinations analogous to the notion of a concept in Formal Concept Analysis. Currently, the model is generalized to account for three-mode or two-mode data of the type $x_{ijk} \in \{-1, 0, 1\}$.

6. References

ARABIE, P., BOORMAN, S.A., and LEVITT, P.R. (1978): Constructing blockmodels: How and why. *Journal of Mathematical Psychology, 17, 21–63.*

CARROLL, J.D., and ARABIE, P. (1980): Multidimensional scaling. *Annual Review of Psychology, 31, 607–649.*

CARROLL, J.D., and ARABIE, P. (1983): Indclus: An individual differences generalization of the Adclus model and the Mapclus algorithm. *Psychometrika, 48, 157–169.*

CARROLL, J.D., and CHANG, J.J. (1970): Analysis of individual differences in multidimensional scaling via an n–way generalization of 'Eckart–Young' decomposition. *Psychometrika, 35, 283–319.*

DE BOECK, P., and ROSENBERG, S. (1988): Hierarchical classes: Model and data analysis. *Psychometrika, 53, 361–381.*

DeSARBO, W.S. (1982): Gennclus: New models for general nonhierarchical clustering analysis. *Psychometrika, 47, 449–475.*

ECKES, T. (1991): *Strukturen der alltagssprachlichen Kategorisierung von Personen, Situationen und Person-Situations-Kombinationen.* DFG–Abschlussbericht, Universität des Saarlandes, Fachrichtung Psychologie.

ECKES, T., and ORLIK, P. (1991): An agglomerative method for two-mode hierarchical clustering. In: H.H. Bock and P. Ihm (eds.): *Classification, data analysis and knowledge organization. Models and methods with applications.* Springer, Berlin.

FISKE, S.T., and NEUBERG, S.L. (1990): A continuum of impression formation, from category-based to individuating processes: Influence of information and motivation on attention and interpretation. In: M.P. Zanna (ed.): *Advances in experimental social psychology.* Vol. 23, Academic Press, New York.

GANTER, B. (1987): Algorithmen zur Formalen Begriffsanalyse. In: B. Ganter, R. Wille, and K.E. Wolff (1987): *Beiträge zur Begriffsanalyse.* B.I.–Wissenschaftsverlag, Mannheim.

HARTIGAN, J.A. (1976): Modal blocks in definition of west coast mammals. *Systematic Zoology, 25, 149–160.*

HORST, P. (1965): *Factor analysis of data matrices.* Holt, Rinehart & Winston, New York.

KELLY, G.A. (1955): *The psychology of personal constructs.* Norton, New York.

KRUSKAL, J.B. (1984): Multilinear methods. In: H.G. Law, C.W. Snyder, J.A. Hattie, and R.P. McDonald (eds.): *Research methods for multimode data analysis.* Praeger, New York.

McCORMICK, W.T., SCHWEITZER, P.J., and WHITE, T.W. (1972): Problem decomposition and data reorganization by a clustering technique. *Operations Research, 20, 993–1009.*

ORLIK, P. (1980): Das Summax–Modell der dreimodalen Faktorenanalyse mit interpretierbarer Kernmatrix. *Archiv für Psychologie, 133,189–218.*

WILLE, R. (1982): Restructuring lattice theory: An approach based on hierarchies of concepts. In: I. Rival (ed.): *Ordered sets.* Dordrecht–Boston, Reidel.

Block-relaxation Algorithms in Statistics

Jan de Leeuw

UCLA Statistics
Mathematical Sciences Building
405 Hilgard Avenue, LA, CA 90024-1555

1. Introduction

Many algorithms in recent computational statistics are variations on a common theme. In this paper we discuss four such classes of algorithms. Or, more precisely, we discuss a single class of algorithms, and we show how some well-known classes of statistical algorithms fit in this common class. The subclasses are, in logical order,

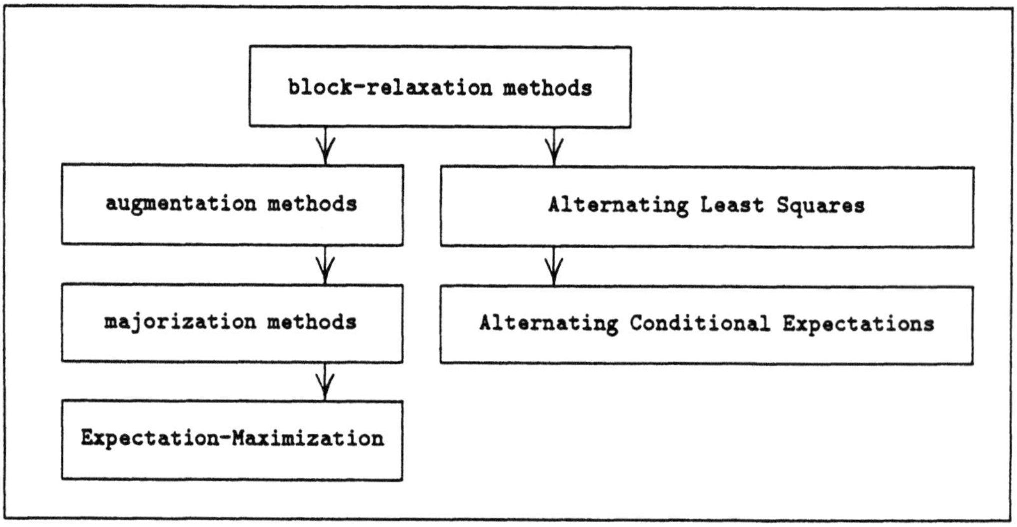

We discuss the general principles and results underlying these methods.

All the methods are special cases of what we shall call *block-relaxation methods,* although other names have also been used. There are many areas in applied mathematics where these methods have been discussed. Mostly, of course, in optimization and mathematical programming, but also in control and numerical analysis, and in differential equations. Bellman's theory of quasi-linearization [4] is closely related to what we call *augmentation* and *majorization.* We cannot give an extensive review of the literature in this paper, but a much more complete list of references is given in [12].

There is not much statistics in this paper. It is almost exclusively about deterministic optimization problems (although we shall optimize a likelihood function or two). Some of our results have been derived in the more restricted context of maximizing a likelihood function by Jensen, Johansen, and Lauritzen [21]. They develop their own results, not relying on the existing results in the optimization literature. More or less the same applies to much of the literature on convergence of the EM algorithm, starting with Dempster, Laird, and Rubin [14]. Because we want to cover a much more general class of algorithms, we need more general results than this.

One thing we shall not discuss, at least not in this version of the paper, is stochastic extensions. But of course the integrals in the majorization algorithms can be approximated by Monte Carlo, functions can be optimized by simulated annealing, and the expected value of the posterior distribution approximates the maximum likelihood estimate (and can obviously be written as an integral). Incorporating this material into this paper would take us too far astray.

2. Block relaxation

Let us thus consider the following general situation. We minimize a real-valued function ψ defined on the product-set $\Omega = \Omega_1 \otimes \Omega_2 \otimes \cdots \otimes \Omega_p$, with $\Omega_s \subseteq \mathcal{R}^{n_s}$. In order to minimize ψ over Ω we use the following iterative algorithm.

[Starter]	Start with $\omega^{(0)} \in \Omega$.
[Step k.1]	$\omega_1^{(k+1)} \in \operatorname{argmin}_{\omega_1 \in \Omega_1} \psi(\omega_1, \omega_2^{(k)}, \cdots, \omega_p^{(k)})$.
[Step k.2]	$\omega_2^{(k+1)} \in \operatorname{argmin}_{\omega_2 \in \Omega_2} \psi(\omega_1^{(k+1)}, \omega_2, \omega_3^{(k)}, \cdots, \omega_p^{(k)})$.
...	...
[Step $k.p$]	$\omega_p^{(k+1)} \in \operatorname{argmin}_{\omega_p \in \Omega_p} \psi(\omega_1^{(k+1)}, \cdots, \omega_{p-1}^{(k+1)}, \omega_p)$.
[Motor]	$k \leftarrow k+1$ and go to k.1

We assume that the minima in the substeps exist (although they need not be unique, i.e. the argmin's can be point-to-set maps). We set $\omega^{(k)} \triangleq (\omega_1^{(k)}, \cdots, \omega_p^{(k)})$, and $\psi^{(k)} \triangleq \psi(\omega^{(k)})$. Also $\Omega_0 \triangleq \{\omega \in \Omega \mid \psi(\omega) \leq \psi^{(0)}\}$. For this method we have our first (trivial) convergence theorem.

Theorem: If ψ is bounded below on Ω, then the sequence $\{\psi^{(k)}\}$ converges. If Ω_0 is compact, then $\{\omega^{(k)}\}$ has a convergent subsequence.

In the special case in which blocks consist of only one coordinate we speak of the *coordinate relaxation method* or the *cyclic coordinate descend* method. Classical papers, with applications to systems of equations, quadratic programming, and convex programming are Schechter [33], [34],[35], Hildreth, D'Esopo [15], Ortega and Rheinboldt [28], [29], Elkin [17], Céa [9], [7], [8], and Auslender [2],[3]. Many of these papers present the method as a nonlinear generalization of the Gauss-Seidel method of solving a system of linear equations. Modern papers on block-relaxation are by Abatzoglou and O'Donnell [1] and by Bezdek et al. [5]. Statistical applications to mixed linear models, with the parameters describing the mean structure collected in one block and the parameters describing the dispersion collected in the second block, are in Oberhofer and Kmenta [27]. Applications to exponential family likelihood functions, cycling over the canonical parameters, are in Jensen et al. [21].

We give a simple statistical application. Let

$$\mathcal{L}(\theta) = \sum_{k=1}^{K} n_k \log \lambda_k(\theta) - \lambda_k(\theta),$$

be a Poisson-likelihood with

$$\lambda_k(\theta) = \exp \sum_{j=1}^{m} x_{kj}\theta_j.$$

Here $\{x_{kj}\}$ is a design-type matrix, with elements equal to 0 or 1. Let

$$\mathcal{K}_j = \{k \mid x_{kj} = 1\}.$$

Then the likelihood equations are

$$\sum_{k\in\mathcal{K}_j} n_k = \sum_{k\in\mathcal{K}_j} \lambda_k(\theta).$$

Solving each of these in turn is cyclic-coordinate descent, but also the *iterative proportional fitting* algorithm. We have, using e_j for the coordinate directions,

$$\lambda_k(\theta + \tau e_j) = \begin{cases} \lambda_k(\theta) & \text{if } k \notin \mathcal{K}_j, \\ \mu\lambda_k(\theta) & \text{if } k \in \mathcal{K}_j, \end{cases}$$

with $\mu = \exp\tau$. Thus the optimal μ is simply

$$\hat{\mu} = \frac{\sum_{k\in\mathcal{K}_j} n_k}{\sum_{k\in\mathcal{K}_j} \lambda_k(\theta)}.$$

3. Generalized block-relaxation methods

If there are more than two blocks, we can move through them in various ways. In analogy with linear methods such as Gauss-Seidel and Gauss-Jacobi, we distinguish *cyclic* and *free-steering* methods. We could select the block, for instance, that seems most in need of improvement. We can pivot through the blocks (A, B, C) as $\{A, B, C, B, A, B, C, B, A, \cdots\}$ or $\{A, B, B, B, C, A, B, B, B, C, \cdots\}$. We can even choose blocks in random order.
We give a formalization of these generalizations, due to Fiorot and Huard [18]. Suppose Δ_s are p point-to-set mappings of Ω into $\mathcal{P}(\Omega)$, the set of all subsets of Ω. We suppose that $\omega \in \Delta_s(\omega)$ for all $s = 1, \cdots, p$. Also define

$$\Gamma_s(\omega) \triangleq \text{argmin}\{\psi(\varpi) \mid \varpi \in \Delta_s(\omega)\}.$$

There are now two versions of the generalized block-relaxation method which are interesting. In the free-steering version we set

$$\omega^{(k+1)} \in \cup_{s=1}^{p} \Gamma_s(\omega^{(k)}).$$

This means that we select, from the p subsets defining the possible updates, one single update before we go to the next cycle of updates. In the cyclic method we set

$$\omega^{(k+1)} \in \otimes_{s=1}^{p} \Gamma_s(\omega^{(k)}).$$

In a little bit more detail this means

$$\begin{aligned} \omega^{(k,0)} &= \omega^{(k)}, \\ \omega^{(k,1)} &\in \Gamma_s(\omega^{(k,0)}), \\ \cdots &\in \cdots, \\ \omega^{(k,p)} &\in \Gamma_s(\omega^{(k,p-1)}), \\ \omega^{(k+1)} &= \omega^{(k,p)}. \end{aligned}$$

Since $\omega \in \Delta_s(\omega)$, we see that, for both methods, if $\xi \in \Gamma(\omega)$ then $\psi(\xi) \leq \psi(\omega)$. This implies that the trivial convergence theorem above continues to apply to this generalized block relaxation method.

A simple example of the Δ_s is the following. Suppose the G_s are arbitrary mappings defined on Ω. They need not even be real-valued. Then we can set

$$\Delta_s(\omega) \stackrel{\Delta}{=} \{\xi \in \Omega \mid G_s(\xi) = G_s(\omega)\}.$$

Obviously $\omega \in \Delta_s(\omega)$ for this choice of Δ_s. There are some interesting special cases. If G_s projects on a subspace of Ω, then $\Delta(\omega)$ is the set of all ξ which project into the same point as ω. By defining the subspaces using blocks of coordinates, we recover the usual block-relaxation method discussed in the previous section. In a statistical context, in combination with the EM algorithm, functional constraints of the form $G_s(\overline{\omega}) = G_s(\omega)\}$ were used by Meng and Rubin [24].

4. Some counterexamples

She shall now strengthen our trivial convergence theorem, by imposing additional conditions on the problem. Some simple examples show that such a strengthening is necessary. We also list some examples which illustrate later results.

Convergence need not be towards a minimum. Take the function

$$\psi(\omega, \xi) = (\omega - \xi)'(\omega - \xi) - 2\omega'\xi.$$

Clearly it does not have minima (on $\omega = \xi$ we have $\psi(\omega, \omega) = -2\|\omega\|^2$). The only stationary point is the saddle $\omega = \xi = 0$, and block-relaxation convergences to that saddle from any starting point.

Convergence need not be towards a minimum, even if the function is convex. This example is from [1]. Let

$$\psi(\omega, \xi) = \max_{x \in [0,1]} \mid x^2 - \omega - \xi x \mid .$$

Start with $\xi = 0$. The optimal ω for this ξ is $1/2$. The optimal ξ for this ω is 0, which means we have convergence. But the best Chebyshev approximation to $f(x) = x^2$ is $g(x) = x + \frac{1}{8}$, and not g(x)=$1/2$.

Coordinate descend may not converge at all, even if the function is differentiable. This is a nice example, due to Powell [32]. It is somewhat surprising that Powell does not indicate what the source of the problem is, using Zangwill's convergence theory. The reason seems to be that the mathematical programming community has decided, at an early stage, that linearly convergent algorithms are not interesting and/or useful. The recent developments in statistical computing suggest that this is simply not true. Powell's example involves three variables, and the function

$$\begin{aligned} \psi(x, y, x) &= -xy - yz - zx + \\ &+ (x-1)_+^2 + (-x-1)_+^2 + (y-1)_+^2 + (-y-1)_+^2 + (z-1)_+^2 + (-z-1)_+^2, \end{aligned}$$

where

$$(x-c)_+^2 = \begin{cases} 0, & \text{if } x \leq c, \\ (x-c)^2, & \text{if } x \geq c. \end{cases}$$

Powell does not tells us that the last part of the function, with the truncated squares, is actually the squared distance of (x, y, z) to the cube $\{\pm 1, \pm 1, \pm 1\}$.

A little analysis shows that the function does not have any minima on the outside of the cube, and it also does not have minima in the interior of the cube. The only points where the derivatives vanish are saddle points. Thus the only place where there can be minima is on the surface of the cube.

Let us apply coordinate descend. A search along the x-axis finds the optimum at

$$\hat{x} = \begin{cases} +1 + {}^1\!/\!_2(y+z) & \text{if } y+z>0, \\ -1 + {}^1\!/\!_2(y+z) & \text{if } y+z<0, \\ \text{anywhere in } [-1,+1] & \text{if } y+z=0. \end{cases}$$

This guarantees that the partial derivative with respect to x is zero. The other updates are given by symmetry. Thus, if we start from $(-1-\epsilon, 1+\frac{1}{2}\epsilon, -1-\frac{1}{4}\epsilon)$, with ϵ some small positive number, then we generate the following sequence.

$$\begin{array}{lll} (+1+\frac{1}{8}\epsilon, & +1+\frac{1}{2}\epsilon, & -1-\frac{1}{4}\epsilon) \\ (+1+\frac{1}{8}\epsilon, & -1-\frac{1}{16}\epsilon, & -1-\frac{1}{4}\epsilon) \\ (+1+\frac{1}{8}\epsilon, & -1-\frac{1}{16}\epsilon, & +1+\frac{1}{32}\epsilon) \\ (-1-\frac{1}{64}\epsilon, & -1-\frac{1}{16}\epsilon, & +1+\frac{1}{32}\epsilon) \\ (-1-\frac{1}{64}\epsilon, & +1+\frac{1}{128}\epsilon, & +1+\frac{1}{32}\epsilon) \\ (-1-\frac{1}{64}\epsilon, & +1+\frac{1}{128}\epsilon, & -1-\frac{1}{256}\epsilon) \end{array}$$

But the sixth point is of the same form as the starting point, with ϵ replaced by $\frac{\epsilon}{64}$. Thus the algorithm will cycle around six edges of the cube. At these edges the gradient of the function is bounded away from zero, in fact two of the partials are zero, the other is ± 2. The function value is $+1$. The other two edges of the cube, i.e. $(+1,+1,+1)$ and $(-1,-1,-1)$ are the ones we are looking for, because there the function value is -3, the global minimum. At these two points all three partials are ± 2. Powell gives some additional examples which show the same sort of cycling behaviour, but are somewhat smoother.

Convergence can be sublinear.

$$\begin{aligned} \psi(\omega,\xi) &= (\omega-\xi)^2+\omega^4, \\ \mathcal{D}_1\psi(\omega,\xi) &= 2(\omega-\xi)+4\omega^3, \\ \mathcal{D}_2\psi(\omega,\xi) &= -2(\omega-\xi), \\ \mathcal{D}_{11}\psi(\omega,\xi) &= 2+12\omega^2, \\ \mathcal{D}_{12}\psi(\omega,\xi) &= -2, \\ \mathcal{D}_{22}\psi(\omega,\xi) &= 2. \end{aligned}$$

It follows that coordinate ascent updates $\omega^{(k)}$ by solving the cubic

$$\omega - \omega^{(k)} + 2\omega^3 = 0.$$

The sequence converges to zero, and by l'Hopitâl's rule

$$\lim_{k\to\infty} \frac{\omega^{(k+1)}}{\omega^{(k)}} = 1.$$

This leads to very slow convergence. The reason is that the matrix of second derivatives of ψ is singular at the origin.

5. Global convergence

In order to prove global convergence (i.e. convergence from any initial point) we use the general theory developed initially by Zangwill [39],[40] (and later by Polak [31], R.R. Meyer [26], G.G.L. Meyer [25], and others). The best introduction and overview is perhaps the volume edited by Huard [16].

The theory studies iterative algorithms with the following properties. An algorithm works in a space Ω. It consists of a triple $(\mathcal{A}, \psi, \mathcal{P})$, with $\mathcal{A}$ a mapping of Ω into the set of nonempty subsets of Ω, with ψ is real-valued continuous function on Ω, and with $\mathcal{P}$ a subset of Ω. We can $\mathcal{A}$ the *algorithmic map*, ψ the *evaluation function*, and $\mathcal{P}$ the *desirable points*. The algorithm works as follows.

1) start at an arbitrary $\omega^{(0)} \in \Omega$,
2) if $\omega^{(k)} \in \mathcal{P}$, then we stop,
3) otherwise we construct the *successor* by the rule $\omega^{(k+1)} \in \mathcal{A}(\omega^k)$,

We study properties of the sequences $\omega^{(k)}$ generated by the algorithm, in particular their convergence.

Theorem: (Zangwill [39]) If

- $\mathcal{A}$ is *uniformly compact* on Ω, i.e. there is a compact $\Omega_0 \subseteq \Omega$ such that $\mathcal{A}(\omega) \subseteq \Omega_0$ for all $\omega \in \Omega$,
- $\mathcal{A}$ is *upper-semicontinuous* or *closed* on $\Omega - \mathcal{P}$, i.e. if $\xi_i \in \mathcal{A}(\omega_i)$ and $\xi_i \rightarrow \xi$ and $\omega_i \rightarrow \omega$ then $\xi \in \mathcal{A}(\omega)$,
- $\mathcal{A}$ is *strictly monotonic* on $\Omega - \mathcal{P}$, i.e. $\xi \in \mathcal{A}(\omega)$ implies $\psi(\xi) < \psi(\omega)$ if ω is not a *desirable point.*

then all accumulation points of the sequence $\{\omega^{(k)}\}$ generated by the algorithm are desirable points.

Proof: Compactness implies that $\{\omega^{(k)}\}$ has a convergent subsequence. Suppose its index-set is

$$\mathcal{K} = \{k_1, k_2, \cdots\}$$

and that it converges to $\omega_{\mathcal{K}}$. Since $\{\psi(\omega^{(k)})\}$ converges to, say, ψ_∞, we see that also

$$\{\psi(\omega^{(k_1)}), \psi(\omega^{(k_2)}), \cdots\} \rightarrow \psi_\infty.$$

Now consider $\{\omega^{(k_1+1)}, \omega^{(k_2+1)}, \cdots\}$, which must again have a convergent subsequence. Suppose its index-set is $\mathcal{L} = \{\ell_1 + 1, \ell_2 + 1, \cdots\}$ and that it converges to $\omega_{\mathcal{L}}$. Then $\psi(\omega_{\mathcal{K}}) = \psi(\omega_{\mathcal{L}}) = \psi_\infty$.

Assume $\omega_{\mathcal{K}}$ is not a fixed point. Now

$$\{\omega^{(\ell_1)}, \omega^{(\ell_2)}, \cdots\} \rightarrow \omega_{\mathcal{K}}$$

and

$$\{\omega^{(\ell_1+1)}, \omega^{(\ell_2+1)}, \cdots\} \rightarrow \omega_{\mathcal{L}},$$

with $\omega^{(\ell_j+1)} \in \mathcal{A}(\omega^{(\ell_j+1)})$. Thus, by usc, $\omega_{\mathcal{L}} \in \mathcal{A}(\omega_{\mathcal{K}})$. If $\omega_{\mathcal{K}}$ is not a fixed point, then strict monotonicity gives $\psi(\omega_{\mathcal{L}}) < \psi(\omega_{\mathcal{K}})$, which contradicts our earlier $\psi(\omega_{\mathcal{K}}) = \psi(\omega_{\mathcal{L}})$. **Q.E.D.**

The concept of closedness of a map can be illustrated with the following picture, showing a map which is not closed at at least one point.

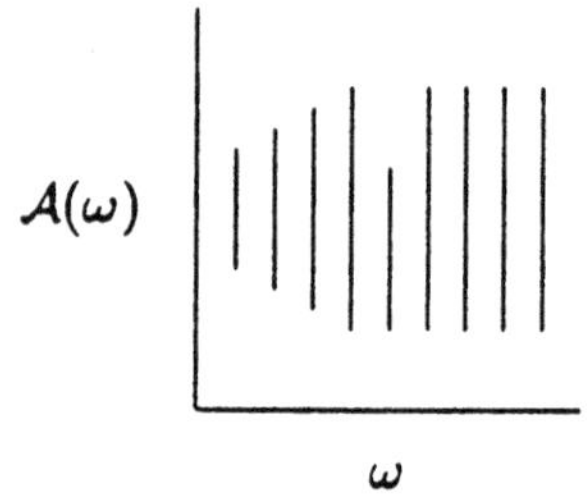

We have already seen another example: Powell's coordinate descend example shows that the algorithm map is not closed at six of the edges of the cube $\{\pm 1, \pm 1, \pm 1\}$.

It is easy to see that desirable points are generalized fixed points, in the sense that $\omega \in \mathcal{P}$ is equivalent to that $\omega \in \mathcal{A}(\omega)$. According to Zangwill's theorem each accumulation point is a generalized fixed point. This, however, does not prove convergence, because there can be many accumulation points. If we redefine fixed points as points such that $\mathcal{A}(x) = \{x\}$, then we can strengthen the theorem.

Theorem: (Meyer, [26]) Suppose the conditions of Zangwill's theorem are satisfied for the stronger definition of a fixed point, i.e. $\xi \in \mathcal{A}(\omega)$ implies $\psi(\xi) < \psi(\omega)$ if ω is not a fixed point, then in addition to what we had before $\{\omega^{(k)}\}$ is *asymptotically regular,* i.e.

$$\|\omega^{(k)} - \omega^{(k+1)}\| \rightarrow 0.$$

Proof: Use the notation in the proof of Zangwill's theorem. Suppose $\|\omega^{(\ell_i+1)} - \omega^{(\ell_i)}\| > \delta > 0$. Then $\|\omega_{\mathcal{L}} - \omega_{\mathcal{K}}\| \geq \delta$. But $\omega_{\mathcal{K}}$ is a fixed point (in the strong sense) and thus $\omega_{\mathcal{L}} \in \mathcal{A}(\omega_{\mathcal{K}}) = \{\omega_{\mathcal{K}}\}$, a contradiction. **Q.E.D.**

It follows (from a result of Ostrowski [30]) that either $\{\omega^{(k)}\}$ converges, or $\{\omega^{(k)}\}$ has a continuum of accumulation points (all with the same function value). This is still not actual convergence, but it is close enough for all practical purposes.

6. Global convergence of block methods

We can now apply this theory to block-relaxation methods. We concentrate on the cyclic methods. The free-steering methods are interesting, but inherently more complicated. Details on free-steering can be found in [18]. Obviously block-relaxation is monotonic if we choose the evaluation function equal to the function we are minimizing, and if we assume that the minima exist. If we assume that the minima of the subproblems are always unique (for instance, if they are least squares projections on convex sets), then Meyer's theorem applies. Actually, we have the following result for generalized block methods.

Theorem: (Fiorot and Huard, [18]) If

- $\omega \in \Delta_s(\omega)$ for all ω and s,
- Δ_s is continuous on Ω, i.e. both upper-semicontinuous and lower-semicontinuous,
- ψ has a unique minimum over $\Delta_s(\omega)$ for all ω and s,
- $\Omega_0 = \{\omega \in \Omega \mid \psi(\omega) \leq \psi(\omega^{(0)})\}$ is compact,

then

- the sequence $\omega^{(k)}$ is asymptotically regular,
- each accumulation point of the sequence is a fixed point of each of the Γ_s.

A fixed point $(\omega_1, \cdots, \omega_p)$ is by definition a point such that ω_s is the unique minimum of $\psi(\omega_1, \cdots, \omega_{s-1}, \omega, \omega_{s+1}, \cdots, \omega_p)$ over $\omega \in \Omega_s$ for all s. This does not imply that the point is a local minimum of ψ on Ω unless we impose extra conditions such as convexity. Actually, convexity is not enough, as the Chebyshev approximation example in section 4 shows.

If we drop the assumption that the partial minima of the subproblems are unique (which is of course basically an identification condition, similar to assumptions needed for consistency) then fixed points must be replaced by generalized fixed points. Also, accumulation points are no longer generalized fixed points of *all* Γ_s. In fact each accumulation point ω_∞ has an associated index set $S(\omega_\infty)$ such that $s \in S(\omega_\infty)$ if the operation of maximizing over Δ_s occurs an infinite number of times in the subsequence. For the six edges in the Powell example, these index sets consist of a single element.

Theorem: ((Fiorot and Huard, [18]) If

- $\omega \in \Delta_s(\omega)$ for all ω and s,
- Δ_s is continuous on Ω, i.e. both upper-semicontinuous and lower-semicontinuous,
- if $\xi \in \Delta_s(\omega)$ then $\Delta_s(\xi) = \Delta_s(\omega)$,
- $\Omega_0 = \{\omega \in \Omega \mid \psi(\omega) \leq \psi(\omega^{(0)})\}$ is compact,

then for every $s \in S(\omega_\infty)$ we have $\omega_\infty \in \Gamma_s(\omega_\infty)$ and $\omega_\infty \in \Gamma_{s+1}(\omega_\infty)$.

7. Quantitative convergence theory

We now switch from the qualitative or global theory of convergence to the quantitative or local theory. We look into the question of convergence speed. To get this more specific information on convergence, we again have to make stronger assumptions. To be able to compute the rate, we need to be able to differentiate ψ sufficiently many times. Also, the solution of the subproblems needs to be unique in a neighborhood of the true value. Thus we forget all references to point-to-set maps, and to free-steering, because our techniques here simply cannot cope with that much freedom. The basic result we use is due to Ostrowski [30].

Theorem: If

- the iterative algorithm $\omega^{(k+1)} = \mathcal{A}(x^{(k)})$, converges to ω_∞,
- $\mathcal{A}$ is differentiable at ω_∞,
- $0 < \rho = \|\mathcal{D}A(\omega_\infty)\| < 1$,

then the algorithm is linearly convergent with rate ρ.

The norm in the theorem is the spectral norm, i.e. the modulus of the maximum eigenvalue. Let us call the derivative of A the *iteration matrix* and write it as $\mathcal{M}$. In general block relaxation methods have linear convergence, and the linear convergence can be quite slow. In cases where the accumulation points are a continuum we usually have sublinear rates. The same things is true if the local minimum is not strict, or if we are converging to a saddle point.

In order to study the rate of convergence of block relaxation, we study the nonlinear system

$$\begin{aligned}\mathcal{D}_1(\omega_1,\xi_2,\xi_3,\cdots,\xi_p) &= 0,\\ \mathcal{D}_2(\omega_1,\omega_2,\xi_3,\cdots,\xi_p) &= 0,\\ \cdots &= \cdots,\\ \mathcal{D}_p(\omega_1,\omega_2,\omega_3,\cdots,\omega_p) &= 0,\end{aligned}$$

which defines the new solution ω in terms of the old solution ξ. The $\mathcal{D}_s$ are the partials of ψ with repect to the blocks. We assume that the assumptions for the implicit function theorem are satisfied at the solution. Differentiating these equations again, and solving for the derivatives, we find the iteration matrix

$$\mathcal{M} = -\begin{pmatrix}\mathcal{D}_{11} & 0 & 0 & \cdots & 0\\ \mathcal{D}_{21} & \mathcal{D}_{22} & 0 & \cdots & 0\\ \cdots & \cdots & \cdots & \cdots & \cdots\\ \mathcal{D}_{s1} & \mathcal{D}_{s2} & \mathcal{D}_{s3} & \cdots & \mathcal{D}_{ss}\end{pmatrix}^{-1}\begin{pmatrix}0 & \mathcal{D}_{12} & \mathcal{D}_{13} & \cdots & \mathcal{D}_{1s}\\ 0 & 0 & \mathcal{D}_{23} & \cdots & \mathcal{D}_{2s}\\ \cdots & \cdots & \cdots & \cdots & \cdots\\ 0 & 0 & 0 & \cdots & 0\end{pmatrix}.$$

If there are only two blocks this simplifies to

$$\mathcal{M} = -\begin{pmatrix}\mathcal{D}_{11}^{-1} & 0\\ -\mathcal{D}_{22}^{-1}\mathcal{D}_{21}\mathcal{D}_{11}^{-1} & \mathcal{D}_{22}^{-1}\end{pmatrix}\begin{pmatrix}0 & \mathcal{D}_{12}\\ 0 & 0\end{pmatrix} = \begin{pmatrix}0 & -\mathcal{D}_{11}^{-1}\mathcal{D}_{12}\\ 0 & \mathcal{D}_{22}^{-1}\mathcal{D}_{21}\mathcal{D}_{11}^{-1}\mathcal{D}_{12}\end{pmatrix}.$$

Thus, in a local minimum, we find that the largest eigenvalue of $\mathcal{M}$ is the largest squared canonical correlation ρ of the two sets of variables, and is consequently less than or equal to one. We also see that a sufficient condition for local convergence to a stationary point of the algorithm is that $\rho < 1$. This precludes having more than one accumulation point, and it is always true for an isolated local minimum. If $\mathcal{D}^2\psi$ is singular at the solution, we find a canonical correlation equal to $+1$, and we do not have linear convergence. Similar calculations can also be carried out in the case of constrained optimization, i.e. when the subproblems optimize over differentiable manifolds. We then use the implicit function calculations on the Langrangean conditions, which makes them a bit more complicated, but essentially the same.

The result for block-relaxation can also derived from a similar result for generalized block relation, that has been used in an EM context by Meng [23]. We minimize ψ over ω under the condition that $G_s(\omega) = G_s(\xi)$, where ξ is the current solution. Once again we can differentiate the stationary equations to find that

$$\frac{\partial\omega}{\partial\xi} = T_s^{-1}H_s'(H_sT_s^{-1}H_s')^{-1}H_s,$$

where H_s is the Jacobian of G_s at the solution, and where

$$T_s = \mathcal{D}^2\psi + \sum_{r=1}^{m}\lambda_{sr}\mathcal{D}^2 g_{rs}.$$

Here g_{rs} is the $r-$th restriction in the $s-$th system, and the λ_{sr} are the corresponding Lagrange multipliers. If the G_s are linear, the second term disappears, and all T_s are equal to the Hessian of ψ at the solution. If we use a generalized block method that cycles over the constraints G_s, then the matrix we need to find the spectral norm of is simply

$$\mathcal{M} = \prod_{s=1}^{p} T_s^{-1} H_s' (H_s T_s^{-1} H_s')^{-1} H_s.$$

In the case of ordinary block relaxation the G_s are linear, because they are the indicator matrices selecting the blocks that do not change in a subproblem. For the first subproblem $G_1 = (0 \mid \mathcal{I})$, and we find

$$\mathcal{M}_1 = \begin{pmatrix} 0 & \mathcal{D}^{12}(\mathcal{D}^{22})^{-1} \\ 0 & \mathcal{I} \end{pmatrix},$$

with the $\mathcal{D}^{st}$ the blocks of the inverse of $\mathcal{D}^2\psi$. If we substitute the G_s for the H_s, we find an alternative expression for the iteration matrix as a product of simpler matrices.

8. Alternating least squares

We now go into the history of block-relation in statistics and data analysis. Alternating Least Squares (ALS) methods were first used systematically in *Optimal Scaling* (OS). Optimal scaling is discussed in detail in the book by Gifi [19]. We only give a brief introduction here.

Suppose we have n observations on two sets of variables x_i and y_i. We want to fit a model of the form

$$F_\theta(\Phi(x_i)) \approx G_\xi(\Psi(y_i))$$

where the unknowns are the structural parameters θ and ξ and the transformations Φ and Ψ. In ALS we measure loss-of-fit by

$$\sigma(\theta, \xi, \Phi, \Psi) = \sum_{i=1}^{n} [F_\theta(\Phi(x_i)) - G_\xi(\Psi(y_i))]^2$$

This loss function is minimized by starting with initial estimates for the transformations, minimizing over the structural parameters, keeping the transformations fixed at their current values, and then minimizing over the transformations, with structural values kept fixed at their new values. These two minimizations are alternated, which produces a nonincreasing sequence of loss function values, bounded below by zero, and thus convergent. This is a version of the trivial convergence theorem.

The first ALS example is due to Kruskal [22]. We have a factorial ANOVA, with, say, two factors, and we minimize

$$\sigma(\phi, \mu, \alpha, \beta) = \sum_{i=1}^{n} \sum_{j=1}^{m} [\phi(y_{ij}) - (\mu + \alpha_i + \beta_j)]^2.$$

Kruskal required ϕ to be monotonic. Minimizing loss for fixed ϕ is just doing an analysis of variance, minimizing loss over ϕ for fixed μ, α, β is doing a *monotone regression*. Obviously also some normalization reuqirement is needed to exclude trivial zero solutions.

This general idea was extended by De Leeuw, Young, Takane around 1975 to

$$\sigma(\phi; \psi_1, \cdots, \psi_m) = \sum_{i=1}^{n} [\phi(y_i) - \sum_{s=1}^{p} \psi_j(x_{ij})]^2.$$

This ALSOS work, in the period 1975-1980, is summarized in [38]. Subsequent work, culminating in the book by Gifi [19], generalized this to ALSOS versions of principal component analysis, path analysis, canonical analysis, discriminant analysis, MANOVA, and so on. The classes of transformations over which loss was minimized were usually step-functions, splines, monotone functions, or low-degree polynomials. To illustrate the use of more sets in ALS, consider

$$\sigma(\psi_1,\cdots,\psi_m;\alpha,\beta)=\sum_{i=1}^{n}\sum_{j=1}^{m}(\psi_j(x_{ij})-\sum_{s=1}^{p}\alpha_{is}\beta_{js})^2.$$

This is principal component analysis (or partial singular value decomposition) with optimal scaling. We can now cycle over three sets, the transformations, the component scores α_{is} and the component loadings β_{js}. In the case of monotone transformations this alternates monotone regression with two linear least squares problems.

The ACE methods, developed by Breiman and Friedman [6], "minimize" over all "smooth" functions. A problem with ACE is that smoothers, at least most smoothers, do not really minimize a loss function (except for perfect data). In any case, ACE is less general than ALS, because not all least squares problems can be interpreted as computing conditional expectations. Another obviously related area in statistics is the Generalized Additive Models discussed extensively by Hastie and Tibshirani [20].

It is easy to apply the general results from the previous sections to ALS. The results show that it is important that the solutions to the subproblems are unique. The least squares loss function has some special structure in its second derivatives which we can often exploit in a detailed analysis. If

$$\sigma(\omega,\xi)=\sum_{i=1}^{n}(f_i(\omega)-g_i(\xi))^2,$$

then

$$\mathcal{D}^2\sigma=\begin{pmatrix}S_1 & 0\\ 0 & S_2\end{pmatrix}+\begin{pmatrix}G'G & -G'H\\ -H'G & H'H\end{pmatrix},$$

with G and H the Jacobians of f and g, and with S_1 and S_2 weighted sums of the Hessians of the f_i and g_i, with weights equal to the least squares residuals at the solution. If S_1 and S_2 are small, because the residuals are small, or because the f_i and g_i are linear or almost linear, we see that the rate of ALS will be the canonical correlation between G and H.

9. Augmentation methods

We take up the historical developments. Alternating Least Squares was useful for many problems, but it some cases it was not powerful enough to do the job. In order to solve some additional least squares problems, we can use *augmentation*. We first illustrate this with some examples.

If we want to fit a factorial ANOVA model to an unbalanced two-factor design, we minimize

$$\sigma(\mu,\alpha,\beta)=\sum_{i=1}^{I}\sum_{j=1}^{J}\sum_{k=1}^{K}w_{ijk}(y_{ijk}-(\mu+\alpha_i+\beta_j))^2,$$

where the weights w_{ijk} are either one (there) or zero (not there). Instead of this we can also minimize

$$\sigma(\mu,\alpha,\beta,z)=\sum_{i=1}^{I}\sum_{j=1}^{J}\sum_{k=1}^{K}(z_{ijk}-(\mu+\alpha_i+\beta_j))^2,$$

with

$$z_{ijk} = \begin{cases} y_{ijk}, & \text{if } w_{ijk} = 1 \\ \text{free}, & \text{otherwise.} \end{cases}$$

Minimizing this by ALS is due to Yates an others, see Wilkinson [37] for references. Augmentation reduces the fitting to the balanced case (where we can simply use row, column, and cell means), with an additional step to *impute* the missing y_{ijk}. The idea of adding variables that augment the problem to a simpler one is very general. It is also at the basis, for instance, of the Lagrange multiplier method.

In LS factor analysis we want to minimize

$$\sigma(A) = \sum_{i=1}^{m} \sum_{j=1}^{m} w_{ij}(r_{ij} - \sum_{s=1}^{p} a_{is}a_{js})^2,$$

with

$$w_{ij} = \begin{cases} 0, & \text{if } i = j, \\ 1, & \text{if } i \neq j. \end{cases}$$

We augment by adding the *communalities,* i.e. the diagonal elements of R as variables, and by using ALS over A and the communalities. For a complete R, minimizing over A just means computing the p dominant eigenvalues-eigenvectors. This algorithm dates back to the thirties, were it was proposed by Thomson and others.

A final example, less trivial in a sense. Suppose we want to minimize

$$\sigma(X) = \sum_{i=1}^{m} \sum_{j=1}^{m} (\delta_{ij} - d_{ij}^2(X))^2,$$

with $d_{ij}^2(X) = (x_i - x_j)'(x_i - x_j)$ squared Euclidean distance. This can be augmented to

$$\sigma(X, \eta) = \sum_{i=1}^{m} \sum_{j=1}^{m} \sum_{k=1}^{m} \sum_{\ell=1}^{m} (\eta_{ijk\ell} - (x_i - x_j)'(x_k - x_\ell))^2,$$

where of course $\eta_{ijij} = \delta_{ij}$ and the others are free. After some computation, ALS again leads to a sequence of eigenvalue-eigenvector problems. This shows that augmentation is an art (like integration). The augmentation is in some cases not obvious, and there are no mechanical rules.

Formalizing augmentation is easy. Suppose ϕ is a real valued function, defined for all $\omega \in \Omega$, where $\Omega \subseteq \mathcal{R}^n$. Suppose there exists another real valued function ψ, defined on $\Omega \times \Xi$, where $\Xi \subseteq \mathcal{R}^m$, such that

$$\phi(\theta) = \min\{\psi(\theta, \xi) \mid \xi \in \Xi\}.$$

We also suppose that minimizing ϕ over Θ is *hard,* while minimizing ψ over Θ is *easy* for all $\xi \in \Xi$. And we suppose that minimizing ψ over $\xi \in \Xi$ is also *easy* for all $\theta \in \Theta$. This last assumption is not too far-fatched, because we already know what the value at the minimum is.

I am not going to define *hard* and *easy.* What may be easy for you, may be hard for me. Anyway, by augmenting the function we are in the block-relaxation situation again, and we can apply our general results on global convergence and linear convergence. Augmentation is used in other areas of statistics [36], where integration is used instead of minimization. If it is difficult to sample from $p(\omega)$ and easy to sample from $p(\omega, \xi)$, then we sample from the joint distribution and integrate out the ξ by summation.

We give another, more serious, example from the area of mixed-model fitting. This is from a paper of De Leeuw and Liu [13], which describes the algorithm in detail. We simply give a list of results that show augmentation at work.

Lemma: If $A = B + TCT'$, with $B, C > 0$,

$$y'A^{-1}y = \min_x (y - Tx)'B^{-1}(y - Tx) + x'C^{-1}x.$$

Lemma: If $A = B + TCT'$, with $B, C > 0$,

$$\log | A | = \log | B | + \log | C | + \log | C^{-1} + T'B^{-1}T | .$$

Theorem: If $A = B + TCT'$, then

$$\begin{aligned} \log | A | + y'A^{-1}y &= \min_x \log | B | + \log | C | + \\ &+ \log | C^{-1} + T'B^{-1}T | + \\ &+ (y - Tx)'B^{-1}(y - Tx) + x'C^{-1}x. \end{aligned}$$

Lemma: If $T > 0$, then

$$\log | T | = \min_{S>0} \log | S | + \operatorname{tr} S^{-1}T - p,$$

with the unique minimum attained at $S = T$.

Theorem:

$$\begin{aligned} \log | A | + y'A^{-1}y &= \min_{x,S>0} \log | B | + \log | C | + \\ &+ \log | S | + \operatorname{tr} S^{-1}(C^{-1} + T'B^{-1}T) + \\ &+ (y - Tx)'B^{-1}(y - Tx) + x'C^{-1}x. \end{aligned}$$

Minimize over x, S, B, C using block-relaxation. The minimizers are

$$\begin{aligned} S &= C^{-1} + T'B^{-1}T, \\ C &= S^{-1} + xx', \\ B &= TS^{-1}T' + (y - Tx)(y - Tx)', \\ x &= (T'B^{-1}T + C^{-1})^{-1}T'B^{-1}y. \end{aligned}$$

10. Majorization methods

The next step (history again) was to find *systematic ways* to do augmentation (which is an art, remember). We start with examples. The first is an algorithm for MDS, developed by De Leeuw [10]. We want to minimize

$$\sigma(X) = \sum_{i=1}^{m} \sum_{j=1}^{m} w_{ij}(\delta_{ij} - d_{ij}(X))^2,$$

with $d_{ij}(X)$ again Euclidean distance, i.e. $d_{ij}(X) = \sqrt{(x_i - x_j)'(x_i - x_j)}$, and thus, by Cauchy-Schwarz,

$$d_{ij}(X) \geq \frac{(x_i - x_j)'(y_i - y_j)}{d_{ij}(Y)}.$$

This implies

$$\begin{aligned} \sigma \quad (X) \leq \eta(X,Y) \triangleq \sum_{i=1}^{m}\sum_{j=1}^{m} w_{ij}\delta_{ij}^2 - \\ + \; 2\sum_{i=1}^{m}\sum_{j=1}^{m} w_{ij}\frac{\delta_{ij}}{d_{ij}(Y)}(x_i - x_j)'(y_i - y_j) + \\ + \; \sum_{i=1}^{m}\sum_{j=1}^{m} w_{ij} d_{ij}(X)^2. \end{aligned}$$

Here is another example: Suppose we want to maximize $\phi(\omega) = \log \int \eta(\omega, x)dx$. By Jensen's inequality

$$\begin{aligned} \log \quad & \frac{\int \eta(\omega,x)dx}{\int \eta(\xi,x)dx} = \log \frac{\int \eta(\xi,x)\frac{\eta(\omega,x)}{\eta(\xi,x)}dx}{\int \eta(\xi,x)dx} \geq \\ \geq \quad & \frac{\int \eta(\xi,x)\log\frac{\eta(\omega,x)}{\eta(\xi,x)}dx}{\int \eta(\xi,x)dx} = \\ = \quad & \frac{\int \eta(\xi,x)\log\eta(\omega,x)dx}{\int \eta(\xi,x)dx} - \frac{\int \eta(\xi,x)\log\eta(\xi,x)dx}{\int \eta(\xi,x)dx}. \end{aligned}$$

It follows that

$$\phi(\omega) \geq \phi(\xi) + \kappa(\omega,\xi) - \kappa(\xi,\xi),$$

Maximizing the right-hand-side by block relaxation is the EM algorithm [14].

As before, we now stop and wonder what these two examples have in common. We have a function $\phi(\omega)$ on Ω, and a function $\psi(\omega,\xi)$ on $\Omega \otimes \Omega$ such that

$$\begin{aligned} \phi(\omega) &\leq \psi(\omega,\xi) \quad \forall \omega, \xi \in \Omega, \\ \phi(\omega) &= \psi(\omega,\omega) \quad \forall \omega \in \Omega. \end{aligned}$$

This is just another way of saying

$$\phi(\omega) = \min_{\xi\in\Omega} \psi(\omega,\xi),$$

and thus we are in the ordinary block relaxation situation. We say that ψ *majorizes* ϕ, and we call the block relaxation algorithm corresponding with a particular majorization function a *majorization algorithm*. It is a special case of our previous theory, because $\Omega = \Xi$ and because $\xi(\omega) = \omega$. This implies that $cD_2(\omega,\omega) = 0$ for all ω, and consequently $\mathcal{D}_{12} = -\mathcal{D}_{22}$. Thus $\mathcal{M} = -\mathcal{D}_{11}^{-1}\mathcal{D}_{12}$. The E-step of the EM algorithm, in our terminology, is the construction of a new majorization function. We prefer a nonstochastic description of EM, because maximizing integrals is obviously a more general problem.

Again, to some extent, finding a majorization function is an art. Many of the classical inequalities can be used (Cauchy-Schwarz, Jensen, Hölder, AM-GM, and so on). Here are some systematic ways to find majorizing functions.

1) If ϕ is concave, then $\phi(\omega) \leq \phi(\xi) + \eta'(\omega - \xi)$, with $\eta \in \partial\phi(\xi)$, the subgradient of ϕ at ξ. Thus concave functions have a linear majorizer.

2) If $\mathcal{D}^2\phi(\xi) \leq D$ for all $\xi \in \Omega$, then

$$\phi(\omega) \leq \phi(\xi) + (\omega - \xi)'\nabla\phi(\xi) + {}^1\!/\!_2(\omega - \xi)'D(\omega - \xi).$$

Let $\eta(\xi) = \xi - D^{-1}\nabla\phi(\xi)$, then

$$\begin{aligned}\phi(\omega) &\leq \phi(\xi) - {}^1\!/\!_2\nabla\phi(\xi)'D^{-1}\nabla\phi(\xi) + \\ &+ {}^1\!/\!_2(\omega - \eta(\xi))'D(\omega - \eta(\xi)).\end{aligned}$$

Thus here we have quadratic majorizers.

3) For d.c. functions (differences of convex functions) such as $\phi = \alpha - \beta$ we can write $\phi(\omega) \leq \alpha(\omega) - \beta(\xi) - \eta'(\omega - \xi)$, with $\eta \in \partial\beta(\xi)$. This gives a convex majorizer. Interesting, because basically all continuous functions are d.c.

We close with a final example. Suppose ψ is a convex and differentiable function defined on the space of all correlation matrices R between m random variables $x_1, \cdots, x_m$. Suppose we want to maximize $\psi(R(\eta_1(x_1), \cdots, \eta_m(x_m)))$ over all transformations η_j. Now

$$\psi(R) \geq \psi(S) + \text{ tr } \nabla\psi(S)'(R - S).$$

Collect the gradient in the matrix G. A majorization algorithm can maximize

$$\sum_{i=1}^{m}\sum_{j=1}^{m} g_{ij}(S)\mathbf{E}\,(\eta_i\eta_j),$$

over all standardized transformations, which we do with block relaxation using m blocks. In each block we must maximize a linear function under a quadratic constraint (unit variance), which is usually very easy to do. This algorithm generalizes ACE, CA, and many other forms of MVA with OS. It was proposed first by De Leeuw [11], with many variations. The function ψ can be based on multiple correlations, eigenvalues, determinants, and so on.

References

[1] T . Abatzoglou and B. O'Donnell. Minimization by coordinate descent. *Journal of Optimization Theory and Applications*, 36:163–174, 1982.

[2] A. Auslender. Méthodes numériques pour la décomposition et la minimisation de fonctions non differentiables. *Numerische Mathematik*, 18:213–223, 1971.

[3] A. Auslender and B. Martinet. Méthodes de decomposition pour la minimisation d'une fonctionelle sur un espace produit. *Comptes Rendus Académie Sciences Paris*, 274:632–635, 1972.

[4] R. E. Bellman and R. E. Kalaba. *Quasilinearization and nonlinear boundary-value problems*. RAND Corporation, Santa Monica, CA, 1965.

[5] J. C. Bezdek, R. J. Hathaway, R. E. Howard, C. A. Wilson, and M. P. Windham. Local convergence analysis of a grouped variable version of coordinate descend. *Journal of Optimization Theory and Applications*, 54:471–477, 1987.

[6] L. Breiman and J. H. Friedman. Estimating optimal transformations for multiple regression and correlation. *Journal of the American Statistical Association*, 80:580–619, 1985.

[7] J. Céa. Les méthodes de "descente" dans la theorie de l'optimisation. *Revue Francaise d'Automatique, d'Informatique et de Recherche Opérationelle*, 2:79–102, 1968.

[8] J. Céa. Recherche numérique d'un optimum dans un espace produit. In *Colloquium on Methods of Optimization*, Lecture notes im mathematics, Berlin, Germany, 1970. Springer-Verlag.

[9] J. Céa and R. Glowinski. Sur les méthodes d'optimisation par rélaxation. *Revue Francaise d'Automatique, d'Informatique et de Recherche Opérationelle*, 7:5–32, 1973.

[10] J. de Leeuw. Applications of convex analysis to multidimensional scaling. In B. van Cutsem et al., editor, *Recent advantages in Statistics*, Amsterdam, Netherlands, 1977. North Holland Publishing Company.

[11] J. de Leeuw. Multivariate analysis with optimal scaling. In S. Das Gupta and J. Sethuraman, editors, *Progress in Multivariate Analysis*, Calcutta, India, 1990. Indian Statistical Institute.

[12] J. de Leeuw. Block-relaxation methods in statistical computation. Preprint, UCLA Statistics, Los Angeles, CA, 1993.

[13] J. de Leeuw and G. Liu. Augmentation methods for mixed model fitting. Preprint, UCLA Statistics, Los Angeles, CA, 1993.

[14] A. P. Dempster, N. M. Laird, and D. B. Rubin. Maximum Likelihood for Incomplete Data via the EM Algorithm. *Journal of the Royal Statistical Society*, B39:1–38, 1977.

[15] D. A. D'Esopo. A convex programming procedure. *Naval Research Logistic Quarterly*, 6:33–42, 1959.

[16] P. Huard (ed). *Point-to-set maps and mathematical programming*. Mathematical Programming Study #10. North Holand Publishing Company, Amsterdam, Netherlands, 1979.

[17] R. M. Elkin. Convergence theorems for Gauss-Seidel and other minimization algorithms. Technical Report 68-59, Computer Sciences Center, University of Maryland, College Park, MD, 1968.

[18] J. Ch. Fiorot and P. Huard. Composition and union of general algorithms of optimization. *Mathematical Programming Study*, 10:69–85, 1979.

[19] A. Gifi. *Nonlinear multivariate analysis*. Wiley, Chichester, England, 1990.

[20] T. J. Hastie and R. J. Tibshirani. *Generalized Additive Models*. Chapman and Hall, London, England, 1990.

[21] S. T. Jensen, S. Johansen, and S. L. Lauritzen. Globally convergent algorithms for maximizing a likelihood function. *Biometrika*, 78:867–877, 1991.

[22] J. B. Kruskal. Analysis of factorial experiments by estimating monotone transformations of the data. *Journal of the Royal Statistical Society*, B27:251–263, 1965.

[23] X.-L. Meng. On the rate of convergence of the ECM algorithm. Technical report, Department of Statistics, University of Chicago, Chicago, IL, 1993.

[24] X.-L. Meng and D.B. Rubin. Maximum likelihood estimation via the ECM algorithm. *Biometrika*, 80, 1993. In press.

[25] G. G. L. Meyer. A systematic approach to the synthesis of algorithms. *Numerische Mathematik*, 24:277–289, 1975.

[26] R. R. Meyer. Sufficient conditions for the convergence of monotonic mathematical programming algorithms. *Journal of Computer and System Sciences*, 12:108–121, 1976.

[27] W. Oberhofer and J. Kmenta. A general procedure for obtaining maximum likelihood estimates in generalized regression models. *Econometrica*, 42:579–590, 1974.

[28] J. M. Ortega and W. C. Rheinboldt. Monotone iterations for nonlinear equations with application to Gauss-Seidel methods. *SIAM Journal of Numerical Analysis*, 4:171–190, 1967.

[29] J. M. Ortega and W. C. Rheinboldt. Local and global convergence of generalized linear iterations. In J. M. Ortega and W. C. Rheinboldt, editors, *Numerical solution of nonlinear problems.* Society of Inductrial and Applied Mathematics, Philadelphia, PA, 1970.

[30] A. M. Ostrowski. *Solution of Equations and Systems of Equations.* Academic Press, New York, N.Y., 1966.

[31] E. Polak. On the convergence of optimization algorithms. *Revue Francaise d'Automatique, d'Informatique et de Recherche Opérationelle*, 3:17–34, 1969.

[32] M. J. D. Powell. On search directions for minimization algorithms. *Mathematical Programming*, 4:193–201, 1973.

[33] S. Schechter. Iteration methods for nonlinear problems. *Transactions American Mathematical Society*, 104:179–189, 1962.

[34] S. Schechter. Relaxation methods for convex problems. *SIAM Journal Numerical Analysis*, 5:601–612, 1968.

[35] S. Schechter. Minimization of a convex function by relaxation. In J. Abadie, editor, *Integer and nonlinear programming.* North Holland Publishing Company, Amsterdam, Netherlands, 1970.

[36] M. A. Tanner. *Tools for statistical Inference. Observed data and data augmentation methods.* Lecture notes in statistics, #10. Spinger-Verlag, New York, N.Y., 1991.

[37] G. N. Wilkinson. Estimation of missing values for the analysis of incomplete data. *Biometrics*, 14:257–286, 1958.

[38] F. W. Young. Quantitative analysis of qualitative data. *Psychometrika*, 46:357–388, 1981.

[39] W. I. Zangwill. Convergence conditions for nonlinear programming algorithms. *Management Science*, 16:1–13, 1969.

[40] W. I. Zangwill. *Nonlinear Programming: a Unified Approach.* Prentice-Hall, Englewood-Cliffs, N.J., 1969.

Multidimensional Scaling with ℓ_p–Distances, a Unifying Approach

Rudolf Mathar

Institut für Statistik, RWTH Aachen,
Wüllnerstraße 3, D-5100 Aachen, Germany

Summary: In a unifying approach, this note deals with three different methods to find the best embedding of n objects in an ℓ_p–space, $p \geq 1$, if only pairwise dissimilarities are given and fitting is measured by weighted least squares. The procedures are based on (1) a nested algorithm with an inner linear constrained problem, (2) a generalized eigenvector procedure resembling inverse iteration, and (3) majorization. All resulting algorithms are quite similar, though the optimization problem is approached from different viewpoints. This paper explains why, by interpreting (2) and (3) as a relaxed version of a first order subgradient method.

1. Introduction

Assume that pairwise dissimilarities δ_{ij}, $1 \leq i, j \leq n$, are given for a set of n objects or individuals. $\delta_{ij} = \delta_{ji} \geq 0$ are supposed to be symmetric, nonnegative, and $\delta_{ii} = 0$ for all $i, j = 1, \ldots, n$. A basic problem in Multidimensional Scaling (MDS) is to find n points $x_1, \ldots, x_n$ in a metric space $(\mathcal{X}, d)$ such that the interpoint distances fit the given dissimilarities as well as possible. This means to determine a configuration $\mathbf{X} = (x_1, \ldots, x_n) \in \mathcal{X}^n$ which represents the given data under a metric structure. Most times $\mathcal{X} = \mathbb{R}^k$ is chosen, for some fixed dimension k, such that configurations may be represented as $(n \times k)$ matrices $\mathbf{X} = (x_1, \ldots, x_n)' \in \mathbb{R}^{n \times k}$. We will concentrate on this case, and furthermore on Minkowski ℓ_p–metrics, $p \geq 1$, i.e.

$$d(a, b) = \Big(\sum_{\ell=1}^{k} |a_\ell - b_\ell|^p\Big)^{\frac{1}{p}}, \quad a, b \in \mathbb{R}^k.$$

In the following, by $d_{ij}(\mathbf{X})$ we denote the distance between points x_i and x_j of a configuration $\mathbf{X}$ with respect to the metric d,

$$d_{ij}(\mathbf{X}) = d(x_i, x_j) = \Big(\sum_{\ell=1}^{k} |x_{i\ell} - x_{j\ell}|^p\Big)^{\frac{1}{p}}.$$

With these notations, least squares scaling means to minimize STRESS, i.e.

$$\text{minimize} \sum_{i,j=1}^{n} w_{ij}(\delta_{ij} - d_{ij}(\mathbf{X}))^2 \quad \text{over all} \quad \mathbf{X} \in \mathbb{R}^{n \times k}. \tag{1}$$

w_{ij} are nonnegative weights which express the relative importance or accuracy of data. They also incorporate the case of missing values by setting $w_{ij} = 0$. We assume that the matrix $W = (w_{ij})$ is irreducible, i.e. problem (1) may not be decomposed into smaller problems of the same type.

Problem (1) has been addressed by several authors, see e.g. de Leeuw (1977, 1988, 1992), de Leeuw and Heiser (1980), and Mathar (1989), mainly in the Euclidian case

$p = 2$, and for general p, Mathar and Groenen (1991), Mathar and Meyer (1992), Groenen, Mathar, and Heiser (1993). Two different methods have been applied to find at least local minima of (1): reformulation of (1) as a constrained convex optimization problem and subgradient based (local descent) procedures, and a more direct approach via stepwise majorizing the target function by a simpler, easy to minimize function.

For special cases, especially $k = 1$ (where all ℓ_p–metrics coincide) or $p = 1$ (the city-block metric), there exist more efficient methods based on combinatorial optimization, see Hubert and Arabie (1986), Heiser (1989), and Hubert, Arabie and Hesson-McInnis (1992).

The purpose of this note is to give a short overview on existing methods and algorithms, and to show how majorizing functions are easily obtained from subgradients, thus unifying approaches from convex analysis and majorization. Moreover, a relaxed version of a nested algorithm derived by Mathar and Groenen (1991) may be obtained in the same framework.

2. Subgradients

The target function in (1) is easily transformed to

$$\begin{aligned}\sigma^2(\mathbf{X}) &= \sum_{i,j=1}^{n} w_{ij}\delta_{ij}^2 - 2\sum_{i,j=1}^{n} w_{ij}\delta_{ij}d_{ij}(\mathbf{X}) + \sum_{i,j=1}^{n} w_{ij}d_{ij}^2(\mathbf{X}) \\ &= \text{const} - 2r(\mathbf{X}) + s^2(\mathbf{X}). \qquad (2)\end{aligned}$$

It is easy to see that $r(\mathbf{X})$ and $s(\mathbf{X})$ are nonnegative, positively homogeneous, and convex on $\mathbb{R}^{n\times k}$. Thus, $\sigma^2(\mathbf{X})$ is a difference of convex functions, a so called d.c. function (cf. Horst and Tuy (1993)). $r(\mathbf{X})$ and $s(\mathbf{X})$ are nondifferentiable at configurations where at least two points coincide, except the case $p \geq 2$ when $s(\mathbf{X})$ is differentiable everywhere. But because of convexity, subgradients exist everywhere.

$\mathbf{Z} \in \mathbb{R}^{n\times k}$ is called a subgradient of a real valued function f at $\mathbf{X} \in \mathbb{R}^{n\times k}$, if

$$f(\mathbf{Y}) \geq f(\mathbf{X}) + \langle \mathbf{Z}, \mathbf{Y} - \mathbf{X}\rangle \quad \text{for all} \quad \mathbf{Y} \in \mathbb{R}^{n\times k}. \qquad (3)$$

This definition carries over to arbitrary Hilbert spaces with inner product $\langle\cdot,\cdot\rangle$, but in this paper we will only consider the Euclidian scalar product on $\mathbb{R}^{n\times k}$: $\langle \mathbf{X}, \mathbf{Y}\rangle = \text{trace}(\mathbf{XY}')$. The set of all subgradients at $\mathbf{X}$ is called subdifferential of f at $\mathbf{X}$, and is denoted by $\partial f(\mathbf{X})$. If f is differentiable at $\mathbf{X}$ then $\partial f(\mathbf{X})$ consists of just one element, namely the derivative of f at $\mathbf{X}$.

For notational convenience we introduce the columnwise vectorized version of a matrix $\mathbf{X} = (x_{ij})_{1\leq i\leq n, 1\leq j\leq k} \in \mathbb{R}^{n\times k}$:

$$\mathbf{x} = \text{vec}(\mathbf{X}) = (x_{11}, \ldots, x_{n1}, x_{12}, \ldots, x_{n2}, \ldots, x_{1k}, \ldots, x_{nk})' \in \mathbb{R}^{nk}.$$

Now, for $\ell = 1, \ldots, k$ let

$$r_{ij}^{(\ell)}(\mathbf{x}) = \begin{cases} -2\, w_{ij}\, \delta_{ij}\, d_{ij}^{1-p}(\mathbf{x})\, |x_{i\ell} - x_{j\ell}|^{p-2}, & \text{if } 1 \leq p < \infty, \\ -2\, w_{ij}\, \delta_{ij}\, \varepsilon_{\ell,m(\mathbf{x}_i,\mathbf{x}_j)} |x_{i\ell} - x_{j\ell}|^{-1}, & \text{if } p = \infty, \end{cases} \quad i \neq j,$$

$$s_{ij}^{(\ell)}(\mathbf{x}) = \begin{cases} -2\, w_{ij}\, d_{ij}^{2-p}(\mathbf{x})\, |x_{i\ell} - x_{j\ell}|^{p-2}, & \text{if } 1 \leq p < \infty, \\ -2\, w_{ij}\, d_{ij}(\mathbf{x})\, \varepsilon_{\ell,m(\mathbf{x}_i,\mathbf{x}_j)} |x_{i\ell} - x_{j\ell}|^{-1}, & \text{if } p = \infty, \end{cases} \quad i \neq j,$$

where $m(\mathbf{x}_i, \mathbf{x}_j) = \min\{s \in \{1, \ldots, k\} \mid |x_{is} - x_{js}| = \max_{1 \le \ell \le k} |x_{i\ell} - x_{j\ell}|\}$ for $1 \le i, j \le n$, and $\varepsilon_{\ell,s} = 1$ if $\ell = s$ and 0 otherwise.

Define $\mathbf{R}_\ell(\mathbf{x}) = (r_{ij}^{(\ell)}(\mathbf{x}))_{1 \le i,j \le n}$ such that $\mathbf{R}_\ell(\mathbf{x})\mathbf{1} = \mathbf{0}$, and $\mathbf{S}_\ell(\mathbf{x})$ analogously. Furthermore, let

$$\mathbf{R}(\mathbf{x}) = \text{blockdiag}[\mathbf{R}_1(\mathbf{x}), \ldots, \mathbf{R}_k(\mathbf{x})], \quad \text{and} \quad \mathbf{S}(\mathbf{x}) = \text{blockdiag}[\mathbf{S}_1(\mathbf{x}), \ldots, \mathbf{S}_k(\mathbf{x})].$$

These notations yield the following convenient representations of $r(\mathbf{x})$, $s^2(\mathbf{x})$, and corresponding subgradients. $r(\mathbf{x})$ and $s(\mathbf{x})$ are defined in an obvious way as functions of the vectorized configurations.

Proposition 1. *Let $r(\mathbf{x})$ and $s^2(\mathbf{x})$ be defined as above. Then,*

$$r(\mathbf{x}) = \mathbf{x}'\mathbf{R}(\mathbf{x})\mathbf{x} \quad \textit{and} \quad s^2(\mathbf{x}) = \mathbf{x}'\mathbf{S}(\mathbf{x})\mathbf{x},$$

and subgradients of $r(\mathbf{x})$ and $s^2(\mathbf{x})$ in $\mathbf{R}^{nk}$ are given by

$$\mathbf{R}(\mathbf{x})\mathbf{x} \in \partial r(\mathbf{x}) \quad \textit{and} \quad 2\,\mathbf{S}(\mathbf{x})\mathbf{x} \in \partial s^2(\mathbf{x}).$$

For a proof of this see Mathar and Meyer (1992).

3. Majorization and generalized eigenvalues

Let $\varphi(\mathbf{x})$ be a real valued function to be minimized over its domain $\mathcal{D}$. A function $\hat{\varphi}(\mathbf{x}, \mathbf{y})$, $\mathbf{x}, \mathbf{y} \in \mathcal{D}$, with the properties

$$\varphi(\mathbf{x}) \le \hat{\varphi}(\mathbf{x}, \mathbf{y}) \quad \text{and} \quad \varphi(\mathbf{x}) = \hat{\varphi}(\mathbf{x}, \mathbf{x}) \quad \text{for all } \mathbf{x}, \mathbf{y} \in \mathcal{D}$$

is called a majorizing function. Now, for fixed $\mathbf{y}$ let

$$\bar{\mathbf{x}} = \arg\min_{\mathbf{x} \in \mathcal{D}} \hat{\varphi}(\mathbf{x}, \mathbf{y}) \tag{4}$$

denote a minimum point. Then immediately we have the following chain of inequalities,

$$\varphi(\bar{\mathbf{x}}) \le \hat{\varphi}(\bar{\mathbf{x}}, \mathbf{y}) \le \hat{\varphi}(\mathbf{y}, \mathbf{y}) = \varphi(\mathbf{y}), \tag{5}$$

which is named the *sandwich* inequality by de Leeuw (1992). A corresponding algorithm which updates "$\bar{\mathbf{x}} \leftarrow \mathbf{x}$" of course yields a decreasing sequence of function values.

Now consider the d.c. function (2), using vec notation. Since $r(\mathbf{x}) \ge r(\mathbf{y}) + \mathbf{z}'(\mathbf{x} - \mathbf{y})$ for any subgradient $\mathbf{z} \in \partial r(\mathbf{y})$ and all $\mathbf{x} \in \mathbb{R}^{nk}$, from Proposition 1 we obtain for all $\mathbf{x}$ that

$$\begin{aligned} \sigma^2(\mathbf{x}) &= \text{const} + s^2(\mathbf{x}) - 2r(\mathbf{x}) \\ &\le \text{const} + s^2(\mathbf{x}) - 2(r(\mathbf{y}) + \mathbf{z}'(\mathbf{x} - \mathbf{y})) \\ &= \text{const} + \mathbf{x}'\mathbf{S}(\mathbf{x})\mathbf{x} - 2\mathbf{y}'\mathbf{R}(\mathbf{y})\mathbf{y} - 2\mathbf{y}'\mathbf{R}(\mathbf{y})(\mathbf{x} - \mathbf{y}) \\ &= \text{const} + \mathbf{x}'\mathbf{S}(\mathbf{x})\mathbf{x} - 2\mathbf{y}'\mathbf{R}(\mathbf{y})\mathbf{x} \\ &= \hat{\sigma}(\mathbf{x}, \mathbf{y}). \end{aligned} \tag{6}$$

$\hat{\sigma}(\mathbf{x}, \mathbf{y})$ is a majorizing function of $\sigma^2(\mathbf{x})$, since $\hat{\sigma}(\mathbf{x}, \mathbf{x}) = \sigma^2(\mathbf{x})$ for all $\mathbf{x} \in \mathbb{R}^{nk}$. This type of majorization is analogous to what de Leeuw (1992) calls *type I majorization*.

The algorithm described in (4) needs minimizing $\mathbf{x}'\mathbf{S}(\mathbf{x})\mathbf{x} - 2\mathbf{y}'\mathbf{R}(\mathbf{y})\mathbf{x}$ over all $\mathbf{x}$. This seems to be a quite complicated task because of the dependence of $\mathbf{S}$ on $\mathbf{x}$. The following upper bound on $\mathbf{S}(\mathbf{x})$ leads to a quadratic function in $\mathbf{x}$ which can be minimized easily.

Proposition 2. *Let $1 \le p \le 2$, and $\mathbf{y}$ such that $y_{i\ell} \neq y_{j\ell}$ for all $i,j = 1,\dots,n$, $\ell = 1,\dots,k$. Then it holds that*

$$\mathbf{x}'\mathbf{S}(\mathbf{x})\mathbf{x} \le \mathbf{x}'\mathbf{S}(\mathbf{y})\mathbf{x} \quad \textit{for all } \mathbf{x} \in \mathbb{R}^{nk}.$$

Proof. By Hölder's inequality

$$\Big(\sum_{\ell=1}^{k} |x_{i\ell} - x_{j\ell}|^p\Big)^{\frac{2}{p}} \Big(\sum_{\ell=1}^{k} |y_{i\ell} - y_{j\ell}|^p\Big)^{\frac{p-2}{p}} \le \sum_{\ell=1}^{k} |x_{i\ell} - x_{j\ell}|^2 |y_{i\ell} - y_{j\ell}|^{p-2} \tag{7}$$

holds for all $p \le 2$ (thus $p/2 \le 1$) and positive $|y_{i\ell} - y_{j\ell}|$. Deviding by $\big(\sum_{\ell=1}^{k} |y_{i\ell} - y_{j\ell}|^p\big)^{\frac{p-2}{p}}$, multiplication of both sides of (7) by w_{ij}, and summing over all $i,j = 1,\dots,n$ yields

$$\begin{aligned}\sigma^2(\mathbf{x}) &= \sum_{i,j=1}^{n} w_{ij} d_{ij}^2(\mathbf{x}) = \sum_{i,j=1}^{n} w_{ij} \Big(\sum_{\ell=1}^{k} |x_{i\ell} - x_{j\ell}|^p\Big)^{\frac{2}{p}} \\ &\le \sum_{\ell=1}^{k} \sum_{i,j=1}^{n} w_{ij} \frac{|x_{i\ell} - x_{j\ell}|^2 |y_{i\ell} - y_{j\ell}|^{p-2}}{d_{ij}^{p-2}(\mathbf{y})} = \mathbf{x}'\mathbf{S}(\mathbf{y})\mathbf{x}.\end{aligned}$$

□

From Proposition 2 and (6) we get a majorizing function of $\sigma^2(\mathbf{x})$ as follows:

$$\sigma^2(\mathbf{x}) \le \text{const} + \mathbf{x}'\mathbf{S}(\mathbf{y})\mathbf{x} - 2\mathbf{y}'\mathbf{R}(\mathbf{y})\mathbf{x} = \tilde{\sigma}(\mathbf{x},\mathbf{y}) \tag{8}$$

for all $\mathbf{x}$ and $\mathbf{y}$ with $y_{i\ell} \neq y_{j\ell}$ for all i, j, ℓ.

$\tilde{\sigma}(\mathbf{x},\mathbf{y})$, as a quadratic function, is easy to minimize over $\mathbf{x}$ for fixed $\mathbf{y}$. The minimum is attained at $\mathbf{x}^*$ satisfying the stationary equations $2\mathbf{S}(\mathbf{y})\mathbf{x}^* - 2\mathbf{R}(\mathbf{y})\mathbf{y} = 0$. It follows that

$$\arg\min_{\mathbf{x}} \tilde{\sigma}(\mathbf{x},\mathbf{y}) = \mathbf{S}^+(\mathbf{y})\mathbf{R}(\mathbf{y})\mathbf{y},$$

where $\mathbf{S}^+(\mathbf{y})$ denotes the Moore–Penrose inverse of $\mathbf{S}(y)$. Any solution of the stationary equations may as well be used to obtain a decreasing sequence of function values. Thus, $\mathbf{S}^+(\mathbf{y})$ may be substituted by any generalized inverse $\mathbf{S}^-(\mathbf{y})$ of $\mathbf{S}(y)$.

Now (5) leads to the following algorithm to minimize STRESS:

$$\mathbf{x}_{m+1} = \mathbf{S}^+(\mathbf{x}_m)\mathbf{R}(\mathbf{x}_m)\mathbf{x}_m, \quad m \in \mathbb{N}_0. \tag{9}$$

This is exactly the same Groenen, Mathar and Heiser (1992) arrived at, and it has all desirable properties proved in that paper.

On the other hand, Mathar and Meyer (1992) transformed (2) to a constrained optimization problem over the level set of a convex function.

$$\text{maximize } r(\mathbf{x}) \quad \text{such that} \quad s(\mathbf{x}) \le 1. \tag{10}$$

They showed that

$$\mathbf{v} = \lambda\mathbf{z}, \text{ for some } \mathbf{v} \in \partial r(\mathbf{x}^*),\ \mathbf{z} \in \partial s^2(\mathbf{x}^*), \text{ and } \lambda \in \mathbf{R}, \tag{11}$$

is a necessary condition for an optimum point $\mathbf{x}^*$. Using the representation of the subgradients of $r(\mathbf{x})$ and $s^2(x)$ in Proposition 1, (11) may be seen as a generalized eigenvalue problem. Resembling inverse iteration (see e.g. Peters and Wilkinson (1971)) Mathar and Meyer (1992) gave an algorithm to solve (10), which is nearly the same as (9). The only difference is renormalization to the constraining set in each step. Analogous convergence properties also hold for this version of the algorithm.

4. Nested algorithms and majorization revisited

Mathar and Groenen (1991) proposed a nested algorithm to solve the MDS–problem (10):

$$\text{choose } \mathbf{z}_m \in \partial r(\mathbf{x}_m), \quad \mathbf{x}_{m+1} = \arg \max_{s(\mathbf{x})=1} \langle \mathbf{x}, \mathbf{z}_m \rangle, \quad m \in \mathbb{N}_0, \tag{12}$$

where the scalar product in the inner optimization problem is the Euclidian one in $\mathbb{R}^{nk}$. The generated sequence $\{\mathbf{x}_m\}$, $m \in \mathbb{N}_0$, is proved to be monotone, each accumulation point of $\{\mathbf{x}_m\}$ satisfies a necessary condition for an optimizing point, and if the inner optimization problem has a unique solution then even it holds that $\|\mathbf{x}_{m+1} - \mathbf{x}_m\| \to 0$ as $m \to \infty$ (cf. Mathar and Groenen (1991)). The inner optimization $\max_{s(\mathbf{x})=1} \langle \mathbf{x}, \mathbf{z}_m \rangle$ is a convex problem which may be solved by standard tools of convex programming, but this in general needs additional computing time and makes the algorithm slow.

Procedure (12) already yields an increasing sequence $\{r(\mathbf{x}_m)\}$, if the updates $\mathbf{x}_{m+1}$ satisfy

$$s(\mathbf{x}_{m+1}) = 1 \quad \text{and} \quad \langle \mathbf{x}_{m+1}, \mathbf{z}_m \rangle \geq \langle \mathbf{x}_m, \mathbf{z}_m \rangle \text{ for all } m \in \mathbb{N}. \tag{13}$$

This may easily be seen from the proof of Theorem 2.5 in Mathar and Groenen (1991). Now choose

$$\mathbf{y}_{m+1} = \mathbf{S}^+(\mathbf{x}_m)\mathbf{R}(\mathbf{x}_m)\mathbf{x}_m \quad \text{and} \quad \mathbf{x}_{m+1} = \frac{\mathbf{y}_{m+1}}{s(\mathbf{y}_{m+1})}. \tag{14}$$

By Proposition 1 we may choose $\mathbf{z}_m = \mathbf{R}(\mathbf{x}_m)\mathbf{x}_m$ to obtain for $1 \leq p \leq 2$

$$\langle \mathbf{x}_{m+1}, \mathbf{z}_m \rangle = \mathbf{x}'_{m+1}\mathbf{R}(\mathbf{x}_m)\mathbf{x}_m \geq \mathbf{x}'_m\mathbf{R}(\mathbf{x}_m)\mathbf{x}_m = \langle \mathbf{x}_m, \mathbf{z}_m \rangle. \tag{15}$$

This inequality corresponds to (18) in Mathar and Meyer (1992). (15) shows that the above relaxed version of the nested algorithm (12) leads to the same result as majorization and the generalized eigenvalue procedure based on (11) do.

But even the nested algorithm (12) has its counterpart in the majorization approach. Consider again the majorizing function $\hat{\sigma}(\mathbf{x}, \mathbf{y}) = \text{const} + \mathbf{x}'\mathbf{S}(\mathbf{x})\mathbf{x} - 2\mathbf{y}'\mathbf{R}(\mathbf{y})\mathbf{x}$. Minimizing $\hat{\sigma}(\mathbf{x}, \mathbf{y})$ over $\mathbf{x}$ for fixed $\mathbf{y}$ is as hard as (up to a one–dimensional differentiable optimization problem with an explicit solution)

$$\max \; \mathbf{y}'\mathbf{R}(\mathbf{y})\mathbf{x} \quad \text{such that} \quad \mathbf{x}'\mathbf{S}(\mathbf{x})\mathbf{x} = 1. \tag{16}$$

We have already used this approach in (10). If $\mathbf{y} = \mathbf{x}_m$ is the actual point in a corresponding majorization algorithm, then by the representations of Proposition 1 we may write (16) as

$$\max \; \mathbf{y}'_m\mathbf{x} \quad \text{such that} \quad s(\mathbf{x}) = 1,$$

where $\mathbf{y}_m \in \partial r(\mathbf{x}_m)$. This is exactly (12), and shows that an algorithm based on the (not easy to minimize) function $\hat{\sigma}$ basically is equivalent to the nested convex programming approach (12).

5. Conclusions

We have shown that all algorithms to solve the least squares scaling problem are quite similar. Approaches by majorization or inverse iteration via Hölder's inequality are shown to be simply relaxed versions of the more basic procedure in Mathar and Groenen (1991). The majorizing function (8) may be improved by the one in (6), which should yield faster convergence. As is shown, this is obtained at the price of a nontrivial inner optimization problem which in general has to be solved iteratively.

The figure below gives a compact survey of how the different methods are related to each other. Horizontal arrows mean "*as hard as*", or "*equivalent after a slight transformation*", whereas arrows from top to bottom say that methods of the upper layer lead to problems in the lower one. Numbers in brackets refer to the corresponding displayed equations in the text.

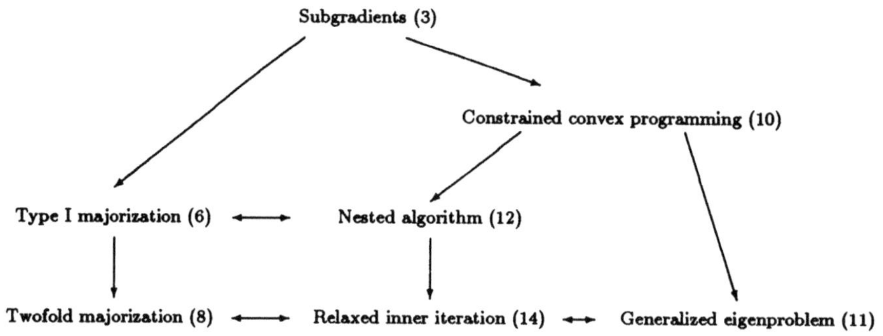

References:

DE LEEUW, J. (1977): Applications of Convex Analysis to Multidimensional Scaling. In: J.R. Barra, F. Brodeau, G. Romier and B. van Cutsem (eds.): *Recent Developments in Statistics*. North-Holland, Amsterdam, 133–145.

DE LEEUW, J. (1988): Convergence of the Majorization Method for Multidimensional Scaling. *Journal of Classification, 5, 163–180.*

DE LEEUW, J. (1992): Fitting Distances by Least Squares. Unpublished report, Los Angeles, UCLA.

DE LEEUW, J., and HEISER, W.J. (1980): Multidimensional Scaling with Restrictions on the Configuration. In: P.R. Krishnaiah (ed.): *Multivariate analysis V*. North–Holland, Amsterdam, 501–522.

GROENEN, P.J.F., MATHAR, R., and HEISER, W.J. (1993): The Majorization Approach to Multidimensional Scaling for Minkowski Distances. Submitted for publication, University of Leiden and RWTH Aachen.

HEISER, W.J. (1989): The City–block Model for Three–way Multidimensional Scaling. In: R. Coppi and S. Bolasco (eds.): *Multiway Data Analysis*. North–Holland, Amsterdam, 395–404.

HORST, R., TUY, H. (1993): *Global Optimization, Deterministic Approaches.* Springer, Berlin.

HUBERT, L.J., and ARABIE, P. (1986): Unidimensional Scaling and Combinatorial Optimization. In: J. De Leeuw, W.J. Heiser, J. Meulman, and F. Critchley (eds.): *Multidimensional Data Analysis.* DSWO Press, Leiden, 181–196.

HUBERT, L.J., ARABIE, P., and HESSON-MCINNES, M. (1992): Multidimensional Scaling in the City-Block Metric: a Combinatorial Approach. *Journal of Classification, 9, 211–236.*

MATHAR, R. (1989): Algorithms in Multidimensional Scaling. In: O. Opitz (ed): *Conceptual and Numerical Analysis of Data.* Springer, Berlin, 159–177.

MATHAR, R., and GROENEN, P.J.F. (1991): Algorithms in Convex Analysis Applied to Multidimensional Scaling. In: E. Diday, and Y. Lechevallier (eds.): *Symbolic–Numeric Data Analysis and Learning.* Nova–Science, Commack, New York, 45–56.

MATHAR, R., and MEYER, R. (1992): Algorithms in Convex Analysis to Fit l_p–Distance Matrices. Submitted for publication, RWTH Aachen.

PETERS, G., and WILKINSON, J.H. (1971): The calculation of specified eigenvectors by inverse iteration. In: J.H. Wilkinson and C. Reinsch (eds.): *Handbook for Automatic Computation, Vol. II, Linear Algebra.* Springer, Berlin, 418–439.

Universal Optimality of Rank Constrained Matrix Approximation

Renate Meyer

Institut für Medizinische Informatik und Biometrie, RWTH Aachen, Pauwelsstr. 30, D-52057 Aachen, Germany

Summary: The major results of this paper consist of an extension of optimality properties concerning the best rank-r matrix approximation (cf. Eckart and Young (1936), Mirsky (1960)), and of the solution of classical MDS (cf. Mathar (1985)), in either case from the class of orthogonally invariant norms to a wider class of approximation criteria. Observing that many problems in multivariate analysis exhibit symmetries under a group of transformations, and, furthermore, realizing the relevance of preorderings for matrix approximation problems in multivariate statistics and MDS leads to a natural approach towards proving universal optimality of their solution via invariant preorderings of matrices.

1. Introduction and Notation

In some respect, this paper may be regarded as a stronger version of Meyer (1991a) concerning the embedding of Correspondence Analysis (CA) as well as many other multivariate analysis (MVA) methods in the framework of classical multidimensional scaling (MDS). A survey of the literature reveals that many authors have emphasized the mutual relationships between various MVA methods. E.g. Meulman (1986) shows that the approximation problem underlying CA may be reduced to the approximation of a given Euclidean distance matrix (EDM) by an EDM of lower embedding dimension with respect to the Frobenius norm on the doubly centered distance matrices. This resulted in the development of *nonlinear* CA, the approximation of the corresponding EDM with respect to the so called STRESS loss function (i.e. weighted least squares on the distances). In Meyer (1991a) this approximation result is established by considering orthogonally invariant norms on the doubly centered distance matrices. The two main theorems of this paper (see also Mathar and Meyer (1993)) underpin these interrelationships between MVA techniques and classical MDS.

As is generally known, the key problem underlying most MVA methods is the approximation of a given $n \times k$ matrix by another one of lower rank r. Its solution, based on the singular value decomposition, is well known since the famous work of Eckart and Young (1936) which is stated in the following theorem.

Theorem 1.1: Let $\mathbf{A} \in \mathbb{R}^{n \times k}$, $1 \leq k \leq n$, and $\mathbf{A} = \mathbf{U}\text{diag}(\sigma_{[1]}, \ldots, \sigma_{[k]})\mathbf{V}'$ denote the singular value decomposition of $\mathbf{A}$ with $\mathbf{U} \in \mathbb{R}^{n \times k}$, $\mathbf{U}'\mathbf{U} = \mathbf{I}_k$, $\mathbf{V} \in \mathcal{O}(k)$, and $\sigma_{[1]} \geq \ldots \geq \sigma_{[k]} \geq 0$. For $1 \leq r \leq rk(\mathbf{A})$ define $\mathbf{A}_{(r)} = \mathbf{U}\text{diag}(\sigma_{[1]}, \ldots, \sigma_{[r]}, 0, \ldots, 0)\mathbf{V}'$ and let $||\mathbf{A}|| = tr(\mathbf{A}\mathbf{A}')^{1/2}$ denote the Frobenius norm of $\mathbf{A}$. Then

$$||\mathbf{A} - \mathbf{A}_{(r)}|| \leq ||\mathbf{A} - \mathbf{G}|| \qquad \forall\, \mathbf{G} \in \mathbb{R}^{n \times k},\ rk(\mathbf{G}) \leq r.$$

$\mathbf{A}_{(r)}$ turned out to be quite "robust" with respect to the choice of the approximation criterion, as it was extended by Mirsky (1960) to the class of orthogonally invariant norms. The approach towards proving universal optimality of the solution, implying its optimality with respect to various approximation criteria, is motivated by two findings:

1. On the one hand, many problems in multivariate statistical analysis show symmetries under certain transformations, which may be described by a certain transformation group acting on the linear space of $n \times k$ matrices. Thus, the principle of "*reduction by invariance*" might facilitate the problem by dropping insignificant aspects.

2. On the other hand, one should realize the importance of inequalities for assessing goodness-of-fit of matrix approximations. Instead of using ad hoc methods to establish such inequalities, the consideration of order relations between matrices is a far more powerful tool, often leading to a deeper insight and giving rise to possible generalizations. The well-known majorization of vectors, as discussed in the comprehensive book of Marshall and Olkin (1979) may serve as an example, where the study of a preordering plays a fundamental role in deriving inequalities. Of course, the larger the number of "natural" applications and the wider the class of order preserving functions, the more useful is the preordering under consideration.

As a synthesis, therefore, it is only natural to examine preorderings of matrices which are invariant under the operation of a group $\mathcal{G}$ on $\mathbb{R}^{n \times k}$.

2. Rank Constrained Approximations in MVA

As a characteristic and well-known example of an MVA method let us consider Correspondence Analysis, since CA may be considered as a special case of many well-known MVA techniques, such as principal components analysis, canonical correlation analysis, linear discriminant analysis or simultaneous linear regression, as surveyed e.g. by Meyer (1991b).

CA proceeds with N realizations of a two-dimensional categorical random vector $(V, W)'$ with n resp. k categories which may be represented by an $n \times k$ contingency table, i.e. a matrix $\mathbf{P} = (p_{ij})$ where $0 \leq p_{ij} \leq 1$ denotes the observed relative frequency of the combination of the i-th category of the row variable V and the j-th category of the column variable W. Furthermore, the row– and column totals $p_{i.} = \sum_{j=1}^{k} p_{ij}$ and $p_{.j} = \sum_{i=1}^{n} p_{ij}$ are assumed to be positive for all $i = 1, \ldots, n$ and all $j = 1, \ldots, k$. The conditional relative frequencies of the categories of the column variable in a row i, the so called *row profiles* $\mathbf{r}_i := \left(\frac{p_{i1}}{p_{i.}}, \ldots, \frac{p_{ik}}{p_{i.}}\right)'$ $i = 1, \ldots, n$, the rows of $\mathbf{R}^{-1}\mathbf{P}$ with $\mathbf{R} = \operatorname{diag}(p_{1.}, \ldots, p_{n.})$, may be conceived as n points in the weighted k–dimensional Euclidean space $(\mathbb{R}^k, \mathbf{C})$ with scalar product defined by $\langle \mathbf{x}, \mathbf{y} \rangle := \mathbf{x}'\mathbf{C}^{-1}\mathbf{y}$ for $\mathbf{x}, \mathbf{y} \in \mathbb{R}^k$ where $\mathbf{C} = \operatorname{diag}(p_{.1}, \ldots, p_{.k})$. In general, each row category has a different relative frequency, so each row profile should be weighted by its corresponding relative frequency $p_{i.}$. If one wants to check this point cloud for some unanticipated pattern, clusters, etc., the most effective approach is a visual inspection of a graphical display. But if the dimension k is high, one has to reduce dimensionality first. This is done — following the lines of projection pursuit — by projecting the points onto an optimal lower–dimensional, say r–dimensional hyperplane, optimal in the sense that the weighted sum of squared distances of the given points to the corresponding projection points is minimized. Evidently, the optimal hyperplane must pass through the centroid, which is $\mathbf{c} = (p_{.1}, \ldots, p_{.k})'$. Thus, the problem is reduced to the projection of the centered Euclidean configuration onto an optimal r-dimensional linear subspace $\mathcal{H}_r$. The numerical problem is to minimize $\sum_{i=1}^{n} p_{i.}(\mathbf{r}_i - \mathbf{c} - \hat{\mathbf{r}}_i)'\mathbf{C}^{-1}(\mathbf{r}_i - \mathbf{c} - \hat{\mathbf{r}}_i)$ over all $\hat{\mathbf{r}}_i \in \mathcal{H}_r$, or in matrix notation with $\mathbf{A} = \mathbf{R}^{-1/2}(\mathbf{P} - \mathbf{rc}')\mathbf{C}^{-1/2}$:

$$tr[(\mathbf{A} - \mathbf{G})(\mathbf{A} - \mathbf{G})'] \longrightarrow \min_{\mathbf{G} \in \mathbb{R}^{n \times k}, rk(\mathbf{G}) \leq r} . \tag{1}$$

For any matrix $\mathbf{G}$ of rank $\leq r$ may be decomposed into the product $\mathbf{G} = \mathbf{UV}'$ with a $k \times r$ matrix $\mathbf{V} = (\mathbf{v}_1, \ldots, \mathbf{v}_r)$ with orthonormal columns forming an orthonormal basis of an r-dimensional linear subspace and an $n \times r$ matrix $\mathbf{U}$ whose rows are the coordinates of n points with respect to this basis.

As will be immediately agreed upon, any rotation or reflection of the configuration should have no influence on the goodness-of-fit. This leads to consider preorderings on $\mathbb{R}^{n \times k}$ which are invariant with respect to orthogonal transformations of the rows of a matrix $\mathbf{A}$.

The duality concept of CA is well known. Of course, one might have defined k column profiles $\mathbf{c}_j$ in n-dimensional weighted Euclidean space $(\mathbb{R}^n, \mathbf{R})$ in an analogous way as rows of the matrix $\mathbf{C}^{-1}\mathbf{P}'$ with centroid $\mathbf{r} = (p_{1.}, \ldots, p_{n.})'$. Projecting these onto an optimal r-dimensional hyperplane again leads to the approximation problem (1) with $\mathbf{A}$ substituted by $\mathbf{A}'$. Hence, the preordering should at the same time be invariant with respect to orthogonal transformations of the columns of a matrix $\mathbf{A}$.

These arguments eventually give rise to consider the topological transformation group $\mathcal{G} = \mathcal{O}(n) \times \mathcal{O}(k) = \{(\mathbf{U}, \mathbf{V});\ \mathbf{U} \in \mathbb{R}^{n \times n}, \mathbf{U}'\mathbf{U} = \mathbf{I}_n, \mathbf{V} \in \mathbb{R}^{k \times k}, \mathbf{V}'\mathbf{V} = \mathbf{I}_k\}$ acting on $\mathbb{R}^{n \times k}$ by matrix multiplication from the left and the right, and to specify the following requirement any sensible preordering $\preceq$ on $\mathbb{R}^{n \times k}$ should meet:

$$\mathbf{A} \preceq \mathbf{B} \quad \Longrightarrow \quad \mathbf{UAV} \preceq \mathbf{UBV} \qquad \forall (\mathbf{U}, \mathbf{V}) \in \mathcal{G} \tag{2}$$

for all matrices $\mathbf{A}, \mathbf{B} \in \mathbb{R}^{n \times k}$.

In a first approach we consider the so called *group-induced* preordering or *$\mathcal{G}$-majorization*, induced by $\mathcal{G}$ via the following general definition.

Definition 2.1: Let $x, y \in \mathcal{X}$, $\mathcal{X}$ a finite-dimensional Hilbert space and $\mathcal{G}$ a subgroup of the unitary group $\mathcal{U}(\mathcal{X})$. Then x is *$\mathcal{G}$-majorized by y*, $x \preceq_{\mathcal{G}} y$, if x is an element of the convex hull of the $\mathcal{G}$-orbit of y, i.e. $x \in \text{conv}\{gy;\ g \in \mathcal{G}\}$.

It is easy to see that this definition actually specifies a preordering, i.e. a transitive and reflexive (not necessarily antisymmetric) relation on $\mathcal{X}$.

The well known majorization of vectors is a specimen of a group-induced preordering. It is defined for $\mathbf{x}, \mathbf{y} \in \mathbb{R}^n$ by (cf. Marshall and Olkin (1979)):

$$\mathbf{x} \preceq \mathbf{y} \quad \Longleftrightarrow \quad \sum_{j=1}^{i} x_{[j]} \leq \sum_{j=1}^{i} y_{[j]} \quad \text{for } i = 1, \ldots, n-1 \text{ and } \sum_{j=1}^{n} x_{[j]} = \sum_{j=1}^{n} y_{[j]}$$

where $\mathbf{x}_{de} = (x_{[1]}, \ldots, x_{[n]})'$ denotes a permutation of $\mathbf{x}$ with components $x_{[1]} \geq \ldots \geq x_{[n]}$ in decreasing order. By a classical result of Hardy, Littlewood and Pólya (cf.

Marshall and Olkin (1979), p. 18) this is equivalent to the existence of a doubly stochastic matrix $\mathbf{D}$ such that $\mathbf{x} = \mathbf{Dy}$. As however, by the famous theorem of Birkhoff (cf. Marshall and Olkin (1979), p. 19), the doubly stochastic matrices just constitute the convex hull of the permutation matrices, majorization is indeed a group–induced preordering, induced by the group of permutation matrices.

Now, it still has to be verified that the group–induced preordering, induced by $\mathcal{G} = \mathcal{O}(n) \times \mathcal{O}(k)$ is $\mathcal{G}$–invariant, i.e. meets (2). However, this is a consequence of the following lemma which is due to Eaton (1984). It also provides an easily verifiable criterion to establish whether an element x is $\mathcal{G}$–majorized by another element y.

Lemma 2.2: Let $\mathcal{G}$ be a subgroup of $\mathcal{U}(\mathcal{X})$, $\mathcal{X}$ is a finite–dimensional Hilbert space, and $x, y \in \mathcal{X}$. Let $\mathcal{C}(x) = \mathrm{conv}\{gx;\ g \in \mathcal{G}\}$ and $m(z\,|\,\mathcal{C}) = \sup_{c \in \mathcal{C}} \langle z, c\rangle$ denote the support functional of a convex set $\mathcal{C} \subset \mathcal{X}$. Then

$$x \preceq_{\mathcal{G}} y \quad \Longleftrightarrow \quad \mathcal{C}(x) \subset \mathcal{C}(y).$$

If, in addition, $\mathcal{G}$ is compact, then

$$x \preceq_{\mathcal{G}} y \quad \Longleftrightarrow \quad m(z\,|\,\mathcal{C}(x)) \leq m(z\,|\,\mathcal{C}(y)) \qquad \forall\, z \in \mathcal{X}.$$

As $\mathcal{G} = \mathcal{O}(n) \times \mathcal{O}(k)$ is compact, Lemma 2.2 can be employed to characterize the group–induced preordering on $n \times k$ matrices in terms of the weak majorization of the corresponding vectors of singular values. Prior to stating this characterization in Lemma 2.4, weak and componentwise majorization on $\mathbb{R}^k$ will be defined.

Definition 2.3: For $\mathbf{x} \in \mathbb{R}^k$ let $|\mathbf{x}| = (|x_1|, \ldots, |x_k|)'$ and $|x|_{[i]}$ denote the i-th largest component of $|\mathbf{x}|$. For $\mathbf{x}$, $\mathbf{y} \in \mathbb{R}^k$ define

(a) $\mathbf{x} \preceq_c \mathbf{y} \quad \Longleftrightarrow \quad |x|_{[i]} \leq |y|_{[i]}$ for $i = 1, \ldots, k$.

(b) $\mathbf{x} \preceq_w \mathbf{y} \quad \Longleftrightarrow \quad \sum_{i=1}^{l} |x|_{[i]} \leq \sum_{i=1}^{l} |y|_{[i]}$ for $l = 1, \ldots, k$.

Lemma 2.4: Let $\mathbf{A}, \mathbf{B} \in \mathbb{R}^{n \times k}$ and $\mathcal{G} = \mathcal{O}(n) \times \mathcal{O}(k)$. Then

$$\mathbf{A} \preceq_{\mathcal{G}} \mathbf{B} \quad \Longleftrightarrow \quad \sigma_{de}(\mathbf{A}) \preceq_w \sigma_{de}(\mathbf{B}).$$

Proof: This result is shown by applying Lemma 2.2 and using a result of Fan (1951):

$$\sup_{\mathbf{U} \in \mathcal{O}(n), \mathbf{V} \in \mathcal{O}(k)} tr(\mathbf{UAVB}) = \sum_{i=1}^{k} \sigma_{[i]}(\mathbf{A}) \sigma_{[i]}(\mathbf{B}) \quad \forall\, \mathbf{A}, \mathbf{B} \in \mathbb{R}^{n \times k}.$$

A detailed proof may be found in Eaton (1984). □

If $\mathcal{G} = \mathcal{O}(n) \times \mathcal{O}(k)$ let us therefore denote $\preceq_{\mathcal{G}}$ *by* $\preceq_w$. Recalling the approximation result of Mirsky (1960), it becomes evident that

$$||\mathbf{A} - \mathbf{A}_{(r)}|| \leq ||\mathbf{A} - \mathbf{G}|| \qquad \forall\, \mathbf{G} \in \mathbb{R}^{n \times k},\ rk(\mathbf{G}) \leq r,\ \forall \text{ orth. inv. norms } ||\cdot||$$

is equivalent to

$$\mathbf{A} - \mathbf{A}_{(r)} \preceq_w \mathbf{A} - \mathbf{G} \qquad \forall\, \mathbf{G} \in \mathbb{R}^{n \times k},\ rk(\mathbf{G}) \leq r.$$

For every orthogonal invariant norm $||\mathbf{A}||$ is generated by a symmetric gauge γ of the corresponding vector of singular values (cf. Marshall and Olkin (1979), p. 273), i.e. $||\mathbf{A}|| = \gamma(\sigma_{de}(\mathbf{A}))$, and γ is Schur convex and thereby order–preserving w.r.t. $\preceq_w$ on $\mathbb{R}^{k+}$. On the other hand, $\gamma_i(\sigma_{de}(\mathbf{A})) = \sum_{j=1}^{i} \sigma_{[j]}(\mathbf{A})$ is an orthogonal invariant norm for all $i = 1, \ldots, n$.

Consequently, $\mathbf{A}_{(r)}$ is the best rank r approximation to $\mathbf{A}$ simultaneously w.r.t. all real–valued and $\preceq_w$–monotone loss functions $\psi \in \Psi_w$. However, this result can be extended to a larger class of loss functions. To this end, instead of the group–induced preordering a stronger preordering on $\mathbb{R}^{n\times k}$ will be considered, which still meets (2). First, the notion of preorderings induced by a maximal invariant function will be introduced.

Definition 2.6: Let $\mathcal{G}$ be a transformation group on a set $\mathcal{X}$. A function f on $\mathcal{X}$ is said to be *invariant under* $\mathcal{G}$ if $f(gx) = f(x)$ for all $g \in \mathcal{G}$, $x \in \mathcal{X}$. An invariant function f is *maximal invariant under* $\mathcal{G}$, if $f(x) = f(y)$ implies $y = gx$ for some $g \in \mathcal{G}$.

Definition 2.7: Let $\mathcal{G}$ be a transformation group on a set $\mathcal{X}$ and $x, y \in \mathcal{X}$. Let $M : \mathcal{X} \to \mathcal{Y}$ be a maximal invariant function whose range $\mathcal{Y}$ is preordered by $\preceq_{\mathcal{Y}}$. Then x is *$(M, \preceq_{\mathcal{Y}})$–majorized by y*, $x \preceq_{(M,\preceq_{\mathcal{Y}})} y$, if $M(x) \preceq_{\mathcal{Y}} M(y)$.

It is easy to see that the preordering $\preceq_{(M,\preceq_{\mathcal{Y}})}$ is $\mathcal{G}$–invariant and that the function $M : \mathbb{R}^{n\times k} \to \mathbb{R}^k, \mathbf{A} \mapsto \sigma_{de}(\mathbf{A})$ is maximal invariant w.r.t. $\mathcal{G} = \mathcal{O}(n) \times \mathcal{O}(k)$. As shown in Lemma 2.2, the group induced preordering $\preceq_{\mathcal{G}}$ is equivalent to the preordering induced by M and weak majorization on $\mathbb{R}^k$, i.e. $\mathbf{A} \preceq_{\mathcal{G}} \mathbf{B} \iff \mathbf{A} \preceq_{(M,\preceq_w)} \mathbf{B}$. Now, let us consider componentwise instead of weak majorization on $\mathbb{R}^k$. For ease of notation, the induced preordering $\preceq_{(M,\preceq_c)}$ on $\mathbb{R}^{n\times k}$ is abbreviated by $\preceq_c$, so that $\mathbf{A} \preceq_c \mathbf{B} \iff \sigma_{de}(\mathbf{A}) \preceq_c \sigma_{de}(\mathbf{B})$. Obviously, $\mathbf{A} \preceq_c \mathbf{B}$ implies $\mathbf{A} \preceq_{\mathcal{G}} \mathbf{B}$. The proof of the approximation result w.r.t. this stronger preordering is based on the following inequality of Fan (1951).

Lemma 2.8: For $\mathbf{B}, \mathbf{C} \in \mathbb{R}^{n\times k}$, $1 \le k \le n$, $r, s > 0$ and $r + s + 1 \le k$:

$$\sigma_{[r+s+1]}(\mathbf{B} + \mathbf{C}) \le \sigma_{[r+1]}(\mathbf{B}) + \sigma_{[s+1]}(\mathbf{C}).$$

Theorem 2.9: Given $\mathbf{A} \in \mathbb{R}^{n\times k}$, let $\mathbf{A}_{(r)}$ be defined as in Theorem 1.1 and $\Psi_c = \{\psi : \mathbb{R}^{n\times k} \to \mathbb{R};\ \psi \preceq_c \text{-monotone}\}$. Then, for all $\mathbf{G} \in \mathbb{R}^{n\times k}$ with $rk(\mathbf{G}) \le r$ it holds that

$$\mathbf{A} - \mathbf{A}_{(r)} \preceq_c \mathbf{A} - \mathbf{G} \text{ and } \psi(\mathbf{A} - \mathbf{A}_{(r)}) \le \psi(\mathbf{A} - \mathbf{G}) \text{ for all } \psi \in \Psi_c.$$

Proof: Substituting $\mathbf{B}$ by $\mathbf{G}$ and $\mathbf{C}$ by $\mathbf{A} - \mathbf{G}$ in Lemma 2.8 gives $\sigma_{[i]}(\mathbf{A} - \mathbf{G}) \ge \sigma_{[r+i]}(\mathbf{A})$ for $i = 1, \ldots, k - r$ as $\sigma_{[r+1]}(\mathbf{G}) = 0$. Furthermore $\sigma_{[i]}(\mathbf{A} - \mathbf{G}) \ge 0$ for the remaining $i = k - r + 1, \ldots, k$. Together, these inequalities yield

$$\begin{aligned} \sigma_{de}(\mathbf{A} - \mathbf{G}) &= (\sigma_{[1]}(\mathbf{A} - \mathbf{G}), \ldots, \sigma_{[k-r]}(\mathbf{A} - \mathbf{G}), \sigma_{[k-r+1]}(\mathbf{A} - \mathbf{G}), \ldots, \sigma_{[k]}(\mathbf{A} - \mathbf{G}))' \\ &\succeq_c (\sigma_{[r+1]}(\mathbf{A}), \ldots, \sigma_{[k]}(\mathbf{A}), 0, \ldots, 0)' \\ &= \sigma_{de}(\mathbf{A} - \mathbf{A}_{(r)}) \end{aligned}$$

and thereby the assertion. □

Theorem 2.9 provides optimality properties of $\mathbf{A}_{(r)}$ w.r.t. an abundant class of approximation criteria. Using a result of Jensen (1984), theorem 1, the classes Ψ_c and Ψ_w can be characterized in the following way.

Lemma 2.10:

i) $\psi \in \Psi_c$ if and only if there exists a function γ on $\mathbb{R}^{k+}$ which is increasing in each component, such that $\psi(\mathbf{A}) = \gamma(\sigma_{de}(\mathbf{A}))$ for all $\mathbf{A} \in \mathbb{R}^{n\times k}$.

ii) $\psi \in \Psi_w$ if and only if there exists a function γ on $\mathbb{R}^{k+}$ which is increasing in each component and Schur convex, such that $\psi(\mathbf{A}) = \gamma(\sigma_{de}(\mathbf{A}))$ for all $\mathbf{A} \in \mathbb{R}^{n\times k}$.

iii) $\Psi_w \subset \Psi_c$.

Examples of functions which might be considered as sensible loss functions for matrix approximation and which are $\preceq_c$– however not $\preceq_w$–monotone are

i) *gauge–like functions* of the singular values (cf. Rockafellar (1970), p. 133).

ii) *information functionals*, i.e. nonnegative, positively homogeneous, and superadditive functionals on the singular values of a matrix. An important subclass is generated by the generalized means of order $p \in [-\infty, 1]$ defined by

$$\gamma_p(\mathbf{x}) = \begin{cases} \left(\frac{1}{k}\sum_{i=1}^{k}|x_i|^p\right)^{1/p} & \text{for } p \le 1, p \ne 0, p \ne -\infty \\ \left(\prod_{i=1}^{k}|x_i|\right)^{1/k} & \text{for } p = 0, \\ \min_{1\le i\le k}|x_i| & \text{for } p = -\infty. \end{cases}$$

They are the concave analogues of the l_p–norms for $p > 1$, increasing in each component, though not Schur convex.

iii) *the generalized variance of order s*, $1 \le s \le k - r$, defined by $\det^{+}_{(s)}(\mathbf{A}) = \prod_{i=1}^{s}\sigma_{[i]}(\mathbf{A})$, generated by $\gamma(\mathbf{x}) = \prod_{i=1}^{s} x_{[i]}$ on $\mathbb{R}^{k+}$ which is increasing, but not Schur convex.

iv) *an elementary symmetric function* of the singular values as generated by $\gamma_l(\mathbf{x}) = \sum_{i_1+\ldots+i_k=l, i_j\in\{0,1\}} x_1^{i_1}\ldots x_k^{i_k}$ for $1 \le l \le k$. As shown by Marshall and Olkin (1979), p. 81, these are concave, thence not Schur convex.

3. Rank Constrained Approximation in MDS

Finally, we turn to the problem of approximating a given symmetric matrix $\mathbf{A} \in \mathcal{H}^n$ by a matrix $\mathbf{H} \in \mathcal{H}^{n+}_{(r)} = \{\mathbf{H} \in \mathcal{H}^n;\ \mathbf{H} \succeq_L \mathbf{0},\ rk(\mathbf{H}) \le r\}$, the set of positive semidefinite matrices of some lower rank $\le r$. Here, $\succeq_L$ denotes the well-known *Loewner preordering*, i.e. $\mathbf{H} \succeq_L \mathbf{0} \iff \mathbf{x}'\mathbf{H}\mathbf{x} \ge 0 \quad \forall \mathbf{x} \in \mathbb{R}^n$. Let $x^+ = \max\{x, 0\}$ denote the positive and $\mathbf{x}^- = \max\{-x, 0\}$ the negative part of $x \in \mathbb{R}$.

Theorem 3.1: Given the matrix $\mathbf{A} \in \mathcal{H}_n$ with spectral decomposition denoted by $\mathbf{A} = \mathbf{U}\operatorname{diag}\left(\lambda_{[1]}(\mathbf{A}), \ldots, \lambda_{[n]}(\mathbf{A})\right)\mathbf{U}'$ where $\mathbf{U}'\mathbf{U} = \mathbf{I}_n$, and for $1 \le r \le n$ define $\mathbf{A}^{+}_{(r)} =$

$\mathbf{U}\ \mathrm{diag}\left(\lambda^+_{[1]}(\mathbf{A}),\ldots,\lambda^+_{[r]}(\mathbf{A}),0,\ldots,0\right)\mathbf{U}'$. Then, the following statement holds for all $\mathbf{H} \in \mathcal{H}^{n+}_{(r)}$:

$$\mathbf{A} - \mathbf{A}^+_{(r)} \preceq_c \mathbf{A} - \mathbf{H} \text{ and } \psi(\mathbf{A} - \mathbf{A}^+_{(r)}) \leq \psi(\mathbf{A} - \mathbf{H}) \text{ for all } \psi \in \Psi_c.$$

Proof: The proof of this theorem consists of a careful combination of three sets of inequalities:

i) $\sigma_{[i]}(\mathbf{A} - \mathbf{H}) \geq \sigma_{[r+i]}(\mathbf{A})$ for all $\mathbf{H}$ with $rk(\mathbf{H}) \leq r$ and $i = 1,\ldots,n-r$.

ii) the monotonicity of eigenvalues when adding a positive semidefinite matrix, i.e. $\lambda_{[i]}(\mathbf{A} + \mathbf{H}) \geq \lambda_{[i]}(\mathbf{A})$ for all positive semidefinite $\mathbf{H}$ (cf. Marshall and Olkin (1979), p. 510, A.1.6).

iii) some structural properties of the preordering $\preceq_c$ on $\mathbb{R}^n$ (cf. Mathar and Meyer (1993), Lemma 4.1):

   a) $|x|_{[i]} \geq x^-_{[n-i+1]}$ for $i = 1,\ldots,n$,

   b) for $r \in \{0,\ldots,n\}$: $(|x_{[r+1]}|,\ldots,|x_{[n]}|)_{[i]} \leq \max\{|x|_{[r+1]}, -x_{[n-i+1]}\}$ for $i = 1,\ldots,n-r$. □

A problem of this type occurs in classical MDS, where a given dissimilarity matrix

$$\Delta \in \mathbf{\Delta}_n = \{\Delta \in \mathcal{H};\ \Delta = (\delta_{ij})_{1\leq i,j\leq n},\ \delta_{ii} = 0,\ \delta_{ij} \geq 0,\ 1 \leq i,j \leq n\}$$

is to be approximated by an EDM $\mathbf{D}(\mathbf{X})$ of embedding dimension $r < n$, i.e. a matrix from the set

$$\mathcal{D}^n_{(r)} = \{\mathbf{D} = (d_{ij}) \in \mathcal{H}^n;\ \exists \mathbf{X} = (\mathbf{x}_1,\ldots,\mathbf{x}_n)' \in \mathbb{R}^{n\times r} : d_{ij} = (\mathbf{x}_i-\mathbf{x}_j)'(\mathbf{x}_i-\mathbf{x}_j)\ \forall i,j\}.$$

The loss function is any composition of a real valued function ψ and the linearly transformed dissimilarity matrix $\tau^{(\mathbf{W},\mathbf{w})}(\Delta) = -\frac{1}{2}\mathbf{W}^{1/2}(\mathbf{I}_n - \mathbf{1}_n\mathbf{w}')\Delta(\mathbf{I}_n - \mathbf{w}\mathbf{1}'_n)\mathbf{W}^{1/2}$ for some given positive definite $\mathbf{W}$ and weight vector $\mathbf{w} \in \mathbb{R}^n$ with $\mathbf{w}'\mathbf{1} = 1$. If $\mathbf{W} = \mathbf{I}_n$ and $\mathbf{w} = \frac{1}{n}\mathbf{1}_n$, then $\tau^{(\mathbf{W},\mathbf{w})}(\Delta)$ is just the doubly centered dissimilarity matrix. Thus, the approximation problem takes the form

$$\psi[\tau^{(\mathbf{W},\mathbf{w})}(\Delta) - \tau^{(\mathbf{W},\mathbf{w})}(\mathbf{D})] \quad \longrightarrow \quad \min_{\mathbf{D}\in\mathcal{D}^n_{(r)}}.$$

As $\tau^{(\mathbf{W},\mathbf{w})}(\Delta) \in \mathcal{H}^n_0 := \{\mathbf{H} \in \mathcal{H}^n;\ \mathbf{H}\mathbf{W}^{-1/2}\mathbf{w} = \mathbf{0}\}$ and $\tau^{(\mathbf{W},\mathbf{w})}(\mathcal{D}^n_{(r)}) = \{\mathbf{A} \in \mathcal{H}^n_0;\ \mathbf{H} \succeq_L \mathbf{0},\ rk(\mathbf{H}) \geq r\}$, this is actually the approximation of a symmetric matrix in the subspace $\mathcal{H}^n_0$ by a positive semidefinite matrix of rank $\leq r$ from this subspace. The hitherto most general approximation result concerning an orthogonally invariant norm ψ is due to Mathar (1985) and may be restated in terms of weak majorization:

Theorem 3.2: Given a positive definite $\mathbf{W}$, $\mathbf{w} \in \mathbb{R}^n$ with $\mathbf{w}'\mathbf{1}_n = 1$, $\Delta \in \mathbf{\Delta}_n$, $r \in \{1,\ldots,n-1\}$, and the spectral decomposition $\tau^{(\mathbf{W},\mathbf{w})}(\Delta) = \mathbf{U}\mathrm{diag}(\lambda_{[1]},\ldots,\lambda_{[n]})\mathbf{U}'$ with $\mathbf{U} = [\mathbf{u}_1,\ldots,\mathbf{u}_n]$, $\mathbf{U}'\mathbf{U} = \mathbf{I}_n$. Define $\mathbf{X}^+_{(r)} = [\mathbf{u}_1,\ldots,\mathbf{u}_r]\mathrm{diag}(\lambda^+_{[1]},\ldots,\lambda^+_{[r]})^{1/2}$. Then the following statements hold for all $\mathbf{D} \in \mathcal{D}^n_{(r)}$:

$$\tau^{(\mathbf{W},\mathbf{w})}(\Delta) - \tau^{(\mathbf{W},\mathbf{w})}(\mathbf{D}(\mathbf{X}_{(r)})) \preceq_w \tau^{(\mathbf{W},\mathbf{w})}(\Delta) - \tau^{(\mathbf{W},\mathbf{w})}(\mathbf{D}) \text{ and}$$

$$\psi\left[\tau^{(\mathbf{W},\mathbf{w})}(\Delta) - \tau^{(\mathbf{W},\mathbf{w})}(\mathbf{D}(\mathbf{X}_{(r)}))\right] \preceq_w \psi\left[\tau^{(\mathbf{W},\mathbf{w})}(\Delta) - \tau^{(\mathbf{W},\mathbf{w})}(\mathbf{D})\right] \forall \psi \in \Psi_w.$$

In view of Theorem 3.1, this result can be extended to the preordering $\preceq_c$ and accordingly to the larger class Ψ_c.

Coming back to the analysis of the row profiles in CA, the k–dimensional configuration $\mathbf{X} = \mathbf{R}^{-1}\mathbf{P}\mathbf{C}^{-1/2}$ was approximated by an r–dimensional configuration $\mathbf{X}_{(r)}$ w.r.t. a weighted least squares loss function. If instead of $\mathbf{X}$ we approximate the linearly transformed corresponding EDM $\tau^{(\mathbf{W},\mathbf{w})}(\mathbf{D}(\mathbf{X}))$ with $\mathbf{W} = \mathbf{R}$ and $\mathbf{w} = \mathbf{r}$ by a linearly transformed EDM of embedding dimension r, then the r–dimensional configuration $\mathbf{X}^{+}_{(r)}$ turns out to be the same as $\mathbf{X}_{(r)}$ up to reflections, rotations and/or translations. Consequently, in this respect, CA may be conceived as a special case of MDS.

References:

EATON, M.L. (1984): On Group Induced Orderings, Monotone Functions, and Convolution Theorems. In: Y.L. Tong (ed.): *Inequalities in Statistics and Probability.* Institute of Mathematical Statistics, Hayward, CA, 13-25.

ECKART, C. and G. Young (1936): The Approximation of One Matrix by Another of Lower Rank. *Psychometrika, 1, 211–218.*

FAN, K. (1951): Maximum Properties and Inequalities for the Eigenvalues of Completely Continuous Operators. *Proc. Nat. Acad. Sci. U.S.A., 37, 760–766.*

JENSEN, D.R. (1984): Invariant Ordering and Order Preservation. In: Y.L. Tong (ed.): *Inequalities in Statistics and Probability.* Institute of Mathematical Statistics, Hayward, CA, 26–34.

MARSHALL, A.W. and OLKIN, I. (1979): *Inequalities: Theory of Majorization and Its Applications.* Academic Press, New York.

MATHAR, R. (1985): The Best Euclidean Fit to a Given Distance Matrix in Prescribed Dimensions. *Linear Algebra Appl., 67, 1–6.*

MATHAR, R. and MEYER, R. (1993): Preorderings, Monotone Functions, and Best Rank r Approximations with Applications to Classical MDS. *Journal of Statistical Planning and Inference, to appear.*

MEULMAN, J. (1986): A Distance Approach to Nonlinear Multivariate Analysis. DSWO Press, Leiden.

MEYER, R. (1991a): Multidimensional Scaling as a Framework for Correspondence Analysis and its Extensions. In: M. Schader (ed.): *Analyzing and Modeling Data and Knowledge.* Springer, Heidelberg, 63–72.

MEYER, R. (1991b): Canonical Correlation Analysis as a Starting Point for Extensions of Correspondence Analysis. *Statistique et Analyse des Données, 16, 1, 55–77.*

MIRSKY, L. (1960): Symmetric Gauge Functions and Unitarily Invariant Norms. *Quart. J. Math. Oxford, Ser. 2, 11, 50–59.*

ROCKAFELLAR, R.T. (1970): *Convex Analysis.* Princeton University Press, Princeton.

The Analysis of Spatial Data from Marine Ecosystems

Dietmar Pfeifer, Ulrike Schleier–Langer[1], Hans–Peter Bäumer[2]

[1] Fachbereich Mathematik, Carl von Ossietzky Universität Oldenburg, Postfach 25 03, D–26015 Oldenburg, Germany

[2] HRZ – Angewandte Statistik, Carl von Ossietzky Universität Oldenburg, Postfach 25 03, D–26015 Oldenburg, Germany

Summary: In this paper we show how established methods from the fields of point process theory, stochastic geometry, and geostatistics can be applied to analyze spatial data from marine ecosystems. In particular, examples and data sets from the research project "Ökosystemforschung Niedersächsisches Wattenmeer" are presented.

1. Introduction

The use of statistical methods has a very good tradition in applied sciences such as medicine or biology, and seems to play a more and more important role also in related fields like ecology (cf. Richter and Söndgerath (1990).) While the "classical" statistical theory is well–established and largely adapted to the particular problems arising here (cf. Krebs (1985) or Begon, Harper, and Townsend (1990)), more modern advances in stochastic modelling have seemingly not yet found their final way into the field. The recent papers of Rothschild (1992) and Pfeifer, Bäumer, and Albrecht (1992) as well as the forementioned monograph by Richter and Söndgerath (1990) show, however, that applications of point process theory and stochastic geometry may be fruitful to obtain a deeper insight into the highly complicated stochastic structures which are the basis for the analysis of many of the spatial data obtained in biological or ecological field experiments (cf. also Diggle (1983) and Cressie (1991)).

Besides these two disciplines, we want to show in this paper that also geostatistical aspects could well be included in an analysis of such data, especially in marine ecology (cf. also the recent monograph by Haining (1990)).

2. Point patterns and processes: small–scale communities

Among the many species which form typical communities in marine ecosystems there are several whose spatial distribution may by physical reasons supposed to be more or less uniform over certain areas. This assumption can be justified if e.g. the offspring is in larval form, being drifted by hydrodynamical forces and distributed "at random". Since the physical extension of the larger part of meio– and microfauna species is typically in the millimeter range or even less (e.g. the copepode species *Harpacticus obscurus*) we may neglect this aspect of statistical modelling, such that point process theory becomes an appropriate tool for the

description and analysis of distributional patterns created by such species.
A first basic model is the so-called spatial *Poisson process*, denoted by $\xi(\,.\,)$. Finite versions of them are characterized by the following two properties ($d \in \mathbf{N}$ denoting the dimension):

i) For any measurable (i.e. Borel–)set $A \subseteq \mathbf{R}^d$, the random variable $\xi(A)$ is Poisson distributed with mean $\lambda(A) \in [0, \infty)$.
ii) For any countable collection of disjoint measurable sets $A, B, C, \ldots$, the random variables $\xi(A), \xi(B), \xi(C), \ldots$ are independent.

Here $\lambda(\,.\,)$ (also denoted as $E\xi$) is a finite measure (called *intensity measure*) on the Borel σ–field over $\mathbf{R}^d$. A canonical representation of ξ is

$$\xi(A) = \sum_{k=1}^{N} \varepsilon_{X_k}(A) = \sum_{k=1}^{N} \mathbb{1}_A(X_k) \tag{1}$$

where the random variables N and $X_1, X_2, X_3, \ldots$ are independent, N follows a Poisson distribution with mean $E(N) = \lambda(\mathbf{R}^d) = E\xi(\mathbf{R}^d)$, and $X_1, X_2, X_3, \ldots$ follow the distribution Q given by

$$Q(A) = P(X_k \in A) = \frac{E\xi(A)}{E\xi(\mathbf{R}^d)} \tag{2}$$

for all measurable sets $A \subseteq \mathbf{R}^d$ and $k \in \mathbf{N}$, provided $E\xi(\mathbf{R}^d) > 0$ (otherwise there are no points realized by the process). Here $\mathbb{1}_B$ denotes the indicator random variable of the event B, which shows that $\xi(A)$ counts the number of points in the set A. If, in particular, the intensity measure $E\xi$ has the form

$$E\xi(A) = c \cdot m(A \cap \mathcal{X}), \tag{3}$$

for some bounded measurable region $\mathcal{X} \subset \mathbf{R}^d$ and $c > 0$, where $m(\,.\,)$ denotes the Lebesgue measure (i.e. area or volume in case $d = 2$ or $d = 3$, resp.), then ξ is called *homogeneous Poisson process over* $\mathcal{X}$. In this case, the positions of points $\{X_n\}$ are uniformly distributed over $\mathcal{X}$ since here

$$Q(A) = P(X_k \in A) = \frac{E\xi(A)}{E\xi(\mathbf{R}^d)} = \frac{m(A \cap \mathcal{X})}{m(\mathcal{X})} \tag{4}$$

for all Borel sets A, provided $m(\mathcal{X}) > 0$.
In order to test the hypothesis of homogeneity it is sometimes convenient to use the *index-of-dispersion test*, especially if only spatially aggregated data are available. For this purpose, the region $\mathcal{X}$ has to be subdivided into n, say, disjoint observation windows $A_1, \ldots, A_n$ of equal size (i.e. $m(A_k) = m(\mathcal{X})/n$ for all k). Under the assumption of homogeneity the random variables $\xi(A_1), \ldots, \xi(A_n)$ are independent and follow a Poisson distribution with mean $(c/n)m(\mathcal{X})$, which is identical to their variance. Hence the distribution of the test statistic

$$D_n = (n-1)\frac{\sigma_n^2}{\bar{\xi}_n} \tag{5}$$

with

$$\overline{\xi}_n := \frac{1}{n}\sum_{k=1}^{n}\xi(A_k), \quad \sigma_n^2 := \frac{1}{n-1}\sum_{k=1}^{n}\left(\xi(A_k)-\overline{\xi}_n\right)^2$$

is asymptotically independent of c and asymptotically χ^2_{n-1}, by the central limit theorem for Poisson distributed random variables. D_n is called normalized index–of–dispersion. The following figure shows the positions of individuals of the polychaeta species *Arenicola marina* in a sample of size 174, together with the corresponding 6×6 abundance matrix of spatially aggregated data:

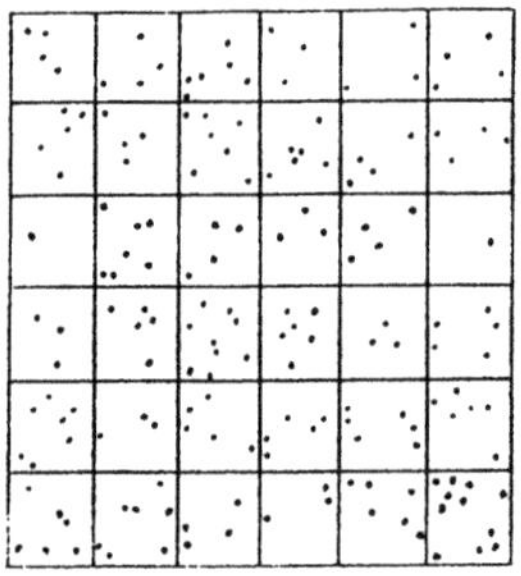

4	4	6	3	3	4
5	4	7	6	4	4
1	7	4	3	4	1
3	5	9	6	3	5
7	3	5	5	6	6
6	7	4	3	6	11

Fig. 1

spatial distribution *aggregated data*

$\overline{\xi}_n = 4.8333 \qquad \sigma_n^2 = 4.0857 \qquad D_n = 29.59$

The corresponding lower and upper 10%–quantiles of the χ^2_{35}–distribution here are $\chi^2_{35;0.10} = 24.80$, $\chi^2_{35;0.90} = 46.05$, hence the test will not reject the homogeneity hypothesis at a significance level of $\alpha = 0.2$ or less. (A similar result would be obtained with the data set of the centipede species *Lithobius crassipes* in Krebs (1985), Figure 10.2, p. 160.)
The following two tables contain the aggregated data of the abundance of the copepode species *Harpacticus obscurus*, taken at two different sites (labeled sites 6 and 8):

95	1	3	0	42
1	1	1	4	2
0	5	8	81	24
11	1	6	71	116
1	5	116	2	10

165	22	1	94	68
11	82	111	97	153
0	0	24	13	15
31	1	46	22	11
2	0	5	8	6

Fig. 2

site 6 *site 8*

$D_n = 1464.20 \qquad D_n = 1490.14$

The index–of–dispersion test here clearly rejects the homogeneity hypothesis for both sites, at all reasonable significance levels α. In this particular case, alternative models for the spatial distribution of individuals should be taken into account. For a discussion of non–homogeneous Poisson processes and, more generally, cluster processes, we refer the reader to Richter and Söndgerath (1990), Stoyan, Kendall and Mecke (1989) or Pfeifer, Bäumer and Albrecht (1992). Geostatistical aspects of such distributional patterns will be treated in more detail in Chapter 5 below.

3. Dynamic point patterns: revitalization & relative stability

One of the most important factors for relative stability in marine ecosystems seems to be the enormous dynamics due to water and wind forces, which guarantee to a certain extent the revitalization potential of the system. It is therefore desirable to introduce a time–dependent dynamic component in stochastic point patterns of the above type, which enables a study of the long–time behaviour of (at least a part of) the entire system. In the recent paper of Pfeifer, Bäumer, and Albrecht (1993) such an attempt is made by considering a family of Poisson point processes $\{\xi_t \mid t \geq 0\}$ which allows for birth, death and movements of points over time. In particular, the counting process $M_t = \xi_t(\mathbf{R}^d)$ of particles at time $t > 0$ forms a Markovian birth–death process with time– and state–dependent birth and death rates. Depending on the choice of system parameters, extinction, explosion and stabilization of the system over time can be modelled. Simulation studies show that the revitalization of (artificially) depopulated areas in the wadden sea by e.g. the gastropode species *Hydrobia ulvae* can very well be described through such models. However, a general framework for dynamic point processes is not yet fully developed, such that only particular models are available at present (cf. also Chapter 5.5.5 in Stoyan, Kendall and Mecke (1989)).

4. A Boolean model: mussel banks

Point processes can of course also serve as a basis for more complicated geometric structures, such as random sets. *Boolean models* are obtained just in this way: the points of a homogeneous Poisson process, say, are the centers of geometric objects such as discs or balls with fixed or random radius. The random sets Ξ, which are created in this manner, are suitable models for patchiness or spatial clustering, such as mussel bank structures. The picture below shows a part of a juvenile *Mytilus edulis* bank.

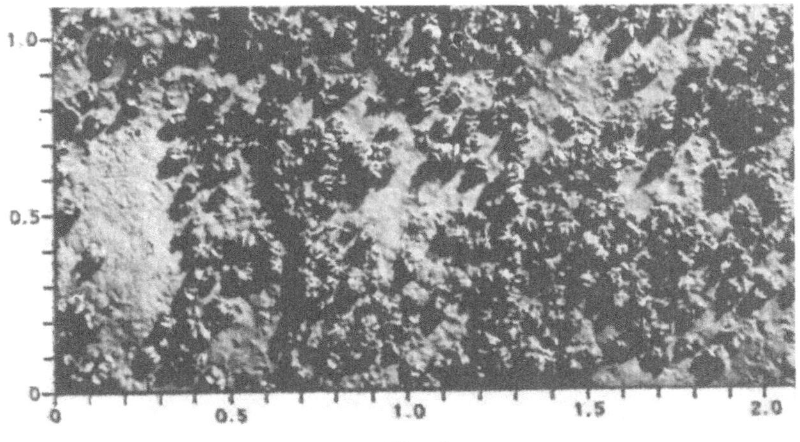

Fig. 3
mussel bank of Mytilus edulis

In applications one of the important parameters of the model is the so-called *area* or *volume fraction* p, i.e. the average area or volume per unit square or unit cube, resp., covered by the image. In the model outlined above it is given by

$$p = 1 - \exp\{-c \cdot E(m(B))\} = \begin{cases} 1 - \exp(-c\pi E(R^2)), & d = 2 \\ 1 - \exp(-\frac{4}{3}c\pi E(R^3)), & d = 3 \end{cases} \tag{6}$$

where R is the random variable describing the (random) radius of the "typical" disc or ball B in the model. This quantity can also be interpreted as the probability that a point of the Poisson process hits the "average" disc or ball with radius $E(R^2)$ or $E(R^3)$, resp. In marine ecology, the knowledge of p for mussel banks is basic to estimate the biomass in the bank, for instance. For stationary and isotropic random sets (which means shift and rotation invariance of the distribution of Ξ, as in the above Boolean model) there are simple and efficient estimators for p, for instance

$$\hat{p} = \frac{1}{n} \sum_{i=1}^{n} \mathbb{1}_{\Xi}(x_i) \tag{7}$$

where $\{x_i\}$ is a grid of n points in $\mathbf{R}^d$, typically larger in extent than the given image. Thus $\hat{p}$ counts the corresponding number of grid points $\{x_i\}$ that hit the image. This estimator is unbiased; under appropriate conditions, its variance is of order $\mathcal{O}(1/n)$ for large n (cf. Stoyan, Kendall and Mecke (1989), Chapter 6.3). For the (complete) mussel bank above, the grid estimate gives values of $\hat{p}$ about 0.4 which is in good coincidence with the value that was obtained by a deterministic technique measuring the covered area by exhaustion with "small" rectangles.

5. Geostatistical models for spatial dependence: benthos data

If we interpret the spatial data as being realizations of a *random field* $\{Z(\boldsymbol{x}) \mid \boldsymbol{x} \in \mathbf{R}^d\}$, then geostatistical approaches could be tried for the data analysis as well. Such techniques have gained increasing importance in environmental sciences over the years (see Haining (1990); in particular in connection with GIS's (Geographical Information Systems)). For instance, for the Harpacticus data (Fig. 2 above), the assumption of a homogeneous spatial distribution is surely inappropriate, as can be seen even with the bare eye. Corresponding observations can be made throughout for a great deal of other benthic species; it seems that such species have a tendency to spatial aggregation in general (cf. also the recent Ph.D. Thesis by Ekschmitt (1993)). A simple but nevertheless efficient way to model such spatial dependence structures is the assumption of a weakly stationary random field with constant mean μ and variance σ^2 in each point $\boldsymbol{x}$, such that information on the spatial dependence is given through the *variogram* function

$$2\gamma(\boldsymbol{h}) = V(Z(\boldsymbol{x}) - Z(\boldsymbol{x} + \boldsymbol{h})) = 2(\sigma^2 - C(\boldsymbol{h})) = 2(C(0) - \mathbf{C}(\boldsymbol{h})) \tag{8}$$

for vectors $\boldsymbol{x}, \boldsymbol{h} \in \mathbf{R}^d$. Here

$$C(\boldsymbol{h}) = Cov(Z(\boldsymbol{x}), Z(\boldsymbol{x} + \boldsymbol{h})) \tag{9}$$

denotes the *covariance* function of the random field. If stationarity and isotropy can be assumed, $C(\boldsymbol{h})$ depends only on the length $\|\boldsymbol{h}\|$. The function $\gamma(\cdot)$ is also called *semi–variogram* function. (Note that in the literature, variograms and semi–variograms are sometimes identified, and that only in the case of normal distributions, this function uniquely determines the distribution of the random field $\{Z(\boldsymbol{x})\}$.) The behaviour of the variogram or semi–variogram function in the neighbourhood of the origin determines the degree of "smootheness" of the random field. For variograms of Gaussian type, i.e.

$$\gamma(\boldsymbol{h}) \sim c \cdot \|\boldsymbol{h}\|^2 \qquad (\boldsymbol{h} \to \boldsymbol{0}) \tag{10}$$

with some positive constant c there exist realizations of a random field with normal marginal distributions that have "smooth", i.e. differentiable paths; in the case

$$\gamma(\boldsymbol{h}) \sim c \cdot \|\boldsymbol{h}\| \qquad (\boldsymbol{h} \to \boldsymbol{0}) \tag{11}$$

the corresponding paths will be continuous, but non–differentiable. If $\gamma(\cdot)$ is not continuous in the origin (so–called *nugget effect*) then the paths of the random field will also not be continuous. Although such an effect cannot be discovered by finite sampling it is sometimes convenient to incorporate it into the model in order to describe *micro–scale variation*, which can be considered as being caused by some white noise process superposed to the underlying "continuous" random field (cf. also the discussion in Cressie (1991), p. 59). Another effect that is occasionally considered is the so–called *hole effect*, which produces some kind of periodic oscillations in the (semi–)variogram, corresponding to spatial correlations at fixed distances.
In a geostatistical analysis, a first goal is to estimate the (semi–)variogram function from data taken at measurement points $\boldsymbol{x}_1, \ldots, \boldsymbol{x}_n \in \mathbf{R}^d$. The "classical" estimator here is

$$\hat{\gamma}(\boldsymbol{h}) = \frac{1}{2m} \sum_{j=1}^{m} \left\{ \hat{Z}(\boldsymbol{x}_{i_j}) - \hat{Z}(\boldsymbol{x}_{i_j} + \boldsymbol{h}) \right\}^2 \tag{12}$$

where $\hat{Z}(\boldsymbol{x})$ denotes the observed value of the random field at the point $\boldsymbol{x} \in \mathbf{R}^d$, and $\boldsymbol{x}_{i_j} - \boldsymbol{x}_{i_{j-1}} = \boldsymbol{h}$ for all j (transsect sampling). In a second step, a curve fitting procedure is applied in order to estimate $\hat{\gamma}(\cdot)$ also at intermediate distances.
A second goal in the geostatistical analysis is a prediction of values of the random field at arbitrary points $\boldsymbol{x} \in \mathbf{R}^d$, on the basis of the complete set of data and the empirical (semi-)variogram observed. With the aid of this *Kriging*[1] *procedure*, contour maps of the underlying random field can be established. In *block Kriging*, the prediction of an averaged value

$$Z_V = \frac{1}{m(V)} \int_V Z(\boldsymbol{x})\, d\boldsymbol{x} \tag{13}$$

over some bounded and measurable region $V \subset \mathbf{R}^d$ is required. *Simple Kriging*, i.e. a pointwise prediction for $\boldsymbol{x}_0 \in \mathbf{R}^d$, is obtained from this by taking the limit $m(V) \to 0$ over regions V which contain $\boldsymbol{x}_0$; then also $Z_V \to Z(\boldsymbol{x}_0)$ if the random field has continuous paths. Usual statistical requirements for a "good" prediction $\hat{Z}_V$ are:

[1] named after the south–african statistician D.G. Krige

a) *Linearity:* $\hat{Z}_V = \sum_{i=1}^{n} \lambda_i Z(\boldsymbol{x}_i)$ with weights $\lambda_1, \ldots, \lambda_n \in \mathbf{R}$;

b) *Unbiasedness:* $E[\hat{Z}_V] = E[Z_V] = \mu$, i.e. $\sum_{i=1}^{n} \lambda_i = 1$;

c) *Minimum Variance:* $E[(\hat{Z}_V - Z_V)^2] \to \min!$ subject to $\sum_{i=1}^{n} \lambda_i = 1$.

A solution of this Lagrangian problem can be obtained from the linear system

$$\boldsymbol{K}\boldsymbol{\lambda} = \boldsymbol{\gamma} \qquad \text{or} \qquad \boldsymbol{\lambda} = \boldsymbol{K}^{-1}\boldsymbol{\gamma}, \tag{14}$$

where

$$\boldsymbol{K} = \left(\begin{array}{c|c} [\gamma_{ij}] & \begin{matrix} 1 \\ \vdots \\ 1 \end{matrix} \\ \hline 1 \cdots 1 & 0 \end{array}\right) \qquad \boldsymbol{\lambda} = \begin{pmatrix} \lambda_1 \\ \vdots \\ \lambda_n \\ \nu \end{pmatrix} \qquad \boldsymbol{\gamma} = \begin{pmatrix} \overline{\gamma}_V(\boldsymbol{x}_1) \\ \vdots \\ \overline{\gamma}_V(\boldsymbol{x}_n) \\ 1 \end{pmatrix}. \tag{15}$$

Here ν is a Lagrangian multiplier, and the matrix $[\gamma_{ij}]$ and the entries $\overline{\gamma}_V(\boldsymbol{x}_i)$ are given by

$$\begin{aligned} \gamma_{ij} &= \gamma(\|\boldsymbol{x}_i - \boldsymbol{x}_j\|) \\ &\qquad [\text{or variogram estimate } \hat{\gamma}(\cdot) \text{ for } \gamma(\cdot), \text{ resp.}] \\ \overline{\gamma}_V(\boldsymbol{x}_i) &= \frac{1}{m(V)} \int_V \gamma(\|\boldsymbol{x}_i - \boldsymbol{y}\|)\, d\boldsymbol{y} \end{aligned} \tag{16}$$

For simple Kriging, i.e. $V = \{\boldsymbol{x}_0\}$, the last terms reduce to

$$\overline{\gamma}_V(\boldsymbol{x}_i) = \gamma(\|\boldsymbol{x}_i - \boldsymbol{x}_0\|). \tag{17}$$

The minimum variance of c) above can be expressed in terms of $\boldsymbol{\lambda}$ and $\boldsymbol{\gamma}$ as follows:

$$E[(\hat{Z}_V - Z_V)^2] = \boldsymbol{\gamma}^{tr}\boldsymbol{\lambda} = \boldsymbol{\gamma}^{tr}\boldsymbol{K}^{-1}\boldsymbol{\gamma}. \tag{18}$$

(For a more thorough discussion of the foundations of geostatistics, we refer the reader to the monographs of Cressie (1991), especially Chapter 3.2, and Journel and Huijbregts (1978) or Haining (1990).)

The following figures show variogram estimates for the Harpacticus data from Fig. 2 (sites 6 and 8), as well as contour maps for the corresponding random fields with simple and block Kriging. In the latter case, blocks V of four neighboured quadrats each were considered. All calculations were performed with the program GEO–EAS (Geostatistical Environmental Assessment Software). For the curve fitting procedure in the semi–variogram, a Gaussian model with nugget effect was used.

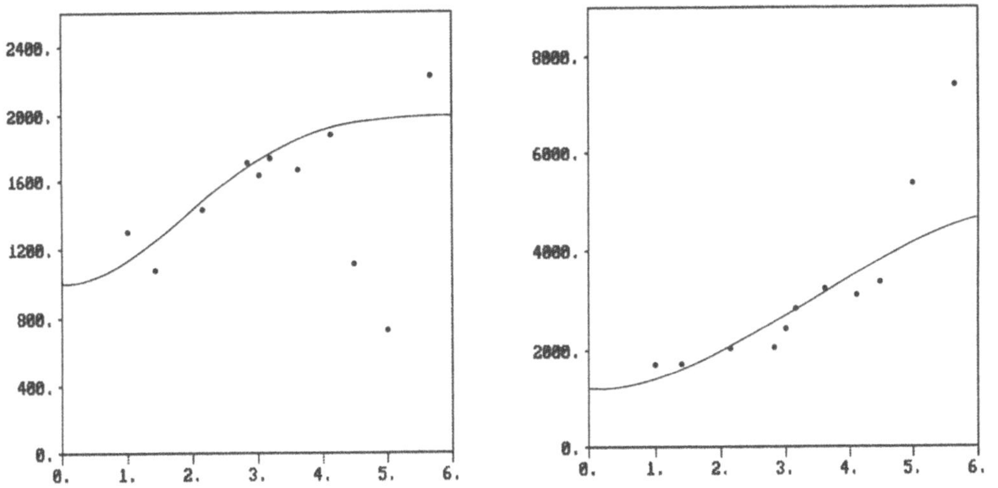

Fig. 4
semi-variogram estimates for Harpacticus obscurus, sites 6 and 8

Seemingly, in the semi-variogram for site 6 a hole-effect can be detected. This is due to the fact that in opposite corners of the data set (north-west and south-east), high count values are observed.

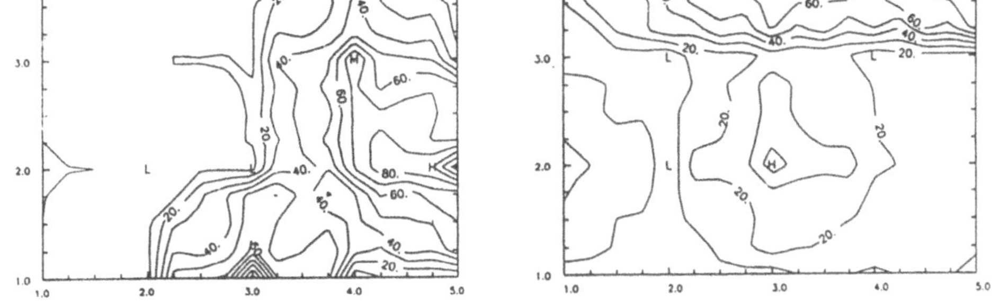

Fig. 5
contour maps for Harpacticus data: simple Kriging, sites 6 and 8

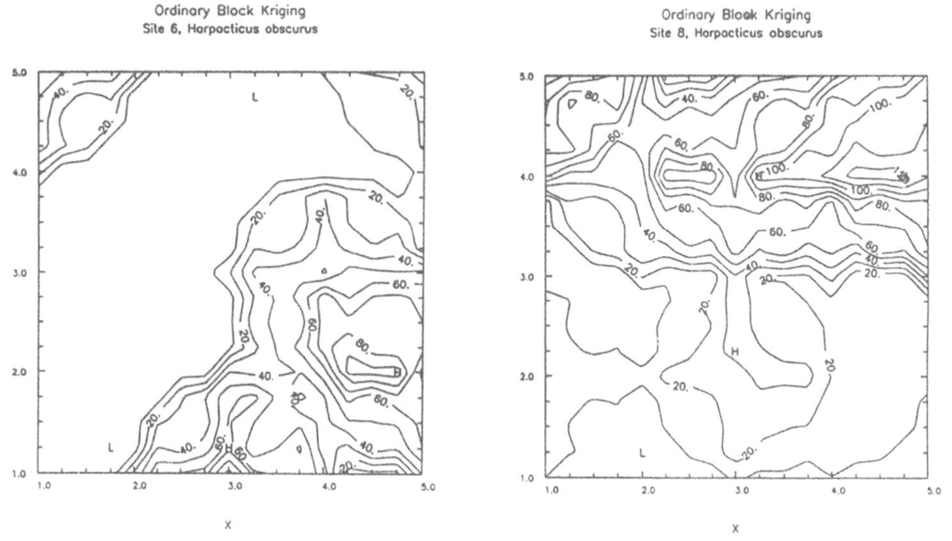

Fig. 6
contour maps for Harpacticus data: ordinary Kriging, sites 6 and 8

The last two figures show 3D–plots produced by a numerical interpolation of the Harpacticus data of the two sites (routine "inverse" of the program SYSTAT). A comparison with the corresponding contour plot above shows that the estimated distribution map is in good coincidence with this numerical data representation.

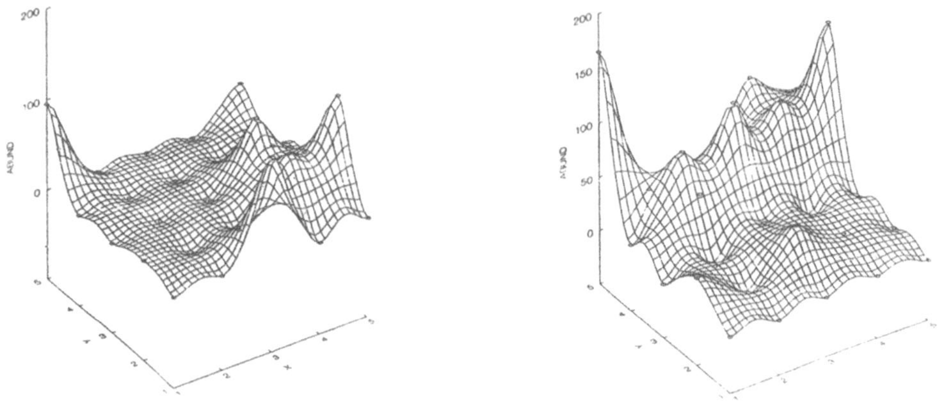

Fig. 7
3D–plots for Harpacticus data, sites 6 and 8

As a final comment it should be pointed out that in the case of count data (like above) the application of geostatistical methods might mathematically not be fully justified since the observed values are realizations of discrete distributions. However, at least in some approximative way, the analysis makes sense in order to obtain information about spatial dependence structures. Such information is e.g. necessary in order to determine *minimal areas* for probe schemes in field experiments, for instance in order to reduce the costs for necessary laboratory analyses. In the above examples, the structure of the semi-variograms shows that a reduction to a four by four probe scheme could be recommended for site 6, while at site 8, a comparable reduction might be inappropriate.

Acknowledgements. A part of the mathematical modelling presented in this paper has been developed and applied in the research project *Ökosystemforschung Niedersächsisches Wattenmeer (Pilotphase): Beiträge der Angewandten Statistik zur Bearbeitung von Maßstabsfragen und zur Versuchsplanung für die Analyse räumlicher Strukturen und dynamischer Vorgänge im Watt*, granted by the *Bundesminister für Forschung und Technologie*, Förderkennzeichen FZ 03F0023F. Special thanks are given to the following colleagues: W. Ebenhöh, ICBM, Universität Oldenburg, who supplied us with photographic material for Fig. 1; B. Hertweck, Forschungsinstitut Senckenberg, Wilhelmshaven, who provided the photograph for Fig. 3; and K.H. van Bernem, Forschungszentrum Geesthacht (GKSS), who contributed lots of data, among them the Harpacticus obscurus data from Fig. 2.

6. References

BEGON, M., HARPER, J.L., and TOWNSEND, C.R. (1990): *Ecology - Individuals, Populations, Communities.* Blackwell Sci. Publ., Oxford.

CRESSIE, N. (1991): *Statistics for Spatial Data.* Wiley, N.Y.

DIGGLE, P.J. (1983): *Statistical Analysis of Spatial Point Processes.* Academic Press, N.Y.

EKSCHMITT, K. (1993): *Über die räumliche Verteilung von Bodentieren. Zur ökologischen Interpretation der Aggregation und zur Probenstatistik.* Ph.D. Thesis, Universität Bremen.

HAINING, R. (1990): *Spatial Data Analysis in the Social and Environmental Sciences.* Camb. Univ. Press, Cambridge.

JOURNEL, A.G. and HUIJBREGTS, C.J. (1978): *Mining Geostatistics.* Academic Press, London.

KREBS, CH.J. (1985): *Ecology: The Experimental Analysis of Distribution and Abundance.* Harper & Row, N.Y.

PFEIFER,D., BÄUMER, H.-P., and ALBRECHT, M. (1992): Spatial point processes and their applications to biology and ecology. *Modeling Geo-Biosphere Processes 1*, 145 - 161.

PFEIFER,D., BÄUMER, H.-P., and ALBRECHT, M. (1993): Moving point patterns - the Poisson case. In: O. Opitz and B. Lausen (eds.): *Information and Classification: Concepts, Methods and Applications.* Springer, N.Y., 248 - 257.

RICHTER, O. and SÖNDGERATH, D. (1990): *Parameter Estimation in Ecology. The Link between Data and Models.* VCH, Weinheim.

ROTHSCHILD, B.J. (1992): Application of stochastic geometry to problems in plankton ecology. *Phil.Trans.R.Soc.Lond. B, 336*, 225 - 237.

STOYAN, D., KENDALL, W.S. and MECKE, J. (1989): *Stochastic Geometry and Its Applications.* Wiley, N.Y.

Automatic Decomposition of Lattice Data including Missing Values and Boundaries

Fritz-Ulrich Siewert[1], Paul O. Degens[2]

[1] Fachbereich Statistik, Universität Dortmund,
Postfach 50 05 00, D-44139 Dortmund, Germany

[2] Berufsgenossenschaftliches Forschungsinstitut für Arbeitsmedizin,
Gilsingstraße 14, D-44789 Bochum, Germany

Summary: Data collected on a regular lattice can be additively decomposed into several components by minimizing a sum of quadratic forms. The automatic construction of suitable quadratic forms which rely on a local definition of the desired properties of the single components is illustrated by the decomposition of data collected on a rectangular system of equidistant mesh points in the plane, as well as the solution of the minimization problem with missing values and the treatment of the lattice boundary points. Furthermore quadratic forms for the decomposition of data measured on a hexagonal lattice will be provided and an appropriate macro for the automated decomposition in the statistical programming language ISP will be presented.

1. Introduction

The objective of this article is to present a method for the automated additive decomposition of data measured on a regular lattice into different components. These data could be, for example, a time series with T equidistant time points, even a multivariate one, or measurements on a lattice in the plane, e.g. heavy metal concentrations in soil.

One possible basis to achieve an additive decomposition of a data vector $z \in \mathbb{R}^T$ into k components $x_i \in \mathbb{R}^T$, so that

$$z = \sum_{i=1}^{k} x_i \tag{1}$$

is valid, is given by the solution of the following minimization problem:

$$\min_{\{x_i \in \mathbb{R}^T \mid \sum x_i = z\}} \quad \sum_{i=1}^{k} x_i{}' Q_i x_i \tag{2}$$

where Q_i denote suitable nnd matrices or quadratic forms. This decomposition method was described by Halekoh and Degens (1986), where inter alia the uniqueness of the solution of (2) under weak regularity conditions is stated. Degens and Halekoh (1990) examined the decomposition of data collected on a rectangular lattice in the plane, while Halekoh and Degens (1991) discussed the decomposition of bivariate time series with this procedure.

The single quadratic forms $x_i' Q_i x_i$ assess how far the desired properties of the corresponding components x_i are fulfilled. In the following section, the typical construction principle will be presented in the case of the the decomposition of real-valued data obtained on a rectangular lattice in the plane. The method is primarily based on the use of difference operators that define locally the desired form of the k components.

The treatment of the boundary points of the lattice will be briefly examined as well.

Since missing values often occur in the data gathering process, this case will also be considered in the decomposition algorithm. It will be seen that the minimization problem can easily be modified for that case.

2. Decomposition of lattice data by minimizing quadratic forms

The results of the measurements can now be written into the data matrix $Z = (z_{ij}) \in \mathbb{R}^{n \times m}$ where (i, j) corresponds to the points of the lattice. The lattice chosen for the demonstration of the decomposition method can be constructed in two ways: Either by subdividing the measurement area with equidistant parallel and rectangular straight lines (see Fig. 1a) and observing one value on every knot of this mesh, or by covering the area with a system of squares and measuring one value in the centre of each individual square (Fig. 1b).

These values z_{ij} should now be decomposed — $k = 2$, $Z = X + Y$ — into a regular ("smooth") component X and an irregular component Y in the following way:

- At first, one has to define the desired properties of the regular component $X \in \mathbb{R}^{n \times m}$ and the irregular component $Y \in \mathbb{R}^{n \times m}$.
- Find an appropriate quadratic form Q_i for the requirements on each component.
- Incorporate these forms into the minimization problem (2).

X should be a smoothed version of the original data while Y should be as small as

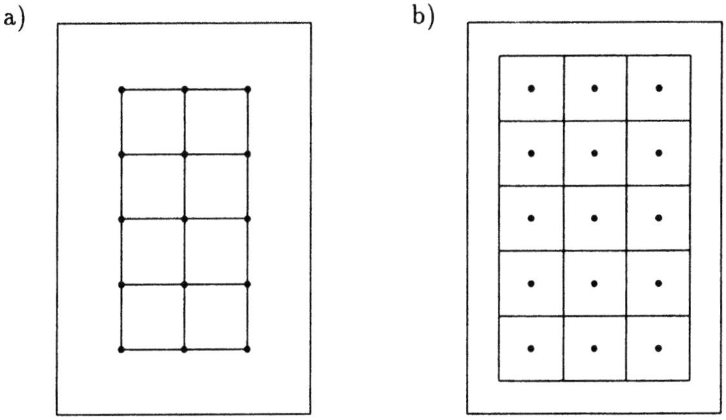

Figure 1: Rectangular lattice

possible. The latter requirement for Y is intended by the sum of squares

$$\mathrm{vec}\,(Y)'\,\mathrm{vec}\,(Y) = \sum\sum y_{ij}^2 \tag{3}$$

where vec: $\mathbb{R}^{n \times m} \ni Y \mapsto \mathrm{vec}\,(Y) \in \mathbb{R}^{n*m}$ assigns a vector of length $n * m$ to the matrix $Y \in \mathbb{R}^{n \times m}$ with the same entries, the first index running faster than the second.

Considering X Wahba (1980) uses the following measure of roughness of a two-dimensional continuous function $g(\xi,\eta): I\!R^2 \to I\!R$ for the computation of a smooth spline function:

$$\int\int \left[\left(\frac{\partial^2}{\partial \xi^2} g(\xi,\eta) \right)^2 + 2 \left(\frac{\partial}{\partial \xi} \frac{\partial}{\partial \eta} g(\xi,\eta) \right)^2 + \left(\frac{\partial^2}{\partial \eta^2} g(\xi,\eta) \right)^2 \right] \mathrm{d}\xi \mathrm{d}\eta. \tag{4}$$

Since here we want to measure the roughness of the "function" X on discrete points it is required to replace the continuous operators by some equivalents.

The equivalent of differentiating functions on vector spaces is the computation of differences on discrete grid points. Therefore, the second order "differentiation" in the vertical direction of the lattice is given by the second difference:

$$\delta_j^i\,(X) := x_{(i-1)j} - 2x_{ij} + x_{(i+1)j} \tag{5}$$

"Differentiation" in the horizontal direction is replaced by:

$$^{i}\delta^{j}\,(X) := x_{i(j-1)} - 2x_{ij} + x_{i(j+1)}, \tag{6}$$

and the mixed partial "differentiation" is given by:

$$^{i}\delta_{j}\,(X) := \left(x_{(i+1)(j+1)} - x_{(i+1)j} \right) - \left(x_{i(j+1)} - x_{ij} \right) \tag{7}$$

These three terms describe locally some type of smoothness of the regular component X. $\delta_j^i\,(X)$ and $^{i}\delta^{j}\,(X)$ measure the roughness of X in the immediate neighbourhood of the point x_{ij}, while $^{i}\delta_{j}\,(X)$ links the points x_{ij}, $x_{(i+1)j}$, $x_{i(j+1)}$ and $x_{(i+1)(j+1)}$ and thus measures the roughness in the neighbourhood of the virtual point $x_{(i+\frac{1}{2})(j+\frac{1}{2})}$. They are symmetric, thus invariant with respect to reflections of the main directions of the lattice. For the smoothness assessment of X, the terms will be shifted individually over the lattice and used on every inner point where the terms are applicable. These three terms will be summed up to the final result for the whole lattice.

Hence for a rectangular grid with $n \times m$ points the following quadratic form is used to measure the roughness of the regular component X:

$$\sum_{i=2}^{n-1} \sum_{j=1}^{m} \left(\delta_j^i\,(X) \right)^2 + 2 \sum_{i=1}^{n-1} \sum_{j=1}^{m-1} \left(^{i}\delta_{j}\,(X) \right)^2 + \sum_{i=1}^{n} \sum_{j=2}^{m-1} \left(^{i}\delta^{j}\,(X) \right)^2 \tag{8}$$

Summation in this discrete problem corresponds to integration in (4).

A matrix representation of the roughness criterion in terms of Kronecker products can be obtained by defining a matrix Δ as follows:

$$\Delta := I_m \otimes P_n' P_n + 2\, F_m' F_m \otimes F_n' F_n + P_m' P_m \otimes I_n \tag{9}$$

with:

$$F_n := \begin{pmatrix} -1 & 1 & 0 & \dots & 0 \\ 0 & -1 & 1 & & \vdots \\ \vdots & & \ddots & \ddots & 0 \\ 0 & \dots & 0 & -1 & 1 \end{pmatrix} \in I\!R^{(n-1)\times n} \tag{10}$$

and

$$P_n := F_{n-1} F_n \tag{11}$$

F_m, P_m similar.

This yields

$$\text{vec}\,(X)'\,\Delta\text{vec}\,(X) \tag{12}$$

as an elegant matrix representation of (8).

The optimum decomposition of the lattice data $Z \in I\!R^{n\times m}$ into a smooth component $X \in I\!R^{n\times m}$ and an irregular component $Y \in I\!R^{n\times m}$ is given by the following minimization problem:

$$\min_{X,Y\in I\!R^{n\times m}} \alpha\,\text{vec}\,(X)'\Delta\text{vec}\,(X) + \text{vec}\,(Y)'\text{vec}\,(Y) \tag{13}$$

with

$$X + Y = Z \tag{14}$$

and a prespecified weight parameter $\alpha > 0$ which controls the amount of roughness of X and the extent of the irregular component Y. Substituting $Y = Z - X$ into (13) results in

$$\min_{X\in I\!R^{n\times m}} \begin{pmatrix}\text{vec}\,(X)\\ \text{vec}\,(Z)\end{pmatrix}' \begin{pmatrix}\alpha\Delta + I & I\\ I & I\end{pmatrix} \begin{pmatrix}\text{vec}\,(X)\\ \text{vec}\,(Z)\end{pmatrix} \tag{15}$$

Taking derivatives with respect to vec (X) and setting the result equal to zero leads to the decomposition:

$$\text{vec}\,(X) = (\alpha\Delta + I)^{-1}\,\text{vec}\,(Z), \quad Y = Z - X \tag{16}$$

3. Treatment of missing values

Due to different causes, e.g. failure of gauges, missing values can occur. Thus, the additive requirement (1) or (14) makes no sense for all z_i or Z_{ij} and the minimization problem (2) or (13) under (14) is undefined.

If Z is defined arbitrary, say 0, at missing value positions, it has been proposed by Halekoh and Degens (1992) to formulate another additivity requirement:

$$I_0\text{vec}\,(X + Y) = I_0\text{vec}\,(Z) \tag{17}$$

where I_0 denotes a matrix with 1's in the main diagonal for each measured Z-value and 0's otherwise. Since the assessment functions (12) and (3) for the two components X and Y are not affected by missing values in Z, we can formulate another minimization problem:

$$\min_{\{X,Y\in I\!R^{n\times m}\,|\,I_0\text{vec}(X+Y)=I_0\text{vec}(Z)\}} \alpha\,\text{vec}\,(X)'\Delta\text{vec}\,(X) + \text{vec}\,(Y)'\text{vec}\,(Y) \tag{18}$$

This can be transformed to:

$$\min_{X\in I\!R^{n\times m}} \begin{pmatrix}\text{vec}\,(X)\\ \text{vec}\,(Z)\end{pmatrix}' \begin{pmatrix}\alpha\Delta + I_0 & I_0\\ I_0 & I_0\end{pmatrix} \begin{pmatrix}\text{vec}\,(X)\\ \text{vec}\,(Z)\end{pmatrix} \tag{19}$$

Taking derivatives with respect to vec (X) yields the solution:

$$\text{vec}\,(X) = (\alpha\Delta + I_0)^{-1}\,I_0\,\text{vec}\,(Z), \quad \text{vec}\,(Y) = I_0\,\text{vec}\,(Z - X) \tag{20}$$

By (18) it is immediately clear that $y_{ij} = 0$ at missing value positions. If the missing value positions are not given in a bad (systematic) order $(\alpha\Delta + I_0)$ is positive definite in general.

In ISP missing values are represented by the symbol "?". In usual formulas and algorithms missing values are treated adequatly i.e. evaluation of a function at a missing value is itself a missing value, in summary statistics missing values are neglected. In our algorithm the missing value symbol "?" indicates the position of a missing value in a data set in a simple manner, but the missing values are treated correctly as described in this section.

4. Treatment of boundaries

The local definition of the desired properties of the separate components leads to the problem that their quality cannot be assessed at all positions. The difference terms cannot be used at boundary sites because no data is available outside of the lattice (Fig. 2a). This may lead to a reduced smoothness of the regular component in the boundaries.

One possible way to circumvent this problem consists in enlarging the lattice by boundary points with missing values z_{ij}, as seen in Fig. 2b. In this example, the

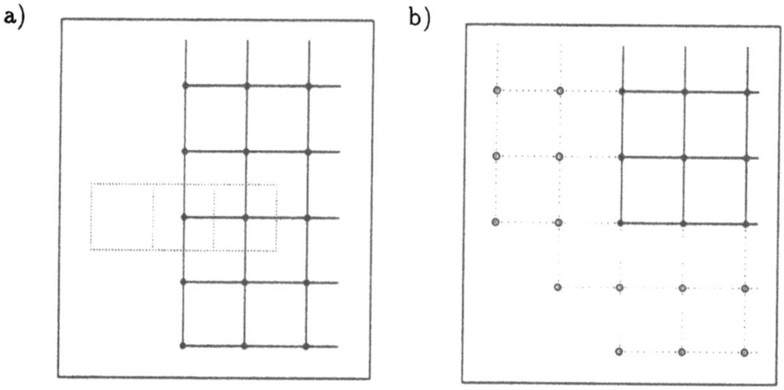

Figure 2: Assessment at the boundary

lattice has been minimally enlarged with missing values "?" as data in such a way that the assessment functions can be evaluated on all boundary points of the original lattice. It is easy to see that a lattice enlarged in such a way has no rectangular shape; this makes algorithms more complicated.

To avoid boundary effects on the regular component, the original data should preferably be embedded in an infinite plane, i.e. the lattice should be enlarged infinitely with missing values as data. In practice, one enlarges the lattice with two or three rows of grid points at each boundary.

5. Hexagonal lattice

In section 2 we have subdivided a 2-dimensional area by equidistant and rectangular straight lines. Alternatively, a symmetric subdivision with straight lines crossing at 60^0 angles can be used (see Fig. 3a). In analogy this subdivision could alternatively be carried out by covering the area with hexagons (Fig. 3b), where a value is observed in the centre of each hexagon.

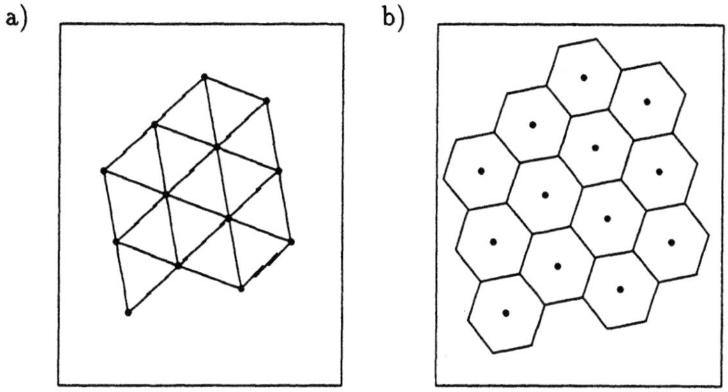

Figure 3: Hexagonal lattice

For the decomposition of data collected on a hexagonal lattice into a smooth and an irregular component a special assessment function for the roughness of the regular component has to be provided, while the irregular component will be assessed with (3). Similary to the three terms (5), (6) and (7) of the rectangular lattice and to the roughness measure (4) of Wahba (1980), it is useful to compute the second derivatives in each of the three main directions of the hexagonal mesh (i.e. to compute the second differences), to compute the mixed partial derivatives in two of the three main directions and to use the results for the construction of an appropriate quadratic form. The terms below measure the local smoothness of the regular component and will be used on every point of the hexagonal lattice they are applicable.

For the development of proper assessment functions, the measured values are labeled in an $n \times m$-array (Fig. 4).

The second differences are computed as follows:

• The "vertical second derivative" is

$$v(X,i,j) := x_{(i-1)j} - 2x_{ij} + x_{(i+1)j} \tag{21}$$

• the "rising 2nd derivative" (one central hexagon becomes linked with the neighbouring hexagon on the left lower side and the hexagon on the upper right side):

$$r(X,i,j) := x_{i(j-1)} - 2x_{ij} + x_{i(j+1)} \tag{22}$$

• the "falling second derivative" (the values of one central hexagon, its adjoining hexagon on the upper left side and its adjoining hexagon on the lower right side are linked):

$$f(X,i,j) := x_{(i-1)(j-1)} - 2x_{ij} + x_{(i+1)(j+1)} \tag{23}$$

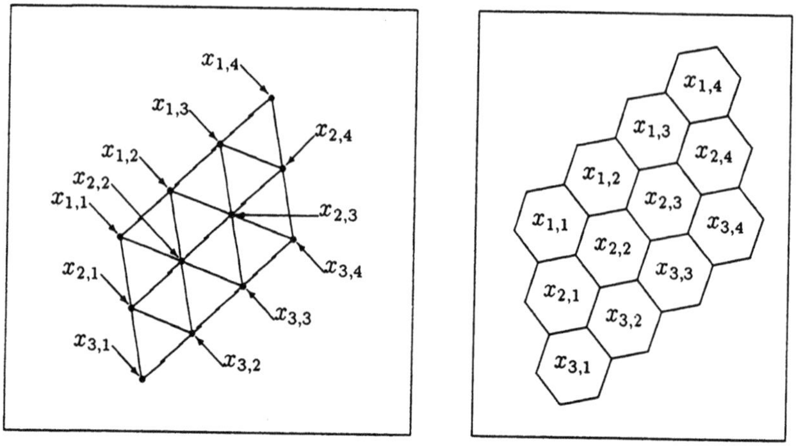

Figure 4: Data structure of the hexagonal lattice

The mixed partial derivatives have the following form:

- Derivation into the "vertical" direction in combination with the "diagonal rising" ($vr(X,i,j)$) and the "diagonal falling" ($vf(X,i,j)$) direction:

$$vr(X,i,j) := x_{(i-1)(j-1)} - x_{i(j-1)} - x_{(i-1)j} + x_{ij} \tag{24}$$
$$vf(X,i,j) := x_{(i-1)(j-1)} - x_{i(j-1)} - x_{ij} + x_{(i+1)j} \tag{25}$$

- "Diagonal rising" in combination with other directions:

$$rv(X,i,j) := x_{(i-1)(j-1)} - x_{(i-1)j} - x_{i(j-1)} + x_{ij} \tag{26}$$
$$rf(X,i,j) := x_{(i-1)(j-1)} - x_{(i-1)j} - x_{ij} + x_{i(j+1)} \tag{27}$$

- "Diagonal falling" in combination with other directions:

$$fr(X,i,j) := x_{(i-1)(j-1)} - x_{ij} - x_{(i-1)j} + x_{i(j+1)} \tag{28}$$
$$fv(X,i,j) := x_{(i-1)(j-1)} - x_{ij} - x_{i(j-1)} + x_{(i+1)j} \tag{29}$$

This leads to the following quadratic form for the assessment of the regular component:

$$\begin{aligned} Q(X) := {} & \sum_{i=2}^{n-1}\sum_{j=1}^{m} v(X,i,j)^2 + \sum_{i=1}^{n}\sum_{j=2}^{m-1} r(X,i,j)^2 + \sum_{i=2}^{n-1}\sum_{j=2}^{m-1} f(X,i,j)^2 \\ & +2\sum_{i=2}^{n-1}\sum_{j=2}^{m} fv(X,i,j)^2 \\ & +2\sum_{i=2}^{n}\sum_{j=2}^{m} vr(X,i,j)^2 + 2\sum_{i=2}^{n}\sum_{j=2}^{m-1} rf(X,i,j)^2 \end{aligned} \tag{30}$$

Thus the decomposition of the hexagonal lattice data can be computed by solving the minimization problem:

$$\min_{\{X,Y\in \mathbb{R}^{n\times m} \mid X+Y=Z\}} Q(X) + \sum_{i=1}^{n}\sum_{j=1}^{m} y_{ij}^2 \tag{31}$$

Due to the complexity of $Q(X)$, a simple matrix representation of this minimization problem as in (13) cannot be provided. It is possible to construct an adequate matrix with an appropriate computer program. This matrix could be used when solving the minimization problem (31).

6. Decomposition macro

It is possible to mechanize the construction of the quadratic forms required for the decomposition as well as the treatment of missing values, the enlargement of the lattice and the solution of the minimization problem. An appropriate macro was developed in the statistical programming language ISP. Its help screen is reproduced in Fig. 5. The macro distinguishes between one-dimensional arrays, i.e. vectors, which

Figure 5: Help screen of a decomposition macro

```
############## (c) F.-U. Siewert, 03.93 ##############
# exec  -- Decompose lattice data
# SYNTAX:
#    exec dat [p] > zer
# INPUTS:
#    dat  [vector(n)|array(n,m)] observations
#         If vector, decompose time series according
#         to Schlicht
#         If array, decompose rectangular lattice data
#    p    integer
#         Length of season in a time series
# PARAMETER:
#    /plo  [n|y]   Default: n If y, data and
#          decomposition will be plotted on screen
#    /enl  integer Default: 0 If >0, the lattice will
#          be enlarged by enl points in each
#          direction
# OUTPUTS:
#    zer  [array(n,3)|array(n,m,2)]
#         decomposition components
# KNOWN BUGS:
#    In time series, the iterative treatment of missing
#    values can lead under rare circumstances to wrong
#    results.
#######################################################
```

will be treated as time series and are decomposed into the three components trend, season and irregular component according to Schlicht (1981), and two-dimensional arrays which will be treated as measurements on a rectangular lattice and are decomposed into a smooth and a irregular component according to the principles mentioned above. The parameter "enl" controls the enlargement of the lattice, while "plo" provides plots of the decomposition.

This presentation is part of a "Diplomarbeit" of F.-U. Siewert. The macro will be further developed in order to cover also hexagonal data and can be obtained by sending an e-mail request to: siewert@amadeus.statistik.uni-dortmund.de.

References:

DEGENS, P.O., and HALEKOH, U. (1990): Bestimmung regulärer und irregulärer Anteile von auf Gittern erhobenen Daten mittels quadratischer Fomen. Vortrag auf dem "2. Herbstkolloquium über Biometrie und Ökologie" der Internationalen Biometrischen Gesellschaft DR in Wuppertal 1989. In: H.D. Quednau (ed.): *Tagungsberichte der AG Biometrie in der Ökologie Heft 1*, 28–42.

HALEKOH, U., and DEGENS, P.O. (1986): Additive Decomposition by Quadratic Forms. In: W. Domschke, W. Krabs, J. Lehn, P. Spelluci (eds.): *Methods of Operations Research 57*, Frankfurt, Athenäum Verlag, 49–50.

HALEKOH, U., and DEGENS, P.O. (1991): Analysis of Data Measured on a Lattice. In: H.H. Bock, P. Ihm (eds.): *Classification, data analysis and knowledge organization*, Proceedings, Jahrestagung der Gesellschaft für Klassifikation, Marburg 1990, Springer Verlag, Heidelberg, 91–98.

HALEKOH, U., and DEGENS, P.O. (1992): Behandlung von Daten mit fehlenden Werten durch Glätter. In: W. Fleischer, M. Nagel, R. Ostermann (eds.): *Interaktive Datenanalyse mit ISP*, Westarp Wissenschaften, Essen, 89–103.

SCHLICHT, E. (1981): A Seasonal Adjustment Principle and a Seasonal Adjustment Method Derived From This Principle. *J. Amer. Statist. Assoc. 76, 374–378.*

WAHBA, G. (1980): Continuous Space Series on the Plane and in *d* Dimensions. In: O. D. Anderson (ed.): *Time Series*. North Holland, Amsterdam, 387–388.

Part III

Genome and Molecular Sequence Analysis

Classification and Data Analysis in Genome Projects: Some Aspects of Mapping, Alignment and Tree Reconstruction

Berthold Lausen

Forschungsinstitut für Kinderernährung Dortmund (FKE)
Heinstück 11, D-44225 Dortmund, Germany

Summary: Some statistical aspects of classification and data analysis in genome projects are reviewed and discussed. The construction of genetic maps involves seriation and hierarchical classification problems. The alignment of molecular data and tree reconstruction with genetic distance data are methodological tasks which can be seen as two steps of a data analysis of DNA fragments sequenced in a genome project. The combination of these two steps and the availability of an enormous amount of data on several levels (primary, secondary order, etc.) provide many challenging problems and applications for classification methods.

1. Introduction

The rapid progress of genome projects provides an enourmous amount of data (cf. e.g. Fuchs et al. (1992) or Chothia (1992)). The questions raised by molecular biologists frequently cause arguments which are based on statistical considerations. For example Weir (1990) introduces underlying principles of genetic data analysis. I concentrate in my paper on statistical aspects of three methodological fields covered by classification and data analysis problems in genome analysis: Mapping, alignment and tree reconstruction.

One of the major tasks in genome projects is to map genes. Mendel's ideas provide the genetical basis for the analysis of recombination data. Statistical methods of linkage analysis are used for the construction of genetic maps (cf. Lathrop and Lalouel (1991) or Ott (1985)) which are described in Section 2.

Discussing for example the question of the number of protein families raised by Chothia (1992) we need an alignment of the protein sequences (e.g. Waterman (1984)) and some classification method to find and to separate the families. The principles and recent approaches of alignment methods for molecular data are viewed in Section 3. I concentrate on a framework for multiple alignments.

Assuming an evolutionary history I focus on hierarchical classifications and consequently the classification can be represented by a tree. Having constructed a tree as hierarchical classification of the operational taxonomic units (OTUs), e.g. proteins or species, it is important to assess the stability of the estimated hierarchical clustering. Three-objects methods provide such an assessment by a variance estimation and bootstrap estimates of the detection probability of an estimated cluster (e.g. Lausen (1991) or Degens et al. (1990)) (Section 4).

In Section 5 I discuss the reviewed aspects of classification and data analysis regarding genome projects and concentrate on recent application problems of these techniques to large data banks.

2. Construction of genetic maps

The task of genetic mapping is to construct chromosome maps for some given set $L = \{u_1, ..., u_m\}$ of marker loci $u_i \in L$. The mapping procedure is based on statistical methods for linkage analysis. Lathrop and Lalouel (1991) give a recent review on various statistical aspects of linkage analysis. In the following I describe the mapping procedure along the proposals given by Lathrop and Lalouel (1991, Sec. 5.5).

The underlying idea is to use the number of recombination events between the given marker locis for an estimation of the order of the loci on the chromosome and moreover to estimate the relative distance between the ordered locis. The proposed method is based on a likelihood model of the pairwise recombination events $r(u_i, u_j)$, which depends on the order and pairwise recombination rates. Consequently, the parameters of the likelihood model are the order, denoted by $u_{()}$, and the series of $(m-1)$ pairwise recombination rates, denoted by $\theta(u_{(i)}, u_{(i+1)}) \in [0, 0.5]$. Assuming stochastical independence of the recombination events between nonoverlapping chromosome intervals defined by different pairs of loci Lathrop and Lalouel (1991, Eq. 5.1) state a likelihood equation, which is in our notation:

$$Pr(u_{()}, \theta; r) = \prod_{i=1}^{m-1} Pr(\theta(u_{(i)}, u_{(i+1)}); r(u_{(i)}, u_{(i+1)})) \ , \tag{1}$$

where $Pr(\theta(u_{(i)}, u_{(i+1)}); r(u_{(i)}, u_{(i+1)}))$ denotes the likelihood for recombination events of the two loci $u_{(i)}$ and $u_{(i+1)}$ (see eq. (2) below). For a discussion of the genetic problems and assumptions like interference and for a review of available solutions see for example Lathrop and Lalouel (1991, Sec. 5). Following their arguments the computation of a maximum likelihood estimate for $(u_{()}, \theta)$ has to be restricted to a search in a given set of possible orders for moderate numbers of loci. The method starts with a given trial map and defines in a first step a set of loci subsets which are distinguished by high numbers of recombination events. In a second step the likelihood for all orders in each subgroup are computed, and a set of orders with high likelihoods is considered in a third step. In the third step the likelihoods for orders between the subgroups are computed and for a set of orders with high likelihoods the maximum is evaluated considering the remaining orders of the subgroups from step two. In the fourth step the maximum defines the new trial order for step one and iterations are continued until convergence occurs. The construction of the first trial map can be seen as seriation problem (Lathrop and Lalouel (1991, Example 5.5.1.)) or as hierarchical classification problem using the pairwise recombination events as distances. Seriation problems are discussed for example by Bock (1989, Chap. 8) and hierarchical classifications in Section 4. The application of a hierarchical clustering algorithm allows also an initial proposal for the choice of subgroups.

Consequently the construction of a genetic map may be seen as an application of a series of pairwise linkage analyses and the corresponding binomial likelihoods for two marker loci v and w

$$Pr(\theta(v, w); r(v, w)) = \frac{n! \theta^{r(v,w)} (1-\theta)^{n-r(v,w)}}{r(v, w)!(n - r(v, w))!} \ , \tag{2}$$

where n denotes the number of informative meioses for the two loci v and w and $r(v,w)$ the number of observed recombinations. Inference of recombinations in humans is based on the analysis of family data (pedigrees). In other higher organisms the inference of recombination can be done by classical Mendelian analysis of crosses between homozygous lines of the organism.

The different approaches of inferring recombination cause various problems of modelling the recombination data, but having the principal problem of ordering marker loci as described above, it is obvious, that the task of genetic mapping is also a challenging problem for the development of new and the adaptation of existing classification and seriation methods (cf. Feingold et al. (1993)).

3. Alignment of molecular data

Another task in genome projects is to infer the sequence structure of the chromosomes. Consequently, it is an aim of the projects to provide detailed DNA sequence information of the genome of the species analysed. Having sequenced a set of DNA fragments of the species or from one individual it is an important problem to compare the DNA fragments. The DNA fragments may be seen as a set of strings $S = \{x_1, ..., x_k\}$ over an alphabet $B = \{A, C, G, T\}$. The algorithm by Needleman and Wunsch (1970) for the alignment of molecular data is discussed for example by Bock (1984), Weir (1990) or Blum (1991). The algorithm computes the optimal alignment by allowing for three different kinds of edit operations transforming sequence x to sequences y, where $x, y \in S$: (i) deleting a character from x; (ii) inserting a character in x; and (iii) substituting a character of x by another character. By giving these operations different weights it is possible to define a cost function of the alignment defined by the series of edit operations. An alignment yielding the minimum of the cost function can be defined as an optimal alignment. For example Blum (1991) and Gotoh (1990) discuss in recent papers different concepts for optimal alignments. The value of the used cost function may be viewed as a genetic distance $d(x,y)$ between the considered sequences x and y. But having k DNA sequences we are interested in a multiple alignment for all DNA sequences which may be seen more formally as (i) the enlargement to the length μ for all sequences, and (ii) a set of edit operations of the single sequence; i.e. a map $Q : B^{\mu_1} \times ... \times B^{\mu_k} \rightarrow B^{k \times \mu}$, where μ_i is the length of the i-th original DNA sequence.

Such multiple alignments can be achieved by assuming or constructing a tree structure defining the relationship between the k sequences (cf. Sankoff and Cedergren (1983) or Felsenstein (1988)). For example Guenoche (1993) suggests multiple aligment methods based on the use of a tree structure and Day and McMorris (1993a) review the problem of discovering consensus molecular sequences and discuss applications regarding multiple sequence alignments (e.g. Vingron and Argos (1991)). For further reading I suggest the recent bibliography by Day and McMorris (1993b).

4. Tree reconstruction with genetic distance data

Considering aligned sequence data there are various methods to reconstruct or estimate a tree. The tree may be interpreted as estimate of the common evolutionary history of the involved operational taxonomic units (OTUs). For example the problem of tree reconstruction is discussed in Felsenstein (1988), Degens et al. (1990), Saitou (1991) and Swofford and Olsen (1990). Recent approaches to the problem of tree reconstruction are also the distance methods of Vach and Degens (1991a, 1991b), the relational approach of Bandelt and Dress (1993) or the consideration of invariants (cf. Day (1991), Navidi et al. (1991, 1993) and Evans and Speed (1991)).

One idea is to use pairwise genetic distance data for tree reconstruction methods. Consequently, it would be possible to use genetic distances computed from pairwise optimal alignments as distance data for the tree reconstruction methods. In contrast the multiple alignment methods based on the computation of tree structures (cf. Section 3) depend directly on the sequence data itself.

Having constructed a tree as hierarchical classification of the OTUs, e.g. proteins, it is important to assess the stability of the estimated hierarchical clustering. Three-objects methods provide such an assessment by a variance estimation and bootstrap estimates of the detection probability of an estimated cluster (Lausen and Degens (1986, 1988), Degens et al. (1990), Lausen (1991) or Wolf and Degens (1991)).

In the following I describe the setup for three-objects methods and I denote by d some measure of genetic distance between the objects of the set of OTUs O. Mathematical formulations of tree structures are the ultrametric d_u for a dendrogram or the additive tree metric d_a for an additive tree. A metric d_u on O is called an ultrametric on O, if the 'ultrametric inequality' holds:

$$d_u(x,y) \leq max\{d_u(x,y), d_u(y,z)\}, \forall x,y,z \in O \tag{3}$$

and a metric d_a on O is called an additive tree metric on O, if the 'additive tree inequality' holds:

$$\begin{aligned} & d_a(u,v) + d_a(x,y) \\ \leq & \; max\{(d_a(u,x) + d_a(v,y)), d_a(u,y) + d_a(v,x))\}, \forall u,v,x,y \in O. \end{aligned} \tag{4}$$

It seems to be straightforward to assume that the given dissimilarity provides some information for the treelike classification by a dendrogram or an additive tree, consequently I assume that the measured genetic distance can be seen as a disturbed ultrametric or additive tree metric:

$$d(\{x,y\}) = d_u(x,y) + e(\{x,y\}), \forall \{x,y\} \subset O \tag{5}$$

or

$$d(\{x,y\}) = d_a(x,y) + e(\{x,y\}), \forall \{x,y\} \subset O, \tag{6}$$

where $e(\{x,y\})$ denotes independent Gaussian distributed errors (noise) for $\{x,y\} \subset O$, $E(e(\{x,y\})) = 0$ and $Var(e(\{x,y\})) = \sigma^2 w(x,y)$, $\sigma^2, w(x,y) > 0$. w denotes given or estimated weights of the measured genetic distance for the objects x and y. The stochastical models above (eq. (5) and (6)) can be justified as a first approximation to the measured genetic distance and allow the choice of certain distance methods for tree reconstruction. On the other hand it is obvious that further knowledge of the stochastical assumptions underlying the measured genetic distance data have to be incorporated in the model (cf. Lausen and Degens (1988) and Lausen (1991)) and may result in other methods for tree reconstruction.

For a set of three objects of O, $\{x',y',z'\} \subset O$, the objects are denoted by x,y,z that for the unknown ultrametric of model (5) holds: $d_u(x,y) \leq d_u(x,z) = d_u(y,z)$ (which is also known as 'ultrametric condition', cf. eq. 3). With $d_{(i)}, i = 1,2,3$, we denote the order statistics of $d(\{x,y\})$, $d(\{x,z\})$ and $d(\{y,z\})$, and w_i denotes the corresponding weights. The generalized agglomerative average linkage method yields the following three-objects estimators of the ultrametric on $\{x',y',z'\}$:

$$\hat{d}_u(x,y) = d_{(1)}, \hat{d}_u(x,z) = \hat{d}_u(y,z) = \frac{w_2 d_{(2)} + w_3 d_{(3)}}{w_2 + w_3}. \tag{7}$$

A three-objects variance estimator is given by:

$$\hat{\sigma^2_{3Ob}}(\{x',y',z'\}) = \frac{w_2 w_3}{w_2 + w_3}(d_{(2)} - d_{(3)})^2. \tag{8}$$

Lausen and Degens (1988) call methods based on the three-objects estimators (7) and (8) "three-objects methods". For example the three-objects variance estimator is defined for the k objects of the set of OTUs O as the arithmetic mean of the three-objects variance estimators of all subsets $\{x',y',z'\} \subset O$ (cf. Lausen (1991)). Another three-object method is the bootstrap estimator of the probability of detection of an estimated cluster (dendrogram) or split (additive tree). This bootstrap assessment of the stability of estimated clusters is based on the three-objects variance estimator and consequently this measure of variability of the used genetic distance data is not computed for the overall tree reconstruction. It is straightforward to apply the three-objects methods to the additive tree situation (cf. Wolf and Degens (1991)).

5. Discussion

The classifications discussed above are based on some measure of genetic distance. The problem of estimating the genetic distance between two objects is controversial in the literature, for example the recent approaches of Weir and Basten (1990), Lausen (1991), Vach (1992) and Schöniger and von Haeseler (1993) assume a known or accepted alignment of the analysed DNA sequences. Consequently, the genetic distances are influenced by the alignment procedure which is often based explictly or informally on some kind of minimum alignment distance (cf. Section 3). A similar minimization problem arises in the context of change point models and in the literature are some statistical approaches (for example cf. Goldstein et al. (1993) or Lausen and Schumacher (1992)). But a feasible statistical approach under reasonable assumptions is still an open problem and the missing and needed link between alignment (Section 3) and tree reconstruction methods (Section 4). Moreover, the distance can be influenced by higher order considerations (e.g. Pascarella and Argos (1992)).

Using large data banks the used distance should be relatively simple and can incorporate the several levels of the data (primary or secondary order, structural information, etc.) in a preliminary and simple way (e.g. Gonnet et al. (1992), Lee (1992)). Afterwards for example the question of the number of protein families discussed by Chothia (1992) can be analysed by hierarchical clustering methods and the constructed hierarchical classification can be assessed by three-objects methods (cf. section 4).

Having described aspects of classification and data analysis in genome projects I would like to mention and emphasize the viewpoint of Weir (1990, p. xi) in his book on Genetic Data Analysis, that "Writing about analyses of molecular genetic data is like aiming at a moving target", but nevertheless the rapid progress of genome projects provides many challenging problems for classification methods.

References:

BANDELT, H.-J., and DRESS, A.W.M. (1993): A Relational Approach to Split Decomposition. In: O. Opitz, B. Lausen, and R. Klar (eds.): *Information and Classification: Concepts, Methods and Applications.* Springer, Berlin Heidelberg, 123-131.

BLUM, N. (1991): *On Locally Optimal Alignments in Genetic Sequences.* Report No. 8567-CS, Institut für Informatik, University Bonn.

BOCK, H.H. (1984): Distanzmaße zum Vergleich von Bäumen, Hierarchien und Sequenzen. In: Bock, H.H. (ed.): *Anwendungen der Klassifikation: Datenanalyse und Numerische*

Klassifikation. Indeks-Verlag, Frankfurt a. M., 52-67.

BOCK, H.H. (1989): Datenanalyse zur Strukturierung und Ordnung von Information. In: Wille, R. (ed.): *Klassifikation und Ordnung (Classification and Order)*. Indeks-Verlag, Frankfurt a. M., 1-22.

CHOTHIA, C. (1992): One Thousand Families for the Molecular Biologist. *Nature, 357*, 543-544.

DAY, W.H.E. (1991): Estimating Phylogenies With Invariant Functions of Data. In: H.-H. Bock and P. Ihm (eds.): *Classification, Data Analysis and Knowledge Organization*. Springer, Berlin Heidelberg, 248-253.

DAY, W.H.E., and MCMORRIS, F.R. (1993a): Discovering Consensus Molecular Sequences. In: O. Opitz, B. Lausen, and R. Klar (eds.): *Information and Classification: Concepts, Methods and Applications*. Springer, Berlin Heidelberg, 394-402.

DAY, W.H.E., and MCMORRIS, F.R. (1993b): Alignment, Comparison and Consensus of Molecular Sequences: A Bibliography. IFCS IV, Paris 1-4, Sept. 1993 (preprint).

DEGENS, P.O., LAUSEN, B. and VACH, W. (1990): Reconstruction of Phylogenies by Distance Data: Mathematical Framework and Statistical Analysis. In: A. Dress and A. von Haeseler (eds.): *Trees and Hierarchical Structures, Lecture Notes in Biomathematics, 84*. Springer, Berlin, 9-42.

EVANS, S.N., and SPEED, T.P. (1991): Invariants of Some Probability Models Used in Phylogenetic Inference. (preprint)

FEINGOLD, E., BROWN, P.O., and SIEGMUND, D. (1993): Gaussian Models for Genetic Linkage Analysis Using Complete High Resolution Maps of Identity-by-descent. *American Journal of Human Genetics, 53, 1*, 234-51.

FELSENSTEIN, J. (1988): Phylogenies From Molecular Sequences: Inference and Reliability. *Annual Review of Genetics, 22*, 521-565.

GONNET, G.H., COHEN, M.A., and BENNER, S.A. (1992): Exhaustive Matching of the Entire Protein Sequence Database. *Science, 256*, 1443-1445.

FUCHS, R., RICE, P., and CAMERON, G.N. (1992): Molecular Biological Databases - Present and Future. *Trends in Biotechnology, 10*, 61-66.

GOTOH, O.: Optimal Sequence Alignments Allowing for Long Gaps. *Bulletin of Mathematical Biology, 52*, 359-373.

GUENOCHE, A. (1993): Alignment and Hierarchical Clustering Method for Strings. In: O. Opitz, B. Lausen, and R. Klar (eds.): *Information and Classification: Concepts, Methods and Applications*. Springer, Berlin Heidelberg, 403-412.

LATHROP, G.M. and LALOUEL, J.M. (1991): Statistical Methods for Linkage Analysis. In: C.R. Rao and R. Chakraborty (eds.): *Handbook of Statistics, Vol. 8, Statistical Methods in Biological and Medical Sciences*. North-Holland, Amsterdam, 81-123.

LAUSEN, B. (1991): Statistical Analysis of Genetic Distance Data. In: H.-H. Bock and P. Ihm (eds.): *Classification, Data Analysis and Knowledge Organization*. Springer, Berlin Heidelberg, 254-261.

LAUSEN, B., and DEGENS, P.O. (1986): Variance Estimation and the Reconstruction of Phylogenies. In: P.O. Degens, H.-J. Hermes, and O. Opitz (eds.): *Die Klassifikation und ihr Umfeld (Classification and its Environment)*. Indeks-Verlag, Frankfurt a.M., 306-314.

LAUSEN, B., and DEGENS, P.O. (1988): Evaluation of the Reconstruction of Phylogenies With DNA-DNA Hybridization Data. In: H.H. Bock (ed.): *Classification and Related Methods of Data Analysis*. North-Holland, Amsterdam, 367-374.

LAUSEN, B., and SCHUMACHER, M. (1992): Maximally Selected Rank Statistics. *Biometrics, 48, 1*, 73-85.

LEE, R.H. (1992): Protein Model Building Using Structural Homology. *Nature, 356*, 543-544.

NAVIDI, W.C., CHURCHILL, G.A., and VON HAESELER, A. (1991): Methods for Inferring Phylogenies From Nucleic Acid Sequences by Using Maximum Likelihood and Linear Invariants. *Molecular Biology and Evolution, 8*, 128-143.

NAVIDI, W.C., CHURCHILL, G.A., and VON HAESELER, A. (1993): Phylogenetic Inference: Linear Invariants and Maximum Likelihood. *Biometrics, 49, 2,* 543-555.

NEEDLEMAN, S.B., and WUNSCH, C.D. (1970): A General Method Applicable to the Search for Similarities in the Amino Acid Sequence of two Proteins. *Journal of Molecular Biology, 48*, 443-453.

OTT, J. (1985): *Analysis of Human Genetic Linkage.* John-Hopkins University Press, Baltimore.

PASCARELLA, S., and ARGOS, P. (1992): A Data Bank Merging Related Protein Structures and Sequences. *Protein Engineering, 5*, 121-137.

SAITOU, N. (1991): Statistical Methods for Phylogenetic Tree Reconstruction. In: C.R. Rao and R. Chakraborty (eds.): *Handbook of Statistics, Vol. 8, Statistical Methods in Biological and Medical Sciences.* North-Holland, Amsterdam , 317-346.

SANKOFF, D., and CEDERGREN, R.J. (1983): Simultaneous Comparison of Three or More Sequences Related by a Tree. In: D. Sankoff and J.B. Kruskal (eds.): *Time Warps, String Edits, and Macromolecules.* Addison-Wesley, Reading/Massachusetts, 253-263.

SCHÖNIGER, M., and VON HAESELER, A. (1993). More Reliable Phylogenies by Properly Weighted Nucleotide Substitutions. In: O. Opitz, B. Lausen, and R. Klar (eds.): *Information and Classification: Concepts, Methods and Applications.* Springer, Berlin Heidelberg, 413-420.

SWOFFORD, D.L., and OLSEN, G.J. (1990): Phylogeny Reconstruction. In: D.M. Hillis and C. Moritz (eds.): *Molecular Systematics.* Sinauer Ass., Massachusetts, 411-501.

VACH, W. (1992): The Jukes-Cantor Transformation and Additivity of Estimated Genetic Distances. In: M. Schader (ed): *Analysing and Modeling Data and Knowledge.* Springer, Berlin Heidelberg, 141-150.

VACH, W., and DEGENS, P.O. (1991a): Least-Squares Approximation of Additive Trees to Dissimilarities - Characterizations and Algorithms. *Computational Statistics Quarterly, 3*, 203-218.

VACH, W., and DEGENS, P.O. (1991b): A new Approach to Isotonic Agglomerative Hierarchical Clustering. *Journal of Classification, 8*, 217-237.

VINGRON, M., and ARGOS, P. (1991): Motif Recognition and Alignment by Consensus. *Journal of Molecular Biology, 218*, 33-43.

WATERMAN, M.S. (1984): General Methods of Sequence Comparisons. *Bulletin of Mathematical Biology, 46, 4,* 473-500.

WEIR, B.S. (1990): *Genetic Data Analysis.* Sinauer Ass., Massachusetts.

WEIR, B.S., and BASTEN, C. (1990): Sampling Strategies for Distances Between DNA Sequences. *Biometrics, 46*, 551-572.

WOLF, K., and DEGENS, P.O. (1991): Variance Estimation in the Additive Tree Model. In: H.-H. Bock and P. Ihm (eds.): *Classification, Data Analysis and Knowledge Organization.* Springer, Berlin Heidelberg, 262-269.

Multiple Alignment of Protein Sequences and Construction of Evolutionary Trees based on Amino Acid Properties - an Algebraic Approach

Willi Schmidt
Max-Delbrück-Center for Molecular Medicine,
Dept. Bioinformatics/Biomathematics,
Robert-Rössle-Str.10, D-13122 Berlin-Buch, Germany

Summary.: This paper presents an essentially new method to construct phylogenetic trees from related amino acid sequences. The method is based on a new distance measure which describes sequence relationships by means of typical steric and physicochemical properties of the amino acids and is advantageous in some essential points. In particular, it avoids negative branches. The method was applied to different sets of protein sequences and the results compared with other well-established methods.

1. Introduction

The interrelationships in a set of amino acid sequences (the primary structures of proteins) are commonly represented as a weighted tree (Fitch and Margoliash 1967; Moore et al. 1973; Sankoff et al. 1982; Penny and Hendy 1986). In such a tree the distance (dissimilarity) between any pair of sequences illustrated is approximately the length of the unique path connecting the corresponding nodes (additivity).
For interpretation purposes, the tree represents the probable evolutionary process which has created the present diversity of the related sequences. Its branching structure is often obtained by means of a stepwise classification procedure aggregating successively those actual objects (i.e. given sequences or postulated predecessors of them) with maximum similarity followed by the re-evaluation of the distances in pairs.
As a rule, the distances to the newly created node will be given by averaging the distances to the aggregated nodes whereas all other distances remain unchanged. The procedure is considered concluded when only one object remains. This procedure must be followed by a second procedure which determines the branch lengths of the constructed tree. However, in general complete additivity is unrealizable. Indeed, for more than three sequences this postulate is equivalent with an overdetermined system of linear equations, i.e. with more equations than unknowns.
As a consequence, the mentioned postulate must be replaced by a weaker postulate which also accepts approximative solutions to the equations.This can be realized, for example, by least square methods.
However, a further problem remains unsolved, i.e. what must be done if negative branch lengths were calculated, which clearly cannot be interpreted biologically? On the one hand, such negative branches can be a consequence of assuming a biologically unacceptable branching order. They can also follow from improper alignments possibly combined with unsuitable gap penalties. Clearly, this part of the problem can be solved by re-analyzing the primary data. However, in general this requires additional information not directly contained in the sequence data. On the other hand, if errors of the even types mentioned cannot be detected there is no guarantee for non-negative branch lengths. In such cases the commonly used computer programs consider alternative trees, i.e. slightly modified trees which do not contain negative branches. In principle, to find such a solution the corresponding tree modifying procedure must be

applied repeatedly (Klotz and Blanken 1981; inter alia). However, we are convinced that solutions of this kind are unsatisfactory from a theoretical point of view. We should therefore look into the more profound reasons for these unacceptable solutions arising in some cases. We have reached the conclusion that the distance measures characterizing the dissimilarities between aligned amino acid sequences are uncertain with respect to deriving evolutionary trees. Indeed, the postulate of additivity of the branch length evaluation stated above seems to be artificial and is equivalent to strong model assumptions about the evolutionary process which are probably not applicable in practice. Nevertheless, additivity is a sound principle when representing the evolutionary process in the form of a binary tree.

Furthermore, the distance measure based on mismatches and gap penalties is a biologically well-founded entity which characterizes dissimilarities between sequences. Thus, to reach a compromise between the seemingly incompatible postulates mentioned above we propose a new distance measure which is similar to the usual one and which possesses advantageous properties. In particular, negative branches are excluded a priori. The idea on which this new measure is based was proposed by Taylor (1986) and Zvelebil et al. (1987). In this new method each amino acid is represented by a vector of steric and physicochemical properties, either present or not present. To do this, Taylor (1986) proposed a total of 11 properties. Due to certain special features, glycine and proline are interpreted as properties per se. Clearly, the number of possible vectors of this kind is much greater than the number of amino acids. Therefore, there are many vectors which cannot be identified with amino acids. The property pattern approach (Bork and Grunwald (1989)) performs protein sequence comparisons on the basis of the aligned amino acids and has been proven to be successful in database searches for well-defined structural and functional motifs (Bork and Grunwald (1989)). As shown in the following, the properties of amino acids can be used to define a well-founded measure describing the degree of similarity of amino acid sequences and, moreover, to derive an algorithm for the construction of evolutionary trees which simultaneously determines branching order and branch length.

2. Mathematical model

The amino acids are denoted by the capitals A,B,C ..., in this section so as not to interpret them as the one-letter codes. Let Ω be the set of all amino acids. Further, for an arbitrary amino acid X we denote the set of properties present for X (i.e. the corresponding components of the property vector are equal to 1) by P_X (associated property set). In our applications described in this paper our work is based on the properties listed in fig.1 in the paper by Bork and Grunwald (1990).

Now, let A be an arbitrary amino acid. Using its associated property set P_A, the amino acid A induces a classification of all amino acids. Let X and Y be any two other amino acids. We then define, X and Y to be equivalent (with respect to A) assuming the equation $P_A \cap P_X = P_A \cap P_Y$ is valid. It is easy to see that this relation between amino acids is reflexive, symmetric and transitive. Consequently, the relation defines a classification of the set Ω, i.e. a partition of Ω into a number of disjunctive subsets in pairs each corresponding to a well-defined common subset of properties of A. We denote this partition by the symbol Π_A.

Properties not present in A are not considered. Trivially, there are a large number of partitions of Ω, however, most of them are not generated by amino acids.

2.1 The lattice of the partitions of a set

In the following we consider the partitions of an arbitrary set S. Let Part(S) be the

total of partitions of a given set S. Part(S) can be partially ordered by introducing the relation "<": Let Π_1 and Π_2 be arbitrary chosen partitions of S. Thus, by definition, $\Pi_1 < \Pi_2$ is equivalent to the statement "every class of Π_1 is completely contained in a (uniquely determined) class of Π_2 ". Obviously, this relation is transitive and from $\Pi_1 < \Pi_2$, $\Pi_2 < \Pi_1$ it follows $\Pi_1 = \Pi_2$.
Furthermore, the set Part(S) can be equipped with an algebraic structure by introducing the operations addition and multiplication. The product $\Pi_1 * \Pi_2$ is defined by the total b1 $\cap$ b2 of non-empty sections of classes b1 of Π_1 and b2 of Π_2. Obviously, this is a partition.
The definition of the sum $\Pi_1 + \Pi_2$ is somewhat more complicated. Qualitatively, the sum $\Pi_1 + \Pi_2$ is the smallest partition Π (in the sense of the relation "<") dominating both Π_1 and Π_2 : $\Pi_1 < \Pi$ and $\Pi_2 < \Pi$.
The detailed theory of the algebraic structure (lattice) defined in this way is described extensively in the monograph by Hartmanis and Stearns (1967). We note without proof some essential facts :
1) The partition '0', defined by the singletons of S is the smallest element of Part(S) and plays the role of the zero with respect to addition and multiplication, i.e. we have for arbitrary $\Pi \in$ Part(S): '0'< Π , Π*'0'= '0' and Π + '0'= Π .
2) The partition '1', consisting of the single class S is the largest element of Part(S) and plays the role of the unity element with respect to multiplication:
Π*'1'= Π and Π + '1'= '1' for all $\Pi \in$ Part(S).
Consequently, we generally have '0'< Π < '1'.
3) The operations addition and multiplication are associative and commutative.
4) Assuming $\Pi_1 < \Pi_2$, it then follows that $\Pi_1 * \Pi_2 = \Pi_1$ and $\Pi_1 + \Pi_2 = \Pi_2$.

2.2 A metric on Part(S)

To discuss further properties of the structure of Part(S) we introduce a distance measure for pairs of partitions, i.e. the structure Part(S) will be equipped with a metric. For definition we assume that S is finite (the following definitions can be generalized but this is irrelevant for our purposes). The first step is to define the covering numbers: Let Π be a partition of S and s an arbitrary element of S. Then $cov(\Pi, s)$ is defined as the number of elements of the (uniquely determined) class of Π containing s. For the second step we define the distance between two partitions Π_1 and Π_2 assuming that the relation $\Pi_1 < \Pi_2$ is valid. Then, distance d is defined as follows:

$$d(\Pi_1, \Pi_2) = \sum_{s \in S}(cov(\Pi_2, s) - cov(\Pi_1, s)).$$

Finally, for arbitrary partitions Π_1 and Π_2 the distance is defined by the equation

$$d(\Pi_1, \Pi_2) = d(\Pi_1, \Pi_1 + \Pi_2) + d(\Pi_2, \Pi_1 + \Pi_2).$$

Note that the second part of this definition is compatible with the first part because of statements 4) (section 2.1) and 5),
However, further statements are required.
5) The values of the distance function d are non-negative integers and $d(\Pi_1, \Pi_2) = 0$ is equivalent to $\Pi_1 = \Pi_2$.
6) The function d is symmetric: $d(\Pi_1, \Pi_2) = d(\Pi_2, \Pi_1)$.
7) Assuming $\Pi_1 < \Pi_2 < \Pi_3$, it then follows that

$$d(\Pi_1, \Pi_3) = d(\Pi_1, \Pi_2) + d(\Pi_2, \Pi_3).$$

8) Let Π_1, Π_2, Π_3 be arbitrary partitions.
The branching triangle inequality holds true as follows:

$$d(\Pi_1,\Pi_3) \leq d(\Pi_1,\Pi_1+\Pi_2+\Pi_3) + d(\Pi_3,\Pi_1+\Pi_2+\Pi_3).$$

In the following section we will show how this structure can be used as a well-suited mathematical framework for the construction of evolutionary trees.

2.3 Constructing trees from partitions generated by property patterns

For further development we interpret the distances between the partitions generated by two amino acids A and B as distances between A and B itself. One of the preconditions to justify this is that the properties used for constructing the partitions are essential for determination of structure and function of the folded protein molecule. But there is no doubt that this is so. On the other hand, the measure is based on the common properties of the amino acids to be compared and, moreover, it considers these properties in the total of all amino acids. This is an important feature because in this way we obtain some weighting of the properties. Indeed, it can be seen from detailed consideration that - according to tendency - rare properties contribute more to the distance measure than more frequent ones. Thus, we expect that the proposed measure will be advantageous from a biological point of view. To clarify the correspondence between an amino acid A and its partition Π_A we provide some additional statements. Let p be an arbitrary property, e.g. p = 'aliphatic'. Then p defines a special partition Π_p of the amino acids, consisting of exactly two classes: One class = all amino acids having this property, the other class = all amino acids which do not have this property. Partitions of this special form are denoted as dichotomies. It can be shown that the associated partitions Π_A can be expressed by the product of all dichotomies derived in this manner from the properties of A: If the following applies $P_A = \{p_1, ..., p_k\}$, then:

$$\Pi_A = \Pi_{p_1} * \Pi_{p_2} * \cdots * \Pi_{p_k}.$$

Let $A_1A_2...A_r$ and $B_1B_2...B_r$ be two aligned amino acid sequences of length r. This includes that some of the A_i and B_j (i,j = 1...r) may be gaps due to insertions and deletions (indels). To denote gaps we use the symbol '-'. To define the distance between these sequences we use the distance measure for amino acids based on the associated partitions of the last section. Obviously, due to the possible gaps, we need a generalization for the symbol '-'. To do this we put $\Pi_- =$ '1'(defined in statement 2), section 2.1).
Clearly, this is arbitrary to a certain extent. Indeed, if it were known from the total alignment that some properties are predominant (or completely conserved) at the considered position then, obviously, the product of the corresponding dichotomies should be taken instead of the partition '1'. Consequently, in the following, amino acid sequences are replaced by partition sequences defined in the set Ω of amino acids. After these preparatory calculations the definition of the distance between the considered sequences is straightforward. This can be defined as follows:

$$d(A_1...A_r, B_1...B_r) = \sum_{i=1...r} d(\Pi_{A_i}, \Pi_{B_i}).$$

To construct evolutionary trees this definition will also be used if there are other arbitrary partitions instead of the associated partitions at the considered sequence positions. For this construction we use the following steps:
1 - determine the minimum distance among the current set of (partition) sequences

2 - replace the corresponding two sequences (assuming $A_1...A_r$ and $B_1...B_r$) by the "sum" $A_1 + B_1...A_r + B_r$
3 - determine the distances of the sequences $A_1...A_r$ and $B_1...B_r$ to the sum sequence $A_1 + B_1...A_r + B_r$
4 - actualize the total distance matrix by replacing the distances of all other aligned sequences to the sequences omitted in step 2 by their distances to the sum sequence
5 - repeat these steps until the number of partition sequences is 1.
The sum sequence defined in step 2 will be interpreted as the ancestor sequence of the two sequences from which it was derived. Note that the sum of two real amino acid sequences will not be generally equivalent to another amino acid sequence. Nevertheless, for matches ($A_i = B_i$) the sum equals the summands at this stage (statement 4, last section).
On the other hand, from A_i ='1' or B_i ='1' it also follows that $A_i + B_i$ ='1'.
As an important byproduct of the tree construction procedure described the branch lengths of this tree can be obtained in a natural manner. Indeed, this is the task of step 3. Moreover, from the distance measure and the general statements in the last section it follows that each branching step is locally additive: the distance of the two sequences is represented as the sum of the distances to their common 'ancestor' sequence. Additionally, the property of additivity is also given along the evolutionary pathways, i.e. the distance of any original sequence to one of its ancestral sequences is given as the sum of all distances between neighbouring intermediate ancestors on the corresponding path.
For all other distances it can be stated that the sum of the branch lengths connecting the sequences in that tree is larger than the corresponding distances, equality being an exception. This follows immediately from the branching triangle inequality and should be interpreted as expressing convergence of the evolutionary process.

3. Results and comparison with other methods

To have well-founded comparisons, the proposed method was applied to a series of multiple alignments and tree constructions published in the literature. In particular, we derived with our new method the evolutionary trees from three aligned groups of amino acid sequences presented and discussed in the paper by Feng and Doolittle (1987). In this paper the trees were compared with the results of the Fitch and Margoliash (1967) procedure and the trees obtained with the "progressive method" of these authors do in fact coincide with the biological expectations. Thus, we assume that these results are a valid basis for testing our new method. In the following, we present the result of the tree reconstruction for a group of 11 hemoglobin sequences. The "progressive method" of Feng and Doolittle was applied to these sequences covering a broad spectrum of types and some significant differences with respect to the tree topologies resulted compared with those trees found from simple binary alignments. Indeed, as pointed out in the paper of Feng and Doolittle, the "progressive" tree is more in line with the probable biological history. As can be seen from Fig.1 the "algebraic" tree is very similar to the "progressive" one, but there are some minor differences. Firstly, the human sequences are now in closer relationship because the hagfish and the lamprey hemoglobin sequences (heha and hbrl) are grouped outside the human subfamily mentioned. Somewhat surprisingly, the human and the gastropod myoglobins are now a subfamily. The second difference between the trees is the obvious greater homogenity of the distances to the common ancestor sequence. The same observation applied to the superoxide dismutase sequences, the tyrosine kinase-like sequences and also for sequence-derived trees obtained by other authors.

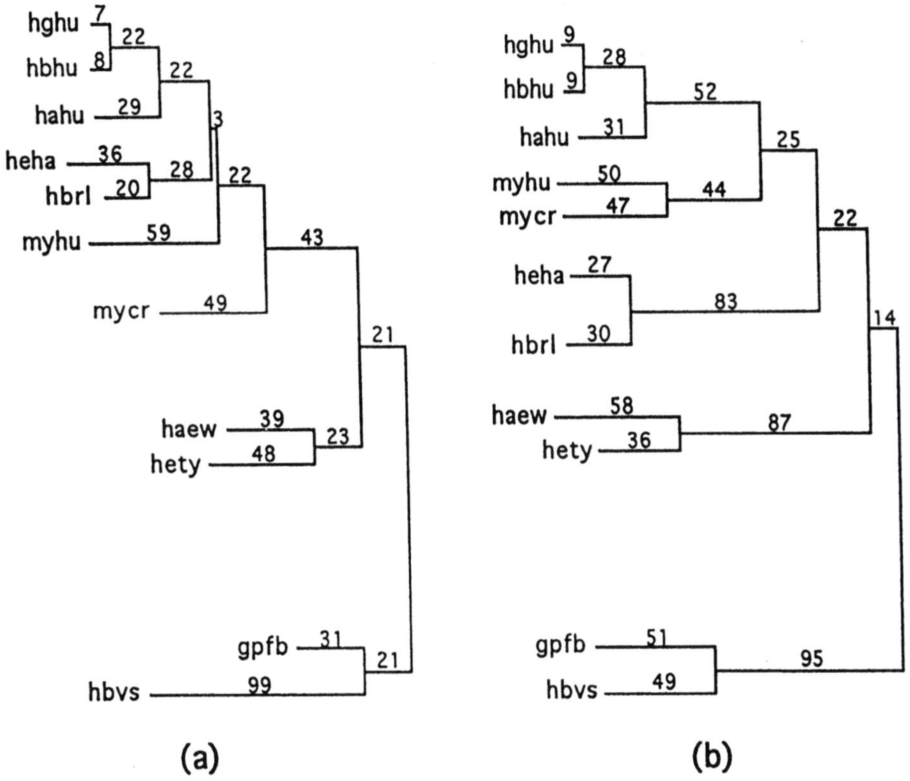

Fig. 1: Phylogenetic trees for 11 globin sequences as determined by (a) the progressive multiple alignment method (Feng and Doolittle (1987)) and (b) the algebraic method proposed in this paper.
Designations: hghu, human globin γ chain; hbhu, human globin β chain; hahu, human globin α chain; heha, hagfish hemoglobin; hbrl, lamprey hemoglobin; myhu, human myoglobin; mycr, gastropod myoglobin; hety, earthworm myoglobin (*Tylorrhynchus*); haew, earthworm hemoglobin (*Lumbricus*); gpfb, kidney bean leghemoglobin; hbvs, bacterial hemoglobin (*Vitreoscilla*).
Tree (a) reproduced from Feng and Doolittle (1987) with permission of the authors.

4. Discussion

The proposed method to construct evolutionary trees from multiple amino acid sequence alignments was derived from the property pattern approach described and successfully applied by the authors Bork and Grunwald (1990). Moreover, the new method is a generalization and refinement of that approach from a mathematical point of view. The amino acid properties induce in a very natural and unique way a new distance measure which describes the dissimilarities between the amino acid sequences.

One of the main advantages of the proposed tree construction method is the fact that determination of the branching order and calculation of the branch lengths are simultaneous steps and negative branch lengths are excluded by the construction principle. Consequently, there will be no need for a posteriori modifications of the

tree topology calculated. Moreover, the branch lengths calculated define, at least locally, an additive evaluation of the tree and the possible global deviations from additivity can be clearly interpreted.
To express the dissimilarity of paired sequences quantitatively within the proposed mathematical framework there is no need to use artificial gap penalties. Indeed, the distance measure describing such sequence positions is uniquely defined on the basis of the property structure within the aligned sequences at this position. This should be considered as a further advantage of the proposed method because choosing penalties is a highly arbitrary decision and, moreover, it is difficult to find plausible biological interpretations of such penalties.
The proposed method to construct evolutionary trees has been applied also to the very actual problem of searching for well-defined motifs, i.e. sequence parts defining structural and/or functional subunits on proteins (e.g. binding sites), in databases (results will be published elsewhere). We remark that in all applications considered up to now our method has a higher level of sensitivity and specificity than conventional searching procedures. On the other hand, we remark that the decision for or against acceptance of a given sequence as a motif in database searches does not depend on probabilistic assumptions. Indeed this is the case if given sequences are interpreted as so-called random sequences. In our opinion, this is a somewhat doubtful notion at least from a biological point of view and, therefore, it seems to be advantageous to have decision rules which avoid reference to such purely probabilistic and biologically meaningless sequence models.

Acknowledgements

I wish to thank Prof. Jens Reich and Dr. Klaus Rohde, Max-Delbrück-Center for Molecular Medicine, Berlin-Buch, for valuable discussions and I am grateful to my colleagues Anita Nothnagel and Gudrun Nürnberg for their excellent technical assistance. Thanks also to the anonymous referees for their detailed and helpful criticisms of the earlier version of this paper.

5. References

BORK, P., and GRUNWALD, C. (1989): A method for property pattern searches in protein sequence databases, demonstrated by detection of GTP binding sites. *Stud. biophys., 129, 231–240.*

BORK, P., and GRUNWALD, C. (1990): Recognition of different nucleotide-binding sites in primary structures using a property pattern approach. *Eur. J. Biochem., 191, 347-358.*

FENG, D.-F., and DOOLITTLE, R. F. (1987): Progressive Sequence Alignment as Prerequisite to Correct Phylogenetic Trees. *J. Mol. Evol., 25, 351–360.*

FITCH, W. M., MARGOLIASH, E. (1967): Construction of phylogenetic trees. *Science, 15, 279–284.*

KLOTZ, L. C., BLANKEN, R. L. (1981): A practical method for calculating evolutionary trees from sequence data. *J. Theor. Biol., 91, 261–272.*

HARTMANIS, J., and STEARNS, R. E. (1966): *Algebraic Structure Theory of Sequential Machines.* Prentice-Hall, Englewood Cliffs. N.J.

MOORE, G. M., et al. (1973): An iterative approach from the standpoint of the additive hypothesis to the dendrogram problem posed by molecular data sets. *J. Theor. Biol., 38,*

423–457.

SANKOFF, D., et al. (1982): A strategy for sequence phylogeny research. *Nucleic Acids Res., 10, 421–431.*

TAYLOR, W. R. (1986): Identification of Protein Sequence Homology by Consensus Template Alignment. *J. Mol. Biol., 188, 233–258.*

PENNY, D., and HENDY, M. (1986): Estimating the reliability of evolutionary trees. *Mol. Biol. Evol., 3, 403–417.*

ZVELEBIL, M. J., et al. (1987): Prediction of Protein Secondary Structure and Active Sites using the Alignment of Homologous Sequences. *J. Mol. Biol., 195, 957–961.*

How to Deal With Third Codon Positions in Phylogenetic Analysis

Michael Schöniger[1], Axel Janke[2] and Arndt von Haeseler[2]

[1] Theoretische Chemie, Technische Universität München, Lichtenbergstraße 4, D-85747 Garching, Germany

[2] Zoologisches Institut, Universität München, Luisenstraße 14, D-80333 München, Germany

Abstract: We investigated all protein coding genes of nine vertebrate mitochondrial genomes. Bootstrap analyses of maximum parsimony trees revealed a tree with bootstrap probabilities of at least 99 % for each branch. If we excluded third codon positions, the branching order of whale, cow and seal changed to a biologically not accepted topology. This observation questions the widespread practice of ignoring third codon positions or only using transversional substitutions at those sites. We explain why considering the changes in third positions is inevitable for the inference of short internal branches of phylogenies under certain evolutionary parameters. We propose a new scheme of analysis: including properly weighted third codon positions. Simulation studies using Kimura's and Hasegawa's model of evolution with parameters inferred from mitochondrial data illustrate our findings.

1. Introduction and the Problem

It is well known that third codon positions in protein coding genes show high substitution rates due to the degeneracy of the genetic code. Thus the rapid accumulation of substitutions may lead to wrong phylogenies, when those positions are subject to a phylogenetic analysis. As a consequence, it is common practice to ignore third codon positions in phylogenetic analyses.

On the other hand, Irwin et al. (1991) and Edwards et al. (1991) showed that it is necessary to consider transversional changes of third codon positions of cytochrome b genes of vertebrate mitochondria in order to obtain a biologically reasonable tree. Edwards et al. (1991) weighted transversions 20 times more than transitions.

We studied the 13 protein coding genes of nine vertebrate mitochondria[1] whale, cow, seal, human, rat, mouse, chicken, frog and opossum (Janke et al. (1993)). In the bootstrapped maximum parsimony tree (PHYLIP version 3.3, Felsenstein (1990)) for the totality of the genes the biologically adequate grouping of cow and whale was detected only when third codon positions were included in the analysis (see fig. 1). However, bootstrap analysis using neighbor–joining as a tree reconstruction method (Saitou and Nei (1987)) provided the true tree no matter if we ignored third positions. This indicated that maximum parsimony is not suited very well for this data set.

Based on estimates from substitution rates and transition–transversion biases from our data, we conducted several simulation studies to elucidate the observation above. Sequences were generated according to Kimura's (1980) or Hasegawa's (Hasegawa et al. (1985)) model of sequence evolution. Distance matrices were then computed using the appropriate weighting and correction schemes (Schöniger and von Haeseler (1993)). Finally a neighbor–joining tree was computed.

[1]Sequences can be retrieved from EMBL under accession numbers: X61145, V00654, X63726, J01415, X14848, J01420, X52392, X02890. The opossum sequence can be obtained from the authors.

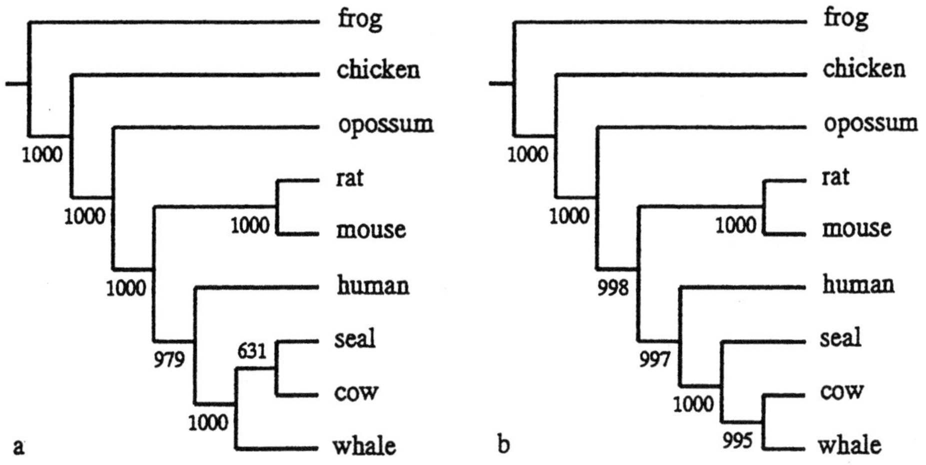

Figure 1: Bootstrapped maximum parsimony trees (PHYLIP version 3.3) from the alignment (gaps excluded) of all 13 protein coding mitochondrial genes of nine vertebrates (length: 3721 codons). Numbers at bifurcations indicate how many times out of 1000 bootstrap replicates the corresponding clustering was observed. (a) Only first and second codon positions were analyzed. (b) All codon positions were analyzed.

2. Methods

2.1 Estimation of Parameters

For Kimura's and Hasegawa's model of sequence evolution it is necessary to obtain estimates for the transition–transversion ratio. Moreover in the latter case it is also necessary to get estimates for the stationary nucleotide distribution.

The stationary nucleotide distribution was inferred from the average nucleotide frequencies of the nine mitochondrial genomes. The rather skewed distribution was computed for each codon position separately (tab. 1). This result correlates with the data reported by Irwin et al. (1991, table 3).

To avoid a severe underestimation of the true transition–transversion bias B, $\hat{B}$ was estimated from the closely related species rat and mouse. B is defined as the ratio of the number of transitions and the total number of substitutions. We obtained the

	1st	2nd	3rd	codon position
A	31.5	19.3	42.0	
G	21.0	12.3	5.0	
C	24.7	26.7	31.6	
T	22.8	41.7	21.4	

Table 1: Average nucleotide composition for nine complete mammalian mt–DNAs (in percent).

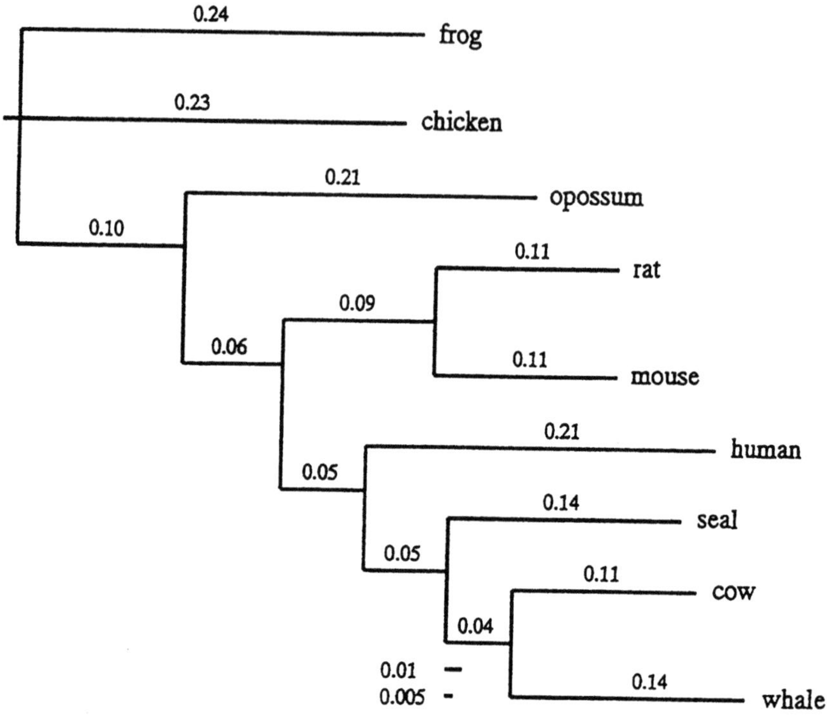

Figure 2: Maximum likelihood tree (PHYLIP version 3.3) for the protein coding genes of nine vertebrate mitochondria. Numbers along the branches are estimated substitution rates per site. This tree was used as model tree for simulation studies. The edge that separates cow and whale was shrunk to 0.01 and 0.005.

following estimates $\hat{B}_1 = 0.641$, $\hat{B}_2 = 0.598$ and $\hat{B}_3 = 0.508$, where the indices refer to the corresponding codon position. $\hat{B}_3$ may already be an underestimate of the true B_3 value since the high substitution rate at those sites can already obscure some transitions.

The relative substitution rates among the codon positions were estimated from the pair rat–mouse to be 0.62 : 0.25 : 2.13 . Similar parameters were obtained using other closely related pairs like cow–whale and cow–seal.

As model tree we used the maximum likelihood tree (PHYLIP version 3.3) obtained from the alignment of the 13 protein genes (fig. 2). The estimated number of substitutions per site was used as value for the evolutionary rate along this branch. Sequences evolved along this tree according to different models of substitutions. The topology of the maximum likelihood tree is identical to the most parsimonious tree.

2.2 Models of Nucleotide Substitutions

2.2.1 Kimura Model: As a simple model of nucleotide substitutions we employed Kimura's (1980) model. A root sequence was generated using pseudorandom numbers with uniformly distributed nucleotides. Along the model tree substitutions were performed using Poisson distributed numbers of changes with means equal to the estimated rates of fig. 2. B was 0.6 for all codon positions and the relative substitution rates were 0.25 : 0.25 : 2.5 for the codon positions. Thus, the probability P_{ij} for substituting nucleotide i by nucleotide j is

$$P_{ij} = \begin{cases} 0.6 & \text{for transition} \\ 0.2 & \text{for transversion} \end{cases} \quad (1)$$

2.2.2 Hasegawa Model: Since the mitochondrial data showed remarkably skewed nucleotide compositions we analyzed the more complex Hasegawa model (Hasegawa et al. (1985)). Now the root sequence was generated according to the estimated average nucleotide composition (cf. tab. 1). Biases B_k and relative substitution rates were used from estimates from subsection 2.1. The probability P_{ij} for substituting nucleotide i by nucleotide j is then

$$P_{ij} = \begin{cases} \alpha\pi_j & \text{for transition} \\ \beta\pi_j & \text{for transversion} \end{cases} \quad (2)$$

where π_j is the frequency of nucleotide j. It can be shown that

$$\frac{\beta}{\alpha} = \frac{(\pi_A\pi_G + \pi_C\pi_T)}{(\pi_A + \pi_G)(\pi_C + \pi_T)} \left(\frac{1}{B} - 1\right). \quad (3)$$

2.3 Weighting Schemes

Schöniger and von Haeseler (1993) showed that weighting nucleotide substitutions by assigning lower weights to more frequently observed changes and higher weights to rare and more informative ones enhances the performance of tree reconstruction methods. For the model tree as illustrated in fig. 2 existential weighting is applicable. Existential weighting suggested by Williams and Fitch (1990) assigns weights for substitutions between nucleotide i and j that are inversely proportional to the number of overall alignment positions containing both i and j. The weights thus obtained were used in the following schemes of pairwise distance calculations:

a) All three codon positions were weighted uniformly: 1+2+3.

b) All three codon positions were weighted existentially: (1+2+3):EX.

c) Codon positions one and two were uniformly whereas the third was existentially weighted: 1+2+(3:EX).

d) Codon positions one and two were uniformly weighted; third positions: only transversions were regarded: 1+2+(3:TRV).

e) Only first and second codon positions weighted uniformly were used: 1+2.

f) Only first and second codon positions weighted existentially were used: (1+2):EX.

2.4 Tree Reconstruction

We used neighbor–joining (Saitou and Nei (1987)) as a fast and efficient distance matrix method.

Percentages of correctly reconstructed model trees cow + whale edge: 0.01					
length:	2400	3000	3600	4200	4800
Kimura model					
1+2+3	83.9	91.9	94.9	96.2	96.3
(1+2+3):EX	85.3	92.4	95.7 *	95.5	97.0 *
1+2+(3:EX)	84.4	92.7 *	95.2	96.5 *	96.4
1+2+(3:TRV)	84.8	88.6 –	90.9 –	93.2 –	95.1
1+2	85.2	91.6	91.6 –	95.2	95.5
(1+2):EX	86.3 *	92.4	91.5 –	94.2	95.6
Hasegawa model					
1+2+3	86.0	93.5	93.1	95.9	95.2
(1+2+3):EX	86.1	95.0 *	93.2	96.1	96.5
1+2+(3:EX)	87.5 *	94.4	93.9 *	96.8 *	96.8 *
1+2+(3:TRV)	87.0	92.4	93.5	94.6	96.1
1+2	84.9	89.2 –	92.6	93.1 –	96.8 *
(1+2):EX	85.7	90.2 –	92.8	94.8-	96.4

Table 2: Results of simulations described in section 2. Schemes 1+2 and (1+2):EX use only 2/3 of the overall sequence lengths. Asterisks indicate the best value of the corresponding column. Percentages differing by more than 3 from the best one are labeled with minus signs.

3. Results

Tabs. 2 and 3 show the percentages of correctly reconstructed trees (P_c) for various weighting schemes, different sequence lengths and two substitution rates ($a = 0.01$ and $a = 0.005$) for the branch separating cow and whale. Since the neighbor–joining method always computed the correct tree when $a = 0.04$ it was necessary to reduce a to the shown values. P_c values are based on 1000 simulations. Only medium sequence lengths were investigated. If sequences are too short sampling problems may render the distance estimation very unreliable. On the other hand, all (P_c) values will approach 100 % if sequences become very long.

The weighting method that provided the largest P_c value for a distinct model and a fixed sequence length is marked with an asterisk. If an alternative method was distinctly worse than the best one, it is labeled with a minus sign.

Although the performance of all schemes differ by at most 10.6 % it is evident that EX weighting is the most robust procedure to infer a tree topology. The best procedure is weighting the third codon positions with EX, and the remaining being uniformly weighted. These results show that it is not advisable to neglect the information contained in third codon positions. Also the strategy of taking into account only transversions among third positions is not superior to EX weighting.

These features are more pronounced if we used a very short branch separating cow and whale ($a = 0.005$).

Percentages of correctly reconstructed model trees cow + whale edge: 0.005					
length:	2400	3000	3600	4200	4800
Kimura model					
1+2+3	60.9 –	72.0	79.3	80.2	80.0
(1+2+3):EX	65.2 *	72.5	82.0 *	80.8 *	80.0
1+2+(3:EX)	64.1	72.7 *	81.0	80.5	82.3 *
1+2+(3:TRV)	64.1	67.0 –	73.4 –	77.5 –	81.3
1+2	63.2	68.8 –	71.7 –	77.4 –	79.0 –
(1+2):EX	64.6	67.6 –	71.4 –	77.4 –	81.5
Hasegawa model					
1+2+3	67.0 *	75.1 *	79.3 *	76.5	82.9
(1+2+3):EX	66.4	73.5	78.7	76.0	81.6
1+2+(3:EX)	66.9	74.5	77.8	77.2 *	83.4 *
1+2+(3:TRV)	65.6	71.7 –	77.8	75.8	79.3 –
1+2	63.4 –	67.5 –	72.9 –	72.6 –	79.1 –
(1+2):EX	66.6	68.4 –	73.8 –	73.7 –	80.0 –

Table 3: Results of simulations described in section 2. Details: see tab. 2.

4. Discussion

Our results indicate that properly weighted substitutions can significantly enhance the reliability of tree reconstruction methods. For the analysis of the mitochondrial data it is recommended to compute EX weighting matrices for the third codon positions and to use uniform weight matrices for the first and second positions. However, the correct choice of the appropriate weighting scheme depends on the data (see Schöniger and von Haeseler (1993)).

In the special situation studied here the correctness of the reconstructed tree crucially depends on the correct recognition of the four species subtree shown in fig. 3 that is a subtree of our model tree. Tab. 4 presents the expected numbers of substitutions along the central edge of that tree assuming a Kimura model of sequence evolution. Assuming a sequence length of 3000 nucleotides, substitution rates and B were taken from 2.2.1. Neighbor-joining correctly infers the clustering of cow(C) and whale(W) versus human(H) and seal(S) if

$$d(\mathrm{C,W}) + d(\mathrm{H,S}) < \min\{d(\mathrm{C,H}) + d(\mathrm{W,S}), d(\mathrm{C,S}) + d(\mathrm{H,W})\},$$

where d is a suitable dissimilarity measure. Ignoring parallel and back mutations it is quite obvious from tab. 4 that for short internal edges the probability to detect an informative change is remarkably small for first and second codon positions only. Including third codon positions provides more phylogenetic informative sites. Hence, the likelihood is increased to distinguish the true tree from artifacts.

So far parallel and back mutations along the outer edges of the tree have not been considered. To keep things simple we computed the expected number of phylogenetic informative sites (in the parsimony sense). If $a = 0.005$ we expect a total of 4.8 informative sites for the first and second codon positions and about 31.1 for the third

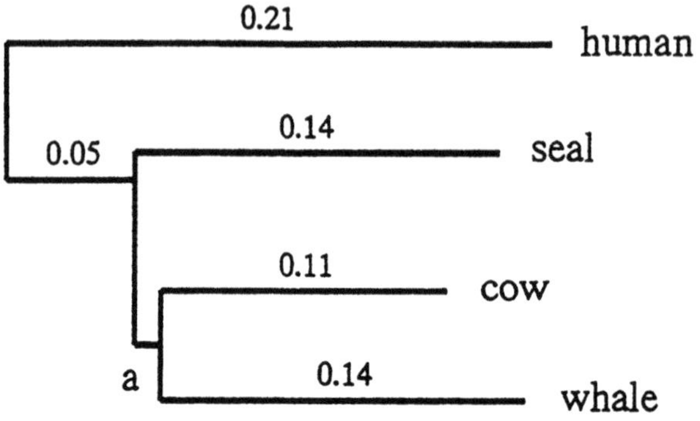

Figure 3: A subtree of the model tree (fig. 2) of four species. The central edge a was set to 0.04, 0.01 and 0.005 in the simulation studies.

codon positions. 2.6 of the informative first and second codon positions support the tree in fig. 3 whereas about 1.1 sites support each of the alternative trees. In case of third codon positions the figures are similar: 12.3 sites support the true tree and 9.4 sites favor one of the remaining trees. The small difference between the number of positions favoring the true tree and the numbers of positions supporting an alternative illustrates why method 1+2, or (1+2):EX performs badly compared to the other methods. A similar argument is valid for the bad performance of weighting scheme 1+2+(3:TRV).

Weighting methods like (1+2+3):EX, 1+2+(3:EX) or simply 1+2+3 are consistently among the best methods. Hence, our conclusion is that it is necessary to consider all three codon positions in this phylogenetic analysis. EX weighting does not dramatically improve the reliability of the neighbor–joining method in this particular instance, since the transition–transversion bias is not very pronounced. If B were larger then EX weighting of all three codon positions should significantly improve the reliability.

Tabs. 2 and 3 do not reveal that in the majority of cases, where we obtained a wrong tree, the maximum parsimony tree of the first two codon positions of the mitochondrial data (fig. 1a) was reconstructed. This justifies our neighbor–joining simulations as explanation of the maximum parsimony results.

In real data it is possible that including positions with high substitution rates increases the variability of estimated distances and thus may outweigh the gain in separability of two closely related groups. In those cases it is indicated to use highly variable positions only to recover the phylogeny of closely related taxa.

We showed that there is no simple method to solve the problem how to analyze third codon positions of genes. The solution rather depends on the data and the question one wants to address. Our findings illustrate the necessity to study each particular data set. Moreover, we showed that simulations using parameters from the data are a helpful tool to understand the outcome of a phylogenetic analysis. Those simulations may also be used to pinpoint pitfalls that may occur in the data.

Kimura model		Expected numbers of substitutions along central branch a			outer branches
	positions	0.005	0.01	0.04	0.65
1 + 2	2 000	2.5	5	20	325
transversions		1	2	8	130
transitions		1.5	3	12	195
3	1 000	12.5	25	100	1 625
transversions		5	10	40	650
transitions		7.5	15	60	975

Table 4: Expected numbers of substitutional events for the four species model tree of fig. 3. We used the Kimura model of sequence evolution with parameters of subsection 2.2.1 and a sequence length of 1000 codons (3000 nucleotides).

Acknowledgments:

We would like to thank G. L. Hofacker and S. Pääbo for stimulating discussions. We thank C. Odorfer for providing us with his program for calculating the parsimony informative sites. We would also like to thank the Genzentrum München for financial support.

References:

EDWARDS, S.V., ARCTANDER, P., and WILSON, A.C. (1991): Mitochondrial resolution of a deep branch in the genealogical tree for perching birds. *Proc. R. Soc. Lond. B, 243, 99-107.*

FELSENSTEIN, J. (1990). PHYLIP (Phylogeny Inference Package) Version 3.3, University of Washington, Seattle.

HASEGAWA, M., KISHINO, H., and YANO, T. (1985): Dating of the human–ape splitting by a molecular clock of mitochondrial DNA. *J. Mol. Evol., 22, 160-174.*

IRWIN, D.M., KOCHER, T.D., and WILSON, A.C. (1991): Evolution of the cytochrome b gene of mammals. *J. Mol. Evol., 32, 128-144.*

JANKE, A., FELDMAIER-FUCHS, G., THOMAS, K., VON HAESELER, A., and PÄÄBO, S. (1993): The marsupial mitochondrial genome and the evolution of placental mammals. Submitted to *Genetics.*

KIMURA, M. (1980): A simple method for estimating evolutionary rates of base substitutions through comparative studies of nucleotide sequences. *J. Mol. Evol. 16, 111–120.*

SAITOU, N., and NEI, M. (1987): The neighbor–joining method: a new method for reconstructing phylogenetic trees. *Mol. Biol. Evol., 4, 406–425.*

SCHÖNIGER, M., and VON HAESELER, A. (1993): A simple method to improve the reliability of tree reconstructions. *Mol. Biol. Evol., 10, 471–483.*

WILLIAMS, P. L., and FITCH, W. M. (1990): Phylogeny determination using dynamically weighted parsimony method. In: R. F. Doolittle (ed.): *Methods in Enzymology Vol. 183.* Academic Press, San Diego, 615–626.

Machine Learning for Protein Structure Prediction

Joachim Selbig

German National Research Center for Computer Science (GMD)
Institute of Methodical Foundations - Molecular Bioinformatics
Schloss Birlinghoven, D-53757 Sankt Augustin, Germany

Summary: The identification of the relationship between the amino acid sequence (1D) and the spatial structure (3D) of proteins belongs to the greatest challenges in modern molecular biology. The understanding of this relationship allows the formation of rules with which parts of the protein 3D-structure may be predicted from the order and type of the occurring amino acids. Information about the spatial structure of proteins determined by X-ray diffraction or NMR have been collected for more than 15 years in databases which form the base to apply statistical methods to discover structural regularities and correlations. But the number of experimentally determined protein structures (about 300) is already small in comparison to the number of known protein sequences (about 30000). Therefore, machine learning methods are used to find „hidden“ structural regularities which are difficult to discover by statistical methods. One of the key questions in this connection is the definition of appropriate representation languages. In this paper we discuss different machine learning methods applied to the outlined problem.

1. Introduction

The necessity to consider computer-based protein structure prediction methods results from the divergence between the number of solved 3D-structures and the number of known amino acid sequences.

There are different prediction methods ranging from simple secondary structure predictions to all-atom free energy calculations, cp. Ouzounis et al. (1993). These methods may be characterized according to different dimensions, e.g. the type of predicted information (secondary structure assignments, backbone atom coordinates, etc.), the relations to biophysical principles (physicochemical properties, solutions effects, etc.), and the algorithmic complexity.

The work of Lim (1974) was the origin of one type of heuristic structure prediction called pattern-based structure prediction. These methods exploit more or less the hierarchical character of protein structure where patterns may be generated „by hand“ on the base of biophysical principles or automatically by machine learning. Of course, also in the latter case biophysical knowledge can be incorporated. Two concepts from the machine learning terminology are of particular interest in connection with the evaluation of the quality of patterns: „completeness“ and „consistency“ meaning that a (class) description of a certain structural element should cover all instances of the considered class and no instance of another one. In our case descriptions of all subsequences are searched for which are related to a certain structural element, e.g. the core of an α-helix. Such relationships may be expressed as rules

$$L(P) \Longrightarrow R(S)$$

where the left hand side $L(P)$ and the right hand side $R(S)$ are expressions over the alphabets P and S, respectively (see below).

Several representation formalisms are used to express patterns to predict secondary and supersecondary structural elements, cp. Cohen et al. (1986), Lathrop et al. (1987), and Rooman et al. (1990). These representation formalisms rest upon biophysical principles and are also used to characterize functional regions, cp. Bork (1989).

The 20 naturally occurring amino acids have different physicochemical and biochemical properties such that one and the same segment of the polypeptide chain can be described by different patterns. Therefore, one of the important factors in deriving patterns is how to find the most characteristic properties to specify a certain position; in particular, if the cooperative nature of the protein structure should be taken into account.

Because the Brookhaven Protein Data Bank (Bernstein et al. (1977)) forms a large data set containing several regularities techniques and methods from the machine learning field may be adopted to identify these regularities as patterns which may be used as left hand sides in rules to predict structural elements.

2. Characterization of learning methods

Learning has been mainly viewed as inducing general concept descriptions from a learning set subdivided into classes. In most cases, two classes are considered containing positive and negative examples of the concept description searched for. In particular, such a description should cover all the positive examples but none of the negative examples. One of the important dimensions for characterizing learning systems is the type of the representation languages used to describe the elements of the learning set and the concepts. These representation languages also determine the generalization and specialization operators applied to modify the concept descriptions if the completeness and consistency criteria are not fulfilled.

The goal of using learning methods in the considered domain is to derive patterns which may be used to predict protein structure. These patterns must be expressed in the language of protein sequences consisting of the 20 amino acids and a large number of amino acid properties.

The nearest neighbour algorithm of Salzberg and Cost (1992) belongs to the class of instance-based learning programs and rests upon the feature space concept. Subsequences of odd length are considered as feature vectors whose class memberships are determined by the secondary structure assignments of the central residues and whose components can take 20 values as illustrated in the following example. [1]

(Q,A,Q,E,V,H,E,K,L,R,G)H

In this example the central residue $H = His = Histidine$ is in the secondary structure (class) $H = Helix$. Because the different amino acid residues on the different positions of the selected subsequences form symbolic feature values, a non-Euclidian distance metric is introduced which rests upon the distance between values of features.

The distance between the feature values $\tilde{x}$ und $\tilde{y}$ is defined by

$$d(\tilde{x},\tilde{y}) = \sum_{i=1}^{n} \left| \frac{\tilde{X}_i}{\tilde{X}} - \frac{\tilde{Y}_i}{\tilde{Y}} \right|^k$$

where n is the number of classes, $\tilde{X}_i$ the frequency of $\tilde{x}$ in class i, $\tilde{X}$ the total frequency of $\tilde{x}$, and k a constant. Therefore, the distance between two feature vectors

[1] Which correspond to the following Brookhaven Protein Data Bank entry; the number in parenthesis indicate the positions of the first feature: 1TIM (179), cp. Bernstein et al. (1977).

$X = (x_1, ..., x_N)$ and $Y = (y_1, ..., y_N)$ is defined by

$$D(X,Y) = w_X w_Y \sum_{j=1}^{N} d(\tilde{x}_j, \tilde{y}_j)^r$$

where $\tilde{x}_j$ and $\tilde{y}_j$ are the values of the feature j of X und Y, respectively. The weights w_X and w_Y are adjusted during the classification, see below.

The method requires two passes through the learning set. During the first pass, the 20×20 feature value distance matrices are computed from the elements of the learning set for each of the features. In the second pass, the program attempts to classify each element from the learning set by computing the distances between the new subsequence and previously stored ones. The new object is assigned the class of the nearest stored object. The feedback from the class comparison is used to adjust the above mentioned weights. Finally, the new object is stored in memory. During the test phase, subsequences are classified in the same way, but no modifications are made to memory or to the feature value distance matrices.

The algorithm requires $O(Nm^2 + Nv^2)$ time for m elements in the learning set, N features, and v values of the features. In most cases, m is much larger than v^2, hence the complexity is $O(Nm^2)$. The classification time during the test phase is at worst $O(Nm)$.

As nearest neighbour algorithms neural net learning systems process subsequences of odd length whose class memberships are determined by the secondary structure assignments of the central residues. Neural net learning systems consist of a network of non-linear processing units that have adjustable connection strengths, cp. Qian and Sejnowski (1988), Holley and Karplus (1989), and Kneller et al. (1990). Each of the 20 different amino acid residues on each of the different positions in the subsequences is associated to an input unit. If a certain amino acid occurs on a certain position the corresponding input unit is activated. Learning consists of altering the weights of the connections between the input units, certain hidden units, and the output units related to the given classes in response to the input and to the information about the correct classification. The goal is to find a good input-output mapping which can then be used to predict elements of the protein structure.

The transmission of the information from the input vectors x via the hidden vectors t to the output vectors s is characterized by

$$t_i = \sigma(\textstyle\sum_k w_{ik} x_k - a_i) \text{ and } s_j = \sigma(\textstyle\sum_i w_{ji} t_i - b_j)$$

where $\sigma(x)$ is the logistic function $\sigma(x) = \frac{1}{1+e^{-x}}$.

The determination of the activation states of the hidden units in correspondence to the input-output mapping is carried out by the back propagation algorithm which is a gradient descent method that propagates error signals back through a multi-layer network. The search for the best weights w_{ik} and w_{ji} is equivalent to the minimization of the error function

$$E = \frac{1}{2} \sum_{l=1}^{m} \sum_{j} (y_j^l - s_j(x^l))^2$$

where y are the output vectors given by the elements of the learning set.

One of the main problems of nearest neighbour and neural net learning systems developed until now for the prediction of protein structural elements seems to consist in the restricted representation language used to describe the input information.

3. Generalizing learning methods

Generalizing learning methods rest upon pattern description languages allowing the derivation of property patterns and providing the possibility to learn explicitly the relationships between physicochemical properties and structural elements.

Starting with the most general rule

$$C_{all} \Longrightarrow S$$

where $C_{all} = \{A, ..., V\}$ is the set of all 20 amino acids [2] and $S = \{C, E, H\}$ is the set of the secondary structural classes coil (C), extended strand (E), and helix (H) King and Sternberg (1990) use the following generalization operators to modify the patterns in the left hand side of the above given rule

- extension of the pattern by introducing another sequence position, i.e. $n_iC_i \rightarrow n_iC_i \wedge n_jC_j$
- extension of the description of a sequence position according to the generalization hierarchy, i.e. $n_iC_i \rightarrow n_iC_i'$ with $C_i \subset C_i'$
- restriction of the description of a sequence position according to the generalization hierarchy, i.e. $n_iC_i \rightarrow n_iC_i''$ with $C_i \supset C_i''$

By considering all the elements of a given learning set rules of the following type are derived using 48 predefined amino acid classes

$$C_{63} \wedge 1C_{29} \wedge 3C_{16} \Longrightarrow C \wedge 1C \wedge 2C \wedge 3C$$

where $C_{63} = C_{hydophobic} \vee C_{small}$, $C_{29} = C_{tiny} \vee (C_{polar} \setminus C_{aromatic}) \vee P$ and $C_{16} = C_{polar} \vee P$ with $C_{hydrophobic} = \{H, W, Y, F, M, L, I, V, C, A, G, T, K\}$, $C_{small} = \{P, V, C, A, G, T, S, N, D\}$, $C_{tiny} = \{A, G, S\}$, $C_{polar} = \{P, T, S, N, D, E, Q, R, K, H, W, Y\}$, and $C_{aromatic} = \{H, W, Y, F\}$.

The works of King and Sternberg (1990) and Schulze-Kremer and King (1992) show that specific rules for the beginning, middle, and end of secondary structural elements should be searched for and that structurally relevant combinations of amino acid properties are often not explicit and occur jumbled together with other features.

In our approach (Kaden et al. (1990), Koch et al. (1992), and Selbig et al. (1992)) we are trying to take into account that a pattern representation language should

- be flexible and capable of being extended,
- have a reasonable physicochemical foundation, and
- allow the derivation of sequentially long-range patterns.

Starting from the contact matrix desciption the data about the spatial structure of the proteins determined by X-ray diffraction are transformed into a graph description G which provides the possibility to define a multitude of pattern graphs for describing secondary and supersecondary structural elements

$$G = (V, E, f, g).$$

[2]The following connection consists between the one letter code, the three letter code, and a frequently used number code: $Ala = A = 1, Arg = R = 2, Asn = N = 3, Asp = D = 4, Cys = C = 5, Glu = E = 6, Gln = Q = 7, Gly = G = 8, His = H = 9, Ile = I = 10, Leu = L = 11, Lys = K = 12, Met = M = 13, Phe = F = 14, Pro = P = 15, Ser = S = 16, Thr = T = 17, Trp = W = 18, Tyr = Y = 19, Val = V = 20$

The n residues of a considered protein are represented by the elements of the node set $V = \{1, ..., n\}$ which are labelled with the corresponding amino acids $f : V \rightarrow C_{all} = \{A, ..., V\}$. The edge set $E \subseteq V \times V$ represents topological neighbourhoods corresponding to specific contact typs (a - no contact, b - backbone/backbone contact, c - backbone/side chain or side chain/backbone or side chain/side chain contact, and d - b and c), i.e. $g : E \rightarrow Pow(B)$ with $B = \{a, b, c, d\}$.

The adjacency matrices of protein graphs G may be understood as discrete forms of the above mentioned contact matrices.
This formalization level allows the derivation of pattern graphs H from specific subgraphs of G

$$H = (U, F, p, q).$$

Specific pattern graph frameworks are defined by the node set $U = \{u_1, ..., u_6\}$, a set of obligatory edges $T \subseteq F \subseteq U \times U$, and a set of forbidden edges $S = \{(u, v) \in F \mid p(u, v) = a\} \subseteq F \setminus T$. These edges are also labelled by specific contact types, i.e. $p : F \rightarrow Pow(B)$. The framework of some pattern graph H occurs in G if there is a mapping $h : U \rightarrow V$ that fulfills the following conditions

- $(u, v) \in F \setminus S \Rightarrow (h(u), h(v)) \in E$ and $p(u, v) \subseteq g(h(u), h(v))$, and
- $(u', v') \in S \Rightarrow (h(u'), h(v')) \notin E$.

Each occurrence of a pattern graph framework defines 6 sequence positions which are related to specific amino acids $f(h(u_1)) \in C_{all}, ..., f(h(u_6)) \in C_{all}$. The number of observed different residue pairs pass into additional edge labels $q : F \rightarrow Pow(C_{all} \times C_{all} \times N)$.

Therefore, learning consists in the application of three generalization operators

- $\forall (u, v) \in S \cup T : (f(h(u)), f(h(v)), w) \rightarrow (f(h(u)), f(h(v)), w + 1)$
- $T \rightarrow T \cup T'$
- $S \rightarrow S \cup S'$

The sets T' and S' correspond to edges which occur in each case and in no case, respectively. Thus, contacts between sequence positions are learnt which are of particular structural significance.

Each of the patterns graphs covers 6 sequence positions and contains information about the pairwise amino acid relations within structurally determined environments which is illustrated in the following figure for the two patterns H_2 ($i, i+1, i+2, i+3, i+4$, and $i+5$) and H_6 ($i-1, i, i+1, i+3, i+4$, and $i+5$) also showing that there are structural relations between the patterns.

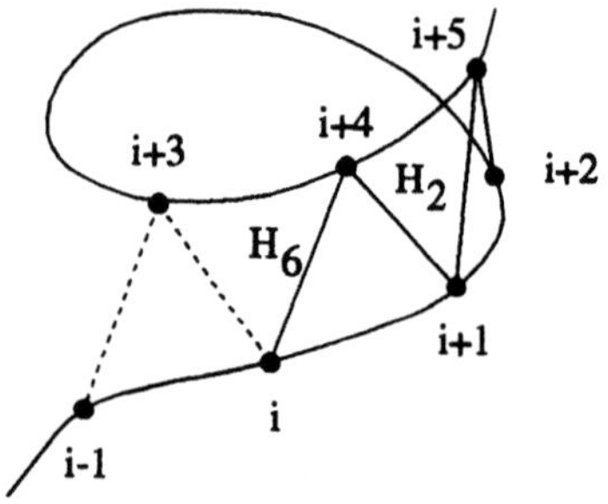

Different patterns graphs are used to characterize the beginning, the continuation, and the end of structure elements. According to the sequential distance of the segments specified by the pattern graph frameworks short-range and long-range patterns are distinguished.

From a set of 67 chains of 64 proteins [3] 68 patterns graphs were derived. Each of these patterns graphs contains about 10 obligatory and forbidden edges from the 15 possible ones describing contacts between the 6 specified sequence positions. Each of these edges is associated to a 20 × 20-matrix containing the frequencies of the observed amino acid pairs defined by the corresponding matrix positions. These 642 20×20-matrices are used in left hand sides of prediction rules, cp. Kaden et al. (1990). In addition to this the pattern graphs and the corresponding 20 × 20-matrices may be used to extend profile-based prediction methods (to be published).

We applied methods from applied graph theory and machine learning to derive pattern graphs from regular contact environments of contact matrices to be used for protein structure prediction.
The mapping of the 3D structure of proteins to graphs provides a powerful simplification by concentrating on the most important characteristics of the structure and eliminating some of the geometrical details.

References:

BERNSTEIN, F.C., KOETZLE, T.F., WILLIAMS, G.J.B., MEYER, E.F. Jr., BRICE, M.D., RODGERS, J.R., KENNARD, O., SHIMANOUCHI, T., and Tasumi, M. (1977): The Protein Data Bank: a computer based archival file for macromolecular structures. *Journal of Molecular Biology, 112, 535–542.*

BORK, P. (1989): Recognition of functional regions in primary structures using a set of property patterns. *FEBS Letters, 257, 191–195.*

COHEN, F.E., ARBABANEL, R.M., KUNTZ, I.D., and FLETTERICK, R.J. (1986): Turn prediction in proteins using a pattern-matching approach. *Biochemistry, 25, 266–275.*

HOLLEY, L.H. and KARPLUS, M. (1989): Protein secondary structure prediction with a neural network. *Proc. Natl. Acad. Sci. USA (Biophysics), 86, 152–156.*

KADEN, F., KOCH, I., and SELBIG, J. (1990): Knowledge-Based Prediction of Protein Structures. *Journal of Theoretical Biology, 147, 85–100.*

KING, R.D. and STERNBERG, M.J.E. (1990): Machine Learning Approach for the Prediction of Protein Secondary Structure. *Journal of Molecular Biology, 216, 441–457.*

KNELLER, D.G., and COHEN, F.E., and LANGRIDGE, R. (1990) Improvements in Protein Secondary Structure Prediction by An Enhanced Neural Network. *Journal of Molecular Biology, 214, 171–182.*

KOCH, I. KADEN, F., and SELBIG, J.: Analysis of Protein Sheet Topologies by Graph Theoretical Methods. *PROTEINS: Structure, Function, and Genetics, 12, 314–323.*

LATHROP, R.H., WEBSTER, T.A., and SMITH, T.M. (1987): ARIANE: Pattern-Directed Inference And Hierarchical Abstraction in Protein Structure Recognition. *Communications*

[3]Corresponding to the Brookhaven Protein Data Bank entries (Bernstein et al. (1977)): 351C, 256B-A, 8ADH, 8ATC-A, 8ATC-B, 2AZA-A, 3B5C, 3BLM, 2CA2, 1CCR, 2CCY-A, 1CD4, 3CLA, 5CPA, 2CPP, 4CPV, 1CSE-E, 1CSE-I, 1CTF, 2CYP, 8DFR, 1ECN, 2ER7-E, 4FD1, 4FXN, 3GAP-A, 2GBP, 1GCR, 1GD1-O, 1GOX, 1GP1-A, 2HLA-B, 1HOE, 1I1B, 4ICD, 1IL8-A, 1L13, 6LDH, 2LIV, 2LTN-A, 2LTN-B, 1LZ1, 1MBD, 2MHR, 2PAB-A, 1PAZ, 4PTP, 1R69, 1RHD, 7RSA, 2RSP-A, 5RXN, 2SGA, 4SGB-I, 2SNS, 2SOD-O, 2SSI, 2STV, 2TMN-E, 1TNF-A, 2TS1, 1UBQ, 1UTG, 2WRP-R, 1WSY-A, 4XIA-A, 1YPI-A

of the ACM, 30, 909–921.

LIM, V.I. (1974): Structural principles of the globular organisation of protein chains: A stereochemical theory of globular protein secondary structure. *Journal of Molecular Biology, 88, 857–872.*

OUZOUNIS, C., SANDER, C., SCHARF, M., and SCHNEIDER, R. (1993): Prediction of protein structure by evaluation of sequence-structure fitness: aligning sequences to contact profiles derived from 3D structures. *Journal of Molecular Biology, 232, 805–825.*

QIAN, N. and SEJNOWSKI, T.J. (1988): Predicting the secondary structure of globular proteins using neural network models. *Journal of Molecular Biology, 202, 865–884.*

ROOMAN, M.J., RODRIGUEZ, J., and WODAK, S.H. (1990): Relations between protein sequence and structure and their significance. *Journal of Molelurar Biology, 213, 337–350.*

SALZBERG, S. and COST, S. (1992): Predicting Protein Secondary Structure with a Nearest-neighbor Algorithm. *Journal of Molecular Biology, 227, 371–374.*

SCHULZE-KREMER, S. and KING, R.D. (1992): IPSA - Inductive Protein Structure Analysis. *Protein Engineering, 5, 377–390.*

SELBIG, J. KADEN, F., and KOCH, I.: Applying machine learning methods for finding significant amino acid properties in proteins. *FEBS Letters, 297, 241–246.*

A Parallel-Processor Implementation of an Algorithm to Delineate Distantly Related Protein Sequences using Conserved Motifs and Neural Networks

Gerhard Vogt[1], Dmitrij Frishman[1,2] and Patrick Argos[1]

[1] European Molecular Biology Laboratory
Postfach 10 22 09
Meyerhofstrasse 1
Germany

[2] Institute of Evolutionary Physiology and Biochemistry
Thorez pr.44
St Petersburg 194223
Russia

Summary: A sensitive technique to recognize distantly related protein sequences has been implemented on a DEC massively parallel computer (MASPAR). The method relies on finding conserved subsequence regions (motifs) in multiple sequence alignments, training neural nets to recognize them, and then searching all segments in large databases of protein sequences to which the nets respond well. The net/motif search has several advantages over other present-day techniques: the sensitivity of neural nets, independence of deletions and insertions as well as gap penalties, and concentration on the most conserved subsegments for searching and scoring. Use of the parallel processors provides ready access to the comparison results as well as quick response for the molecular biological community.

1. Introduction

Comparison and alignment of protein amino acid sequences can result in significant biological knowledge (cf. Argos, 1990) and can substantially reduce experimental effort. In a familial set of multiply matched primary structures, the most conserved subsequence regions (motifs or patterns) become apparent. If the proteins are distantly related, the regions are further isolated and limited in number. The most conserved amino acids within the motifs are often functionally and structurally essential; they may form receptor and nucleic acid binding regions or active sites in enzymes. These constrained spans are very useful for identifying further and very distant members in a molecular family, thereby increasing comprehension of protein structure/function relationships.

For very similar and multiply aligned amino acid sequences the problem of defining a protein pattern becomes trivial as all possible exchanges can be enumerated. For more diverged sequences where most matched residues are not identical, deriving subsequence patterns or motifs becomes a more sophisticated task in pattern recognition. Artificial neural networks represent a powerful technique to deal with noisy patterns as they can extract essential features from a set of variable, imperfect objects (see, for example, Wasserman, 1989). Neural net utilization in molecular biology has already had a long history (for a review see Hirst & Sternberg, 1992). Recent new applications of neural nets to various tasks in primary (Horton & Kamehisa, 1992; Farber, Lapedes & Sirotkin, 1992) and secondary (Pancoska, Blazek & Keiderling, 1992; Hayward & Collins, 1992) structure analysis of biological polymers point to unremitting interest of biological researchers in neural nets.

Frishman and Argos (1992) have developed a system to recognize protein subsequence or motif patterns with neural networks from a set of amino acid sequences known to belong to a family. A trained perceptron-type network is then used to search for additional representatives of the same pattern in the database. From a multiple alignment of protein sequences, the relevant subsequence motifs are first delineated. This is achieved in the following manner. Suppose there are N aligned protein sequences. At each appropriate alignment position, a set of N aligned oligopeptides of given window length is taken. After eliminating each i-th oligopeptide in turn, the remaining N-1 subsequence spans are used to train one of N neural nets. The recognition output of each net is averaged over each i-th span. The process is repeated for all possible multiple alignment sets of given probe length, allowing a sequence conservation profile plot of the average net output *versus* starting alignment position for each set. The subsequence regions that are best identified by their associated neural nets are defined as the conserved motif regions in the overall alignment. The oligopeptides within each of these M conserved regions (as well as randomized versions of them with the same amino acid composition for negative training) are then used to train intensively M neural networks. An entire sequence database can subsequently be searched by submitting every database oligopeptide of given length to the M motif nets and the network outputs noted. Those sequences with sufficient positive response from the individual pattern nets taken in the proper sequential order are then identified as distantly related members of the original family. The motif nets can then be retrained and made more sensitive in the light of the newly found subsequences. A database search is then reinitiated in an attempt to discover even more distant family members. This process can be repeated as often as necessary. These procedures have been implemented in a fully automated computer program written in the C language and originally developed for VAX computers.

Although perceptron-like techniques attempt to imitate distributed data processing in the nervous system to achieve learning and generalization abilities, implementations of these methods have been in most cases essentially serial. Nevertheless, the great efficiency of distributed data processing for biological sequence analysis has been realized (Collins & Coulson, 1984). Comparative tasks associated with large protein and nucleic acid sequence databases can be easily parallelized because the similarity score of the target with each database entry can be computed independently (Coulson et al., 1987; Lander & Mesirov, 1988; Vogt & Argos, 1992; Jones, 1992). Furthermore, the potential of parallel data processing inherent in the neural nets can also be realized on a massively parallel computer (Zhang et al., 1990). Such quick sequence comparison searches will be essential for the human genome project which will not only produce massive sequence data for comparison but will also require each new protein primary structure to be assigned quickly and sensitively to its molecular family from its characteristic sequence motifs. Such classification should allow easy handling of the massive data.

In this work we have used a massively parallel computer to implement the sensitive technique for protein sequence motif recognition and database searches described above (Frishman & Argos, 1992). The great gain in speed over that of more conventional computers allowed us to carry out the comparison task with multiple window lengths, thus achieving peak sensitivity for the method to recognize distant family members.

2. Implementation of the motif/network method on a massively parallel computer

2.1 Hardware

For this project a DECmpp 12000 (hereafter referred to as "mpp") was used. The mpp is a massively parallel SIMD (Single Instruction, Multiple Data) computer consisting of a data parallel unit (DPU) and an ULTRIX workstation for user access (Fig. 1).

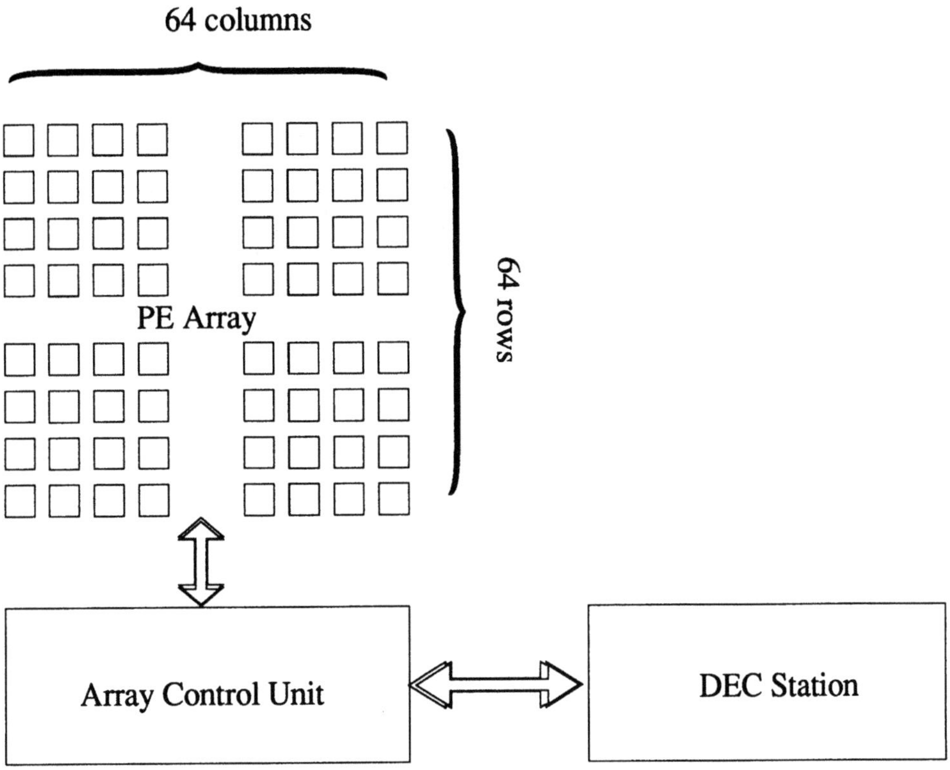

Fig. 1: Schematic of the DECmpp 12000 computer. PE indicates an individual processor element.

The DPU contains the array control unit (ACU) and 4096 processor elements (PEs) which are interconnected in a two-dimensional toroidal-wrapped 64-by-64 matrix. Every PE has 64 kBytes of local data memory; an additional 128 kBytes of ACU memory are globally available for all PEs. The mpp can run time-sliced processes during which PE memory is split between the concurrently running processes, leaving only 16 or 32 kBytes per PE and job. The "front end", a DECstation 5000, is used for mass storage and input/output (I/O) and also acts as a cross development platform. The mpp can either be programmed in Fortran or in MPL which is essentially a parallel C dialect. Programs typically consist of a host part running on the front end and a parallel part running on the mpp. The DEC station mainly performs user tasks related to I/O, the initialization of the mpp and data supply while the mpp performs the actual calculations.

SIMD machines execute the same instruction stream on all processors simultaneously.

The data which is evaluated can however be different in every processor. MPL offers the data type modifier "plural" to specify that a separate instance of a given variable resides on every PE. Operations with plural operators are evaluated independently on every PE. Plural variables can also take part in conditional statements. In this case individual processors are switched on or off depending on an evaluation. If a conditional statement contains a true as well as a false alternative and if the condition evaluates to true on some PEs and to false on others, then both are executed serially. Program loops are similar as their code is repeatedly executed until the condition to "leave" is true in all PEs. It is possible that most processors merely wait for just a few others which are calculating demanding loops.

Each PE can work on individual data which must be loaded into the PE array. Every PE has an individual identification number between 0 and 4095 as well as its x/y coordinates within the processor array. With these numbers in conditional statements, the local memory of every PE can be accessed individually. Additionally, blocks of data can be moved between front end memory and the PE array and *vice versa.* This movement, which can be initiated from either the program running on the workstation or the DPU, splits front end data in equally sized blocks and distributes these within a program defined subsquare of the PE array. Communication within the PE array can either take place between arbitrary processors using plural variables as the source and target addresses or systematically to equidistant PEs along rows, columns or diagonals using the toroidal topology of the PE network.

2.2 Basic concepts

Neural networks, due to their inherent parallelism, are perfectly suited for multiple processor computers which are also amenable for database searches where similarity scores are evaluated independently for every sequence entry. In reality, however, the achievable benefit is limited and depends on the exact problem as well as on the implementation and the available features of the hardware utilized. For the present project, two parts of the algorithm have been ported to the mpp: that which determines the conserved subsequence motifs in the multiply aligned sequences (conservation profile) and that which performs the actual database search. Additional programs to build up the database files and to post process the mpp generated results have been written as serial C programs on the DECstation. As there is usually no user intervention necessary, all individual programs are started from a Unix shell script.

Several criteria have to be considered if a parallel computer is to be used with efficacy and this is especially the case for a SIMD machine such as the mpp. The major limitation is the amount of available memory while a minor but nonetheless significant issue is the optimal utilization of processing power. The memory required per neural network can easily be calculated. Every position (column) in the multiple sequence alignment corresponds to 20 input nodes, one for every possible amino acid. For a window length of 20 such positions, there are 20 times 20 input nodes. Each of these nodes is connected with each hidden unit of the neural net; if 100 hidden units are used, 40000 weights are needed between input and hidden layer. As every weight is a float variable which requires 4 bytes of memory, roughly 160 kBytes are needed for them alone. Therefore, only part of a network can be stored in the 64 kBytes available in each PE. Distributing networks more than necessary amongst the PEs increases the overhead depending on the degree of parallelization. The overhead consists of communication time and lost processing time while some processors stay idle during the calculation of serial parts of the program. Thus an optimal balance between memory requirements and parallelization must be found which in turn requires different approaches for the two major tasks (Mp1 and Mp3) of the program package described here

2.3 Calculation of the sequence conservation profile

The program module Mpl is used to build a conservation profile from a set of N multiply aligned sequences as described in the introduction. For every possible subalignment of a given window length, neural networks are trained with N-1 oligopeptide sequence spans and neural net output is used as the recognition score for span N to measure the degree of conservation for this oligopeptide set. To reduce the negative effect of some badly aligned sequences as well as to discern the consistent and conserved regions best recognized by neural nets, a jack-knife procedure, where every subsequence in the alignment window is used for recognition by a net trained on the others, is emphasized. The average of the neural net outputs (recognition scores) is stored for each possible window span of given length in the multiple alignment. A plot of the mean net outputs at each central alignment site is referred to as the conservation profile of the considered sequence.

The length of the conservation profile is that of the multiple alignment minus the probe size. For every profile position and every sequence in the alignment one network has to be trained such that, for an exemplary alignment length of 400 residues and 25 sequences, 10000 networks would be utilized. For efficiency, all amino acids processed in a neural network are stored in the local memory of the calculating PE. Every PE evaluates the necessary calculations for one hidden unit. Typically 10 hidden units are used in each net for the calculation of the profile. Therefore 409 networks can be evaluated in one pass with 6 PEs idle.

First, all necessary general data such as the algorithm parametric setup is loaded into DPU memory. Then the entire alignment is stored as one long string in DPU memory as well. The actual calculation takes place in several passes and the data is distributed in such a way that all 409 networks are fully utilized for all except the last pass. Filling PE memory is a time critical factor since loading every PE separately by switching off all others using plural conditions is very slow. The loading is performed in such a way that as many PEs as possible access the same sequence character in DPU memory at the same time. This can be achieved since all networks calculating at the same alignment position but for a different jack-knife sequence load the same data. It is, however, inefficient to load the sequences to the same position in PE memory for all networks since the jack-knife sequence, which is different for each network, does not take part in the training. An expensive plural condition would thus have to switch off one network per iteration. Fig 2 shows an optimal distribution of the sequences for the calculation. The sequence indices change in a circular manner from network to network (Fig. 2). The jack-knife sequence is stored in the n-th array element in all networks.

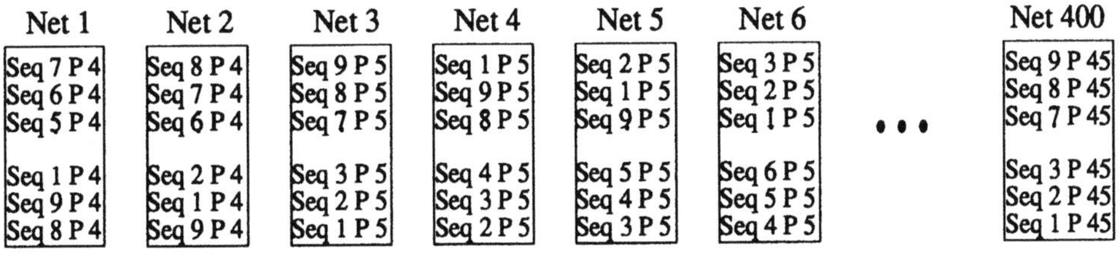

Fig. 2: Distribution of the sequences over the neural networks. The squares represent the arrays in PE memory where the sequences are stored. The example assumes a multiple alignment of 9 sequences. The two numbers for each net element represent the sequence number (Seq) and starting position (P) within the alignment. Because of the jack knife

procedure, 9 networks are required for each position, here for example Net 3 to Net 11 for position 5. For networks with 10 hidden layers, 10 PEs are needed. Therefore 40 full positions can be calculated in every pass. In this example, Net 1 and Net 2 calculate the remainder from the previous pass. The eight bottom sequences are used for training the networks; the top jack-knife sequence is used for the recognition test. This distribution allows usage of the same index in all PEs and avoids the long processing time required for intensive use of plural pointers to access the arrays.

The actual profile calculation begins with the generation of random sequences which are generated with the same amino acid composition as that of the actual alignment and are used as negative examples for the training. It is obviously necessary to have the same random sequence available for all PEs taking part in one particular network. Therefore a random number assigned to each and every PE is initialized with a seed calculated using the PE number modulo, the number of PEs per network and the number of initializations. The random sequences are then determined and the random number generator again initialized using the PE number directly. This guarantees that all PEs use different starting weights associated with each family member. Each oligopeptide sequence in the multiple alignment is given a weight which depends on the closeness of each sequence to all others in the set. The weights are determined by the technique of Sibbald and Argos (1990) and are required so that the trained nets do not overrepresent a series of particularly similar primary structures.

In testing all possible motif spans, the networks are trained a number of times (default or user-defined). The input weights for the hidden layer and for the output layer are both calculated in the hidden unit PEs. The calculation of neural net output score has to be performed serially on the PE with the highest assigned number in every network. After the requested number of training cycles, the jack-knife sequence is tested. The program then reads the score and sequence position from every PE and adds the score to the corresponding cell of an array in DPU memory which is, upon program completion, sent to the DEC station for printing.

In the next two steps, the resulting conservation profile is searched for peaks, higher than an empirically chosen threshold which are indicative of well conserved alignment regions. The oligopeptides corresponding to alignment positions associated with peaks are used to train extensively through many cycles larger networks with many hidden units. This results in a set of networks used for the database search; each network is associated with a given sequence motif that characterizes the sequence family.

2.4 Performing the database search

For the database search, version 24 of the Swissprot protein sequence databank was utilized (Bairoch & Böckman, 1992). The database was reformatted and stored as a binary file on the hard disk of the front-end computer. Reformatting is extremely important in the present application and is therefore described here. The Swissprot 24 release contained 28154 sequences, the longest consisting of 5037 amino acids and the shortest of 3. The entries are not normally sorted by length. For the implementation of the database search, either 256 or 512 sequences are calculated concurrently. If one long sequence and many short ones were examined simultaneously, most of the processors would remain idle for most of the search procedures. Table 1 below exemplifies this effect.

PEs per network	unsorted database	decreasing length
8	14%	91%
16	17%	83%

Table 1: Theoretical limit for the utilization of the mpp for different database sorting arrangements based on sequence length. The results are influenced by the length distribution of sequences concurrently loaded in the mpp.

The database search consists of three steps. For every database sequence, the top n hits for every search network are recorded. The number of search networks depends on the number of peak values in the conservation profile calculated over multiple alignment positions. The location of the hits are then checked for consistency in that they must occur in a sequential order according to their N- to C-terminal appearence in the multiple alignment. The consistent score with the highest square sum of network outputs (Frishman & Argos, 1992) is then maintained for each sequence entry. Finally all the best scores for all database entries are sorted and printed.

Database searches are typically performed with 3 to 10 motif networks, about 100 hidden units and a window length ranging from 7 to 20 amino acids, requiring more then 400 kBytes of memory. Therefore splitting the search task amongst 16 or even 32 PEs per network is reasonable. Fig. 3 shows a network with 96 hidden units split into 16 PEs.

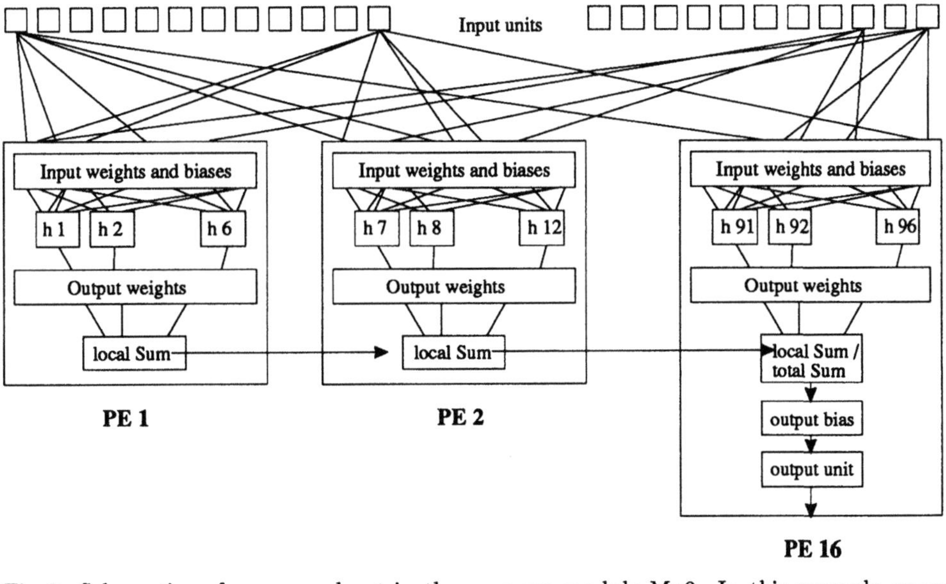

Fig 3. Schematics of one neural net in the program module Mp3. In this example every neural network consists of n input units where n is 20 times the window length chosen, 96 hidden units and 1 output unit. Every net is distributed over 16 processors and thus every processor element (PE) calculates the weights for 6 hidden units. The weighted outputs of the hidden units are then collected in the 16th processor which then calculates the network output.

As previously mentioned, there are very fast blockmove commands for data transfers from the workstation to the PE array. These commands are limited to square submatrices of the PE array. As they are the only method which is sufficiently fast, 64 divided by the number of PEs per network must be integral, otherwise the remaining

PEs cannot be used. The sequences are read one-by-one from the database. Then 16 copies are made and all concatenated to one long string. This string is sent by the front end via blockmove to the PEs. Then the workstation initiates the calculation on the mpp. When completed, the DECstation collects the results from the mpp and writes them to a binary file for later analysis.

2.5 Performance

The performance of the programs Mp1 and Mp3 was compared to the existing serial version of both programs running on a VAX 9000. Table 2 shows the absolute CPU time required for both machines and the relative speedup of the mpp versions. Additionally, the theoretical floating point peak performance of both machines is also shown.

Program	VAX 9000	DECmpp 12000	relative speedup
Peak Performance	60 MFlops	1.3 GFlops	22
Mp1	120 minutes	7 minutes	17
Mp3	12 hours	30 minutes	24

Table 2: Peak performances and required time to calculate a profile based on one window length and to perform a database search using the immunoglobulin family (see Frishman & Argos (1992) for the multiple alignment) on the DECmpp and a VAX 9000. The difference in the speedup factors for Mp1 and Mp3 is due to the more complex and time consuming communication for the profile calculation.

Theoretically, a speedup of 24 for Mp3 should be impossible because parallel programs always suffer from communication overhead and inherently serial parts of the program. This is especially true for SIMD machines where, in contrast to MIMD machines, processors cannnot calculate relatively independant tasks in parallel but are strictly bound to one instruction stream. The numbers also show, that even for programs with a very high fraction of floating point calculations, the peak performance ratios can only be a rough measure of the achievable speed. Though the speed-up can be largely ascribed to the use of the parallel machine, it is clear that the program developed for it has achieved near maximal efficiency.

2.6 Merging the various window results

It is obvious that the window length used for effecting the profile and searching the database should correspond to the length of the motif. In practice, this is difficult since the profile generated from Mp1 is based on only one window size and testing several is computationally expensive. While a window length of 12 yielded good results for previously examined examples (Frishman & Argos, 1992), extra trials showed that other probe sizes sometimes allowed recognition of even more substituted family members with even fewer intervening false positives amongst the top ranking scores. The gain of speed from mpp application provided the advantage required to test many different window lengths. To chose *a priori* the proper window length is typically impossible since the distantly related members are not known. Joining the results from different runs provides the solution if the observations from each window length tested are normalized for proper comparison. For this purpose, all sequences which are not within the top-scoring 1000 are defined as unrelated to the query motifs. This cut-off is chosen since the biggest protein family known to date (the globins) consists of about 700 members. For these sequences the average score and its standard deviation are calculated and used to determine the Z-score for the top hits as a number of standard deviations above or below the mean score. The lists from each window calculation are then joined and sorted on Z-scores. Sequences are only saved in the list according to their highest achievable Z-score at a specific window size. Table 3 shows the results of this process for the immunoglobulin (Williams & Barklay, 1988) and integrase (Argos et al, 1986; Abremski & Hoess, 1992) families. Frishman and Argos (1992) show the inital multiple alignments used as inputs to the neural nets.

Protein Family	Neural networks / Window length					Profilesearch
	8	12	16	20	8, 12, 16, 20	
Immunoglobulin false positives within first 600	79	**52**	106	308	**50**	196
Integrase false positives within first 50	23	23	22	**19**	**19**	25

Table 3. Results from joining the output lists for several window lengths using Z-scores. It is clear that the merged results show the same or better quality than the single optimal window length. The false positives refer to sequences identified in the top-600 scores that are known not to be in the families. Results are given for the PROFILESEARCH technique (Gribskov at al, 1987) which is the most commonly used technique for database searches when multiple alignments are available. Frishman and Argos (1992) discuss this method in greater detail (*vide infra*).

3. Commentary

A fully automatic methodology has been developed to recognize conserved subsequences in a set of multiply aligned sequences with the use of back propagation algorithms. Other cases known to the authors where neural nets were applied to the generation of unknown mappings involve the structural classification of protein sequences using an unsupervised Kohonen algorithm (Ferran & Ferrara, 1992) and protein superfamily classification based on back-propagation networks (Wu et al., 1992). In the latter case the protein sequence database was subdivided into several main enzymatic classes such as transferases, lyases, and the like.

Applying a sliding neural network to each stretch of a multiple alignment in conjunction with a jack-knife procedure affords the possibility to find conserved motifs in alignments of very distant proteins. The motifs discovered in the first stage of the analysis are then used for sensitive searches for related sequences in protein sequence databases. New motifs from the database with the highest similarity scores can then be utilized to retrain networks which can then be applied for further searches in the database. Frishman & Argos (1992) have exemplified this advantage with the immunoglobulin case where retraining reduced the number of false positives by 30%.

The methodology described here is intended for sensitive sequence comparisons where little overall similarity is detectable except for a few conserved regions. For very distantly related proteins, conserved segments "float" atop the background noise of a multiple alignment. In such cases searching a large database with neural networks trained to recognize only motifs results in better recognition of distant sequences as compared with PROFILE-like algorithms (Gribskov et al., 1987). PROFILESEARCH uses the entire sequence alignment for comparison purposes and not just motifs. It is also subject to the gap penalty chosen and allows only limited length for insertions and deletions. Recognition of distant family members is made more difficult as the Z-scores are based on a single high scoring local alignment (Smith & Waterman, 1981) which is typically longer than single motifs and can introduce considerable noise in misaligned segments (likely for distant members). Furthermore, the net/motif approach provides for a full range of intervening sequence length separating motifs, not possible in PROFILESEARCH algorithm as a result of gap penalty constraints. This is not the first indication that neural nets can perform more sensitively in sequence analysis than statistical methods (to which belong PROFILE-like techniques). Lapedes et al. (1990) investigated the effectiveness of various neural net, machine learning and information theory techniques in DNA sequence pattern searches and found that neural nets provided the highest accuracy.

The use of a parallel-processor computer allows the window trials to be executed in reasonable time periods which is not the case for typical scalar machines embodied as workstations or even mainframes such as a VAX 9000. In one example there was even an improvement in the number of false positives for merged window results over that achieved for the optimal window length.

The minimum net output level in the conservation profile to declare a motif region has been recommended as 0.7 which yielded recognition of nearly 75% of the oligopeptides at a given alignment position (Frishman & Argos, 1992). It is nonetheless recommended for completeness that the researcher attempt a variety of thresholds. For example, in the integrase family adding an extra fourth motif resulting from use of a threshold near 0.6, considerably weakened the database search sensitivity where 7 false positives, as juxtaposed to only one using a threshold of 0.7, occurred in the top 25 scores.

The best recognized regions in each searched sequence need not be sequentially consecutive if there is a suspicion that the conserved segments are shuffled. Additionally it is possible to apply restrictions on the lengths of the amino acid stretches separating each of the conserved regions.

Concepts involving the presence of characteristic patterns in protein families or their use to find additional representatives in the database have also been explored by other researchers. There are several methods for delineating these patterns from sets of unaligned and aligned sequences (see Taylor & Jones (1992) for a review). Several collections of protein signatures as well as a number of tools for their automatic generation are available (Bairoch, 1991; Sheridan & Venkataraghvan, 1992; Ogiwara et al., 1992). The concept of flexible protein patterns introduced by Barton & Sternberg (1990) allows specification of sequence distances between pattern elements within a certain length range. Another approach for the description of sequence conservation involves representation of the most reliable parts of a multiple alignment in the form of ungapped blocks as introduced by Henikoff & Henikoff (1991) who also apply these regions to database searches with considerable success. All these techniques are based on the same philosophy: biologically, conserved regions in proteins are likely to be functionally and structurally important and thus represent the essential features of a given family. Since the motifs must be maintained over long evolutionary time periods, they should provide one of the most sensitive approaches to discover very distant but related primary structures. The method presented here is individual in its use of sensitive and retrainable neural nets for pattern delineation and subsequent database searches as well as in its parallel implementation. Researchers interested in obtaining a copy of the programmed net/motif computer algorithm should contact the authors by post or e-mail (VOGT@EMBL-HEIDELBERG.DE on Internet).

D.F. was supported by an Alexander von Humboldt postdoctoral fellowship.

References:

ABREMSKI, K.E. & HOESS, R.H. (1992): Evidence for a second conserved arginine residue in the integrase family of recombination proteins: *Protein Eng. 5, 87-91.*

ARGOS, P. (1990): Computer analysis of protein structure: *Methods Enzymol. 182, 751-776.*

ARGOS, P., LANDY, A., ABREMSKI, K., HAGGARD-LJUNGQUIST, E., Hoess, R.H., Khan, M.L., Kalionis, B., Narayana, S.V.L., Pierson, L.S., III, Sternberg, N. & Leong, J.M. (1986): The integrase family of site-specific recombinases: regional similarities and global diversity: *EMBO J. 5, 433-440.*

BAIROCH, A. (1991): PROSITE: a dictionary of sites and patterns in proteins: *Nucl. Acids Res. 19, 2241-2245.*

BAIROCH, A. & BÖCKMANN, B. (1992): The SWISS-PROT protein sequence data bank: *Nucl. Acids Res. 20, 2019-2022.*

BARTON, G.J. & STERNBERG, M.J.E. (1990): Flexible protein sequence patterns: a sensitive method to detect weak structural similarities: *J. Mol. Biol. 212, 389-402.*

COLLINS, J.F. & COULSON, A.F.W. (1984): Applications of parallel processing algorithms for DNA sequence analysis: *Nucl. Acid Res. 12, 181-192.*

COULSON, A.F.W., COLLINS, J.F., LYALL, A. (1987): Protein and nucleic acid sequence database searching: a suitable case for parallel processing: *Comput. J. 30, 420-424.*

FARBER, R., LAPEDES, A. & SIROTKIN, K. (1992): Determination of eukariotic protein coding regions using neural networks and information theory: *J. Mol. Biol. 226, 471-479.*

FERRAN, E.A. & FERRARA, P. (1992): Clustering proteins into families using artificial neural networks: *Comp. Appl. Biosci. 8, 39-44.*

FRISHMAN, D. & ARGOS, P. (1992): Recognition of distantly related protein sequences using conserved motifs and neural networks: *J. Mol. Biol. 228, 951-962.*

GRIBSKOV, M., MCLACHLAN, A.D. & EISENBERG, D. (1987): Profile analysis: detection of distantly related proteins: *Proc. Nat. Acad. Sci., U.S.A. 84, 4355-4358.*

HAYWARD, S. & COLLINS, J.F. (1992): Limits on a-helix prediction with neural network models: *Proteins 14, 372-381.*

HENIKOFF, S. & HENIKOFF, J.G. (1991): Automated assembly of protein blocks for database searching: *Nucl. Acids Res. 19, 6565-6572.*

HIRST, J.D. & STERNBERG, M.J. (1992): Prediction of structural and functional features of protein and nucleic acid sequences by artificial neural networks: *Biochemistry 31, 7211-7218.*

HORTON, P.B. & KANEHISHA, M. (1992): An assessment of neural network and statistical approaches for prediction of E.coli promoter sites: *Nucl. Acids Res. 20, 4331-4338.*

JONES, R. (1992): Sequence pattern matching on a massively parallel computer: *Comp. Appl. Biosci. 8, 377-383.*

LANDER, E. & MESIROV, J. (1988): Protein sequence comparison on a parallel computer: *Proc. of the 1988 International Conference on Parallel Processing, Penn State* Press, Philadelphia, pp.257- 263.

LAPEDES, A., BARNES, C., BURKS, C., FARBER, R. & SIROTKIN K. (1990): Application of neural networks and other machine learning algorithms to DNA sequence analysis: *Computers and DNA, SFI Studies in the Sciences of Complexity, vol. VII,* (Bell, G. &

Marr ,T, eds.). Addison- Wesley, New York, pp. 157-181.

OGIWARA, A., UCHIYAMA, I., SETO, Y. & KANEHISHA, M. (1992): Construction of a dictionary of sequence motifs that characterize groups of related proteins: *Protein Eng. 5, 479-488.*

PANCOSKA, P., BLAZEK, M. & KEIDERLING, T.A. (1992): Relationships between secondary structure fractions for globular proteins. Neural network analyses of crystallographic data sets: *Biochemistry 31, 10250-10257.*

SHERIDAN, R.P. & VENKATARAGHVAN, R. (1992): A systematic search for protein signiture sequences: *Proteins 14, 16-28.*

SIBBALD, P.R. & ARGOS, P. (1990): Weighting aligned protein or nucleic acid sequences to correct for unequal representation: *J. Mol. Biol. 216, 813-818.*

SMITH, T.F. & WATERMAN, M.S. (1981): Identification of common molecular subsequences: *J. Mol. Biol. 147, 195-197.*

TAYLOR, W.R. & JONES, D.T. (1991): Templates, consensus patterns and motifs: *Curr. Opinion Struct. Biol. 1, 327-333.*

VOGT, G. & ARGOS, P. (1992): Searching for distantly related protein sequences in large databases by parallel processing on a transputer machine: *Comp. Appl. Biosci. 8, 49-55.*

WASSERMAN, P.D. (1989): Neural computing. Theory and practice: *Van Nostrand, New York.*

WILLIAMS, A.F.& BARCLAY, A.N. (1988): The immunoglobulin superfamily - domains for cell surface recognition: *Ann. Rev. Immunol. 6, 381-405.*

WU, C., WHITSON, G., MCLARTY, J., ERMONGKONCHAI, A. & CHANG, T.-C. (1992): Protein classification artificial system: *Protein Sci. 1, 667-677.*

ZHANG, X., MCKENNA, M., MESIROV, J.P. & WALTZ, D.L. (1990): An efficient implementation of the back-propagation algorithm on the connection machine CM-2. In: *Advances in Neural Information Processing Systems 2 (Touretzky, D.S., ed.) Morgan Kaufmann Publishers, San Mateo, California, pp. 801-809.*

Part IV

Applied Data Analysis in Special Fields

- Economy and Marketing
- Archeology
- Screening in Medicine

Knowledge-Based Selection and Application of Quantitative Models of Consumer Behavior

Reinhold Decker

Institut für Entscheidungstheorie und Unternehmensforschung, Universität Karlsruhe, Kollegium am Schloß, D-7500 Karlsruhe, Germany

Summary: Starting from the proposition of a comprehensive set of market diagnostics and a flexible data model, which is well suited for the representation of both these diagnostics and different types of panel data, we shall discuss selected aspects of consumer behavior modeling to motivate a new goal- and data-oriented approach for model selection. The simulation framework presented in this context constitutes an essential extension of traditional applications of quantitative models of consumer behavior. The relationships between panel data, models, and market diagnostics mentioned in this essay can be taken as a starting point for knowledge-based analyses and simulations of consumer buying behavior.

1. Introduction

In many markets - but especially in those of consumer goods - increasing competition, alignment of the product ranges offered by competing suppliers, and an increasing consumer consciousness are becoming more and more serious problems for many firms. Therefore, to be able to strengthen and to improve the actual market position of a product, the marketing management needs detailed information about actual and future buying behavior.

From a data analytical point of view one possible starting point to tackle this problem is the calculation and prediction of meaningful market diagnostics on the basis of (commercial) panel data with the help of a Quantitative Model of Consumer (or buying) Behavior (abbreviated as QMCB in the following).

To follow this course some requirements must be met. For instance, a data analyst must determine which market diagnostic(s) can provide the necessary information. Moreover, it must be known which diagnostic(s) can be calculated from a special QMCB, and which properties the actual panel data should have so that they can be analyzed with the model in mind.

Therewith, we are confronted with a basic conflict. On the one hand, the application of a QMCB requires extensive theoretical knowledge in special areas of panel data analysis, whereas on the other hand, decision support systems which are easy to use are prefered in marketing practice. One possible solution for this problem is the integration of those methods and models which are relevant for panel data analysis in a knowledge-based system.

With regard to the relevant literature concerning the knowledge-based analysis of buying behavior data in general, and panel data in particular, it is noticeable that this area can be regarded as a fairly 'new territory' in consumer behavior research. Most of the knowledge-based systems in marketing (cf., e.g., Gaul and Schader (eds.) (1988), and Decker and Gaul (1990)) deal with the classical marketing disciplines 'product policy', 'pricing policy', 'communications policy', and 'distribution policy', whereas only a few approaches are known from consumer behavior analysis. The systems (and system concepts) listed in Table 1 either deal with selected aspects of quantitative consumer behavior research (cf., e.g., Geyer-Schulz et al. (1988)) or they represent first attempts towards an automatic analysis and interpretation of special types of panel data (cf., e.g., Schmitz et al. (1990), and Rangaswamy et al. (1991)). Although both systems emphasize descriptive data analyses, they still provide useful ideas concerning the treatment of the problem at hand. An interesting contribution to knowledge-based (ex post) analysis of retail panel data is PANELYSER (cf. Beys et al. (1992)). However, according to our knowledge this system was not designed for model-based simulation tasks. Nevertheless, the cited paper supplies some helpful hints with respect to a more user-friendly presentation of results (in panel data analysis), both in graphical and textual form.

Name	Area of Application	Reference
COVERSTORY	Extraction and preparation of relevant information from scanner data	Schmitz et al. (1990)
DANEX	Problem-oriented selection and use of multivariate methods in marketing data analysis	Böckenholt et al. (1989)
DEALMAKER	Assistance in the design of trade deals	McCann and Gallagher (1990)
INFER	Analysis and interpretation of scanner data, and preparation of corresponding reports	Rangaswamy et al. (1991)
M^2KS	Problem-oriented formulation of regression-based marketing models	McCann and Gallagher (1990)
O.N.	Improvement of the performance of stochastic models of consumer behavior	Geyer-Schulz et al. (1988)
PANELYSER	Determination and investigation of market share shifts on the basis of retail panel data	Beys et al. (1992)
SHANEX	Investigation of potential reasons for the changes of a product's market share	Alpar (1991)

Table 1: Knowledge-based approaches in consumer behavior analysis.

The model- and knowledge-based data analysis system KVASS (KaufVerhaltensAnalyse- und SimulationsSystem) developed at our institute within the so-called WIMDAS project (cf. Gaul et al. (1990) as well) is innovative (compared with approaches known from relevant literature), because it not only allows descriptive (ex post) analyses of data but also supports model-based simulations of future buying and market behavior. With the help of the KVASS approach stochastic and econometric models of consumer behavior (which are still reserved to a small group of "specialists in consumer behavior modeling" and which have been applied successfully to different product classes in the past) should become accessible to a wider circle of potential users in market research and marketing management.

An earlier proposal concerning the data-oriented classification and selection of stochastic models of consumer behavior (as an important preliminary work with respect to the development of the actual KVASS knowledge base) was published in Decker and Gaul (1989).

2. Buying behavior data and their representation

The term 'panel data', as used in this paper, covers diary- and scanner-based data recordings within consumer panels, retail panels, and scanner test markets. The latter have gained increasing importance in the last few years. Since panel data collections naturally lead to a large amount of single data, suitable (i.e. goal-oriented) aggregations of the same are indispensable. The result of such a "compression" of data can be a particular diagnostic which on its part supplies concrete diagnostic information about buying and/or market behavior.

The systems designed for panel data analysis encountered until now only support simulations with respect to a few market diagnostics - if at all. In most cases, the simulation facilities of these systems are restricted to market share forecasts. At the same time stochastic models of consumer behavior, for instance, constitute a solid basis for the derivation of a broad spectrum of meaningful market diagnostics. Therefore, we have designed a comprehensive set of diagnostics (cf. Figure 1) for KVASS which are calculable from panel data and/or QMCBs. Such a system of mutually complementing diagnostics can facilitate both the identification and the interpretation of complex market events and trends.

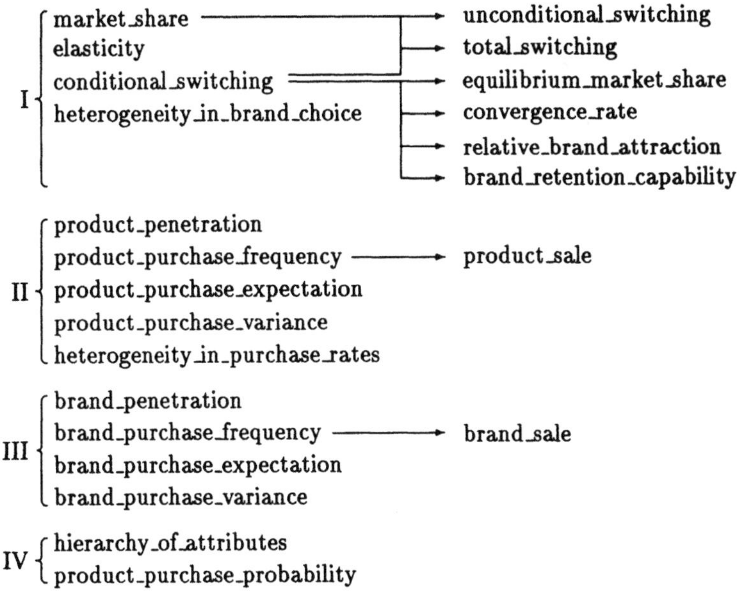

Figure 1: Selected diagnostics of buying behavior in consumer markets.

The market diagnostics can be classified as follows: Those on the left-hand side of Figure 1 have to be calculated directly from panel data or suited QMCBs whereas the diagnostics listed on the right-hand side can be derived from existing diagnostics. The total switching (i.e., the proportion of purchases involving brand switching in a given period of time), for example, can be calculated from market shares and conditional switchings. The latter are corresponding with markovian transition probabilities. Another classification (indicated with Roman numerals) refers to the type of model or method from which the individual diagnostics can be derived (cf. the corresponding remarks in section 3 as well): Those of group I are derivable from brand choice models whereas those of group II can be calculated from purchase incidence models. The calculation of diagnostics of group III even necessitates compound models of purchase incidence and brand choice. Finally, the elements of group IV cannot be assigned to one of the model classes above and have to be calculated with the help of special approaches.

It should be pointed out that the system of market diagnostics introduced above (and the given classification of them respectively) can be used as a measure of the descriptive capabilities of existing QMCBs and as a guide line for the design and development of new approaches in quantitative consumer research as well (cf. Decker (1993) for detailed explanations). Moreover, a well-defined set of diagnostics facilitates the construction of a model selection knowledge base.

To reach a high degree of flexibility, in connection with automated applications of different kinds of QMCBs on different types of panel data, it must be ensured that all relevant data are structured in a way that can be interpreted unambiguously by all system components in each phase of data processing. To this end, a flexible data modeling approach is introduced which will be sketched next.

```
      begin_description
      description            {<general comments concerning the data file>}
      data_type              {purchase; ··· buyer_attribute; parameter;
 I                            market_share; ··· relative_brand_attraction}
      panel_type             {not_defined; individual; ··· test_market}
      type_of_recording      {not_defined; diary; ··· interview}
        ...                    ...
      first_day_of_panel     {<first day of data collection>}
 II   unit_of_time           {not_defined; day; week; month; year}
      no_of_ana_per          {not_defined; <number of analysis periods>}
        ...                    ...
      model                  {not_defined; <model used for calculation>}
 III  estimation_method      {not_defined; least_squares; ··· moments}
      fit_criterion          {not_defined; chi_square; likelihood_ratio}
        ...                    ...
      no_of_rows             {<number of data rows>}
      no_of_columns          {<number of data columns>}
      column_1               {<contents of the first data column>}
        ...                    ...
 IV   scale_type_col_1       {ratio; interval; ordinal; nominal}
        ...                    ...
      range_col_1            {integer; real; <associated strings> *}
        ...                    ...
      values                 {<actual data>}
        ...
      end_description
```

Figure 2: General structure of the data model used in KVASS.

The basic idea of this approach is to model buying behavior data with a given set of predefined descriptors. An abridged description of the data model is given in Figure 2. The present combination of descriptors is the result of different investigations within the WIMDAS project (see above) with respect to the modeling of consumer behavior data, and it was identified to be suitable for both the adequate description of inputs (panel data) and the well-structured representation of interim (e.g. parameter estimators) and final results (market diagnostics) as well. The descriptors employed in this context can be divided into four classes. Class I contains essential aspects for a general description of buying behavior data. The elements of class II are used to characterize temporal aspects of the data whereas class III, if necessary, contains information about the model which was used to calculate the present data. The remaining descriptors are needed for the identification and specification of the contents of individual data columns. To complete the picture it should be mentioned that some of the descriptors (like '`panel_type`') are bound to a predefined set of possible values whereas others (like '`column_1`') are unbound in the sense that they can take any numerical or textual value (corresponding hints are enclosed in '`< >`'). An asterisk identifies lists of values separated by comma. If any descriptor is irrelevant for a special data type, this is indicated with the string 'not_defined'. The data listing itself always starts after the string 'values' and terminates with the string 'end_description'.

The present data model can also be transformed into a relational representation as needed in connection with the use of many commercial data base systems. Consequently large amounts of data can be handled in an efficient way (Panel data sets referring to one product class can easily consist of more than one hundred thousand single data specifying purchase histories, buyer attributes, etc.). Furthermore, a standardized model of buying behavior data facilitates simultaneous time and cost saving developments and modifications of separate system components, since each module is working with a well-defined data interface.

3. Classification of quantitative models of consumer behavior

Since a complete description of all econometric and stochastic models of consumer behavior that are relevant for KVASS would exceed the given frame of this paper it has been restricted to a condensed sketching of some basic aspects of stochastic consumer behavior modeling. A more detailed mathematical discussion of this class of models can be found in Topritzhofer (1974), Wagner and Taudes (1987), and Ehrenberg (1988), while Cooper and Nakanishi (1988) and Ronning (1991) deal with econometric approaches in-depth. Finally, in Decker (1993) both classes of models are examined in view of their applicability to panel data and in view of the spectrum of market diagnostics that can be calculated with their help.

One of the most important aspects of building realistic QMCBs is the adequate incorporation of consumer heterogeneity. In stochastic brand choice models, for example, heterogeneity in buying decision making can be treated in such a way that individual (multinomial) brand choice probabilities are assumed to be Dirichlet distributed over the population. At the same time it should be taken into account (both with respect to the formulation and the selection of models) that individual buying behavior can change over time depending on marketing mix activities and/or seasonal influences. The ability of a model to deal with dynamic buying behavior successfully is a decisive prerequisite for its application to simulation problems. Moreover, buying decisions may be affected by purchase histories as well. The incorporation of this aspect in particular strongly increases the models' formal complexity. Therefore, it is indispensable to substantiate the model selection with a sound pre-investigation of the data to be analyzed. This point will be revisited later. Finally, the number of dimensions of the underlying buying decision process that are included in a model plays a major role in determining the type and spectrum of market diagnostics that can be calculated with its help. Within the stochastic modeling framework, we are able to describe both purchase incidence and brand choice with separate submodels which then can be combined to build so-called compound models.

Figure 3 suggests a classification structure for stochastic approaches which can be used as a guide line for the problem-oriented design (and development) of flexible model bases and the corresponding rule bases for model selection. The lines between different boxes "combine" those aspects which are substantial for realistic modeling and which are tractable from a mathematical point of view. Models depicted in dashboxes are not considered in this paper. The terms in Figure 3 are common in relevant literature and are explained in detail in Wagner and Taudes (1987) and Decker (1993), for example.

One of the most remarkable features of stochastic models of consumer behavior is the wide range of market diagnostics that can be calculated with their help. But unfortunately problems concerning the implementation and estimation of parameters did not always support the acceptance of these models in marketing practice in the past. This is one of the facts that initiated the development of the KVASS approach for panel data analysis described in this paper.

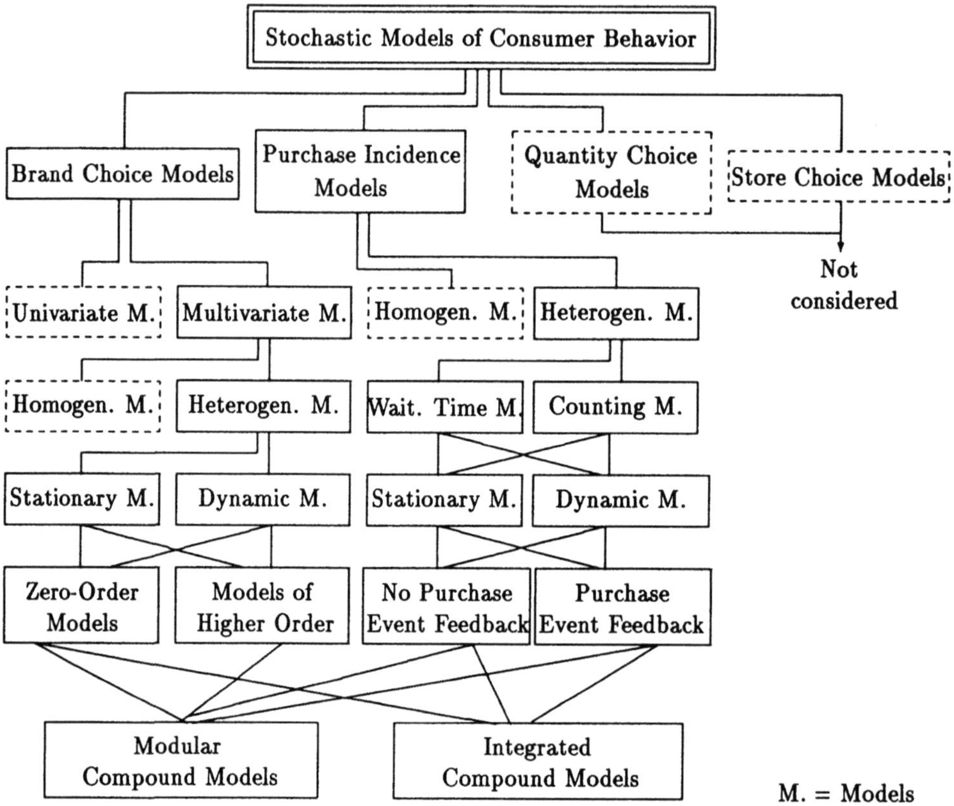

Figure 3: Classification structure for stochastic models of consumer behavior.

4. Goal- and data-oriented analysis of panel data

To answer the general request of marketing management for easy-to-handle decision support systems we have developed a special goal system that covers different areas in the analysis (→ past-oriented) and the simulation (→ future-oriented) of buying behavior. The use of predetermined goal formulations ensures precise specifications of the data analytical problems to be solved by the system, and also enables the knowledge-based determination of those market diagnostics that are relevant for the realization of a goal. Depending on the individual problem - but especially within (ex post) analysis of buying behavior - it is possible that a goal is associated with three, four, or more diagnostics. For example, if we were interested in the '`Identification of Main Competitors of Brand i`', the actual KVASS knowledge base would check if it is possible to calculate market shares, conditional and unconditional switchings, and elasticities as well. If the check was successful, the cited diagnostics could be interpreted (after their calculation) by a knowledge-based reporting system.

Within KVASS the user can specify special constellations of a goal with the help of predefined input masks. The optimization goal depicted exemplarily in Figure 4 represents a typical problem within the simulation of buying behavior and consists of two main parts. First of all, the user has to determine which of the listed market diagnostics should be optimized for which brand. And second, information is needed to specify some necessary constraints in order to guarantee meaningful solutions from an economical point of view.

```
Maximization of [market_share                  ] of brand [ B_1 ]
                [brand_penetration             ]          [ ... ]
                [brand_purchase_expectation    ]          [ B_I ] 1 of I
                [brand_sale                    ]
                [repeat_purchase_probability   ] 1 of 5

    with marketing mix variable [ MMV_1 ] ( between [   and   ] )
                                [  ...  ]
                                [ MMV_L ] l' of L

                          as a(n) [ relative ] value of [ spending ] ,
                                  [ absolute ]          [ receipts ]

                   with an actual unit price of [          ] DM,
                   unit costs in last period of [          ] DM,
                 sales volume in last period of [          ] units,
                         total sales volume of  [          ] units,
                    and an experience effect of [          ] %.
```

Figure 4: Optimization goal (with costs) in KVASS.

Formally, this goal can be written as follows:

$$\text{Max } MD_{i,t+1} = f(\cdots, \eta_{i1}, \cdots, \eta_{iL}, \cdots; \cdots, x_{i1,t+1}, \cdots, x_{iL,t+1}, \cdots)$$

$$\text{with} \qquad x_{il,t+1}^{\min} \leq x_{il,t+1} \leq x_{il,t+1}^{\max}$$

$$\text{and } \sum_{l=1}^{L} \Delta_{2l}(S_{i,t+1}^{\Delta_{1l}} x_{il,t+1} - S_{it}^{\Delta_{1l}} x_{ilt}) - (C_{i,t+1}S_{i,t+1} - C_{it}S_{it}) \geq 0$$

where

$MD_{i,t+1}$	=	market diagnostic (derived from a QMCB) referring to brand i and simulation period $t+1$,
x_{ilt}	=	marketing mix variable l of brand i in last calibration period t,
η_{il}	=	response coefficient,
S_{it}	=	sales volume of brand i in the panel in period t, and
C_{it}	=	unit costs of brand i in period t.

Furthermore, let

$$\Delta_{1l} = \begin{cases} 1, & \text{if } x_{\cdot l \cdot} \text{ refers to one unit} \\ 0, & \text{if } x_{\cdot l \cdot} \text{ refers to the whole panel} \end{cases}$$

$$\Delta_{2l} = \begin{cases} -1, & \text{if } x_{\cdot l \cdot} \text{ is a value of spending} \\ 1, & \text{if } x_{\cdot l \cdot} \text{ is a value of receipts} \end{cases}$$

In view of realistic calculations of $C_{i,t+1}$ we suggest a cost function similar to that one which is used within the so-called 'experience curve' theory (cf. Henderson (1974)):

$$C_{i,t+1} = C_{it} \left(\frac{\tilde{S}_{it}^{acc}}{\tilde{S}_{it}^{acc} + \tilde{S}_{it} \frac{S_{i,t+1}}{S_{it}}} \right)^{\theta}$$

where

$\tilde{S}_{it}^{(acc)}$ = (accumulated) sales volume of brand i in the total market in period t,
θ = coefficient (calculated from the experience effect).

The present cost function avoids the (hardly determinable) so-called 'costs of the first unit of production' used in the original function. Therefore, it can be applied to consumer goods as well. A comprehensive discussion of the experience curve theory with respect to its practical relevance and empirical confirmation can be found in Lange (1984) and Kloock et al. (1987).

The parameter Δ_{1l} is used to separate marketing mix variables referring to one unit (e.g. unit price) from those which are referring to the total market (e.g. total advertising expense of a brand), whereas Δ_{2l} is needed to take into account the effect of an increase, or decrease, of a variable on a brand's profit margin. With the cost function above, a projection of panel results to the total market becomes possible.

The explicit incorporation of costs in KVASS is an important extension of traditional implementations of QMCBs. In this way, modifications to marketing mix variables (in order to improve the value of a market diagnostic) are prevented from implying a worsening of the actual profit margin of a brand. For example, price cuts - as one possible strategy to increase the current market share of a brand - are only permitted, if resulting drops in contribution margin (per unit) can be compensated by a reduction of any value(s) of spending.

An important feature of the KVASS model base is the strict division of each model approach into one calibration module and one diagnostic module. The first is used to carry out parameter estimations whereas the second contains formulas for each market diagnostic (MD_{it}) that can be calculated from the present model. Each of the relevant market diagnostics can then be calculated efficiently on the basis of actual - possibly user-manipulated - marketing mix variables and the parameters estimated beforehand. The independence of model implementation and individual goal formulation enables both the maintenance, and the extension, of the model base and ensures a high degree of goal specification flexibility within the knowledge base.

A basic problem resulting from such an application-oriented data analysis is the connection of (panel) data, application goals (respectively the market diagnostics needed for their realization), and QMCBs.

For this reason, a comprehensive knowledge base was developed which "links" goals and diagnostics (To carry out an optimization of market shares in accordance with the goal sketched above, it is necessary to calculate market shares and brand sales as well.). The knowledge base also assigns market diagnostics to separate QMCBs, and checks the applicability of potentially possible models to the panel data at hand. In the remaining part of this section we shall sketch selected aspects of the knowledge-based data analysis and simulation management of KVASS. A full description can be found in Decker (1993).

The identification of possible combinations of models and diagnostics implies that for all econometric and stochastic approaches used within KVASS, a set of calculable market diagnostics must be identified. The current KVASS knowledge base contains corresponding information about more than 30 different models. With the help of a dynamic model of purchase incidence and brand choice, based on a Gamma-Erlang-2/Dirichlet-Multinomial distribution, for instance (For detailed explanations see Decker (1993).), the system can calculate market shares, elasticities (of market share), brand sales, brand penetrations and several other diagnostics.

The relationships between goals, diagnostics, and models constitute special knowledge that can be used to determine a set of models which are potentially suitable for the calculation of that (those) market diagnostic(s) which is (are) relevant for the current goal.

In the next step, the knowledge base has to check which of these potentially possible models in fact can be applied to the present panel data. To this end, a comprehensive set of criteria for data characterization (tailormade for panel data analysis with the help of econometric and stochastic models of consumer behavior) was developed within the KVASS project. The entire set of criteria can be divided into three classes.

In the first class, we subsume those characteristics that can be taken directly from the standardized (input) data description (cf. Figure 2). The panel type, for example, is of great importance in determining the category of models (econometric or stochastic approaches) that can be applied. The second group of criteria will be acquired within interactive user-system dialogues during the course of a consultation. As an example, reference to the identification of the type of decision-making which is dominant in a product class, and which is of special interest for the selection of econometric choice models can be taken. Depending on the prevailing type of decision-making, models built for simultaneous, or for sequential information processing (i.e. evaluation of product attributes), may be suitable. Finally, a third class of criteria requires explicit pre-analyses of the underlying panel data with the help of statistical tests and/or heuristics. The determination of the order of the brand choice process (cf. Bass et al. (1984)) and the checking of the validity of the so-called IIA-property (cf. Hausman and McFadden (1984)), are typical examples for such pre-analyses of data. A detailed description and discussion of the entire catalogue of criteria can be found in Decker (1993).

5. Concluding remarks

Two repeatedly cited reasons (on the part of the marketing management) for the only modest acceptance of complex (econometric and stochastic) models of consumer behavior in panel data analysis are: the considerable expenditures from mathematical and implementatory points of view (connected with their application), and the fact that in most approaches published until now costs are not considered.

The conceptional framework for model- and knowledge-based analyses and simulations of buying behavior, as presented in this paper, takes into account both of these aspects and - according to our experience - it is an useful approach to make econometric and stochastic models of consumer behavior (which have been used successfully in numerous scientific studies in the past) available to a broader circle of users in panel data analysis and marketing research.

6. References

ALPAR, P. (1991): Knowledge-Based Modeling of Marketing Managers' Problem Solving Behavior. *International Journal of Research in Marketing*, 8, 5-16.

BASS, F.M., GIVON, M.M., KALWANI, M.U., REIBSTEIN, D., and WRIGHT, G.P. (1984): An Investigation into the Order of the Brand Choice Process. *Marketing Science*,

3, 4, 267-287.

BEYS, O., FISCHER, M., TRIPMAKER, S., and MERTENS, P. (1992): Wissensbasierte Analyse von Handelspaneldaten. *Marketing ZFP*, 14, 3, 157-166.

BÖCKENHOLT, I., BOTH, M., and GAUL, W. (1989): A Knowledge-Based System for Supporting Data Analysis Problems. *Decision Support Systems*, 5, 345-354.

COOPER, L.G., and NAKANISHI, M. (1988): *Market-Share Analysis: Evaluating Competitive Marketing Effectiveness.* Kluwer Academic Publishers, Boston.

DECKER, R. (1993): *Analyse und Simulation des Kaufverhaltens auf Konsumgütermärkten - Konzeption eines modell- und wissensorientierten Systems zur Auswertung von Paneldaten.* Dissertation, Universität Karlsruhe (TH).

DECKER, R., and GAUL, W. (1989): Classification and Selection of Consumer Purchase Behaviour Models. In: Opitz, O. (ed.): *Conceptional and Numerical Analysis of Data.* Springer, Berlin, 389-398.

DECKER, R., and GAUL, W. (1990): Einige Bemerkungen über Expertensysteme für Marketing und Marktforschung. *Marketing ZFP*, 12, 4, 257-271.

EHRENBERG, A.S.C. (1988): *Repeat-Buying: Facts, Theory and Applications.* Griffin, London.

GAUL, W., and SCHADER, M. (eds.) (1988): *Data, Expert Knowledge and Decisions.* Springer, Berlin.

GAUL, W., SCHADER, M., and BOTH, M. (1990): Knowledge-Oriented Support for Data Analysis Applications to Marketing. In: Schader, M., and W. Gaul (eds.): *Knowledge, Data and Computer-Assisted Decisions.* Springer, Berlin, 259-271.

GEYER-SCHULZ, A., TAUDES, A., and WAGNER, U. (1988): Exploring the Possibilities of an Improvement of Stochastic Market Models by Rule-Based Systems. In: Gaul, W., and M. Schader (eds.): *Data, Expert Knowledge and Decisions.* Springer, Berlin, 54-66.

HAUSMAN, J., and McFADDEN, D. (1984): Specification Tests for the Multinomial Logit Model. *Econometrica*, 52, 5, 1219-1240.

HENDERSON, B.D. (1974): *Die Erfahrungskurve in der Unternehmensstrategie.* Campus, Frankfurt.

KLOOCK, J., SABEL, H., and SCHUHMANN, W. (1987): Die Erfahrungskurve in der Unternehmenspolitik. *Zeitschrift für Betriebswirtschaft*, Ergänzungsheft, 2, 3-51.

LANGE, B. (1984): Die Erfahrungskurve: Eine kritische Beurteilung. *Zeitschrift für betriebswirtschaftliche Forschung*, 36, 3, 229-245.

McCANN, J.M, and GALLAGHER, J.P. (1990): *Expert Systems for Scanner Data Environments - The Marketing Workbench Laboratory Experience.* Kluwer Academic Publishers, Boston.

RANGASWAMY, A., HARLAM, B.A., and LODISH, L.M. (1991): INFER: An Expert System for Automatic Analysis of Scanner Data. *International Journal of Research in Marketing*, 8, 29-40.

RONNING, G. (1991): *Mikroökonometrie.* Springer, Berlin.

SCHMITZ, J.D., ARMSTRONG, G.D., and LITTLE, J.D.C. (1990): CoverStory - Automated News Finding in Marketing. *Interfaces*, 20, 6, 29-38.

TOPRITZHOFER, E. (1974): *Absatzwirtschaftliche Modelle des Kaufentscheidungsprozesses unter besonderer Berücksichtigung des Markenwahlaspektes.* Österreichische Akademie der Wissenschaften, Wien.

WAGNER, U., and TAUDES, A. (1987): Stochastic Models of Consumer Behaviour. *European Journal of Operational Research*, 29, 1-23.

Goodwill towards Domestic Products as Segmentation Criterion: An Empirical Study within the Scope of Research on Country-of-Origin Effects

Wolfgang Gaul, Ulrich Lutz, Eberhard Aust

Institut für Entscheidungstheorie und Unternehmenforschung, Universität Karlsruhe
Postfach 6980, D-76128 Karlsruhe, Germany

Summary: The phenomenon of market success of domestic products is investigated. We use goodwill towards domestic products as segmentation criterion within an empirical study, describe data collection and methodology, and discuss findings of importance for marketing applications.

1. Introduction

Already a glance at the references of this paper reveals that there is quite an amount of marketing literature in which authors deal with 'attitudes towards foreign products', 'national stereotypes and product evaluations', 'country-of-origin effects of products' or 'the image of products MADE IN a special country'. Against this background, we will use the term 'goodwill towards domestic products' to paraphrase a phenomenon which is of great importance to the area of marketing but has been somewhat neglected in the literature. We discuss indicators and explanations for goodwill towards domestic products and describe an empirical study designed to allow a segmentation of consumers in such a way that characteristics of a 'domestic origin segment' (a group of consumers for which goodwill towards domestic products is (one of) the essential feature(s)) can be evaluated. In this paper, emphasis is not laid on new methodology with respect to classification or segmentation. Instead, we explain how new findings with respect to such a latent variable as goodwill towards domestic products can be yielded by application of data analysis techniques from a classification point of view. Finally, we discuss some managerial implications which goodwill towards domestic products may have in the marketing area.

2. Goodwill towards Domestic Products

In view of the development of the European Internal Market which is reality since January 1993 the phenomenon under study has become even more obvious. Remember the attention given to the removal of physical, technical, and fiscal barriers within Europe in the middle of the eighties (Commision of the European Communities (1985)). If such kinds of trade barriers lose acuteness, mental marketing barriers (Pfeifer (1988, p. 67)) may still remain from a consumer behaviour orientated point of view. Within the scope of European life-style research (Kramer (1991)) one approach (Butter (1988, p. 601)) reveals that an essential feature of a large consumer segment of Europe – the traditional, conservative and nationally-minded 'mainstreamers' – is preference for national products. From an American point of view, one can mention the discussion about 'consumer ethnocentrism' (Shimp and Sharma (1987)) which covers what we call goodwill towards domestic products in this paper.

Although a nationality bias in favour of products from the home country has been observed in many studies on country-of-origin effects (some papers describe results based on preference judgements for products for which price, quality, and design are fixed such that country-of-origin is "the remaining feature", e.g., Nagashima (1970), Darling (1980), Cattin et al. (1982), Hung (1989), while others report on product evaluations on the basis of country-of-origin when special product characteristics are not mentioned, e.g., Reierson (1966), Doeblin and Doeblin (1983), Morello (1984), Festervand et al. (1985), Crawford et al. (1987), Papadopoulos et al. (1990)), goodwill towards domestic products has not been in the centre of this research direction. However, the papers mentioned point to the existence of such a phenomenon as goodwill towards domestic products, and already in the early overview of Bilkey and Nes (1982) on country-of-origin effects the authors conclude that "There is a tendency for consumers to evaluate their own country's products relatively more favourably than do others".

Clearly, the expression "to evaluate more favourably" needs further explanations, e.g., one could start the discussion with the question how goodwill towards domestic products is related to the quotation of prices. Although one might assume that prices of products are lower in the home market than in export markets (see Cateora and Hess (1975, p. 434), Ratnatunga (1987, p. 42), Kublin (1990, p. 57), Cavusgil (1988, p. 57), Jeannet and Hennessey (1988, p. 399), Becker (1990) for so-called international price escalation) opposed findings can readily be presented, e.g. for the automobile market (where Gual (1987, p. 29) and Kirman and Schueller (1990, p. 78) even talk about discrimination against domestic consumers) some researchers have stated that cars are ceteris paribus more expensive in the domestic market than abroad (Mertens and Ginsburgh (1985, p. 163) and Bourdet (1987, p. 188) report about 4 % to 5 % rise in prices which could be confirmed by our own research (Lutz (1992)). Such price-related findings can be understood as willingness of consumers to pay higher prices for products they perceive to be national. More generally, willingness to pay has been related to country-of-origin effects (Dichtl et al. (1983)) which might help to explain higher prices for, e.g., French perfumes, German cars, or Italian shoes.

Another question is whether the fact that domestic producers often show higher or even dominating market shares in their home countries is related to goodwill towards domestic products. Some of the rather obvious reasons, why domestic products may have higher shares in their home markets than foreign products, are: • Firms might concentrate their sales efforts on the domestic market. • Distribution structures – developed over a long period – ensure that the domestic market is efficiently exploited. • Higher brand awareness of domestic products supports their sale. • Products are often geared to fulfil the specific requirements in their respective home country. • As far as public procurement is concerned, authorities tend to keep their purchases and contracts within their own country. • Employees of manufacturing firms identify themselves – to a large extent – with their "own" products. • Additionally, special discounts may be offered to the employees for buying the products of their "own" firm.

Of course, among the explanations for goodwill towards domestic products country-of-origin itself can be considered as relevant to this phenomenon. Here, researchers have made a distinction between cognitive and affective dimensions (Findings by Johansson et al. (1985) indicate that products of domestic origin are not selected in a cognitive way; however, due to usually better knowledge of salient features of domestic products, the perceived risk in buying them is reduced; furthermore, Johansson (1989) distinguishes between affective country stereotypes originating from, e.g., history, political situation, culture, or psychological proximity, and behavioural intentions to meet social norms.). Factors like positive attitude towards one's own nation (which Morello (1984) claims to exist in industrialized countries), national pride, cultural and psychological proximity to local firms, intention to support the domestic economy or intention to avoid social risk are related to goodwill towards domestic products.

The importance of findings of this kind can depend on the product category. As an example, preference towards domestic food products (Kaynak and Cavusgil (1983), Festervand et al. (1985)) seems to be more obvious than towards fashion or luxury products (Garland et al. (1989)). Within the same product category, less sophisticated standard products should be endowed with national identity while niche products with an "own" strong positioning can benefit from this distinctive identity (see Hubbert (1990) for corresponding statements referring to the European car market).

3. Survey and Segmentation

3.1 Data Collection

To gain insight into the structure of consumer segments prefering domestic products, a survey was carried out which dealt with automobile purchasing decisions. The sample in this inquiry consisted of people who had bought at least one new motor-car during the last five years. 161 questionnaire-supported personal interviews were conducted (Friederes (1992)). With respect to criteria such as age, sex, education and percentage of owners of domestically produced cars, respondents were approximately representative of that part of the population involved in automobile purchasing decisions. Three types of data were collected: Demographics, ratings on attitude statements to characterize respondents, and conjoint analysis data to indirectly measure the relative importance of domestic origin in purchasing decisions because respondents – when asked directly – tend to attach comparatively little importance to country-of-origin (Gaul et al. (1992), p. 28).

3.2 Methodology

Already Bilkey and Nes (1982) have appealed to researchers to analyze the importance of the country-of-origin in a multi-attribute context (see Thorelli et al. (1988), Khachaturian et al. (1990), and Wall et al. (1991) who used, e.g., brand, type of store, and guarantee).

For the application of conjoint analysis, after extensive pretests with varying numbers of attributes and levels, and forms of stimuli presentation, a full profile approach was chosen with the following five two level attributes describing a hypothetical motor-car purchasing situation: Origin ("produced domestically/in a foreign country"), price ("some hundred marks more/less than for comparable cars"), guarantee ("equivalent to common conditions/extraordinary good"), design ("distinctive in a positive sense/neither good nor bad") and dealer characteristics ("nearby and well known to provide good service/at a distance and with unknown quality of service"). In contrast to previous research involving conjoint analysis combined with country-of-origin effects (Ettenson et. al. (1988) and Wagner et. al. (1989)) the aim of the present study was to perform a so-called benefit segmentation and not segmentation according to prespecified background variables. To minimize the response burden a reduced orthogonal factorial main effects design (the Addelman (1962) basic plan 1) was employed leading to eight stimuli. As assessment on rating scales – although considerably more demanding – need not lead to better results (Kalish and Nelson (1991)), respondents were asked to rank-order product cards. Individual part worths were estimated in a two step procedure: First, an initial solution was calculated using OLS regression, where the inverted rank was the dependent variable. Secondly, monotonous analysis of variance minimizing Kruskal's (1965) stress was applied. Using the OLS result as starting solution a stress value of 1.4% could be attained. Part worths were standardized within respondents and importances of attributes calculated. While simultaneous procedures are the theoretically adequate approaches to benefit segmentation we used a traditional proceeding for several reasons: Scepticism has been expressed whether simultaneous procedures lead to better results in practice (Green and Helsen (1989)). If – as in our case – there are only two level attributes the Kamakura (1988) approach leads to no improvement. Fuzzy approaches (Hagerty (1985)) were not thought to be adequate due to interpretational difficulties. Finally, Ogawa's (1987) hierarchical clustering algorithm, though specifically aimed at rank-order input data is computationally very demanding in the case of large samples.

Therefore, we resorted to what might be called the traditional approach (Green and Srinivasan (1978)). In the beginning, outliers were eliminated by Single Linkage (Punj and Stewart (1983)) before Ward was applied. The Euclidean metric was employed in both cases as good results have been reported for part worths clustering (Ogawa (1987)). With two level attributes, clustering of attribute importances is equivalent to the use of standardized part worths (if the latter are uni-directional which was the case in our example except for a very small number of respondents). A four cluster solution was suggested by the Elbow criterion.

Principal component factor analysis was used to derive scores on latent personal characteristics where due to missing values the size of the sample was reduced to 154. Four factors, representing a total of 60.3% of variance, were retained by the Mineigen criterion. Due to sharper loadings on the relevant factors the factor pattern after Quartimax rotation was chosen though Varimax led to a quite similar solution. To characterize the clusters average part worths, demographics and factor scores were calculated. The significance of discrimination between the clusters on basis of the demographics and on basis of the latent personal characteristics was assessed by estimating canonical discriminant functions. The importance of single variables of the afore mentioned groups to discriminate between clusters was analyzed using standard t-tests.

3.3 Results

Fig. 1 shows the benefit segmentation structure resulting from the clustering activities sketched before. The upper histogram in Fig. 1 shows the attributes' part-worths aggregated on the whole sample. Origin turned out to be the most important attribute (28.8% of total benefit) of those under consideration. Dealer characteristics attained the second highest value, whereas the importance of guarantee was comparatively small.

Already in the two-cluster-solution the group of respondents who benefit significantly from the domestic origin of a product ("domestic origin segment") separates from the remaining group ("unspecific segment"). The domestic origin segment comprises a little less than 28% of the total number of respondents and will be further investigated in the following.

The four-cluster-solution is the most appropriate one with respect to the Elbow criterion. In this case, there is one cluster the members of which give particular attention to dealer characteristics ("dealer segment"), one cluster in which respondents attach great importance to design ("design segment"), one cluster in which quite similar benefit from different attributes is reported although price is the most important single attribute, hence the name ("price segment"), and the domestic origin segment already mentioned.

Before starting further investigation with respect to the domestic origin segment one should remember that demographic variables such as age, sex, education, income (Schooler (1971), Anderson and Cunningham (1972), Bannister and Saunders (1978), Dickerson (1982), Festervand et al. (1985), Wall and Heslop (1986), Hung (1989)) and latent personal characteristics like nationalism, conservatism, dogmatism, awareness of

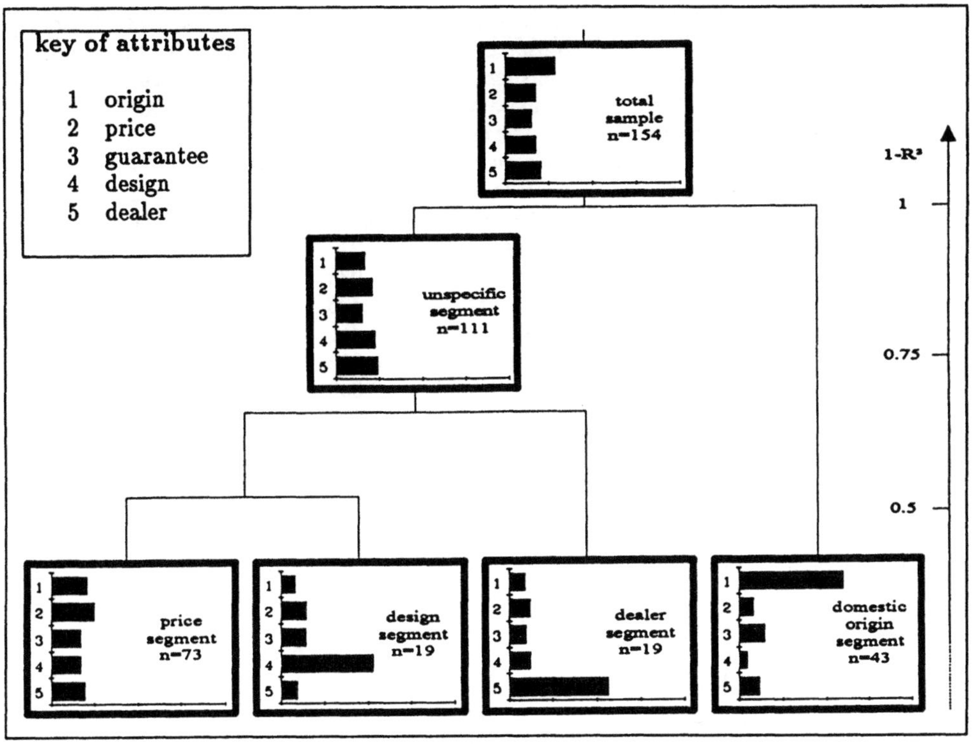

Fig. 1: Structure of benefit segments

status, patriotism, attitude towards big business (Anderson and Cunningham (1972), Shimp and Sharma (1987)) have been used for explanation of country-of-origin effects. In our study demographics as well as information on attitudes towards several statements to determine personal characteristics were collected. As a result respondents in the domestic origin segment are oldest (48.1 years on average), members of the design segment are youngest (32.6 years), and the share of females in the domestic origin segment (27.9%) is less than in the total sample (30.5%). Tab. 1 describes the segments in terms of education level. The corresponding shares in the domestic origin segment decrease monotonously when the level of education increases, while members of the design segment tend to have a higher degree of education. Factor analysis based findings with respect to the statements used for gathering personal characteristics in the questionnaire are given in Tab. 2. Four factors (representing 60.3% of the variance) were selected after Quartimax rotation. Interpretations of these factors ("industry-related nationalism" for factor1, "awareness of status" for factor2, "conformative conservatism" for factor3 and "risk aversion" for factor4) correspond partly to latent personal characteristics employed in the studies referred to above.

Level of Education	domestic origin segment	unspecific segment	price segment	design segment	dealer segment
Total	27.9%	72.1%	47.5%	12.3%	12.3%
Primary School	37.5%	62.5%	37.5%	4.2%	20.8%
9-th Grade Education	37.0%	63.0%	40.7%	11.1%	11.2%
10-th Grade Education	26.0%	74.0%	52.0%	10.0%	12.0%
A-Levels	21.8%	78.2%	56.3%	15.6%	6.3%
University Graduation	19.0%	81.0%	42.9%	23.8%	14.3%

Tab. 1: Description of segments in terms of education level

Short forms of statements in questionnaire	factor1	factor2	factor3	factor4
National pride	0.829	0.116	0.192	-0.027
Enthusiasm about victories of German sportsmen	0.745	0.170	0.167	-0.078
Positive attitude towards national anthem	0.806	0.050	-0.064	0.090
Belief in quality provided by "Made in Germany"	0.802	-0.018	0.005	0.119
Purchase of domestic products to help economy	0.676	0.027	-0.031	0.047
Benefit from a strong German industry	0.481	0.289	-0.163	0.367
Opinion that achievements should be shown	0.310	0.612	0.100	0.021
Belief that wealth is a sign of personal success	0.218	0.705	0.027	0.152
Importance of neighbours' esteem	0.073	0.679	-0.090	-0.087
Distrust of innovations	0.403	0.264	0.624	0.190
Assessment of personal degree of conservatism	0.451	0.264	0.545	0.014
Efforts to attract attention	0.019	0.226	-0.768	-0.043
Purchase of established brands	0.330	0.100	-0.064	0.740
Willingness to take chances	-0.073	0.049	-0.272	-0.815
Variance explained by each factor	33.0%	10.4%	8.6%	8.3%

Tab. 2: Factor pattern of personal attitudes (principal components and Quartimax rotation)

demographic variables	t-test	p-value	latent personal characteristics	t-test	p-value
age	-4.4017	0.0001 (1)	factor1	4.1296	0.0001 (1)
education	1.8425	0.0337 (1)	factor2	2.0356	0.0217 (1)
sex	0.4356	0.6637 (2)	factor3	1.6100	0.0547 (1)
			factor4	2.8367	0.0026 (1)
(1) one-sided test (prior belief suggests a direction for testing)					
(2) two-sided test (results in the literature do not suggest a direction for testing)					

Tab. 3: Discrimination between domestic origin segment and unspecific segment (univariate test statistics)

Some information concerning differences between the domestic origin segment and the unspecific segment are presented in Tab. 3. In the left part, results with respect to some of the demographic variables under consideration are shown. Sex did not discriminate, whereas differences concerning age and education (Degrees of education were represented on a five point integer scale.) proved to be significant, i.e., persons in the domestic origin segment are older and less educated than the rest of the sample. In the right part of Tab. 3 which refers to latent personal characteristics, three out of the four factors determined above discriminate the domestic origin segment from the unspecific segment at values far below the 0.05 level, only factor3 ("conformative conservatism") narrowly exceeds the 0.05 level. These results indicate that members of the domestic origin segment are significantly on a larger scale nationally minded and industry orientated, risk averse, aware of status, and – to some degree – conformative and conservative. The estimated canonical discriminant functions for the domestic origin segment versus the unspecific segment or versus the design segment, the dealer segment, and the price segment were all highly significant (0.0097 against the dealer segment, 0.0001 against the remaining segments).

4. Managerial Implications

Goodwill towards domestic products was investigated within a survey in which attributes concerning motor-car purchasing decisions as well as demographics and personal attitudes, which should help to characterize potential consumers, were used for description of the underlying purchasing situation. Domestic origin proved to be the most important attribute and members of the domestic origin segment, obtained by benefit segmentation, were characterized by age, education, and factors derived by a factor analysis application. Although one could argue that for automobiles the specialist image of German car makes supports the findings in favour of goodwill towards German cars, the results of this survey also give evidence to an existence of goodwill towards domestic products in this field.

This leads us to discuss implications of goodwill towards domestic products in different areas of managerial decision-making. In the mind of consumers the perception of a product's country-of-origin is associated with the location of the manufacturer and/or with places where main business activities (e.g., production, R & D, administration) are performed. However, decisions concerning the location of such main business activities are to such an extent fundamental to the whole strategy of firms and require direct investments of such amounts that they cannot be made only on the basis of image and goodwill. Nevertheless, the extent to which country-of-origin can influence individual purchasing decisions should be considered in locational decisions. International manufacturing can widen the scope of image policy. According to Johansson (1989, p. 54) "hybrid" products (for which the country-of-origin may not be clear) can reduce the importance of country-of-origin effects. If production structures are widespread or not well-known, a brand can succeed in having a national image in many countries (Kreutzer (1989, p. 569) mentions *Nivea* as an example not referring to the automobile industry.). On a long run, subsidiaries in countries of interest may help to create local image for a foreign producer. *Ford* succeeded with such a strategy in European countries even without the use of local brand names. A helpful proceeding to create a national image is to use locally accepted brands or brand names as *General Motors* did in several countries (As an example, *Opel Corsa* is considered to be a German car by German consumers even though it is not necessarily produced in Germany. Additionally, *General Motors* just recently changed the brand name of its *Isuzu*-based off-road motor-cars to *Opel* for the German market.). From all which has been discussed up to now one can see that within marketing activities directed to goodwill towards domestic products communication policy can play an important role. MADE IN-labels, the use of national characteristics with respect to appearance or, at least, packaging of products, and national associations within advertising (Head (1988) gives examples of advertisements with quite sophisticated information concerning country-of-origin.) can be mentioned. With respect to pricing policy, marketing implications have already been stated. When consumers are more price sensitive towards foreign than domestic products it might be appropriate to charge lower prices in foreign markets than at home (Bourdet (1987, p. 185) determined price elasticities for motor-cars in France and reported values of -1.6 for foreign and -0.25 for domestic brands.). However, if price differences become "large enough" problems in form of grey markets can arise (Cavusgil and Sikora (1988) and Weigand (1990)).

5. Conclusion

In a Europe growing together – to end this paper as we have started – different future developments are possible with respect to the phenomenon of goodwill towards domestic products. On the one hand, in a "real" Common European Market the importance of a national origin might be replaced by a European origin (Garland et al. (1989, p. 20)). Even a general decline in the importance of "made in"-labels could be possible (Kirbach (1987)). On the other hand, consumers could perceive a loss of national identity through the Common Market and therefore refuse to give up their preferences for domestic products. (Head (1988, p. 251) proposes to give emphasis to country-of-origin because "the internal market spreads its mantle of national anonymity over the products", Johansson and Nebenzahl (1987) and Papadopolous et al. (1990) also assume that the importance of country-of-origin in purchasing decisions may increase.)

Thus, there seems to be enough evidence that research of the kind described in this paper is of great interest from a marketing point of view. Future activities could collect results based on different product categories, compare results from the same product category investigated in different countries, or consider whether the importance of goodwill towards domestic products changes over time.

References

ADDELMAN, S. (1962): Orthogonal Main Effect Plans for Asymmetrical Factorial Experiments. *Technometrics*, 4, 1, 21–46.

ANDERSON ,W.T., and CUNNINGHAM, W.H. (1972): Gauging Foreign Product Promotion. *Journal of Advertising Research*, 1, 29–34.

BANNISTER, J.P., and SAUNDERS, J.A. (1978): UK Consumers' Attitudes Towards Imports – The Measurement of National Stereotypes, *European Journal of Marketing*, 12, 8, 562–570.

BECKER, H. (1990): Price Escalation in International Marketing. In: H.B. Thorelli and S.T. Cavusgil (eds.): *International Marketing Strategy*. Pergamon Press, Oxford New York, 523–525.

BILKEY, W.J., and NES, E. (1982): Country-of-Origin Effects on Product Evalutations. *Journal of International Business Studies*, Spring/Summer, 89–99.

BOURDET, Y. (1987): *International Integration, Market Structure and Prices*. Lund Economic Studies.

BUTTER, D. (1988): Europa '92 – Die Herausforderung von 320 Mio. Euro-Konsumenten an das Marketing. *Marketing Journal*, No. 6, 600–603.

CATEORA, P.R., and HESS, M.H. (3rd Edition, 1975): *International Marketing*, Richard D. Irwin Inc., Homewood, Illinois.

CATTIN, P., JOLIBERT, A., and LOHNES, C. (1982): A Cross-Cultural Study of "Made in" Concepts. *Journal of International Business Studies*, Winter, 131–141.

CAVUSGIL, S.T. (1988): Unravelling the Mystique of Export Pricing. *Business Horizons*, May-June, 54–63.

CAVUSGIL, S.T., and SIKORA, E. (1988): How Multinationals can Counter Grey Market Imports. *Columbia Journal of World Business*, Winter, 75–85.

COMMISSION of the EUROPEAN COMMUNITIES (1985): Completing the Internal Market. *White Paper from the Commission to the European Council*. Brussels.

CRAWFORD, J.C., GARLAND, B.C., and SCHWEIGER, G. (1987): Austrian, German, and Swiss Product Quality Perceptions of Foreign Goods. *Werbeforschung und Praxis*, 2, 29–34.

DARLING, J.R. (1980): A Longitudinal Study of the Attitudes of Finnish Consumers Toward the Products and Associated Marketing Practices of Various Selected Countries. *The Finnish Journal of Business Economics*, Fall, 252–268.

DICHTL, E., BEESKOW, W., and PULS, S. (1983): Deutsche Erzeugnisse auf japanischen Konsumgütermärkten. *Jahrbuch für Absatz- und Verbraucherforschung*, 29, 221–243.

DICKERSON, K.G. (1982): Imported Versus U.S.-Produced Apparel: Consumer Views and Buying Patterns. *Home Economic Research Journal*, March, No. 3, 241–252.

DOEBLIN, R., and DOEBLIN, J. (1983): Das Image deutscher Produkte in den USA. *Markenartikel*, No. 4, 159ff.

ETTENSON, R., WAGNER, J., and GEATH, G. (1988): Evaluating the Effect of Country of

Origin and the "Made in the USA" Campaign: A Conjoint Approach. *Journal of Retailing*, 64, Spring, 85–100.

FESTERVAND, T.A., LUMPKIN, J.R., and LUNDSTROM, W.J. (1985): Consumers' Perception of Imports: An Update and Extension. *Akron Business and Economic Review*, Spring, 31-36.

FRIEDERES, G. (1992): Die Bevorzugung inländischer Produkte - Eine empirische Studie. *Diploma Thesis*, Institut für Entscheidungstheorie und Unternehmensforschung der Universität Karlsruhe (TH).

GARLAND, B.C., CRAWFORD, J.C., and DALTON, J. (1989): Euro-Consumer Satisfaction with Imported Products. *Der Markt*, 28, 1, 19–24.

GAUL, W., THEUNER, G, and LUTZ, U. (1992): Unternehmen aus den neuen Bundesländern beim Übergang zur Marktwirtschaft – Zusammenfassung von Ergebnissen einer empirischen Studie am Beispiel EDV-Einsatz. *Report*, Institut für Entscheidungstheorie und Unternehmensforschung der Universität Karlsruhe (TH).

GREEN, P.E., and HELSEN, K. (1989): Cross-Validation Assessment of Alternatives to Individual-Level Conjoint Analysis: A Case Study. *Journal of Marketing Research*, 26, August, 346–50.

GREEN, P.E., and SRINIVASAN, V. (1978): Conjoint Analysis in Consumer Research: Issues and Outlook. *Journal of Consumer Research*, 5, September, 103–123.

GUAL, J. (1987): An Econometric Analysis of Price Differentials in the EEC Automobile Market. *Discussion Paper*, IESE, Barcelona.

HAGERTY, M.R. (1985): Improving the Predictive Power of Conjoint Analysis: The Use of Factor Analysis and Cluster Analysis. *Journal of Marketing Research*, 22, May, 168–184.

HEAD, D. (1988): Advertising Slogans and the 'Made-in' Concept. *International Journal of Advertising*, 7, 237–252.

HUBBERT, J. (1990): Strategien für das Fahrzeuggeschäft mit Blick auf den EG-Binnenmarkt. In: E. Zahn (Hrsg.): *Europa 1992 – Wettbewerbsstrategien auf dem Prüfstand*. C.E. Poeschel, Stuttgart, 65–79.

HUNG, C.L. (1989): A Country of Origin Product Image Study: The Canadian Perception and Nationality Biases. *Journal of International Consumer Marketing*, 1, 3, 5–26.

JEANNET, J.-P., and HENNESSEY, H.D. (1988): *International Marketing Management – Strategies and Cases*. Houghton Mifflin Company, Boston.

JOHANSSON, J.K. (1989): Determinants and Effects of the Use of "Made in" Labels. *International Marketing Review*, 6, 1, 47–58.

JOHANSSON, J.K, DOUGLAS, S.P., and NONAKA, I. (1985): Assessing the Impact of Country of Origin on Product Evaluations: A New Methodological Perspective. *Journal of Marketing Research*, 22, 388–396.

JOHANSSON, J.K, and NEBENZAHL, I.D. (1987): Country-of-Origin, Social Norms and Behavioral Intentions. *Advances in International Marketing*, 2, 65–79.

KALISH, S., and NELSON, P (1991): A Comparison of Ranking, Rating and Reservation Price Measurement in Conjoint Analysis. *Marketing Letters*, 2, 4, 327–335.

KAMAKURA, W.A. (1988): A Least Squares Procedure for Benefit Segmentation with Conjoint Experiments. *Journal of Marketing Research*, 25, May, 157–167.

KAYNAK, E., and CAVUSGIL, S.T. (1983): Consumer Attitudes towards Products of Foreign Origin: Do They Vary Across Product Classes? *International Journal of Advertising*, 2, 147–157.

KHACHATURIAN, J.L., and MORGANOSKY, M.A. (1990): Quality Perceptions by Country of Origin. *International Journal of Retail & Distribution Management*, 15, 5, 21–30.

KIRBACH, R. (1987): Den falschen Ruf besiegelt. *Die Zeit*, 7. August, 19.

KIRMAN, A., and SCHUELLER, N. (1990): Price Leadership and Discrimination in the European Car Market. *The Journal of Industrial Economics*, September, 69–91.

KRAMER, S. (1991): *Europäische Life-Style-Analysen zur Verhaltensprognose von Konsumenten.* Verlag Dr. Kovac, Hamburg.

KREUTZER, R.T. (1989): Marketingstrategien im länderübergreifenden Marketing. *Markenartikel*, No. 11, 569–572.

KRUSKAL, J.B. (1965): Analysis of Factorial Experiments by Estimating Monotone Transformations of the Data. *Journal of the Royal Statistical Society Series B*, 27, 251–263.

KUBLIN, M. (1990): A Guide to Export Pricing. *Industrial Management*, May/June, 29–32.

LUTZ, U. (1992): Zur Preisbildung im Europäischen Markt. In: *Operations Research Proceedings.* Springer, Berlin Heidelberg New York, 332–339.

MORELLO, G. (1984): The "made in" Issue. *European Research*, January, 5–21.

MERTENS, Y., and GINSBURGH, V. (1985): Product Differentiation and Price Discrimination in the European Community – The Case of Automobiles. *The Journal of Industrial Economics*, No. 2, December, 151–166.

NAGASHIMA, A. (1970): A Comparison of Japanese and U.S. Attitudes Toward Foreign Products. *Journal of Marketing*, 34, January, 68–74.

OGAWA, K. (1987): An Approach to Simultaneous Estimation and Segmentation in Conjoint Analysis. *Marketing Science*, 6, 73–100.

PAPADOPOULOS, N., HESLOP, L.A., and BERACS, J. (1990): National Stereotypes and Product Evaluations in a Socialist Country. *International Marketing Review*, 7, 1, 32–47.

PFEIFFER, R. (1988): Das neue Marketing für den europäischen Binnenmarkt. In: Rudolf Haufe Verlag (Hrsg.): *EG-Binnenmarkt – Chancen und Risiken für Betriebe.* Rudolf Haufe Verlag, Freiburg, 67–88.

PUNJ, G., and STEWART, D.W. (1983): Cluster Analysis in Marketing Research: Review and Suggestions for Application. *Journal of Marketing Research*, 20, 134–148.

RATNATUNGA, J. (1987): Financial Dimensions of Pricing in International Markets. *Australian Accountant*, July, 41–46.

REIERSON, C. (1966): Are Foreign Products seen as National Stereotypes. *Journal of Retailing*, 33–40.

SCHOOLER, R. (1971): Bias Phenomena Attendant to the Marketing of Foreign Goods in the U.S. *Journal of International Business Studies*, 2, Fall, 71–80.

SHIMP, T.A., and SHARMA, S. (1987): Consumer Ethnocentrism – Construction and Validation of the CETSCALE. *Journal of Marketing Research*, 24, August, 280–289.

THORELLI, H.B., LIM, J., and YE, J. (1988): Relative Importance of Country of Origin – Warranty and Retail Store Image on Product Evaluations. *International Marketing Review*, 6, 1, 35–45.

WAGNER, J., ETTENSON, R., and PARRISH, J. (1989): Vendor Selection among Retail Buyers: An Analysis by Merchandise Division. *Journal of Retailing*, 65, 1.

WALL, M., and HESLOP, L.A. (1986): Consumer Attitudes Toward Canadian-Made Versus Imported Products. *Journal of the Academy of Marketing Science*, 14, 2, 27–36.

WALL, M., LIEFELD, J., and HESLOP, L.A. (1991): Impact of Country-of-Origin Cues on Consumer Judgements in Multi-Cue Situations: A Covariance Analysis. *Journal of the Academy of Marketing Science*, 19, 2 105–113.

WEIGAND, R.E. (1990): Parallel Import Channels – Options for Preserving Territorial Integrity. *Columbia Journal of World Business*, Spring, 53–60.

Merovingian Glass Beads: An Essay of Classification

Peter Ihm[1], Barbara Sasse[2], Claudia Theune[3]

[1] Institut für medizinische Biometrie, Philipps-Universität, D-35033 Marburg

[2] Vorgeschichtliches Seminar, Philipps-Universität, D-35032 Marburg

[3] Institut für Ur- und Frühgeschichte, Albert-Ludwigs-Universität, D-79085 Freiburg

Summary: Until recently Merovingian glass beads were often described only cursorily by colour and shape. New analyses have shown that glass beads, together with other finds, are especially valuable for supplying additional information on chronology as well as economic, cultural and social history. Because of their widespread distribution beads are also useful for the ordering of graves with few other gravegoods. The study of the combination of different bead types per grave has been used for the chronological seriation and the horizontal stratigraphy of the female graves of the Alamannic cemetaries of Weingarten and Eichstetten, respectively.

The 801 graves from the Weingarten cemetery were analysed in three independent Correspondence Analyses (CA) based on: (i) weapons and belt fittings in male graves, (ii) beads in female graves, (iii) other costume ornaments in female graves. Analysis (ii) based on 42 bead types from 101 graves showed seven bead combination groups corresponding to nine fashion phases established by aid of other jewellery and dated by coin finds (Theune-Vogt 1991). A CA with 27 bead types from 51 graves in the Eichstetten cemetary allowed the isolation of five groups, the oldest group associated with special types of brooches, the four younger ones with different types of shoe fittings and earrings (Sasse 1986).

The comparison of bead inventaries of different cemetaries and analysts necessitates a standardized description code. Based on earlier work of Beck (1928) a team of experienced German archaeologists started in 1989 with the elaboration of a hierarchical thesaurus with tree structure allowing the indexation of beads with different degree of detail. The structure of the thesaurus is described in this paper.

1. Introduction

Until recently Merovingian glass beads were often described only cursorily by color and shape. New analyses have shown that glass beads, along with other finds, are especially valuable for supplying additional information on chronology as well as economic, cultural and social history. Because of their widespread distribution beads are also useful for ordering graves with only few other gravegoods. The study of the combination of different bead types per grave has been used for chronological seriation and horizontal stratigraphy of female graves of the Alamannic cemeteries of Weingarten and Eichstetten, respectively.

After a short review of some archaeological results and an explanation of the basic concepts of the thesaurus, the structure of the thesaurus will be described.

2. The Archaeological Background

In South-West-Germany large cemeteries of individual inhumation graves exist in the Merovingian period (450 - 700 a.D.). The people were buried in their costume (Tracht) and other artifacts were usually added. The male graves produced weapons such as one- edged (sax) or two-edged (spatha) swords, spears, shields, arrows and belt-fittings of various types. The women were buried with their cultural dress and accessories: brooches, earrings, belt fittings, shoe fittings, strap ends and bead necklaces with often more than 100 single remaining beads. Other artifacts such as knives, combs or pottery for food destinated for the life after the death can also be found. The outfit of the grave depends on sex, age and social standing.

The gravegoods were subject to changes of function and fashion allowing the establishment of a chronological system. Different methods were used for its establishment (Böhner 1958, Werner 1953 & al.): The methodology ranges from typological comparisons to horizontal stratigraphy to the absolute dating by aid of coins to Correspondence Analysis (CA), which together result in the chronological seriation of the graves.

The archaeologists take much notice in belt fittings of male graves and brooches of female graves since both types of artifacts allow a rather accurate chronological system. Until now, not very much attention was paid to the beads which were described only cursorily by color and shape. It has turned out, however, that because of their abundance, beads are a useful tool for dating a grave, especially in cases where only a few other gravegoods are present. Using beads the number of datable graves may increase up to 50 percent.

Since the combinations of different bead types in the grave varies depending on time, they were used successfully as a tool for datation by Christlein (1966), Koch (1974, 1977, 1982), Sasse (1989, 1993) and Theune-Vogt (1990, 1991). The reliability of a seriation based on beads as compared to one based on jewellery is shown by a high correlation coefficient between the scores x: $r_{Pearson}$=.938,$\rho_{Spearman}$=.889; p=.0001 (n=43, Weingarten cemetery).

As recent studies have shown, glass beads are also especially helpful for supplying information on economic, cultural and social history.

3. Previous Work on Bead Classification

In the publications of Merovingian bead material in the last 40 years each archaeologist used his own typology, e.g. Böhner (1958), Christlein (1966), Koch (1974, 1977, 1982), Sasse (1986), Theune-Vogt (1990). However, the comparison of bead inventories from different cemeteries and analysts requires a standardized description code. Based on the earlier general bead study of Beck (1928) a team of experienced German archaeologists started in 1989 with the elaboration of a partially hierachical thesaurus with a tree structure which allowed the indexation of beads with different degree of detail. This structure differs from the detailed, but nominally scaled, classifications of viking-age beads from Scandinavia (Callmer (1977)), early medieval Hungarian beads (Szilgyi (1987, 1990)) and Indian trade beads of North America (Kidd and Kidd (1970, 1983), Allen (1983) and Karklins (1985)).

4. The Basic Concept of the Thesaurus

Evidently, a one-to-one description of a bead can never be attained and any choice of descriptors will depend on the particular problem the investigator wants to solve. As a consequence, the thesaurus will reflect the actual problems of research. In order to reach a maximum consensus, it is therefore worked out by a greater number of specialists and will be presented for public criticism in order to arrive finally at an implemented version which responds to the requirements of as many Merovingian archaeologists as possible. The thesaurus should contain the following properties:

(i) Results from different excavations and excavators must be comparable. This necessitates a standardization of nomenclature, as it is already in use for other kinds of archaeological finds, and a set of well defined descriptors. Furthermore, this set must be open and allow the addition of new descriptors if new attributes become important.

(ii) The description of the beads must be possible to every desired degree of detail. This means that an archaeologist who is not interested in a thorough investigation may be content with a low level of description; whereas, a bead specialist may find an answer to his particular questions only if he takes also the smallest details into account.

(iii) Data files of different investigations must have the same structure allowing merging and use of a common retrieval program.

(iv) It must be possible to print catalogues in different languages directly from the descriptors of the beads without storage of natural language texts.

(v) A statistical analysis must be possible. The descriptive variates must be defined on scales of the highest possible level; quantitative scales (counts, measurements) should be used where possible. For qualitative variates, ordinal scales are preferred to nominal ones. Therefore, following the revolutionary work of Beck (1928), the shape of the beads will be defined on a multivariate ordinal scale inspired by a simple pattern recognition language. Another requirement is the quantification of color allowing the computation of color similarity.

(vi) Data description must be possible with minimum effort. A hierarchical menu driven program is offered which allows a description to any desired degree of detail.

5. Structure of the Classification

Beads and pendants are defined as objects characterized by one or more thread holes and a maximum diameter of approximately 5 cm. Regarding the vertical section the thread hole of beads is always situated in the center, the thread hole of pendants outside the center (Fig. 1). Beads and pendants usually do not have a practical function. They are jewellery and ornaments used as chains or as costume applications. However archaeological and ethnographic observations have found great evidence of symbolic significance of special bead types.

In the classifcation, a bead or a pendant is defined as a unit characterized by attributes. In other words, it will be dissected. Similarity and dissimilarity of the attributes are responsible for the similarity and dissimilarity of the beads. However, there is a difficulty in the classification of ancient glass beads: they are handmade articles and are, therefore, not only irregular but also partially destroyed by time. In classifying the beads, it is principally assumed that there were ideal types which the beadmaker tried to produce. Therefore, it is not necessary to describe irregularities resulting from manual work. The classification is principally based on attributes which are still possible to observe.

Describing a bead for a catalogue or data base the work begins with:

A - the identification number of the bead
B - remarks, e.g. the status of conservation or earlier classification type.

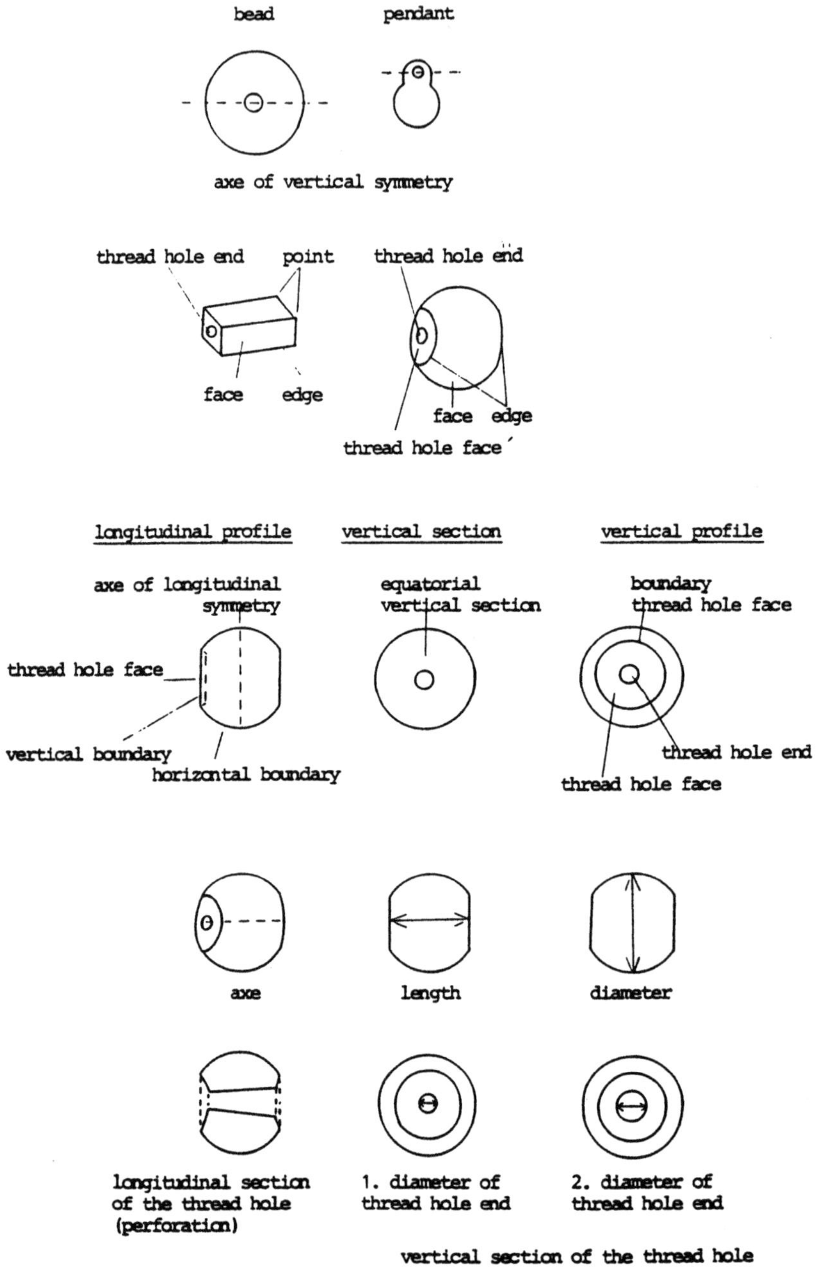

Fig. 1. Terminology of bead description

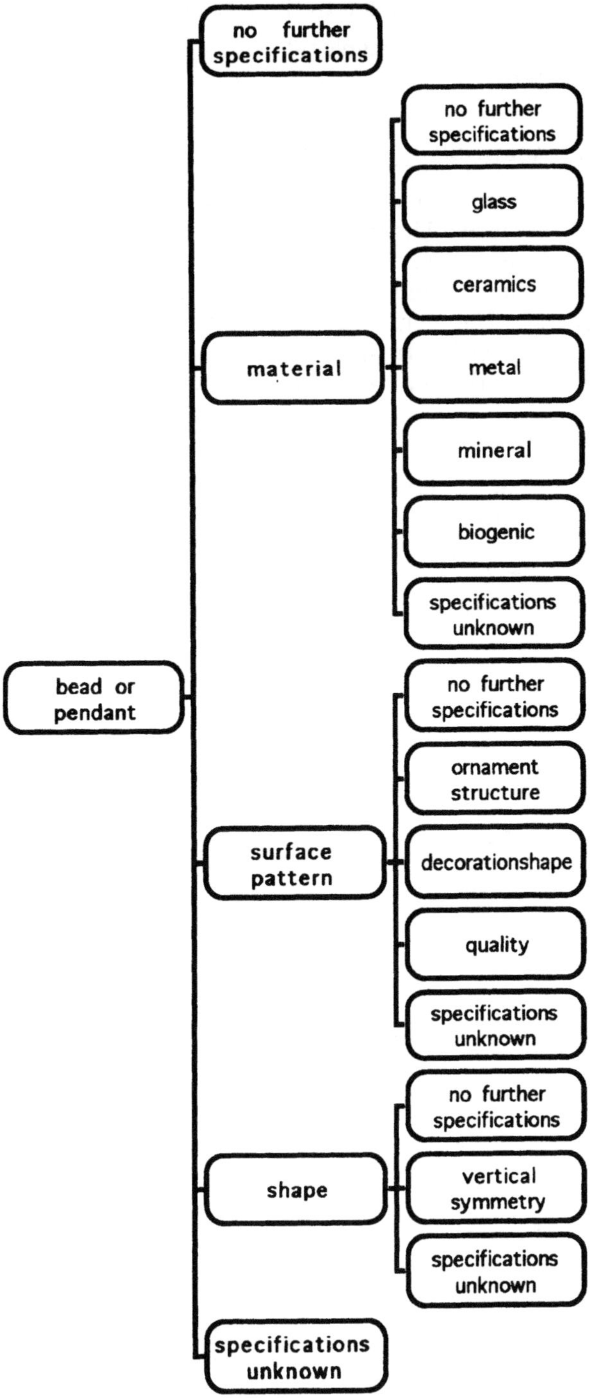

Fig. 2. Part of thesaurus structure

The classification structure is divided into three sections (Fig. 2):

C - Quality (material and technology)
D - Surface pattern
E - Shape

5.1. Quality: Raw Material and Technology

The first category of classification is the raw material of which the bead is made (Callmer 1977). It is organized hierachically in the first steps which lead to the main groups of raw material like glass, ceramics, metal, mineral and organic (biogenic). The technology of production and the kind of raw material depends on the main groups of raw material. The technology of glass bead production is judged by its complexity and ranges from beads with a simple body to beads consisting of more layers of glass, of different threads (reticella beads) or millefiori rods.

5.2. Surface Pattern

After determining the raw material and technology, the surface of the bead is classified. The surface of a glass bead is the sum of all optical signals registered on its surface. It is composed of a different number of elements which are to be described one after the other. Beads without decoration have only one element: the whole surface, the body or the matrix in the terminology of Beck (1928). Decorated beads have more elements which form the background(s) and the ornament(s) (Fig. 3).

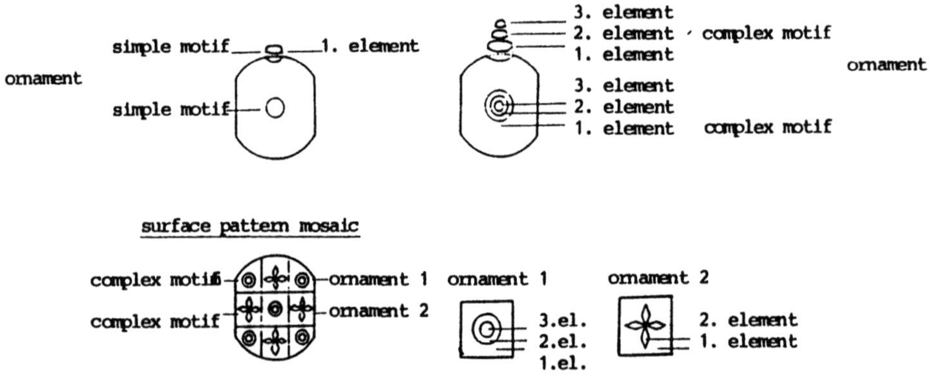

Fig. 3. Description of surface pattern

An ornament consists of a motif, its frequency, arrangement on its background and its quality which includes its material, its color and the structure of the surface. A motif is defined as a picture made by one or more elements. It is possible to reduce the main groups of elements to simple figures like spot and thread and to give them greater complexity using a hierachical system (Fig.4). If motifs are complex that is they have more than one element, there is an element stratification which may be vertical or horizontal.

5.3. Color

Many archaeologists use color as defined on a nominal scale but in doing so lose information on color similarity. Color can be described, however, as a point in a three dimensional Euclidean space. Theory and methods can be found in a number of books, e.g. Küppers (1987), Völz (1990). Roughly speaking, the proportions of yellow (Y), cyan (C) and magenta (M) are used (Y+C+M=100) which allow two coordinates A and B; the third coordinate L is defined by presence or absence of light (white vs. black). Each color can thus be described by its CIELAB-value (CIE = Commission Internationale d'Eclairage). Up to a change of scale, CIELAB-values correspond to those given in the Munsell Book of Color (1966, 1971). Color is measured optimally by reflection spectometry, but most archaeologists will not dispose of the necessary equipment and color tables are necessary. A table containing the colors most frequently occurring in Merovingian glass beads is being prepared. The code is based on the Munsell system, but codes of different systems can be transformed into one another.

5.4 Description of Shape

It is a general experience that decoration is the most important attribute for the classification of Merovingian beads (see e.g. Koch, 1977). Another important criterion for the description of a bead is the shape. The description given in the present paper is inspired by the work of Beck (1928). The main categories of Beck are the transverse section and the profile, that is the sum of lines bordering the longitudinal section of the bead. For the longitudinal section he used a scale with two ordinal and one nominal, nearly ordinal dimension. The first dimension has categories 1.1 disc, 1.2 short, 1.3 standard and 1.4 long beads, the second 2.1 convex, 2.2 straight and 2.3 concave beads. Symmetrical standard beads are subdivided into circular, barrel and others (see list below). Disc, short and long beads are derived from these shapes by change in the longitudinal scale. An attempt was made to describe shape by a syntactic pattern recognition language. In syntactic pattern recognition, according to Kanal, Lambird and Lavine (1982), the pattern is divided into subpatterns which can be further divided, until primitive patterns are reached. Formally, pattern is a sentence in a language specified by a grammar used for the generation of sentences (strings) from given symbols denoting subpatterns or primitives.

Standard primitives and their symbols are shown in Fig. 5. Due to rotational symmetry, a sentence describing the part above the axis (horizontal), from left to right, is sufficient. The standard primitives are defined in the unit square. For drawing a bead, however, more general primitives must be used. They will be generated by multiplying the sides of the square with factors α and β, respecting symmetry and continuity and discontinuity of the derivative of the limiting curve. In other terms: only transformations are allowed such that presence and absence of edges by concatenation of standard primitives will not be altered. In descriptive practice the use of standard primitives is sufficient when diameter and length of the bead are given.

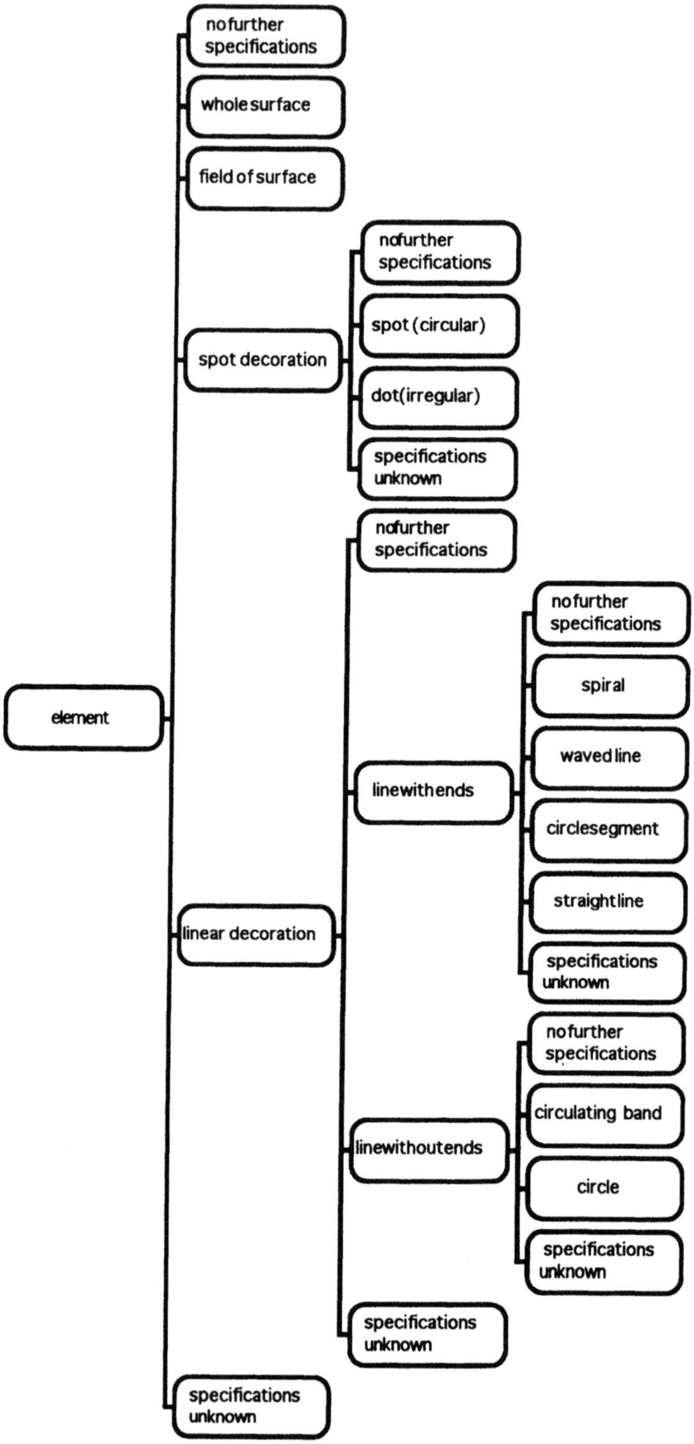

Fig. 4. Part of ornament classification

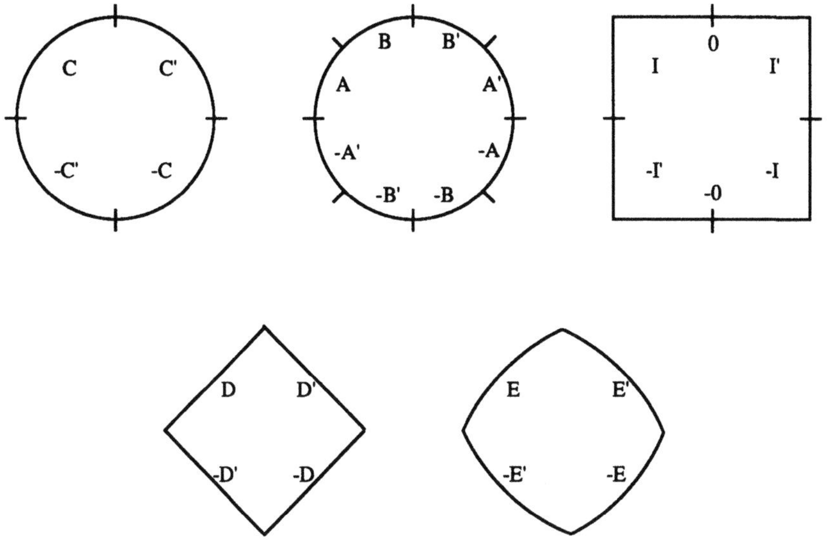

Fig. 5. Standard primitives for shape description

The following algebraic rules will be used for subpatterns X, Y, Z,... (including pattern and primitives as limiting cases):

$$-O = O$$

$$-(-X) = X$$
$$-(XY) = -X - Y$$

$$(X')' = X$$
$$(XY)' = Y'X'$$

Because of symmetry a pattern is a sentence (string) of type XX' or XYX' where Y is a primitive. The generative grammar is as follows: Let X, Y, Z,... be subpatterns and x, y,... be primitives. Then for $y = y'$, $Y = Y'$

$$S \rightarrow xx'$$
$$S \rightarrow y$$
$$XX' \rightarrow XyX'$$
$$Y \rightarrow zYz'$$
$$XYX' \rightarrow XzYz'X'$$

The following list gives most of Beck's standard bead types with the corresponding sentence in the pattern recognition language:

2.1 convex

2.1.1 cicular: CC'

2.1.2 barrel: $IB(IB)'$

2.1.3 convex bicone: EE'

2.1.4 truncated convex bicone: $IE(IE)'$

2.2 straight

2.2.1 -

2.2.2 cylinder: IOI'

2.2.3 bicone: DD'

2.2.4 truncated bicone: $ID(ID)'$

2.3 concave

2.3.1 -

2.3.2 concave: $I-B'(I-B')'$

2.3.3 concave bicone: $-E-E'$

2.3.4 truncated concave bicone: $I-E(I-E)'$

The grammar allows derivation of more complicated patterns. For example, $ID(ID)'$ (truncated bicone) $\rightarrow IDO(ID)'$ (Beck's chamfered cylinder). It is possible to derive notions like 'edged' (e) or 'rounded' (r), e.g.

$$DD' \rightarrow DeD'$$

$$AA' \rightarrow ArA'$$

The vertical section is classified geometrically. Further important attributes of the shape are the longitudinal symmetry, the proportion, that is the value of the length divided by the diameter of the bead and at least the vertical and the longitudinal section of the thread hole.

References:

ALLEN, J.D. (1983): The manufacture of inticate glass canes and a new perspective on the relationsship between chevron-star beads and mosaic millefiori beads. *Proceedings of the 1982 glass trade bead conference. Research records No. 16, 173-191.* New York

BECK, H. C. (1928): Classification and nomenclature of beads and pendants. *Antiquity 77, 1-76*

BÖHNER, K. (1958): *Die fränkischen Altertümer des Trierer Landes. = Germanische Denkmäler der Völkerwanderungszeit Ser. B.1*, Berlin

CALLMER, J. 1977): Trade beads and bead trade in Scandinavia ca. 800-1000 A.D. *Acta Archaeologica Lundensia. Ser. 4, n. 11.* Bonn, Lund

CHRISTLEIN, R. (1966): *Das alamannische Reihengräberfeld von Marktoberdorf im Allgäu. = Materialhefte zur Bayerischen Vorgeschichte Heft 21.* Kallmünz (Oberpfalz)

KANAL, L. N., LAMBIRD, B. A., LAVINE, D. (1982): Structural methods in image analysis and recognition. In: Krishnaiah, P. R., Kanal, L. N. (eds.): *Classification, Pattern Recognition and Reduction of Dimensionality. = Handbook of Statistics, 2, 361-382.* North Holland, New York, Oxford, Amsterdam

KARKLINS, K. (1985): *Glass beads. The Levin catalogue of mid-19th century beads. A sample book of 19th century Venetian beads. Guide to the discription and classification of glass beads.* Ottawa

KIDD, K.E., KIDD, M.A. (1970): Classification system for glass beads for the use of field archaeologists. Canadian historic sites. In: *Occasional papers in archaeology and history*

No. 1, 45-89 Reprinted in: *Proceedings of the 1982 glass trade bead conference. Research records No. 16, 219-257 (1983),* New York

KOCH, U. (1974): Probleme merowingerzeiticher Glasperlen aus Süddeutschland. *Annales du 6. Congrès de l'Association internationale pour l'histoire du verre. Cologne 1. - 7. juillet 1973, 131seq.* Lige

KOCH, U. 1977: *Das Reihengräberfeld bei Schretzheim. = Germanische Denkmäler der Völkerwanderungszeit Ser. A 10.* Berlin

KOCH, U. (1982): *Die fränkischen Gräberfelder von Bargen und Berghausen in Nordbaden. = Forschungen und Berichte zur Vor- und Frühgeschichte Baden-Württembergs 12.* Stuttgart

KÜPPERS, H. (1987): *Farbe: Ursprung - Systematik - Anwendung. 4th* ed. Callwey, München

MUNSELL, A. H. (1966): *The Munsell Book of Color.* Macbeth, Baltimore

MUNSELL, A. H. (1971) *A Color Notation. An Illustrated System Defining all Colors and Their Relations by Measured Scales of Hue, Value, and Chroma.* Macbeth, Baltimore

ROTH, H., THEUNE, C. (1988): *SW I-V: Zur Chronologie merowingerzeitlicher Frauengräber in Südwestdeutschland. = Archäologische Informationen aus Baden-Württemberg 6.* Stuttgart

SASSE, B. (1986): Aufnahme und Auswertung des merowingerzeitlichen Friedhofs von Eichstetten, Kaiserstuhl, mit einem PC. Als Beispiel: Die Perlen. *Archäologische Informationen. Mitteilungen zur Ur- und Frühgeschichte 9(1), 26seq.*

SASSE, B. (1993): *Ein frühmittelalterliches Reihengräberfeld bei Eichstetten am Kaiserstuhl. = Forschungen und Berichte zur Vor- und Frühgeschichte Baden-Württembergs.* Stuttgart

SZILAGYI, K. (1990): Computergestützte Merkmalsanalyse ungarischer Perlen aus dem 10. - 12. Jahrhundert. *Ethnographisch-Archäologische Zeitschrift 28, 87-96.*

SZILAGYI, K. (1990): Die Perlen des Gräberfeldes Halimba-Cseres aus dem 10. - 12. Jahrhundert. *Zeitschrift für Archäologie 24, 79-100.*

THEUNE-VOGT, C. (1990): *Chronologische Ergebnisse zu den Perlen aus dem alamannischen Gräberfeld von Weingarten. Eine Datenanalyse. = Kleine Schriften aus dem Vorgeschichtlichen Seminar der Philipps-Universität 33*

THEUNE-VOGT, C. (1991): Analysis of beads found in the Merovingian cemetary of Weingarten. In: H.-H. Bock, P. Ihm (eds.): *Classification, Data Analysis and Knowledge Organization. Models and Methods with Applications. Proceedings of the 14. Annual Conference of the Gesellschaft für Klassifikation e. V. University of Marburg, March 12-14, 1990, 352-361,* Springer, Berlin, Heidelberg, New York

VÖLZ, H. G. (1990): *Industrielle Farbprüfung. Grundlagen und Methoden. Farbmetrische Testverfahren für Farbmittel in Medien.* VCH, Weinheim, New York, Basel

WERNER, J. (1955): *Das alamannische Gräberfeld von Mindelheim. = Materialhefte zur Bayerischen Vorgeschichte Heft 6.* Kallmünz (Oberpfalz)

WERNER, J. (1953): Das alamannische Gräberfeld von Bülach. *Monographien zur Ur- und Frühgeschichte der Schweiz 9.* Basel

Statistical Measures to Quantify the Benefit from Screening: a Case Study on Cholesterol Screening

Olaf Gefeller[1], Rainer Muche[2]

[1] Abteilung Medizinische Statistik, Universität Göttingen, Humboldtallee 32, D-37073 Göttingen, Germany

[2] Klinische Dokumentation, Universität Ulm, Schwabstraße 13, D-89075 Ulm, Germany

Summary: A variety of cholesterol screening strategies aiming at the identification of persons at high risk of developing some cardiovascular disease have been proposed during the last decade. These proposals, which have been controversially discussed in the medical community, are analyzed in this paper using standard and non-standard methodological techniques of screening evaluation. The intention of the presentation is twofold: (1) to provide a brief comprehensive review of statistical measures employed in this context, (2) to evaluate the predictive properties of six important screening recommendations in the framework of a large cohort study. The results of the analysis add critical caveats to the discussion about what can be expected from cholesterol screening on the population level.

1. Introduction

Although mortality rates from cardiovascular diseases have declined steadily during the last decade in many industrialized countries, this disease entity accounts for a substantial proportion of all deaths even today. Identified as the leading cause of death in most national mortality statistics of developed countries, research directed towards finding effective strategies to prevent deaths from cardiovascular diseases is thus of major public health concern. Besides primary prevention activities, screening (as part of secondary prevention) for persons at high risk of developing some cardiovascular disease forms an integral part of the current activities aimed at reducing the disease burden on the population level.

A number of prospective observational epidemiologic studies have identified important risk factors for cardiovascular diseases. Due to the outstanding role of elevated cholesterol levels, a variety of screening proposals have been suggested during the last years in which differentiated ascertainment of the lipoprotein profile plays a dominating role in the screening decision. These "cholesterol screening strategies" differ, however, considerably in the cut-points used for dichotomizing the lipoprotein information measured continously and the number of additional cardiovascular risk factors (for example, hypertension, smoking, obesity, familial disposition etc.) included for modifying the screening classification.

Despite the growing demand for standardization of these campaigns, no comparative evaluation of the screening strategies using appropriate data sets can be found in the literature. Hence, the present discussion about this important topic in cardiovascular disease prevention is based on the analysis of single proposals (e.g. Bush and Riedel (1991)) and subjective evaluation considerations of benefits and costs (e.g. Steinberg (1989); Schmidt (1990)). The aim of this paper is therefore twofold: (1) to provide a brief comprehensive review of statistical measures employed in screening evaluation (including non–standard concepts as the critical value approach and measures quantifying the gain in certainty), (2) to evaluate the predictive properties of six different cholesterol screening strategies with respect to the development of myocardial infarction in the framework of a large prospective cohort study. The results of this analysis will add relevant information to the present controversial discussion about what can be expected from cholesterol screening on the population level.

2. Statistical measures for screening evaluation

The usual situation in screening evaluation of a dichotomous classification rule can be characterized as follows: in the population of interest all persons are classified according to the outcome of the screening procedure as screening–positive (T) or screening–negative ($\overline{T}$). In a subsequent validation of the screening classification the true disease status of all screening participants as being diseased (D) or healthy ($\overline{D}$) is obtained by some "gold standard" assumed to be without error. The corresponding joint probability distribution is given in table 1.

Tab. 1: Fourfold table of joint probability distribution of dichotomous screening classification and true disease status

true disease status	screening classification positive (T)	 negative ($\overline{T}$)	Σ
diseased (D)	$P(D \cap T)$	$P(D \cap \overline{T})$	$P(D)$
healthy ($\overline{D}$)	$P(\overline{D} \cap T)$	$P(\overline{D} \cap \overline{T})$	$P(\overline{D})$
Σ	$P(T)$	$P(\overline{T})$	1

The standard measures for screening evaluation are sensitivity ($Sens$), specificity ($Spec$), and positive ($PosPV$) as well as negative predictive value ($NegPV$) which are defined as follows:

$$Sens := \frac{P(D \cap T)}{P(D)} = P(T \mid D) \tag{1}$$

$$Spec := \frac{P(\overline{D} \cap \overline{T})}{P(\overline{D})} = P(\overline{T} \mid \overline{D}) \tag{2}$$

$$PosPV := \frac{P(D \cap T)}{P(T)} = P(D \mid T) \tag{3}$$

$$NegPV := \frac{P(\overline{D} \cap \overline{T})}{P(\overline{T})} = P(\overline{D} \mid \overline{T}) \tag{4}$$

Whereas the sensitivity provides the information to what extent the disease of interest is actually detected by the screening test, the specificity quantifies the proportion of subjects classified correctly as screening–negative among healthy screening participants. Hence, sensitivity and specificity describe the fundamental characteristics of a screening test and are viewed as test–specific constants. On the contrary, the predictive values depend critically on the prevalence of the disease under study in the population to be screened. Positive (negative) predictive values offer the important information for the screening participant how sure he can be that his classification as screening–positive (screening–negative) is correct. Even if the test–specific values of sensitivity and specificity are very high indicating an effective screening test in identifying truely diseased and healthy persons, the positive predictive value is surprisingly low when screening for a rare disease is performed. These well–known relationships between the measure defined above have been extensively discussed in a variety of textbooks, see, for example, Galen and Gambino (1975) and Kõbberling et al. (1991).

In addition to these standard measures in screening evaluation, two other concepts of describing the efficiency of a screening test are employed in this paper: (i) measures of gain in certainty, (ii) critical values.

The first group of measures tries to quantify the gain in information derived from the screening test compared with what was known before the screening. To this end, these measures consider the (absolute or relative) differences between the predictive values and the corresponding prevalences in the population. More precisely, the following (unweighted) measures have been used in the literature.

Absolute gains:

$$\begin{aligned} PosG &:= PosPV - P(D) \\ &= \frac{P(D)\cdot P(\overline{D})\cdot [Sens + Spec - 1]}{P(D)\cdot Sens + P(\overline{D})\cdot [1 - Spec]} \end{aligned} \tag{5}$$

$$\begin{aligned} NegG &:= NegPV - P(\overline{D}) \\ &= \frac{P(D)\cdot P(\overline{D})\cdot [Sens + Spec - 1]}{P(D)\cdot [1 - Sens] + P(\overline{D})\cdot Spec} \end{aligned} \tag{6}$$

Relative gains (relative to the maximal attainable gain):

$$\begin{aligned} PosG\% &:= \frac{PosPV - P(D)}{P(\overline{D})} \\ &= \frac{P(D)\cdot [Sens + Spec - 1]}{P(D)\cdot Sens + P(\overline{D})\cdot [1 - Spec]} \end{aligned} \tag{7}$$

$$\begin{aligned} NegG\% &:= \frac{NegPV - P(\overline{D})}{P(D)} \\ &= \frac{P(\overline{D})\cdot [Sens + Spec - 1]}{P(D)\cdot [1 - Sens] + P(\overline{D})\cdot Spec} \end{aligned} \tag{8}$$

Additionally, it is important to consider not only the magnitude of the gain in certainty reflected by the measures defined above but also the probability that the particular gain will occur. These probabilities, $P(T)$ for $PosG$ / $PosG\%$ and $P(\overline{T})$ for $NegG$ / $NegG\%$, respectively, can be employed to define weighted measures describing the expected gain in certainty from a positive and negative screening test. As $P(T)$ can be expressed as $P(D) \cdot Sens + P(\overline{D}) \cdot [1 - Spec]$ and $P(\overline{T})$ analogously as $P(D) \cdot [1 - Sens] + P(\overline{D}) \cdot Spec$, the expected gains can be simply calculated from (5) - (8) as

$$\begin{aligned} E(PosG) &= P(D) \cdot P(\overline{D}) \cdot [Sens + Spec - 1] = E(NegG) \\ E(PosG\%) &= P(D) \cdot [Sens + Spec - 1] \\ E(NegG\%) &= P(\overline{D}) \cdot [Sens + Spec - 1] \end{aligned}$$

It may also be reasonable to combine these expected gains from positive and negative screening tests by adding the corresponding figures as suggested by Connell and Koepsell (1985). This results in:

$$E(G) = 2 \cdot P(D) \cdot P(\overline{D}) \cdot [Sens + Spec - 1] \tag{9}$$

$$E(G\%) = Sens + Spec - 1 \tag{10}$$

Incidentally, the last formula (10), the expected relative gain from the screening test, can also be found as "Youden index" in the literature, initially proposed by Youden (1950) using a quite different motivation. Further discussion of this approach quantifying the gain in certainty from screening has been provided by Klar & Reuter (1983) and Klar (1985).

Another perspective when evaluating the efficiency of a screening test has been introduced by Brecht (1985). He defines the so-called "critical value" of a screening test as the inverse odds of the positive predictive value:

$$\begin{aligned} CV &:= \frac{1 - PosPV}{PosPV} = \frac{P(\overline{D} \cap T)}{P(D \cap T)} \\ &= \frac{1 - Spec}{Sens} \cdot \frac{P(\overline{D})}{P(D)} \end{aligned} \tag{11}$$

The critical value gives the ratio of the false-positives and the true-positives. Hence, CV can be interpreted as the expected number of screening participants classified erroneously as diseased per each correctly discovered diseased person who can potentially benefit from the screening. If a screening strategy leads to a high critical value, this points to an inefficient screening enterprise as the number of unnecessarily worried participants who subsequently undergo a process of diagnostic validation of the incorrect screening classification is high compared to those who can profit from screening.

3. Evaluation of cholesterol screening based on data from the GRIPS study

A variety of screening strategies for identifying persons at high risk of developing some cardiovascular disease have been proposed during the last decade. In this paper we will focus on the following:

- guidelines of the expert panel of the American National Cholesterol Education Program, abbreviated: NCEP (National Cholesterol Education Program Expert Panel (1988))
- recommendations of the Euopean Atherosclerosis Society, abbreviated: EAS (European Atherosclerosis Society Study Group (1988))
- the strategic paper of the German National Cholesterol Initiative, abbreviated: NatCI (Nationale Cholesterin–Initiative (1990))
- recommendations of the expert panel "Gesundheitsökonomie" at the German ministry of research and technology, abbreviated: BMFT (Materialien zur Gesundheitsforschung (1990))
- information of the Austrian heart fund, abbreviated: Austria (Weiss et al. (1990))
- recommendations of the Swiss Society for Cardiology, abbreviated: Swiss (Arbeitsgruppe Lipide der Schweizerischen Stiftung für Kardiologie (1989))

The first two constitute the most important and influential proposals on the international level, whereas the last four have been included because of their relevance for the German speaking countries. An explanation of details concerning these cholesterol screening strategies is beyond the scope of this paper, the reader is referred to the original papers and additional commentaries. A common element in all strategies is that they use dichotomized information on total cholesterol and/or lipoprotein fractions depending on the presence of other cardiovascular risk factors to classify screening participants. With the exception of the EAS proposal all other screening guidelines lead to a classification of participants into three groups: the "high risk group" needing dietary and drug therapy, the "medium risk group" requiring only dietary recommendations, and "low risk group" being without need for further intervention. For the purpose of screening evaluation using the statistical measures introduced in the last section, this screening classification has to be reduced to a dichotomy according to one of the following two rules: (A) only members of the high risk group are considered as screening–positive (T), whereas members of the other two groups are screening–negative ($\overline{T}$), (B) screening–positive is defined as belonging to the high or medium risk group, only the low risk group forms the screening–negatives.

The evaluation of the cholesterol screening strategies presented in the following is based on data of the **G**öttingen **R**isk, **I**ncidence and **P**revalence **S**tudy (GRIPS), a prospective cohort study on 6029 male German employees at one industrial plant aged 40 – 60 years (Cremer et al. (1991); Muche et al. (1991)). The baseline examination of the cohort members in spring 1982 is taken as a hypothetical screening visit. The subsequent five year follow–up period can be viewed as providing information on the natural course of disease in the cohort members since no intervention has taken place in GRIPS, the design of the study was strictly observational. During the follow–up period, in which only 4.7% of the study participants have been lost, myocardial infarction has been observed in 107 cohort members. It is the intention of the evaluation to analyze how the different cholesterol screening strategies perform in (i) identifying these 107 cohort members as screening–positive in the baseline examination and (ii) excluding the other cohort members as screening–negative on the same occasion.

Tab. 2: Estimated values of the standard measures for screening evaluation for six cholesterol screening strategies employing two rules of dichotomization (A, B) based on data of the GRIPS study

screening strategy	*Sens* (%)	*Spec* (%)	*PosPV* (%)	*NegPV* (%)
NCEP (A)	80.37	57.22	3.77	99.29
NCEP (B)	80.37	55.88	3.77	99.27
EAS	90.65	36.48	2.89	99.47
NatCI (A)	78.50	67.01	4.73	99.34
NatCI (B)	83.18	54.51	3.67	99.36
BMFT (A)	77.57	68.90	4.95	99.33
BMFT (B)	83.18	54.51	3.67	99.36
Austria (A)	47.66	83.23	5.60	98.68
Austria (B)	91.59	35.60	2.88	99.51
Swiss (A)	67.29	77.93	5.98	99.13
Swiss (B)	91.59	35.60	2.88	99.51

Table 2 presents the standard measures of evaluation (sensitivity, specificity, positive and negative predictive values) for all strategies and using both classification rules (A and B) whenever possible. Although substantial heterogeneity in the results can be observed, the overall picture is quite clear: the sensitivity of the strategies reaches mostly a satisfactory level of about 80%, whereas the specificity, taking values in the range from 36% up to 83% with a centre around 55%, is in most cases far away from an acceptable level. Consequently, the positive predictive values yield 2.9% – 6% meaning that only one of about 25 cohort members classified as screening–positive in the baseline examination develops a myocardial infarction during the follow–up.

Table 3 provides additional information on the prognostic properties of the cholesterol screening strategies. Absolute, relative (to maximal attainable) and the combined expected gains in certainty as well as the critical values are presented for all screening proposals. Due to the fact that the cumulative incidence of myocardial infarction during the study period has been rather low ($P(D) = 2\%$), absolute and relative (to maximal attainable) gains in certainty from a positive screening classification are nearly identical. Both measures reveal that the prognostic gain in this situation is negligible for all competitors. From a screening–negative classification the absolute gain in certainty is also very low, however, as 2% is the maximal attainable gain in certainty here, the relative figures are much higher indicating that a substantial proportion of the maximal gain is actually achieved by the strategies. The combined expected gains in certainty point to the fact all screening proposals add somehow to

Tab. 3: Estimated values of measures of gain in certainty and critical value for screening evaluation for six cholesterol screening strategies employing two rules of dichotomization (A, B) based on data of the GRIPS study

screening strategy	*PosG* (%)	*PosG%* (%)	*NegG* (%)	*NegG%* (%)	*E(G%)* (%)	*CV*
NCEP (A)	1.73	1.77	1.33	65.20	37.59	25.5
NCEP (B)	1.73	1.77	1.31	64.22	36.25	25.5
EAS	0.85	0.87	1.51	74.02	27.13	33.6
NatCI (A)	2.69	2.75	1.38	67.65	45.51	20.1
NatCI (B)	1.63	1.66	1.40	68.63	37.69	26.2
BMFT (A)	2.91	2.97	1.37	67.16	46.47	19.2
BMFT (B)	1.63	1.66	1.40	68.63	37.69	26.2
Austria (A)	3.56	3.63	0.72	35.29	30.89	16.9
Austria (B)	0.84	0.86	1.55	75.98	27.19	33.7
Swiss (A)	3.94	4.02	1.17	57.35	45.22	15.7
Swiss (B)	0.84	0.86	1.55	75.98	27.19	33.7

the knowledge about the future risk of developing myocardial infarction (if the pre–screening knowledge remains unchanged, the value of $E(G\%)$ would be 0), however, no screening strategy reaches the 50% level. Hence, all screening recommendations are far away from an optimal solution to the goal of screening characterized by high values for sensitivity and specificity simultaneously leading to a combined expected gain in certainty above 70–80%. Critical values in the range 16 – 34 further indicate that a remarkably high number of healthy persons would be unnecessarily worried per each correctly classified case of myocardial infarction in the screening.

4. Discussion

The predictive properties of six important cholesterol screening strategies with respect to identifying persons at high risk for developing myocardial infarction have been analyzed in the framework of a large prospective cohort study. The results revealed that all proposals currently discussed in the medical community are inappropriate tools for secondary prevention on the level of the general population. Primarily, this finding is due to the high number of false–positive screening classifications, which is reflected in a low specificity and positive predictive value as well as in a high critical value and a negligible gain in diagnostic certainty resulting from the screening. Even after taking the limitations of the present analysis into account (e.g. unrepresentativeness of the GRIPS cohort for the general population, limited length of only five years of follow-up time), it has to be concluded that the populationwide (continuing) realization of current cholesterol screening recommendations seems unjustified. Large amounts of money and health care professionals are shifted towards the healthy population for doubtful success while serious other problems of medical care remain unsolved. It should be conceded by protagonists of cholesterol screening that the current knowledge of the etiology of myocardial infarction does not offer the opportunity to propose a screening strategy for the general population yielding high sensitivity and specificity simultaneously. This does not mean that the formulation of such a strategy is an impossible task for all times. However, other cardiovascular risk factors presently not known have to be identified to achieve this goal of improving the performance of future screening strategies.

Finally, it should be recognized that the analysis of the performance of the cholesterol screening strategies as diagnostic tests for myocardial infarction in a preclinical stage deals only with one dimension of screening evaluation. The efficacy of screening measures is also conditioned by the compliance with the screening programme on the population level and by the effectiveness of early treatment of those identified as screening-positive. In addition, a variety of other issues have to be considered before introducing a screening procedure, for example, the availability of resources in the health care system and cost-benefit considerations. As the focus of this paper has been on the description of the properties of cholesterol screening strategies as diagnostic tests, these further aspects of screening evaluation have been neglected, for a detailed list of criteria summarizing the principle ideas and prerequisites of early disease detection see Wilson and Jungner (1968).

Acknowledgements

The authors wish to express their grateful thanks to Prof. Dr. D. Seidel and Dr. P. Cremer, Munich, for the permission to use the data of the GRIPS study for this analysis.

References

ARBEITSGRUPPE LIPIDE DER SCHWEIZERISCHEN STIFTUNG FÜR KARDIOLOGIE (1989): Lipide und die Prävention der koronaren Herzkrankheit: Diagnostik und Maßnahmen. *Schweizer Ärztezeitung*, 7, 1279–1292.

BRECHT, J.G. (1985). The utility of cancer screening – a decision help by critical values. In: E. Walter and A. Neiß (eds.): *Methodical Problems in Early Detection Programmes.* Springer-Verlag, Berlin, 110–115.

BUSH, T.L., and RIEDEL, D. (1991): Screening for total cholesterol: do the NCEP recommendations detect individuals at high risk of CHD? *Circulation*, 83, 1287–1293.

CONNELL, F.A., and KOEPSELL, T.D. (1985): Measures of gain in certainty from a diagnostic test. *American Journal of Epidemiology*, 121, 744–753.

CREMER, P., NAGEL, D., LABROT, B., et al. (1991): *Göttinger Risiko-, Inzidenz- und Prävalenzstudie (GRIPS).* Springer-Verlag, Berlin.

EUROPEAN ATHEROSCLEROSIS SOCIETY STUDY GROUP (1988): The recognition and management of hyperlipidemia in adults: a policy statement of the European Atherosclerosis Scociety. *European Heart Journal*, 9, 571–600.

GALEN, R.S., and GAMBINO, S.R. (1975): *Beyond normality: the predictive value and efficiency of medical diagnosis.* John Wiley & Sons, New York.

KLAR, R. (1985): The relation between Bayesian and information theoretical approaches in the validation of diagnostic tests. In: J.H. van Bemmel, F. Gremy, and J. Zvarova (eds.): *Medical Decision Making.* North-Holland, Amsterdam, 204–207.

KLAR, R., and REUTER, R. (1983): Vergleich zweier unterschiedlicher Ansätze zur Bestimmung von a-posteriori-Kenngrößen für die Bewertung diagnostischer Tests. In: J. Berger and J. Höhne (Hrsg.): *Methoden der Statistik und Informatik in Epidemiologie und Diagnostik.* Springer-Verlag, Berlin, 407–413.

KÖBBERLING, J., RICHTER, K., TRAMPISCH, H.J., and WINDELER, J. (1991): *Methodologie der medizinischen Diagnostik.* Springer-Verlag, Berlin.

MATERIALIEN ZUR GESUNDHEITSFÖRDERUNG (1990): *Gesundheitsökonomische Aspekte der Früherkennung von Herz-Kreislauf-Krankheiten.* Schriftenreihe zum Programm "Forschung und Entwicklung im Dienste der Gesundheit", Band 18, Bonn.

MUCHE, R., GEFELLER, O., and CREMER, P. (1991): Der Einfluß der Lipoproteine auf den Myokardinfarkt: Risikoquantifizierung und Screening-Überlegungen in der GRIPS-Studie. *Zeitschrift für Präventivmedizin und Gesundheitsförderung,* 3, 14–20.

NATIONAL CHOLESTEROL EDUCATION PROGRAM EXPERT PANEL (1988): Report of the NCEP Expert Panel on detection, evaluation and treatment of high blood cholesterol in adults. *Archives of Internal Medicine,* 148, 33–63.

NATIONALE CHOLESTERIN-INITIATIVE (1990): Strategie-Papier zur Erkennung und Behandlung von Hyperlipidämien. *Deutsches Ärzteblatt,* 87, 1358–1382.

SCHMIDT, J.G. (1990): Wie sinnvoll ist ein Cholesterin-Screening? *Zeitschrift für Allgemeinmedizin,* 66, 789–794.

STEINBERG, D. (1989): The cholesterol controversy is over – why did it take so long? *Circulation,* 80, 1070–1078.

WEISS, K., EGGER, S., HIRMANN, P., WIDHALM, K., and SINZINGER, H. (1990): Risikofaktoren für Atherosklerose und Risikobewußtsein bei Mitarbeitern einer Wiener Versicherung. *Wiener Medizinische Wochenschrift,* 140, 513–519.

WILSON, J.M.G., and JUNGNER, G. (1968): Principles and practice of screening for disease. *Public Health Paper,* 34, World Health Organization, Geneva.

YOUDEN, W.J. (1950): Index rating for diagnostic tests. *Cancer,* 3, 32–35.

Evaluation of Screening in Case–Control Studies: an Alternative to Randomized Controlled Trials ?

Olaf Gefeller[1], Jürgen Windeler[2], Hans Joachim Trampisch[2]

[1] Abteilung Medizinische Statistik, Universität Göttingen, Humboldtallee 32, D-37073 Göttingen, Germany

[2] Abteilung Medizinische Informatik und Biomathematik, Universität Bochum, Overbergstraße 17, D-44780 Bochum, Germany

Summary: Numerous methodologic problems of evaluating screening programs in case–control studies have been subject to controversial discussion in recent years. On the other hand, the growing number of published papers presenting results of case–control studies on the evaluation of cancer screening demonstrate that this design has rapidly gained popularity in cancer epidemiology, although the methodologic criticism has not been solved until yet and is usually not reflected in the interpretation of the study results. This paper reviews the problematic issues relating to the application of the case–control design to screening situations. From this review it is concluded that no satisfactory solutions of the methodologic problems of this design applied to screening evaluation are available, at least at present. Hence, randomized controlled trials remain the only appropriate study design to demonstrate conclusively the efficacy of screening measures.

1. Introduction

Screening for the detection of early disease has become a popular concept in health care. During the last two decades screening programs have increased in number and dissemination and are now established in a wide variety of settings. A major reason for this development has been the experience that methods directed towards primary prevention as well as treatment of these diseases have remained unsatisfactory until yet. Another argument is the compelling intuitive appeal of the screening premise that detection of early disease will be rewarding (Cole and Morrison (1980)).

However, there is an ongoing critical discussion about the efficacy of screening measures. The efficacy is conditioned by: (a) performance of the screening procedure to identify preclinical diseased individuals, (b) efficacy of early treatment to prevent the most adverse effects of the disease, (c) compliance with screening on the population level. Simply the existence of some diagnostic test procedure with a high sensitivity for detecting the disease of interest in symptom-free but diseased subjects and a high specificity for excluding the disease in healthy persons does by no means guarantee that implementing this diagnostic test as a screening procedure will show any benefit. For example, if the treatment offered to screen-detected early stage cases is not more beneficial than the usual therapeutic regime for symptom-detected cases, screening becomes senseless. Other aspects to be considered before introducing a screening procedure relate to available resources of the health care system, acceptability of the screening procedure for the general population, knowledge about the biology and natural history of the disease to be screened for and cost-benefit analyses. A detailed list of criteria summarizing the principle ideas and prerequisites of early disease detection has been given by Wilson and Jungner (1968) and Whitby (1974).

The most straightforward evaluation of efficacy is provided by a well-designed randomized controlled trial (RCT) in which the mortality rates from the disease in question (or the incidence of some other surrogate outcome sufficiently advanced to ensure its detection even in the absence of screening) are compared between two groups: those offered screening and those not. The RCT reaches a definite, methodologically sound answer to the question whether screening shows the hoped-for benefit. Several authors have claimed, however, that RCTs in this context suffer from many logistic and ethical problems, most important the huge number of participants as well as the long time of follow-up required to complete the evaluation and the ethical inadvisability of randomization after screening has become widely available.

In this situation, the case-control design, which has long been popular in other fields of epidemiologic research, has been adopted to the screening situation, especially for cancer screening. Although Weiss correctly stated as late as 1983, "Case-control studies of the efficacy of a rapid diagnosis of cancer, or of any disease, are at present more a theoretical possibility than reality", today case-control studies *are* the reality of current attempts to evaluate the efficacy of screening. Numerous studies concerning cancer of the breast, cervix, lung, stomach, rectum, prostate as well as malignant melanoma have been published in recent years (for reviews see Friedman and Dubin (1991) and Moss (1991)) and document the rapid development in this area.

Simultaneously, several methodologic papers have discussed severe drawbacks of the case-control design in studying the efficacy of screening (Sasco et al. (1986); Dubin et al. (1987); Connor et al. (1991); Knox (1991); Sasco (1991); Weiss et al. (1992)). These refer to the problem of defining cases and adequate controls in this situation, self selection bias, healthy screenee bias, lead time bias, and length bias. This paper critically reviews these problematic issues. Primarily, it is the intention to raise caution against the naive practical application of the case-control design and the overoptimistic interpretation of results derived from such studies devoted to screening evaluation.

2. Screening evaluation: basic considerations

Studying the efficacy of screening needs some information about the natural history of the disease to be screened for (see figure 1). The total preclinical phase in the disease process is the time period between onset of disease and onset of symptoms (in the example of figure 1: 20 years). Screening attempts to advance the detection of disease by recognizing early precursors of the disease manifestation. However, it is too optimistic to assume that screening can "save" the total preclinical phase (Morrison (1985)). Within this period the disease of interest will be detectable only during some time in the latter part of the interval, which is referred to as detectable preclinical phase (in the example of figure 1: 10 years). The length of this phase depends on the actual screening test to be used.

In a case-control study of screening the study subjects are identified on the basis of outcome (i.e. death from the disease or late-stage disease) rather than exposure (i.e. screening status). Thus, the investigator's point of access to the exposure-disease relationship is at the end of the sequence rather than at the start. The screening status can only be measured retrospectively (by questionnaire or medical records) and is not determined by design but by various factors not under the investigator's control. These principal features of the case-control design lead to several (general and screening-specific) methodologic problems in screening evaluation which are discussed in the following sections.

Fig. 1: Example for the natural history of some disease in the presence of a screening procedure

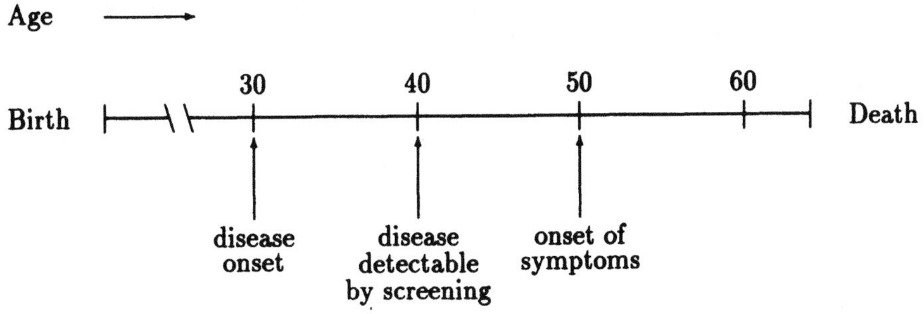

3. The general dilemma of case–control studies

The problem of establishing causality in observational epidemiologic studies has been extensively discussed and is not recapitulated here, see, for example, Feinstein (1985), and Miettinen (1989). Even by proponents of the opinion that causality can be established using evidence from observational epidemiologic studies it is conceded that especially the practical realizations of the case–control design are prone to many forms of bias potentially resulting in misleading conclusions (Sackett (1979); Kopec and Esdaile (1990)). Except for breast cancer screening no results of RCTs are available in the context of cancer screening and causality in the sense that screening prevents any invasive disease (or mortality from the disease) is not yet established. Taking this reality of screening evaluation into consideration, the commonly readable prerequisite for case–control studies on cancer screening topics, for example in Sasco et al. (1986), "once experimental studies have established the benefit of screening", is a farce.

Uncontrolled confounding and/or effect-modification are pervasive problems in any observational epidemiologic study and the validity of all measures of effect depends on proper control of these disturbing phenomenons. Neither careful subject selection nor measuring known confounders, followed by an appropriate adjustment in the analysis, will ever provide the level of comparability that is attainable in RCTs. A specific problem in evaluating screening programs by non-randomized trials is that many authors have pointed to the fact that social class is a strong confounder in these studies. Confounding by social class leads to enormous practical problems, especially when considering women, as readily available social class indicators have been demonstrated to possess only a poor validity for women (Liberatos et al. (1988)).

An additional problem relates to misclassification of study variables which results in the so-called "information bias". While this constitutes a universal problem in any type of investigation irrespective of its design, its potential implications with respect to biasing effect estimates are much more pronounced in case–control studies than in RCTs as the relevant exposure information (i.e. screening status) is here ascertained retrospectively by questionnaire or medical records rather than determined by the randomized design. Thus, the potential for exposure misclassification is only present in case–control studies.

4. Screening–specific problems of case–control studies

The intention of this paper is to examine more closely the methodologic problems of the case–control design in evaluating screening programs. Besides the general drawbacks of this observational design sketched in the preceding section, a number of screening-specific issues have to be taken into account when assessing the advantages and disadvantages of the case-control design in screening evaluation. It is beyond the scope of this article to discuss all these problems in detail, however, some remarks on important aspects are summarized in the following subsections.

4.1 Subject selection

The framework for the selection of study subjects for a case-control study, i.e. the definition of eligibility criteria and sampling mechanisms for cases and controls, forms an essential part of the design of any case-control study. Especially for the purpose of studying screening efficacy the matter of subject selection is of outstanding importance for the validity of the study and closely related to at least some of the other problems discussed in subsequent sections.

Concerning the definition of cases the following procedure is widely accepted: a case of the study population should be a person who died of the disease in question (or a person with a diagnosis of invasive disease, if the aim of the screening measure is to prevent invasive disease) some time after the screening program has been introduced (Morrison (1982)). The problem of the study cases being representative of cases in the source population is not addressed in most screening papers. It should be noted that there is an ongoing discussion about whether and how screen-detected cases should be included in the study population which is of importance for some of the bias problems (see 4.4 and 4.5).

Concerning the selection of controls, too, there seems to be some kind of agreement: neither dead controls (Gordis (1982)), nor solely people with the disease in question at an earlier stage should be included. Although stating that regarding the latter "there seems to be a certain intuitive appeal" Weiss (1983) argued that controls should be taken from the whole population under risk, cases excluded. The reason is that selecting controls from patients with early stage disease and matching for time of diagnosis leads to an overoptimistic estimate of the screening effect due to lead time bias (see 4.4).

4.2 Self selection bias

It is a well-known phenomenon that screening participants differ from non-participants in several aspects, for example, in socioeconomic indicators as education and occupation or in their attitude towards health-related behaviour. The crucial question is, however, whether these differences affect disease outcome, or, to put it in an epidemiologic wording, whether confounding is present. For some cancer locations it is known that socioeconomic factors relate to both, high risk of disease and low screening compliance resulting in an overoptimistic estimate of the screening benefit if confounding by these factors is not properly taken into account. In general, however, the direction of this bias is not clear at all, it may vary with different cancer locations. In fact, even for one location the results of randomized trials show negligible (Shapiro et al. (1971)) as well as serious self-selection bias (Andersson et al. (1988)) in favour of screening.

4.3 Healthy screenee bias

Analogous to the problem of the healthy worker effect in occupational epidemiology, in screening evaluation one has to cope with the "healthy screenee bias". This bias occurs if the frequency of screening depends on the disease status of the subject. Correction procedures are necessary, for example, to ensure that the screening benefit is not only the result of not further screening an individual when his/her diagnosis is established. As Friedman and Dubin (1991) recently pointed out, it is not clear what the results of a possible correction procedure ("truncation") will be. They conclude that "all case–control studies to date may be biased, with the direction and extent of the bias unknown". Further research seems to be necessary.

4.4 Lead time bias

The duration by which the diagnosis of disease is advanced among screen–detected cases due to the screening participation is referred to as lead time. The actual lead time for a screening participant is an unobservable part of the detectable preclinical phase (see figure 1) and not estimable without strong assumptions. Irrespective of any potentially existing screening benefit, the effect of the lead time is that the observed survival times of screen–detected cases *after diagnosis* are longer than those of symptom–detected cases, which results in the so–called lead time bias. The lead time problem is not at all solved for the case–control design. The major controversies in the discussion relates to the question whom to select as controls and how to count subjects with screen–detected disease. As pointed out earlier in section 4.1 the current practice of selecting controls from the whole population at risk which gave rise to the cases is advocated as the solution to avoid lead time bias. Arguments provided in favour of this opinion have not addressed, however, two basic questions: (i) since the lead time bias results from using patients with early disease as controls and since patients with early diseases are members of the population at risk, isn't the problem of lead time bias only diluted by this way of selecting controls?, (ii) additionally, do people who always had negative screening tests and never suffered from the disease in question provide any information about the benefit of screening whatsoever?

4.5 Length bias

Length bias refers to the observation that cancers detected at regular screening are of a less aggressive type than those symptom–detected and hence have a better prognosis. Additionally, screening will detect some cancers that would never have become apparent in patients' lives, i.e. that also have a good prognosis. Consequently, no matter how beneficial screening actually may be, the method of sampling in itself leads to an overoptimistic effect estimation. This important methodologic problem has been neglected in all case–control evaluations of screening efficacy. No practical measures that could be used to reduce or allow for length bias in case–control studies have been developed and thus it is not sufficiently clear how this problem can be dealt with in practice.

5. Discussion

The growing popularity of the case–control design for studying screening efficacy results primarily from its cheapness, the shortness of time required to yield a result, and its feigned simplicity of application. The preceding discussion of methodologic issues has revealed that enormous problems with respect to the interpretation of results from case–control studies evaluating screening exist. These problems have mostly been ignored in the actual publications of study results. One of the most disturbing features in screening evaluation by case–control studies relates to the semantic inaccuracy of many statements which can be found throughout methodologic and applied papers on this topic during the last ten years. For example, Sasco et al. (1986), although correctly stating that "case–control studies cannot replace randomized controlled trials" at the very beginning of their paper, claim at the end of the same paper that a case-control evaluation is "the method of choice when ... screening tests have been performed over a period of years ... but no proper control evaluation has been undertaken", which means nothing else but that case–control studies may substitute the evaluation by RCTs. This contradiction is symptomatic of a fairly ambiguous attitude which leads to an inflational use of such words as "protect" and "prevent" in the majority of articles focussing on this topic. For nearly all cancer locations no RCTs have been undertaken until yet. Hence, there is absolutely no justification at present to refer to "screening benefits" and imply causal relationships by using such terms (Gefeller and Windeler (1991)). Apart from our sustaining doubts that case–control studies will ever be able to allow conclusions about causal relationships, such an assertion as, for example, "Pap smears protect against cervical neoplasia" (LaVecchia et al. (1984)) is misleading in another way: screening tests definitely cannot protect people who are free from disease, the vast majority of screening participants. Screening in itself is not a therapy. The potential protective element lies only in the treatment following the screening test provided that the test result is truely positive (Knox (1991)).

Looking at the unsolved methodologic problems combined with the naive practical application, it has to be doubted that the current practice of the case–control design offers an appropiate framework to yield valid answers regarding the efficacy of screening. Additional methodologic research resulting in practical improvements of the design of these studies is required in order to refute such critical opinions as "the benefits indicated by case–control studies are partly or even totally factitious" (Rodgers (1991). Since an appropriate research design, the RCT, is already available to reach a definite answer concerning the efficacy of a screening programme, it seems to us, however, that these efforts (with uncertain success) are of limited value. Case–control studies of screening efficacy can never replace the RCT, they can only supplement it by providing additional information. Consequently, when attempting to improve the current situation of screening evaluation, preference should be given to activities directed towards implementing well-planned RCTs rather than finding sophisticated methods to cope with the many forms of bias in case–control studies.

We speculate that one reason for the present tendency to promote the use of case–control studies in the evaluation of screening stems from the fact that those primarily engaged in this field seem to be not very familiar with clinical trial methodology. They preferred to adopt a research design known to them and with a proven usefulness in other epidemiologic areas rather than to become acquainted with a "new" research design as the RCT. The similarities between screening evaluation and clinical studies on the efficacy of some medical intervention (drug therapy, surgical procedure, preventive measures etc.) are, however, much more pronounced than between screening evaluation and conventional epidemiologic studies identifying risk factors for some disease of interest. Consequently, those involved in screening evaluation should learn from the experiences made in thousands of clinical trials during this century and discontinue their doubtful enterprise to evaluate screening efficacy in a non–randomized design as the case–control study.

References:

ANDERSSON, I., ASPEGREN, K., JANZON, L., et al. (1988): Mammographic screening and mortality from breast cancer: the Malmö mammographic screening trial. *British Medical Journal*, 297, 943–948.

COLE, P., and MORRISON, A.S. (1980): Basic issues in population screening for cancer. *Journal of the National Cancer Institute*, 64, 1263–1272.

CONNOR, R.J., PROROK, P.C., and WEED, D.L. (1991): The case–control design and the assessment of the efficacy of cancer screening. *Journal of Clinical Epidemiology*, 44, 1215–1221.

DUBIN, N., FRIEDMAN, D.R., TONIOLO, P.G., et al. (1987): Breast cancer detection centers and case–control studies of the efficacy of screening. *Journal of Chronic Diseases*, 40, 1041–1050.

FEINSTEIN, A.R. (1985): Experimental requirements and scientific principles in case–control studies. *Journal of Chronic Disease*, 38 ,127–133.

FRIEDMAN, D.R., and DUBIN, N. (1991): Case–control evaluation of breast cancer screening efficacy. *American Journal of Epidemiology*, 133, 974–984.

GEFELLER, O., and WINDELER, J. (1991): Risk factors for cervical cancer: comments on attributable risk calculations and the evaluation of screening in case–control studies (letter). *International Journal of Epidemiology*, 20, 1140–1141.

GORDIS, L. (1982): Should dead cases be matched to dead controls? *American Journal of Epidemiology*, 115, 1–5.

KNOX, G. (1991): Case–control studies of screening procedures. *Public Health*, 105, 55–61.

KOPEC, J.A., and ESDAILE, J.M. (1990): Bias in case–control studies. A review. *Journal of Epidemiology and Community Health*, 44, 179–186.

LA VECCHIA, C., FRANCESCHI, S., DECARLI, A., et al. (1984): Pap smear and the risk of cervical neoplasia: quantitative estimates from a case–control study. *Lancet*, ii, 779–782.

LIBERATOS, P., LINK, B.G., and KELSEY, J.L. (1988): The measurement of social class in epidemiology. *Epidemiologic Reviews*, 10, 87–121.

MIETTINEN, O.S. (1989): The clinical trial as a paradigm for epidemiologic research. *Journal of Clinical Epidemiology*, 42, 491–496.

MORRISON, A.S. (1982): Case definition in case–control studies of the efficacy of screening. *American Journal of Epidemiology*, 115, 6–8.

MORRISON, A.S. (1985): *Screening in chronic disease.* Oxford University Press, New York.

MOSS, S.M. (1991): Case–control studies of screening. *International Journal of Epidemiology*, 20, 1–6.

RODGERS, A. (1991): Breast cancer screening: the current position. *British Medical Journal*, 302, 1401.

SACKETT, D.L. (1979): Bias in analytic research. *Journal of Chronic Diseases*, 32, 51–63.

SASCO, A.J. (1991): Validity of case–control studies and randomized controlled trials of screening. *International Journal of Epidemiology*, 20, 1143–1144.

SASCO, A.J., DAY, N.E., and WALTER, S.D. (1986): Case–control studies for the evaluation of screening. *Journal of Chronic Diseases*, 39, 399–405.

SHAPIRO, S., STRAX, P., and VENET, L. (1971): Periodic breast cancer screening in reducing mortality from breast cancer. *Journal of the American Medical Association*, 215, 1777–1785.

WEISS, N.S. (1983): Control definition in case–control studies of the efficacy of screening and diagnostic testing. *American Journal of Epidemiology*, 118, 457–460.

WEISS, N.S., MCKNIGHT, B., and STEVENS, N.G. (1992): Approaches to the analysis of case–control studies of the efficacy of screening for cancer. *American Journal of Epidemiology*, 135, 817–823.

WHITBY, L.G. (1974): Screening for disease. Definitions and criteria. *Lancet*, ii, 819–822.

WILSON, J.M.G., and JUNGNER, G. (1968): Principles and practice of screening for disease. *Public Health Paper*, 34, World Health Organization, Geneva.

Subject Index

List of Authors

GPSR Compliance

The European Union's (EU) General Product Safety Regulation (GPSR) is a set of rules that requires consumer products to be safe and our obligations to ensure this.

If you have any concerns about our products, you can contact us on ProductSafety@springernature.com

In case Publisher is established outside the EU, the EU authorized representative is:

Springer Nature Customer Service Center GmbH
Europaplatz 3
69115 Heidelberg, Germany

Batch number: 10838062

Printed by Printforce, the Netherlands